The Project Physics Course

Handbook

Directors

F. James Rutherford
Gerald Holton
Fletcher G. Watson

Concepts of Motion

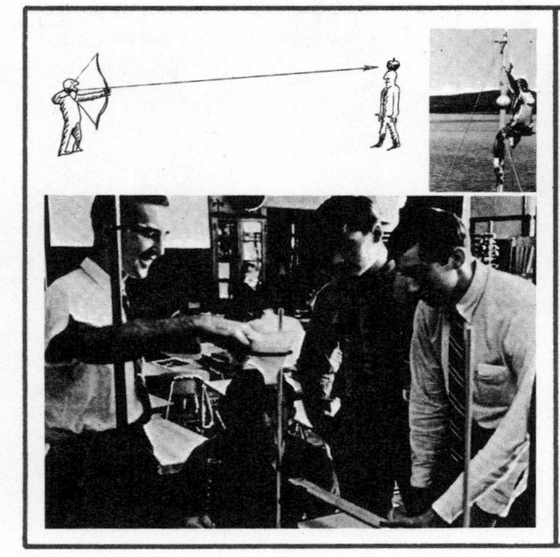

Motion in the Heavens

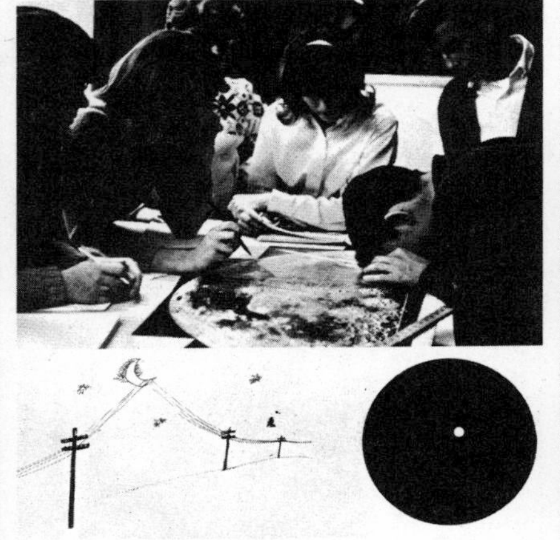

The Triumph of Mechanics

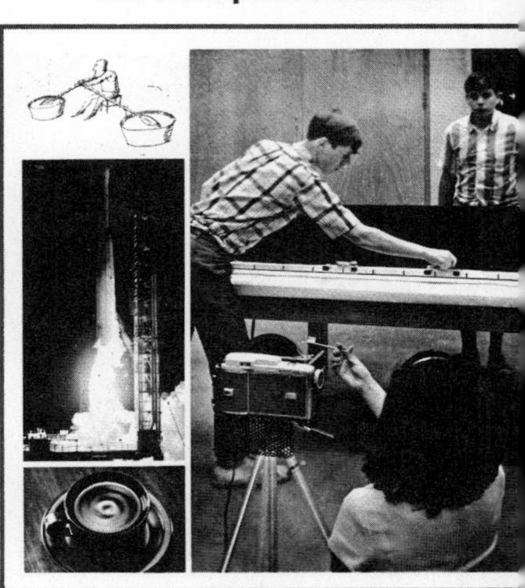

The Project Physics Course

Handbook

Light and Electromagnetism

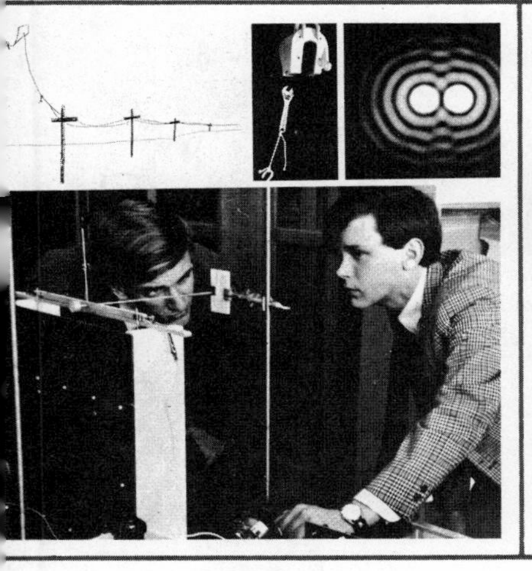

Models of the Atom

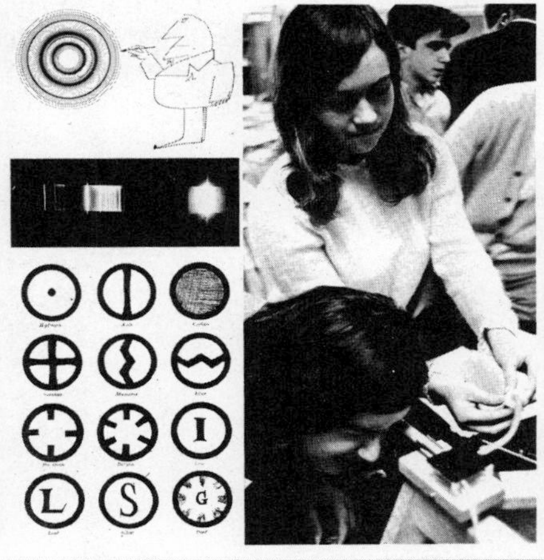

The Nucleus

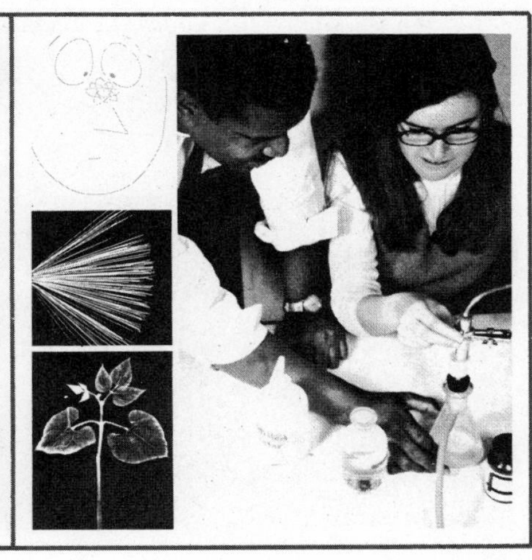

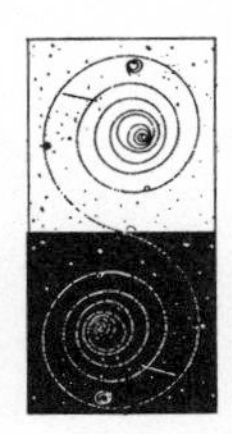

Published by HOLT, RINEHART and WINSTON, Inc. New York, Toronto

Directors of Harvard Project Physics

Gerald Holton, Department of Physics, Harvard University
F. James Rutherford, Capuchino High School, San Bruno, California, and Harvard University
Fletcher G. Watson, Harvard Graduate School of Education

Special Consultant to Project Physics

Andrew Ahlgren, Harvard Graduate School of Education

A partial list of staff
and consultants to Harvard Project Physics
appears in the Text on page 13 of the Appendix.

This Handbook is one of the many instructional materials developed for the Project Physics Course. These materials include Texts, Handbooks, Teacher Resource Books, Readers, Programmed Instruction booklets, Film Loops, Transparencies, 16mm films, and laboratory equipment.

ISBN 0-03-084801-6
34567890 039 98765

Acknowledgments

P. 7 The table is reprinted from *Solar and Planetary Longitudes for Years −2500 to +2500* prepared by William D. Stalman and Owen Gingerich (University of Wisconsin Press, 1963).

Pp. 259, 260 Smedlie, S. Raymond, *More Perpetual Motion Machines,* Science Publications of Boston, 1962.

Pp. 275, 277 "Science and the Artist," *Chemistry,* January 1964, pp. 22–23.

P. 263 I. F. Stacy, *The Encyclopedia of Electronics* (Charles Susskind, Ed.), Reinhold Publishing Corp., N.Y., Fig. 1, p. 246.

Pp. 341, 350, and 357 The three resource letters beginning on these pages are courtesy of the authors and of the *American Journal of Physics.*

Picture credits for the Handbook appear on pages 2, 70, 138, 232, 280 and 314.

Science is an adventure of the whole human race to learn to live in and perhaps to love the universe in which they are. To be a part of it is to understand, to understand oneself, to begin to feel that there is a capacity within man far beyond what he felt he had, of an infinite extension of human possibilities . . . .

I propose that science be taught at whatever level, from the lowest to the highest, in the humanistic way. It should be taught with a certain historical understanding, with a certain philosophical understanding, with a social understanding and a human understanding in the sense of the biography, the nature of the people who made this construction, the triumphs, the trials, the tribulations.

I. I. RABI
Nobel Laureate in Physics

Preface

Background The Project Physics Course is based on the ideas and research of a national curriculum development project that worked in three phases. First, the authors—a high school physics teacher, a university physicist, and a professor of science education—collaborated to lay out the main goals and topics of a new introductory physics course. They worked together from 1962 to 1964 with financial support from the Carnegie Corporation of New York, and the first version of the text was tried out in two schools with encouraging results.

These preliminary results led to the second phase of the Project when a series of major grants were obtained from the U.S. Office of Education and the National Science Foundation, starting in 1964. Invaluable additional financial support was also provided by the Ford Foundation, the Alfred P. Sloan Foundation, the Carnegie Corporation, and Harvard University. A large number of collaborators were brought together from all parts of the nation, and the group worked together for over four years under the title *Harvard Project Physics.* At the Project's center, located at Harvard University, Cambridge, Massachusetts, the staff and consultants included college and high school physics teachers, astronomers, chemists, historians and philosophers of science, science educators, psychologists, evaluation specialists, engineers, film makers, artists and graphic designers. The teachers serving as field consultants and the students in the trial classes were also of vital importance to the success of Harvard Project Physics. As each successive experimental version of the course was developed, it was tried out in schools throughout the United States and Canada. The teachers and students in those schools reported their criticisms and suggestions to the staff in Cambridge, and these reports became the basis for the subsequent revisions of the course materials. In the Preface to Unit 1 *Text* you will find a list of the major aims of the course.

We wish it were possible to list in detail the contributions of each person who participated in some part of Harvard Project Physics. Unhappily it is not feasible, since most staff members worked on a variety of materials and had multiple responsibilities. Furthermore, every text chapter, experiment, piece of apparatus, film or other item in the experimental program benefitted from the contributions of a great many people. On the preceding pages is a partial list of contributors to Harvard Project Physics. There were, in fact, many other contributors too numerous to mention. These include school administrators in participating schools, directors and staff members of training institutes for teachers, teachers who tried the course after the evaluation year, and most of all the thousands of students who not only agreed to take the experimental version of the course, but who were also willing to appraise it critically and contribute their opinions and suggestions.

The Project Physics Course Today. Using the last of the experimental versions of the course developed by Harvard Project Physics in 1964–68 as a starting point, and taking into account the evaluation results from the tryouts, the three original collaborators set out to develop the version suitable for large-scale publication. We take particular pleasure in acknowledging the assistance of Dr. Andrew Ahlgren of Harvard University. Dr. Ahlgren was invaluable because of his skill as a physics teacher, his editorial talent, his versatility and energy, and above all, his commitment to the goals of Harvard Project Physics.

We would also especially like to thank Miss Joan Laws whose administrative skills, dependability, and thoughtfulness contributed so much to our work. The publisher, Holt, Rinehart and Winston, Inc. of New York, provided the coordination, editorial support, and general backing necessary to the large undertaking of preparing the final version of all components of the Project Physics Course, including texts, laboratory apparatus, films, etc. Damon, a company located in Needham, Massachusetts, worked closely with us to improve the engineering design of the laboratory apparatus and to see that it was properly integrated into the program.

In the years ahead, the learning materials of the Project Physics Course will be revised as often as is necessary to remove remaining ambiguities, clarify instructions, and to continue to make the materials more interesting and relevant to students. We therefore urge all students and teachers who use this course to send to us (in care of Holt, Rinehart and Winston, Inc., 383 Madison Avenue, New York, New York 10017) any criticism or suggestions they may have.

F. James Rutherford
Gerald Holton
Fletcher G. Watson

Contents THE PROJECT PHYSICS COURSE: HANDBOOK

The Project Physics Course

Handbook **1**

Concepts of Motion

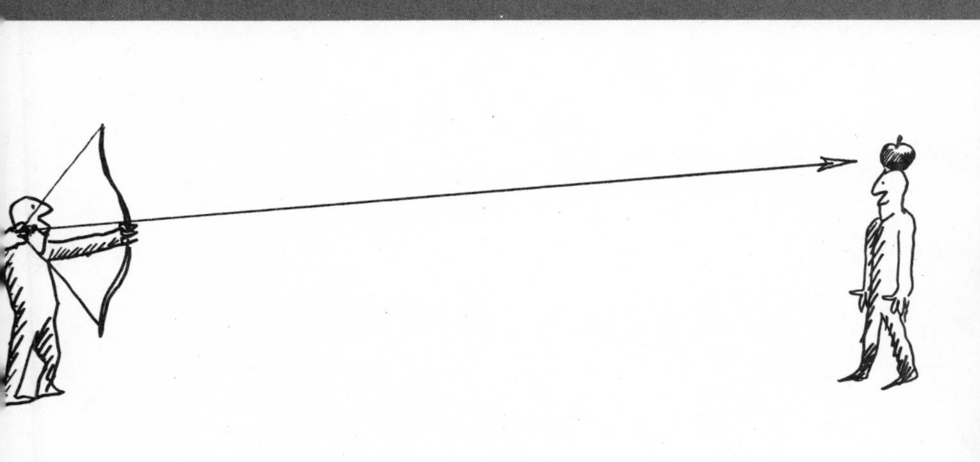

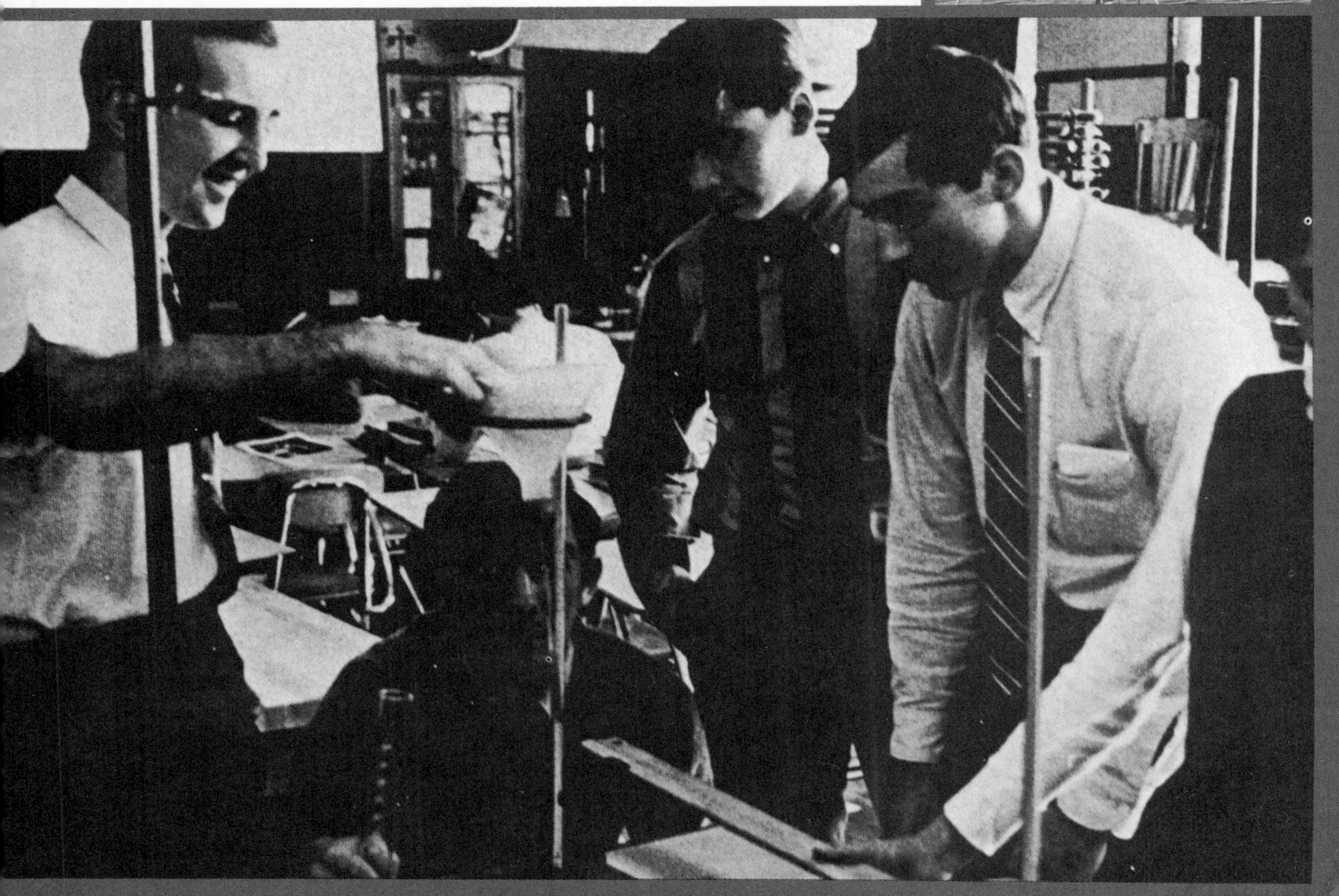

Picture Credits, Unit 1

Cover: (top left) Cartoon by Charles Gary Solin and reproduced by his permission only; (top right) from the film loop Galilean Relativity I—Ball Dropped from Mast of Ship.

P. 11 Isogonic chart through the courtesy of the Environmental Sciences Services Administration, Coast and Geodetic Survey.

Pp. 11, 14, 19, 41, 42, 47, 59, 63 (cartoons). By permission of Johnny Hart and Field Enterprises Inc.

P. 28 Photography unlimited by Ron Church from Rapho Guillumette Pictures, New York.

P. 60 (water drop parabola) Courtesy of Mr. Harold M. Waage, Palmer Physical Laboratory, Princeton University.

P. 61 (water drop parabola—train) Courtesy of Educational Development Center, Newton, Mass.

All photographs used with film loops courtesy of National Film Board of Canada.

Photographs of laboratory equipment and of students using laboratory equipment were supplied with the cooperation of the Project Physics staff and Damon Corporation.

Contents HANDBOOK, Unit 1

This *Handbook* is your guide to observations, experiments, activities, and explorations, far and wide, in the realms of physics.

Prepare for challenging work, fun and some surprises. One of the best ways to learn physics is by *doing* physics, in the laboratory and out. Do not rely on reading alone. Also, this *Handbook* is different from laboratory manuals you may have worked with before. Far more projects are described here than you alone can possibly do, so you will need to pick and choose.

Although only a few of the experiments and activities will be assigned, do any additional ones that interest you. Also, if an activity occurs to you that is not described here, discuss with your teacher the possibility of doing it. Some of the most interesting science you will experience in this course will be the result of the activities which you choose to pursue beyond the regular assignments of the school laboratory.

This *Handbook* contains a section corresponding to each chapter of the *Text*. Usually each section is divided further in the following way:

> The *Experiments* contain full instructions for the investigations you will be doing with your class.
>
> The *Activities* contain many suggestions for construction projects, demonstrations, and other activities you can do by yourself.
>
> The *Film Loop* notes give instructions for the use of the variety of film loops that have been specially prepared for the course.

In each section, do as many of these things as you can. With each, you will gain a better grasp of the physical principles and relationships involved.

Keeping Records

Your records of observations made in the laboratory or at home can be kept in many ways. Your teacher will show you how to write up your records of observations. But regardless of the procedure followed, the key question for deciding what kind of record you need is this: "Do I have a clear enough record so that I could pick up my lab notebook a few months from now and explain to myself or others what I did?"

Here are some general rules to be followed in every laboratory exercise. Your records should be neatly written without being fussy. You should organize all numerical readings in tables, if possible, as in the sample lab write up on pages 6 and 7. You should always identify the units (centimeters, kilograms, seconds, etc.) for each set of data you record. Also, identify the equipment you are using, so that you can find it again later if you need to recheck your work.

In general, it is better to record more rather than less data. Even details that may seem to have little bearing on the experiment you are doing—such as the temperature and whether it varied during the observations, and the time when the data were taken—may turn out to be information that has a bearing on your analysis of the results.

If you have some reason to suspect that a particular datum may be less reliable than other data—perhaps you had to make the reading very hurriedly, or a line on a photograph was very faint—make a note of the fact. But don't erase a reading. When you think an entry in your notes is in error, draw a single line through it—don't scratch it out completely or erase it. You may find it was significant after all.

There is no "wrong" result in an experiment, although results may be in considerable error. If your observations and measurements were carefully made, then your result will be correct. What ever happens in nature, including the laboratory, cannot be "wrong." It may have nothing to do with your investigation. Or it may be mixed up with so many other events you did not expect that your report is not useful. Therefore, you must think carefully about the interpretation of your results.

Finally, the cardinal rule in a laboratory is to choose in favor of "getting your hands dirty" instead of "dry-labbing." In 380 B.C., the Greek scientist, Archytas, summed this up pretty well:

> In subjects of which one has no knowledge, one must obtain knowledge either by learning from someone else, or by discovering it for oneself. That which is learnt, therefore, comes from another and by outside help; that which is discovered comes by one's own efforts and independently. To discover without seeking is difficult and rare, but if one seeks, it is frequent and easy; if, however, one does not know how to seek, discovery is impossible.

EXPERIMENT A. SEPT. 1969

This experiment is to see how a rubber band stretches under the influence of forces

rubber band

weight

YOUR OWN SKETCH WILL ALWAYS HELP YOUR MEMORY

I put different masses on the end of the rubber band and recorded the position of the top of the hook that holds the weight

Room Temperature 26°C

Position of the top of rubber band 36.3 cm

INCLUDE ANY DATA YOU THINK MAY BE RELEVANT

mass (g)	Force (N)	Pts of Bottom (cm)	Extension (cm)	
(weights from set, no error) 0	0	44.0 ±.1	0	ALWAYS SHOW UNITS OF TABULATED QUANTITIES
10	.098	45.1 "	1.1 ±.2	ESTIMATE THE ERROR OF EVERY QUANTITY YOU MEASURE
20	.196	45.8	1.8	
30	.294	46.8	2.8	
50	.490	49.6	5.6	
60	.588	51.5	7.5	second 20 g weight is missing from set
70	.686	53.7	9.7	INCLUDE COMMENTS ON YOUR DATA
80	.784	56.1	12.1	
100	.980	60.6	16.6	KEEP DATA IN NEAT TABLES
80	.784	56.2	12.2	recheck

On these two pages is shown an example of a student's lab notebook report. The table is used to record both observed quantities (mass, scale position) and calculated quantities (force, extension of rubber band). The graph shows at a glance how the extension of the rubber band changes as the force acting on it is increased. The notes in capital letters are comments.

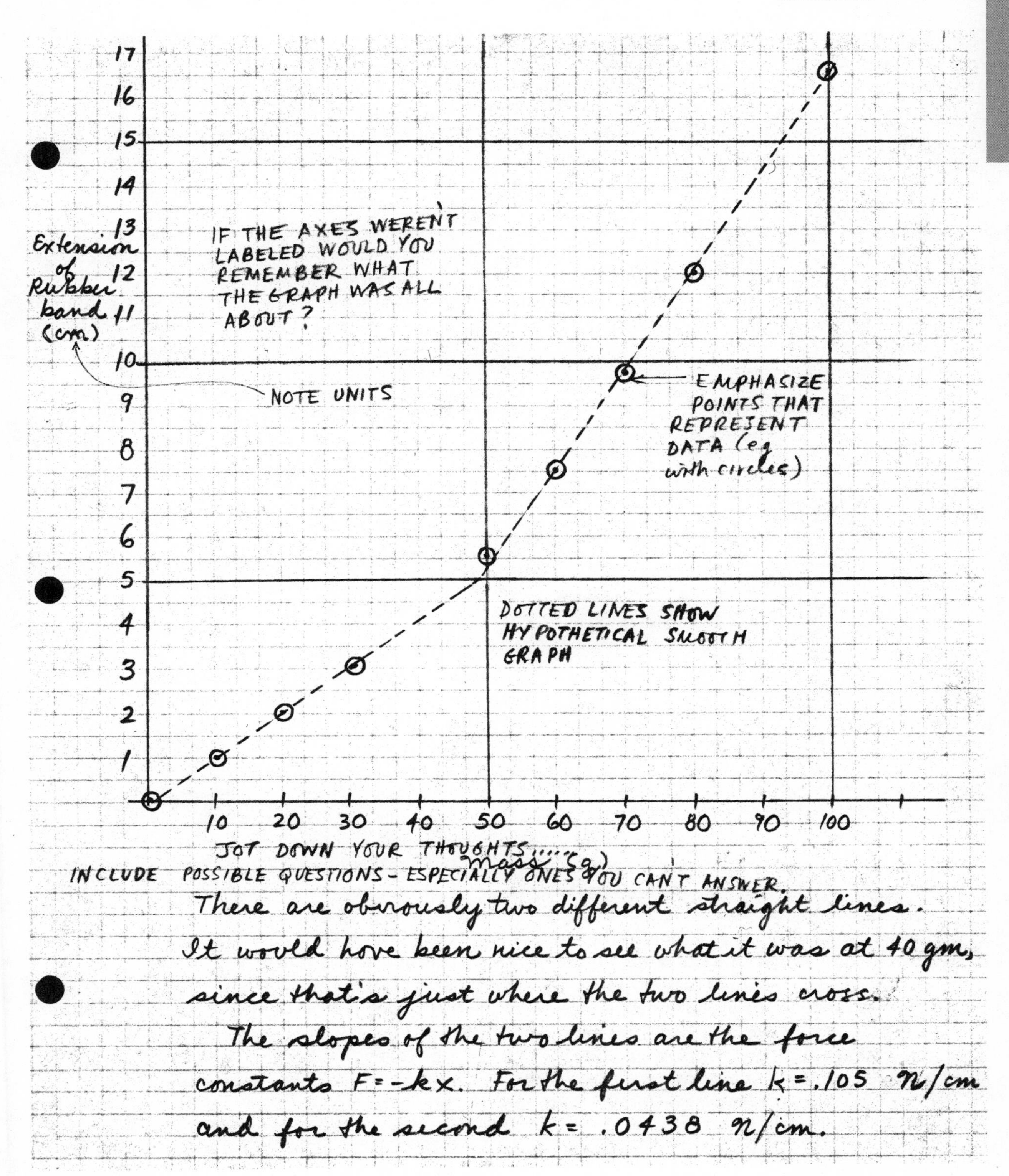
17
16
15
14
13
12
11
10
9
8
7
6
5
4
3
2
1
Extension of Rubber band (cm)
IF THE AXES WEREN'T LABELED WOULD YOU REMEMBER WHAT THE GRAPH WAS ALL ABOUT?
NOTE UNITS
EMPHASIZE POINTS THAT REPRESENT DATA (eg with circles)
DOTTED LINES SHOW HYPOTHETICAL SMOOTH GRAPH
10
20
30
40
50
60
70
80
90
100
mass (g)
JOT DOWN YOUR THOUGHTS.....
INCLUDE POSSIBLE QUESTIONS - ESPECIALLY ONES YOU CAN'T ANSWER.
There are obviously two different straight lines.
It would have been nice to see what it was at 40 gm, since that's just where the two lines cross.
The slopes of the two lines are the force constants F=-kx. For the first line k=.105 n/cm and for the second k= .0438 n/cm.

Using the Polaroid Land Camera

You will find the Polaroid camera is a very useful device for recording many of your laboratory observations. Section 1.3 of your textbook shows how the camera is used to study moving objects. In the experiments and activities described in this *Handbook,* many suggestions are made for photographing moving objects, both with an electronic stroboscope (a rapidly flashing xenon light) and with a mechanical disk stroboscope (a slotted disk rotating in front of the camera lens). The setup of the rotating disk stroboscope with a Polaroid camera is shown below.

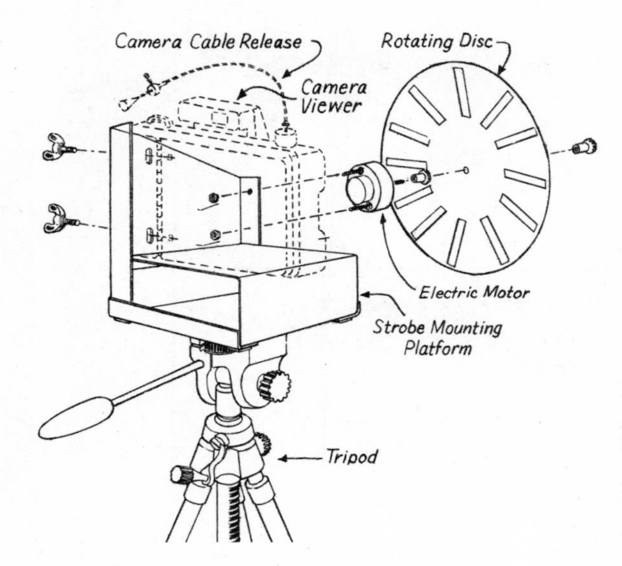

Below is a check list of operations to help you use the modified Polaroid Land camera model 210. For other models, your teacher will provide instructions.

1. Make sure that there is film in the camera. If no white tab shows in front of the door marked "4" you must put in new film.
2. Fasten camera to tripod or disk strobe base. If you are using the disk strobe technique, fix the clip-on slit in front of the lens.
3. Check film (speed) selector. Set to suggested position (75 for disk strobe or blinky; 3000 for xenon strobe).
4. If you are taking a "bulb" exposure, cover the electric eye.
5. Check distance from lens to plane of object to be photographed. Adjust focus if necessary. Work at the distance that gives an image just one-tenth the size of the object, if possible. This distance is about 120 cm.
6. Look through viewer to be sure that whatever part of the event you are interested in will be recorded. (At a distance of 120 cm the field of view is just under 100 cm long.)
7. Make sure the shutter is cocked (by depressing the number 3 button).
8. Run through the experiment a couple of times without taking a photograph, to accustom yourself to the timing needed to photograph the event.
9. Take the picture: keep the cable release depressed only as long as necessary to record the event itself. Don't keep the shutter open longer than necessary.
10. Pull the white tab all the way out of the camera. Don't block the door (marked "4" on the camera).
11. Pull the large yellow tab straight out—all the way out of the camera. Begin timing development.
12. Wait 10 to 15 seconds (for 3000-speed black-and-white film).
13. Ten to 15 seconds after removing film from the camera, strip the white print from the negative.
14. Take measurements immediately. (The magnifier may be helpful.)
15. After initial measurements have been taken, coat your picture with the preservative supplied with each pack of film. Let this dry thoroughly, label it on the back for identification and mount the picture in your (or a partner's) lab report.
16. The negative can be used, too. Wash it carefully with a wet sponge, and coat with preservative.
17. Recock the shutter so it will be set for next use.
18. Always be careful when moving around the camera that you do not inadvertently kick the tripod.
19. Always keep the electric eye covered when the camera is not in use. Otherwise the batteries inside the camera will run down quickly.

The *Readers*

Your teacher probably will not often assign reading in the *Project Physics Reader*, but you are encouraged to look through it for articles of interest to you. In the Unit 1 *Reader* most students enjoy the chapter from Fred Hoyle's science fiction novel, *The Black Cloud*. This chapter, "Close Reasoning," is fictional, but nevertheless accurately reflects the real excitement of scientists at work on a new and important problem.

Since different people have very different interests, nobody can tell you which articles you will most enjoy. Those with interests in art or the humanities will probably like Gyorgy Kepes' article "Representation of Movement." If you are interested in history and in the role science plays in historical development, you are urged to read the Butterfield and Willey articles.

The *Reader* provides several alternative treatments of mechanics which either supplement or go beyond the Unit 1 *Text*. For those seeking a deeper understanding of mechanics, we particularly recommend the article from the *Feynman Lectures on Physics*. For articles that deal with applications of physics, you can turn to Stong on "The Dynamics of the Golf Club," Kirkpatrick on "Bad Physics in Athletic Measurements," and DuBridge on "Fun in Space."

Practice the art of browsing! Don't decide from the titles alone whether you are interested, but read portions of articles here and there, and you may well discover something new and interesting.

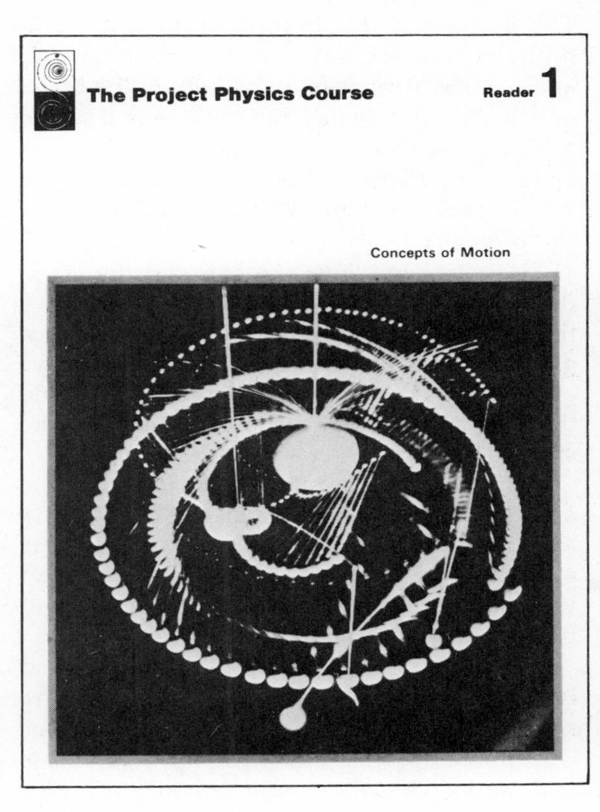

The Project Physics Course
Reader 1
Concepts of Motion

EXPERIMENTS

EXPERIMENT 1 NAKED EYE ASTRONOMY

The purpose of this first experiment is to familiarize you with the continually changing appearance of the sky. By watching the heavenly bodies closely day and night over a period of time, you will begin to understand what is going on up there and gain the experience you will need in working with Unit 2, *Motion in the Heavens*.

Do you know how the sun and the stars, the moon and the planets, appear to move through the sky? Do you know how to tell a planet from a star? Do you know when you can expect to see the moon during the day? How do the sun and planets move in relation to the stars?

The Babylonians and Egyptians knew the answers to these questions over 5000 years ago. They found them simply by watching the everchanging sky. Thus, astronomy began with simple observations of the sort you can make with your unaided eye.

You know that the earth appears to be at rest while the sun, stars, moon, and planets are seen to move in various paths through the sky. Our problem, as it was for the Babylonians, is to describe what these paths are and how they change from day to day, from week to week, and from season to season.

Some of these changes occur very slowly. In fact, this is why you may not have noticed them. You will need to watch the motions in the sky carefully, measuring them against fixed points of reference that you establish. You will need to keep a record of your observations for at least four to six weeks.

Choosing References

To locate objects in the sky accurately, you first need some fixed lines or planes to which your measurements can be referred, just as a map maker uses lines of latitude and longitude to locate places on the earth.

For example, you can establish a north-south line along the ground for your first reference. Then with a protractor held horizontally, you can measure the position of an object in the sky around the horizon from this north-south line. The angle of an object around the horizon from a north-south line is called the object's *azimuth*. Azimuths are measured from the north point (0°) through east (90°) to south (180°) and west (270°) and around to north again (360°or 0°).

To measure the height of an object in the sky, you can measure the angle between the object and a horizontal plane, such as the horizon, for your second reference. This plane can be used even when the true horizon is hidden by trees or other obstructions. The angle between the horizontal plane and the line to an object in the sky is called the *altitude* of the object.

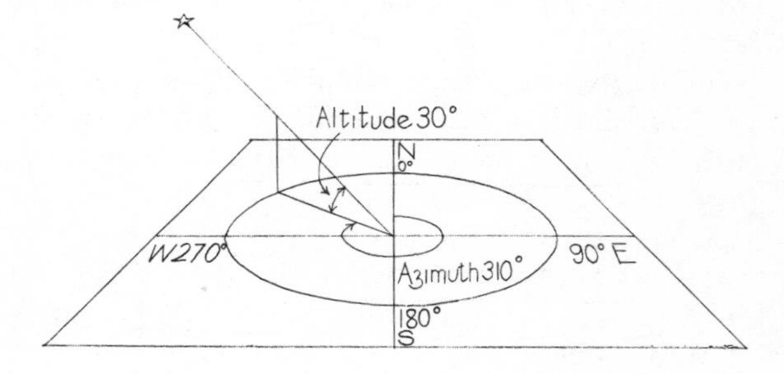

Establishing References

You can establish your north-south line in several different ways. The easiest is to use a compass to establish magnetic north but this may not be the same as true north. A magnetic compass responds to the total magnetic effect of all parts of the earth, and in most localities the compass does not point true north. The angle between magnetic north and true north is called *the angle of magnetic declination*. At some places the magnetic declination is zero, and the compass points toward true north.

At places east of the line where the declination is zero, the compass points west of true north; at places west of the line, the compass points east of true north. You can find the angle of declination and its rate of change per year for your area from the map below.

At night you can use the North Star (Polaris) to establish the north-south line. Polaris is the one fairly bright star in the sky that moves least from hour to hour or with the seasons. It is almost due north of an observer anywhere in the Northern Hemisphere.

To locate Polaris, first find the "Big Dipper" which on a September evening is low in the sky and a little west of north. (See the star map, Fig. 1–1 page 12.) The two stars forming the end of the dipper opposite the handle are known as the "pointers," because they point to the North Star. A line passing through them and extended upward passes very close to a bright star, the last star in the handle of the "Little Dipper." This bright star is the Pole Star, Polaris. On September 15 at 8:30 P.M. these constellations are arranged about as shown in the diagram at the top of page 13.

B.C. **By John Hart**

By permission of John Hart and Field Enterprises, Inc.

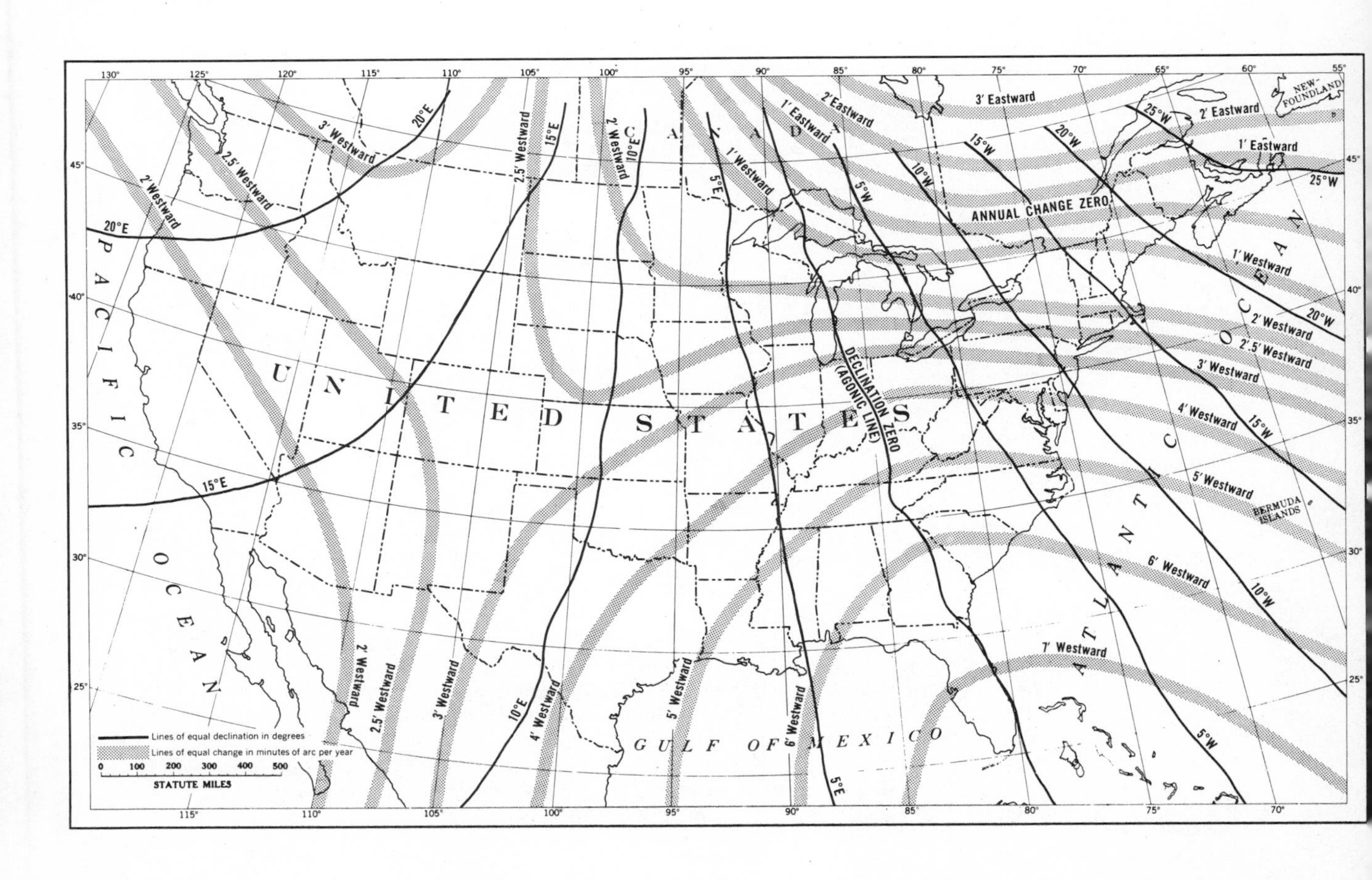

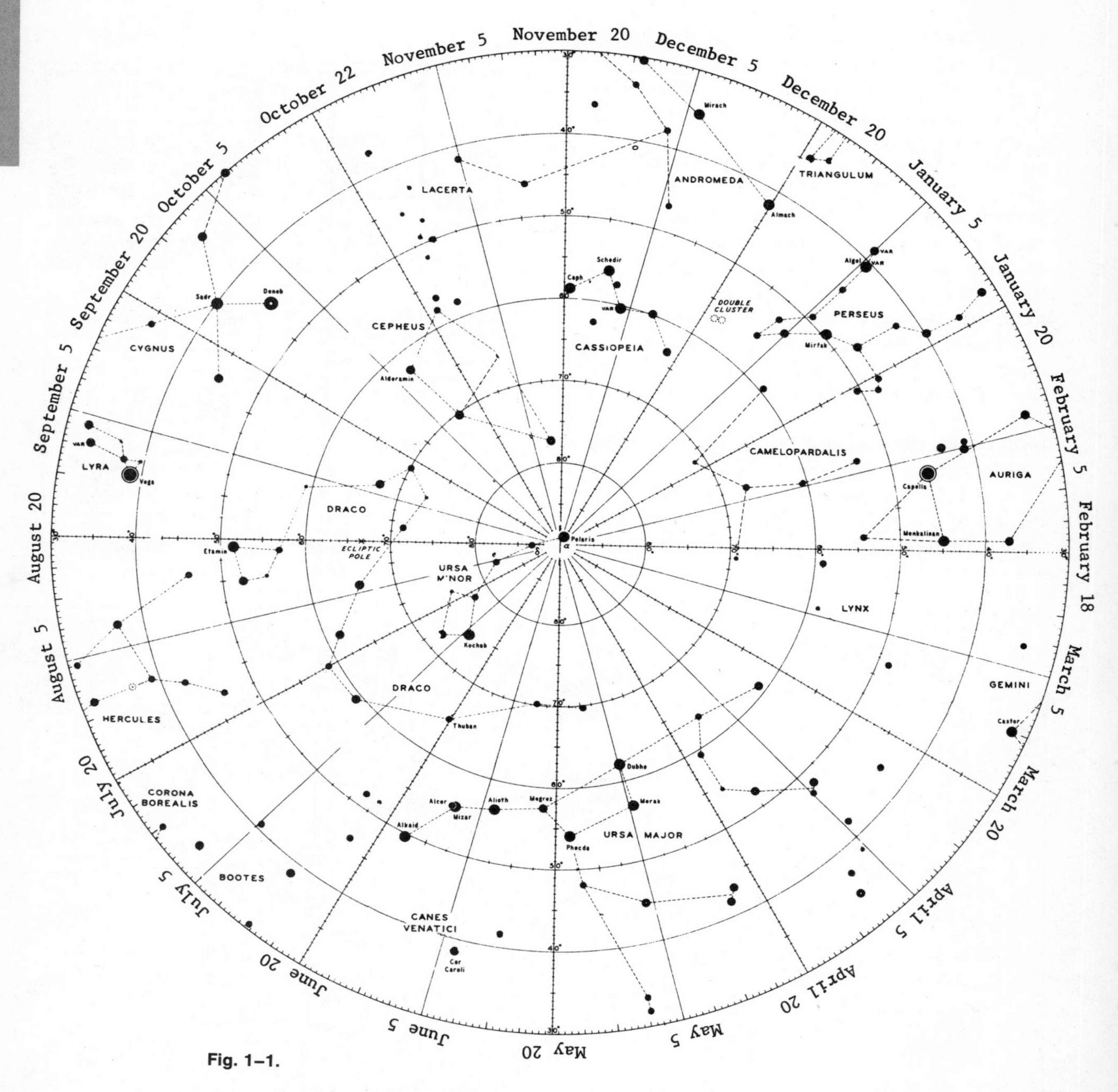

Fig. 1–1.

This chart of the stars will help you locate some of the bright stars and the constellations. To use the map, face north and turn the chart until today's date is at the top. Then move the map up nearly over your head. The stars will be in these positions at 8 P.M. For each hour *earlier* than 8 P.M., rotate the chart 15 degrees (one sector) clockwise. For each hour *later* than 8 P.M., rotate the chart counter-clockwise. If you are observing the sky outdoors with the map, cover the glass of a flashlight with fairly transparent red paper to look at the map. This will prevent your eyes from losing their adaptation to the dark when you look at the map.

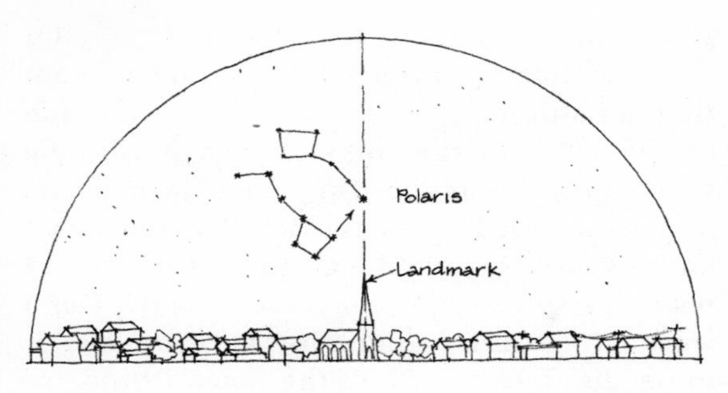

Imagine a line from Polaris straight down to the horizon. The point where this line meets the horizon is nearly due north of you.

Now that you have established a north-south line, either with a compass or from the North Star, note its position with respect to fixed landmarks, so that you can use it day or night.

You can establish the second reference, the plane of the horizon, and measure the altitude of objects in the sky from the horizon, with an astrolabe, a simple instrument you can obtain easily or make yourself, very similar to those used by ancient viewers of the heavens. Use the astrolabe in your hand or on a flat table mounted on a tripod or on a permanent post. A simple hand astrolabe you can make is described in the Unit 2 *Handbook,* in the experiment dealing with the size of the earth.

Sight along the surface of the flat table to be sure it is horizontal, in line with the horizon in all directions. If there are obstructions on your horizon, a carpenter's level turned in all directions on the table will tell you when the table is level.

Turn the base of the astrolabe on the table until the north-south line on the base points along your north-south line. Or you can obtain the north-south line by sighting on Polaris through the astrolabe tube. Sight through the tube of the astrolabe at objects in the sky you wish to locate and obtain their altitude above the horizon in degrees from the protractor on the astrolabe. With some astrolabes, you can also obtain the azimuth of the objects from the base of the astrolabe.

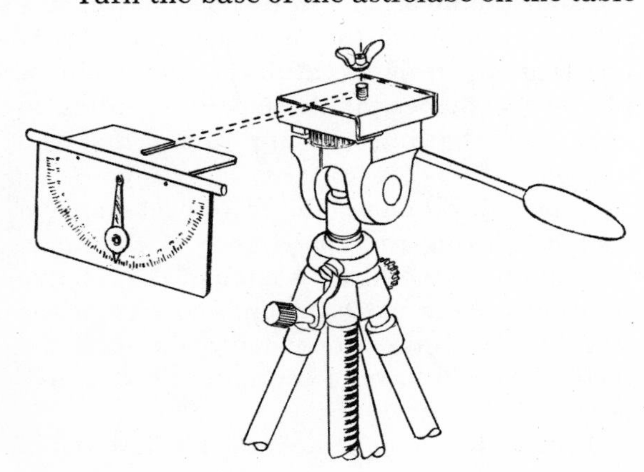

To follow the position of the sun with the astrolabe, slip a large piece of cardboard with a hole in the middle over the sky-pointing end of the tube. (CAUTION: NEVER look directly at the sun. It can cause permanent eye damage!) Standing beside the astrolabe, hold a small piece of white paper in the shadow of the large cardboard, several inches from the sighting end of the tube. Move the tube about until the bright image of the sun appears through the tube on the paper. Then read the altitude of the sun from the astrolabe, and the sun's azimuth, if your instrument permits.

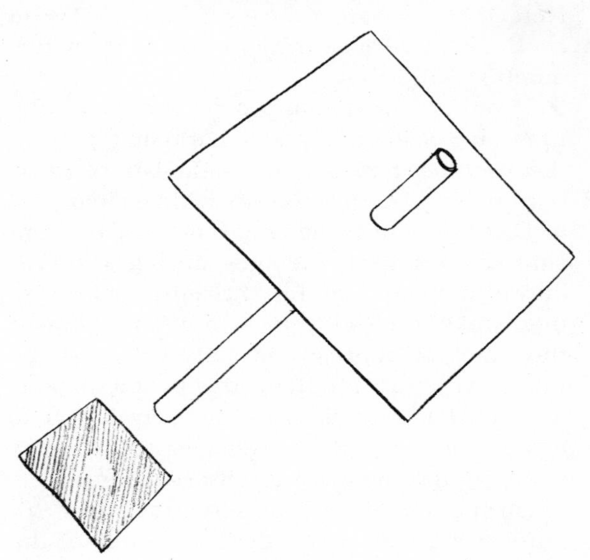

Observations

Now that you know how to establish your references for locating objects in the sky, here are suggestions for observations you can make on the sun, the moon, the stars, and the planets. Choose at least one of these objects to observe. Record the date and time of all your observations. Later compare notes with classmates who concentrated on other objects.

A. *Sun*

CAUTION: NEVER look directly at the sun; it can cause permanent eye damage. Do not depend on sun glasses or fogged photographic film for protection. It is safest to make sun observations on shadows.

1. Observe the direction in which the sun sets. Always make your observation from the same observing position. If you don't have an unobstructed view of the horizon, note where the sun disappears behind the buildings or trees in the evening.
2. Observe the time the sun sets or disappears below your horizon.
3. Try to make these observations about once a week. The first time, draw a simple sketch on the horizon and the position of the setting sun.
4. Repeat the observation a week later. Note if the position or time of sunset has changed. Note if they change during a month. Try to continue these observations for at least two months.
5. If you are up at sunrise, you can record the time and position of the sun's rising. (Check the weather forecast the night before to be reasonably sure that the sky will be clear.)
6. Determine how the length of the day, from sunrise to sunset, changes during a week; during a month; or for the entire year. You might like to check your own observations of the times of sunrise and sunset against the times as they are often reported in newspapers. Also if the weather does not permit you to observe the sun, the newspaper reports may help you to complete your observations.
7. During a single day, observe the sun's azimuth at various times. Keep a record—of the azimuth and the time of observation. Determine whether the azimuth changes at a constant rate during the day, or whether the sun's apparent motion is more rapid at some times than at others. Find how fast the sun moves in degrees per hour. See if you can make a graph of the speed of the sun's change in azimuth.

Similarly, find out how the sun's angular altitude changes during the day, and at what time its altitude is greatest. Compare a graph of the speed of the sun's change in altitude with a graph of its speed of change in azimuth.

8. Over a period of several months—or even an entire year—observe the altitude of the sun at noon—or some other convenient hour. (Don't worry if you miss some observations.) Determine the date on which the noon altitude of the sun is a minimum. On what date would the sun's altitude be a maximum?

By permission of John Hart and Field Enterprises, Inc.

B. *Moon*

1. Observe and record the altitude and azimuth of the moon and draw its shape on successive evenings at the same hour. Carry your observations through at least one cycle of phases, or shapes, of the moon, recording in your data the dates of any nights that you missed.

For at least one night each week, make a sketch showing the appearance of the moon and another "overhead" sketch of the relative positions of the earth, moon, and sun. If the sun is below the horizon when you observe the moon, you will have to estimate the sun's position.

2. Locate the moon against the background of

the stars, and plot its position and phase on a sky map supplied by your teacher.

3. Find the full moon's maximum altitude. Find how this compares with the sun's maximum altitude on the same day. Determine how the moon's maximum altitude varies from month to month.

4. There may be a total eclipse of the moon this year. Consult Table 1 on page 16, or the *Celestial Calendar and Handbook,* for the dates of lunar eclipses. Observe one if you possibly can.

C. *Stars*

1. On the first evening of star observation, locate some bright stars that will be easy to find on successive nights. Later you will identify some of these groups with constellations that are named on the star map in Fig. 1–1, which shows the constellations around the North Star, or on another star map furnished by your teacher. Record how much the stars have changed their positions compared to your horizon after an hour; after two hours.

2. Take a time exposure photograph of several minutes of the night sky to show the motion of the stars. Try to work well away from bright street lights and on a moonless night. Include some of the horizon in the picture for reference. Prop up your camera so it won't move during the time exposures of an hour or more. Use a *small* camera lens opening (large f-number) to reduce fogging of your film by stray light.

3. Viewing at the same time each night, find whether the positions of the star groups are constant in the sky from month to month. Find if any new constellations appear after one month; after 3 or 6 months. Over the same periods, find out if some constellations are no longer visible. Determine in what direction and how much the positions of the stars shift per week and per month.

This multiple exposure picture of the moon was taken with a Polaroid Land camera by Rick Pearce, a twelfth-grader in Wheat Ridge, Colorado. The time intervals between successive exposures were 15 min, 30 min, 30 min, and 30 min. Each exposure was for 30 sec using 2000-speed film. Which way was the moon moving in the sky?

A time exposure photograph of Ursa Major (The Big Dipper) taken with a Polaroid Land camera on an autumn evening in Cambridge, Massachusetts.

D. *Planets and meteors*

1. The planets are located within a rather narrow band across the sky (called the ecliptic) along which the sun and the moon also move. For details on the location of planets, consult Table 1 on page 16, or the *Celestial Calendar and Handbook,* or the magazine *Sky and Telescope.* Identify a planet and record its position in the sky relative to the stars at two-week intervals for several months.

2. On almost any clear, moonless night, go outdoors away from bright lights and scan as much of the sky as you can see for meteors. Probably you will glimpse a number of fairly bright streaks of meteors in an hour's time. Note how many meteors you see. Try to locate on a star map like Fig. 1–1 where you see them in the sky.

Look for meteor showers each year around November 5 and November 16, beginning around midnight. Dates of other meteor showers are given in Table 2 on page 17. Remember that bright moonlight will interfere with meteor observation.

Additional sky observations you may wish to make are described in the Unit 2 *Handbook.*

TABLE 1
A GUIDE FOR PLANET AND ECLIPSE OBSERVATIONS
Check your local newspaper for eclipse times and extent of eclipse in your locality.

Year	Mercury	Venus	Mars	Jupiter	Saturn	Lunar	Solar
	Visible for about one week around stated time.	Visible for several months around stated time.	Very bright for one month on each side of given time. Observable for 16 months surrounding given time.	Especially bright for seven months beyond stated time.	Especially bright for two months on each side of given time. Visible for 13 months.	Eclipses	Eclipses
	Mercury and Venus are best viewed the hour before dawn when indicated as a.m. and the hour after sunset when indicated as p.m.						
1970	mid Feb.: a.m. late Apr.: p.m. early June: a.m. mid Aug.: p.m. late Sept.: a.m. early Dec.: p.m.	early Nov.: p.m. mid Dec.: a.m.		late May: overhead at midnight	early Dec.: overhead at midnight	Feb. 21 Aug. 17	Mar. 7: total in Fla., partial in eastern and southern U. S.
1971	mid Jan.: a.m. late Mar.: p.m. mid May: a.m. late July: p.m. mid Sept.: a.m. late Nov.: p.m.		early Sept.: overhead at midnight	late June: overhead at midnight	late Dec.: overhead at midnight	Feb. 10	
1972	early Jan.: a.m. mid Mar.: p.m. early May: a.m. mid July: p.m. late Aug.: a.m. early Nov.: p.m. mid Dec.: a.m.	mid May: p.m. early Aug.: a.m.		late July: overhead at midnight		Jan. 30 July 26	July 10 partial in northern U.S.
1973	late Feb.: p.m. late Apr.: a.m. late June : p.m. early Aug.: a.m. mid Oct.: p.m. early Dec.: a.m.	late Dec.: p.m.	late Nov.: overhead at midnight	early Sept.: overhead at midnight	early Jan.: overhead at midnight	Dec. 10	
1974	mid Feb.: p.m. late Mar.: a.m. early June: p.m. mid July: a.m. late Sept.: p.m. early Nov.: a.m.	early Mar.: a.m.		mid Oct.: overhead at midnight	late Jan.: overhead at midnight	June 4 Nov. 29	
1975	late Jan.: p.m. early Mar.: a.m. mid May: p.m. early July: a.m. mid Sept.: p.m. late Oct.: a.m.	mid-late July: p.m. early Oct.: a.m.		early Nov.: overhead at midnight	early Feb.: overhead at midnight	May 25 Nov. 18	
1976	mid Jan.: p.m. late Feb.: a.m. early May: p.m. mid June: a.m. late Aug.: p.m. early Oct.: a.m. mid Dec.: p.m.		late Jan.: overhead at midnight	early Dec.: overhead at midnight	late Feb.: overhead at midnight		
1977	early Feb.: a.m. early Apr.: p.m. late May: a.m. mid Aug.: p.m. late Sept.: a.m.	early Mar.: p.m. mid Apr.: a.m.				Apr. 4	

TABLE 2
FAVORABILITY OF OBSERVING METEOR SHOWERS
THE BEST TIME FOR VIEWING METEOR SHOWERS IS BETWEEN MIDNIGHT AND 6 A.M., IN PARTICULAR DURING THE HOUR DIRECTLY PRECEDING DAWN.

Quadrantids Jan. 3-5 Virgo	Lyrids Apr. 19-23 Lyra	Perseids July 27-Aug. 17 Perseus	Orionids Oct. 15-25 Orion	Leonids Nov. 14-18 Leo	Geminids Dec. 9-14 Gemini	
Rises in the east around 2 a.m., upper eastern sky at 5 a.m.	Rises in the east around 10 p.m., western sky at 5 a.m.	Rises in the east at 10 p.m., towards the west at 5 a.m.	Rises in the east at midnight, directly overhead at 5 a.m.	Rises in the east at 2 a.m., upper eastern sky at 5 a.m.	Rises in the east at 8 p.m., towards the far west at 5 a.m.	
Good	Poor	July 27-Aug.11	Oct. 18-25	Poor	Poor	1970
Good	Good	July 27-Aug. 2 Aug. 7-17	Good	Good	Good	1971
Good	Good	Aug. 2-17	Oct. 15-20	Nov. 14-16	Good	1972
Good	Apr. 21-23	July 27-Aug. 9	Good	Good	Poor	1973
Poor	Good	Aug. 7-17	Good	Good	Good	1974
Good	Good	Good	Oct. 21-25	Poor	Dec. 9-12	1975
Good	Good	July 27-Aug. 5 Aug. 12-17	Good	Good	Good	1976
Poor	Good	Aug. 3-17	Oct. 15-21	Good	Good	1977

EXPERIMENT 2 REGULARITY AND TIME

You will often encounter regularity in your study of science. Many natural events occur regularly—that is, over and over again at equal time intervals. But if you had no clock, how would you decide how regularly an event recurs? In fact, how can you decide how regular a clock is?

The first part of this exercise is intended merely to show you the regularity of a few natural events. In the second part, you will try to measure the regularity of an event against a standard and to decide what is really meant by the word "regularity."

Part A

You work with a partner in this part. Find several recurring events that you can time in the laboratory. You might use such events as a dripping faucet, a human pulse, or the beat of recorded music. Choose one of these events as a "standard event." All the others are to be compared to the standard by means of the strip chart recorder.

One lab partner marks each "tick" of the standard on one side of the strip chart recorder tape while the other lab partner marks each "tick" of the event being tested. After a long run has been taken, inspect the tape to see how the regularities of the two events compare. Run for about 300 ticks of the standard. For each 50 ticks of the standard, find on the tape the number of ticks of the other phenomenon, estimating to $\frac{1}{10}$ of a tick. Record your results in a table something like this:

STANDARD EVENT	*TEST EVENT*
First 50 ticks	________ ticks
Second 50 ticks	________ ticks
Third 50 ticks	________ ticks
Fourth 50 ticks	________ ticks

The test event's frequency is almost certain to be different from test to test. The difference could be a real difference in regularity, or it could come from your error in measuring.

Q1 If you think that the difference is larger than you would expect from human error, then which of the two events is *not* regular?

Part B

In this part of the lab, you will compare the regularity of some devices specifically designed to be regular. The standard here will be the time recording provided by the telephone company or Western Union. To measure two peri-

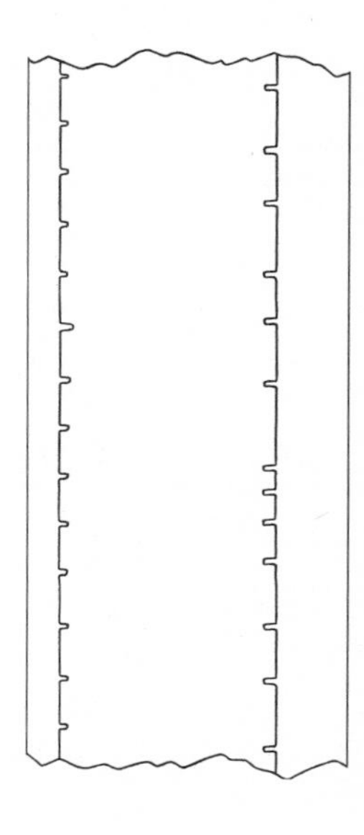

ods of time, you will have to make three calls to the time station, for example, 7 P.M., 7 A.M., and 7 P.M. again. Agreement should be reached in class the day before on who will check wall clocks, who will check wristwatches, and so on. Watch your clock and wait for the recording to announce the exact hour. Tabulate your results something like this:

TIME STATION

Time	Period
"7 P.M. exactly" }	12:00:00 hr
"7 A.M. exactly" }	
"7 P.M. exactly" }	12:00:00 hr

ELECTRIC WALL CLOCK

7: : ____ }	___ : ___ : ___
7: : ____ }	
7: : ____ }	___ : ___ : ___

In Part A, you found that to test regularity you need a standard that is consistent, varying as little as possible. The standard is understood, *by definition,* to be regular.

Q2 What is the standard against which the time station signal is compared? Call to find out what this standard is. Try to find the final standard that is used to define regularity—the time standard against which all other recurring events are tested. How can we be sure of the regularity of *this* standard?

B.C. By John Hart

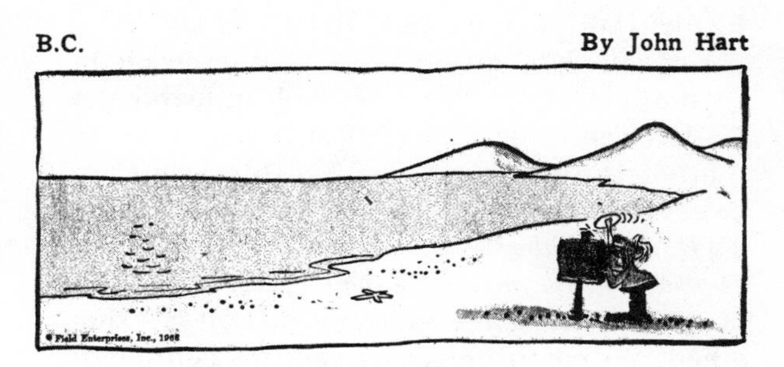

By permission of John Hart and Field Enterprises, Inc.

EXPERIMENT 3 VARIATIONS IN DATA

If you count the number of chairs or people in an ordinary sized room, you will probably get exactly the right answer. But if you measure the length of this page with a ruler, your answer will have a small margin of uncertainty. That is, *numbers read from measuring instruments do not give the exact measurements* in the sense that *one* or *two* is exact when you count objects. Every measurement is to some extent uncertain.

Moreover, if your lab partner measures the length of this page, he will probably get a different answer from yours. Does this mean that the length of the page has changed? Hardly! Then can you possibly find the length of the page without any uncertainty in your measurement? This lab exercise is intended to show you why the answer is "no."

Various stations have been set up around the room, and at each one you are to make some measurement. Record each measurement in a table like the one shown here. When you have completed the series, write your measurements on the board along with those of your classmates. Some interesting patterns should emerge if your measurements have not been influenced by anyone else. Therefore, do not talk about your results or how you got them until everyone has finished.

TYPE OF MEASUREMENT	REMARKS	MEASUREMENT

Chapter 1 The Language of Motion

EXPERIMENT 4 MEASURING UNIFORM MOTION

If you roll a ball along a level floor or table, eventually it stops. Wasn't it slowing down all the time, from the moment you gave it a push? Can you think of any things that have uniform motion in which their speed remains constant and unchanging? Could the dry-ice disk pictured in Sec. 1.3 of the *Text* really be in uniform motion, even if the disk is called "frictionless"? Would the disk just move on forever? Doesn't everything eventually come to a stop?

In this experiment you check the answers to these questions for yourself. You observe very simple motion, like that pictured below, and make a photo record of it, or work with similar photos. You measure the speed of the object as precisely as you can and record your data in tables and draw graphs from these data. From the graphs you can decide whether the motion was uniform or not.

Your decision may be harder to make than you would expect, since your experimental measurements can never be exact. There are likely to be ups and downs in your final results. Your problem will be to decide whether the ups and downs are due partly to real changes in speed or due entirely to uncertainty in your measurements.

If the speed of your object turns out to be constant, does this mean that you have produced an example of uniform motion? Do you think it is possible to do so?

Doing the Experiment

Various setups for the experiment are shown on pages 21 and 22. It takes two people to photograph a disk sliding on a table, or a glider

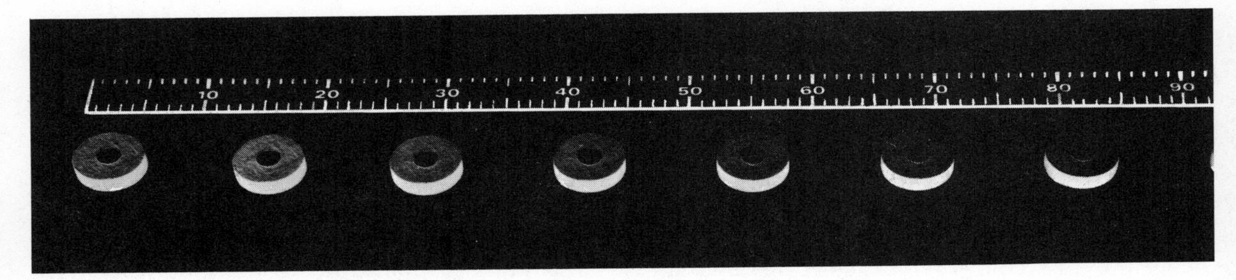

Fig. 1–1. Stroboscopic photograph of a moving CO_2 disk.

on an air track, or a steadily flashing light (called a blinky) mounted on a small box which is pushed by a toy tractor. Your teacher will explain how to work with the set up you are using. Excellent photographs can be made of any of them.

If you do not use a camera at all, or if you work alone, then you may measure a transparency or a movie film projected on the chalk board or a large piece of paper.

Or you may simply work from a previously prepared photograph such as Fig. 1–1, above. If there is time, you might try several of these methods.

One setup uses for the moving object a disk made of metal or plastic. A few plastic beads sprinkled on a smooth, dust-free table top (or a sheet of glass) provide a surface for the disk to slide with almost no friction. Make sure the surface is quite level, too, so that the disk will not start to move once it is at rest.

Set up the Polaroid camera and the stroboscope equipment according to your teacher's instructions. Instructions for operating the Polaroid model 210, and a diagram for mounting this camera with a rotating disk stroboscope is shown on page 8. A ruler need not be included in your photograph as in the photograph above. Instead, you can use a magnifier with a scale that is more accurate than a ruler for measuring the photograph.

Either your teacher or a few trials will give you an idea of the camera settings and of the speed at which to launch the disk, so that the images of your disk are clear and well-spaced in the photograph. One student launches the disk while his companion operates the camera. A "dry run" or two without taking a picture will probably be needed for practice before you get a good picture. A good picture is one in which there are at least five sharp and clear images of your disk far enough apart for easy measuring on the photograph.

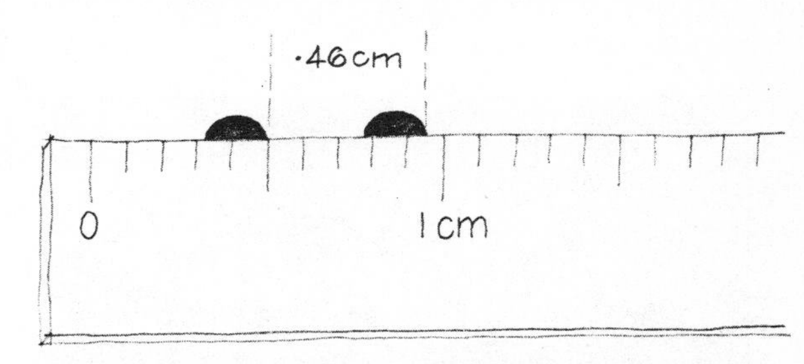

Fig. 1–2. Estimating to tenths of a scale division.

Making Measurements

Whatever method you have used, your next step is to measure the spaces between successive images of your moving object. For this, use a ruler with millimeter divisions and estimate the distances to the nearest tenth of a millimeter, as shown in Fig. 1–2 above. If you use a magnifier with a scale, rather than a ruler, you may be able to estimate these quite precisely. List each measurement in a table like Table 1.

Since the intervals of time between one image and the next are equal, you can use that interval as a unit of time for analyzing the event. If the speed is constant, the distances of travel would turn out to be all the same, and the motion would be uniform.

Q1 How would you recognize motion that is not uniform?

Q2 Why is it unnecessary for you to know the time interval in seconds?

TABLE 1

TIME INTERVAL	DISTANCE TRAVELED IN EACH TIME INTERVAL
1st	0.48 cm
2nd	0.48
3rd	0.48
4th	0.48
5th	0.48
6th	0.48

Table 1 has data that indicate uniform motion. Since the object traveled 0.48 cm during each time interval, the speed is 0.48 cm per unit time.

It is more likely that your measurements go up and down as in Table 2, particularly if you measure with a ruler.

TABLE 2

TIME INTERVAL	DISTANCE TRAVELED IN EACH TIME INTERVAL
1st	0.48 cm
2nd	0.46
3rd	0.49
4th	0.50
5th	0.47
6th	0.48

Q3 Is the speed constant in this case? Since the distances are *not* all the same, you might well say, "No, it isn't." Or perhaps you looked again at a couple of the more extreme data in Table 2, such as 0.46 and 0.50 cm, checked these measurements, and found them doubtful. Then you might say, "The ups and downs are because it is difficult to measure to 0.01 cm with the ruler. The speed really *is* constant as nearly as I can tell." Which statement is right?

Look carefully at the divisions or marks on your ruler. Can you read your ruler accurately to the nearest 0.01 cm? If you are like most people, you read it to the nearest mark of 0.1 cm (the nearest whole millimeter) and *estimate* the next digit between the marks for the nearest tenth of a millimeter (0.01 cm), as illustrated in Fig. 1–2 at the left.

In the same way, whenever you read the divisions of any measuring device you should read accurately to the nearest division or mark and then estimate the next digit in the measurement. Then probably your measurement, including your estimate of a digit between divisions, is not more than half of a division in error. It is not likely, for example, that in Fig. 1–2 on page 22 you would read more than half a millimeter away from where the edge being measured comes between the divisions. In this case, in which the divisions on the ruler are millimeters, you are at most no more than 0.5 mm (0.05 cm) in error.

Suppose you assume that the motion really is uniform and that the slight differences between distance measurements are due only to the uncertainty in reading the ruler. What is then the best estimate of the constant distance the object traveled between flashes?

Usually, to find the "best" value of distance you must average the values. The average for Table 2 is 0.48 cm, but the 8 is an uncertain measurement.

If the motion recorded in Table 2 really is uniform, the measurement of the distance traveled in each time interval is 0.48 cm plus or minus 0.05 cm, written as 0.48±0.05 cm. The ±0.05 is called the *uncertainty* of your measurement. The uncertainty for a single measurement is commonly taken to be half a scale division. With many measurements, this uncertainty may be less, but you can use it to be on the safe side.

Now you can return to the big question: Is the speed constant or not? Because the numbers go up and down you might suppose that the speed is constantly *changing*. Notice though that in Table 2 the changes of data above and below the average value of 0.48 cm are always smaller than the uncertainty, 0.05 cm. Therefore, the ups and down *may* all be due to the difficulty in reading the ruler to better than 0.05 cm—and the speed may, in fact, be constant.

Our conclusion from the data given here is that the *speed is constant to within the uncertainty of measurement, which is 0.05 cm per unit time*. If the speed goes up or down by less than this amount, we simply cannot reliably detect it with our ruler.

CHAPTER 1

EXPERIMENT 4 SEPT. 1969

In this experiment one compares the distances travelled by a moving object during equal time intervals to see if the motion is uniform

Set up

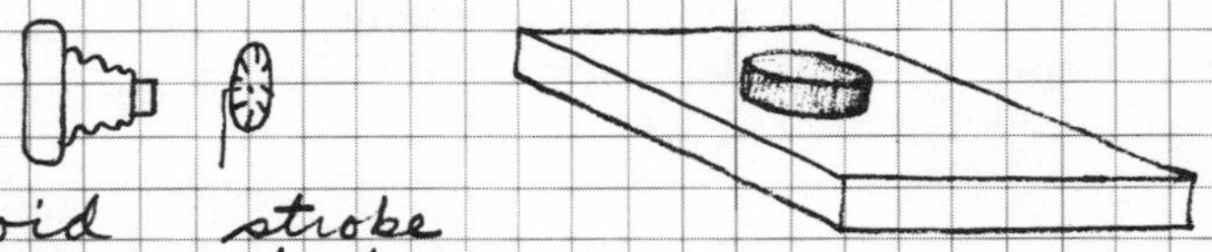

polaroid camera

strobe disk

puck on "frictionless" surface

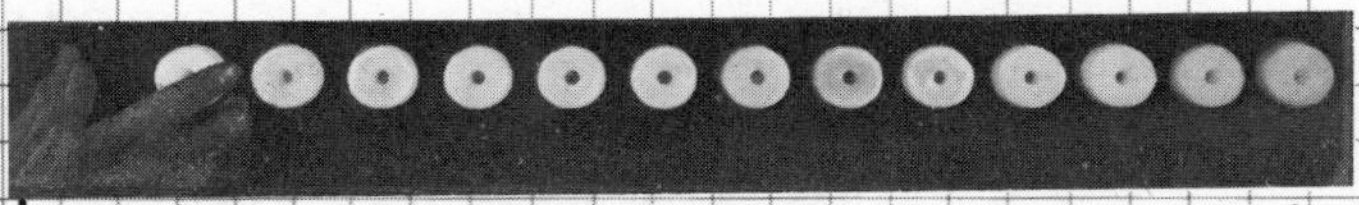

Interval number 1 2 3 4 5 6

Data

Interval number	distance interval (cm)	total distance (cm)	
1	0.48		
2	0.46	.94	
3	0.49	1.43	estimated uncertainty
4	0.50	1.93	of each distance
5	0.47	2.40	measurment
6	0.48		
overall		2.88	± 0.05 cm

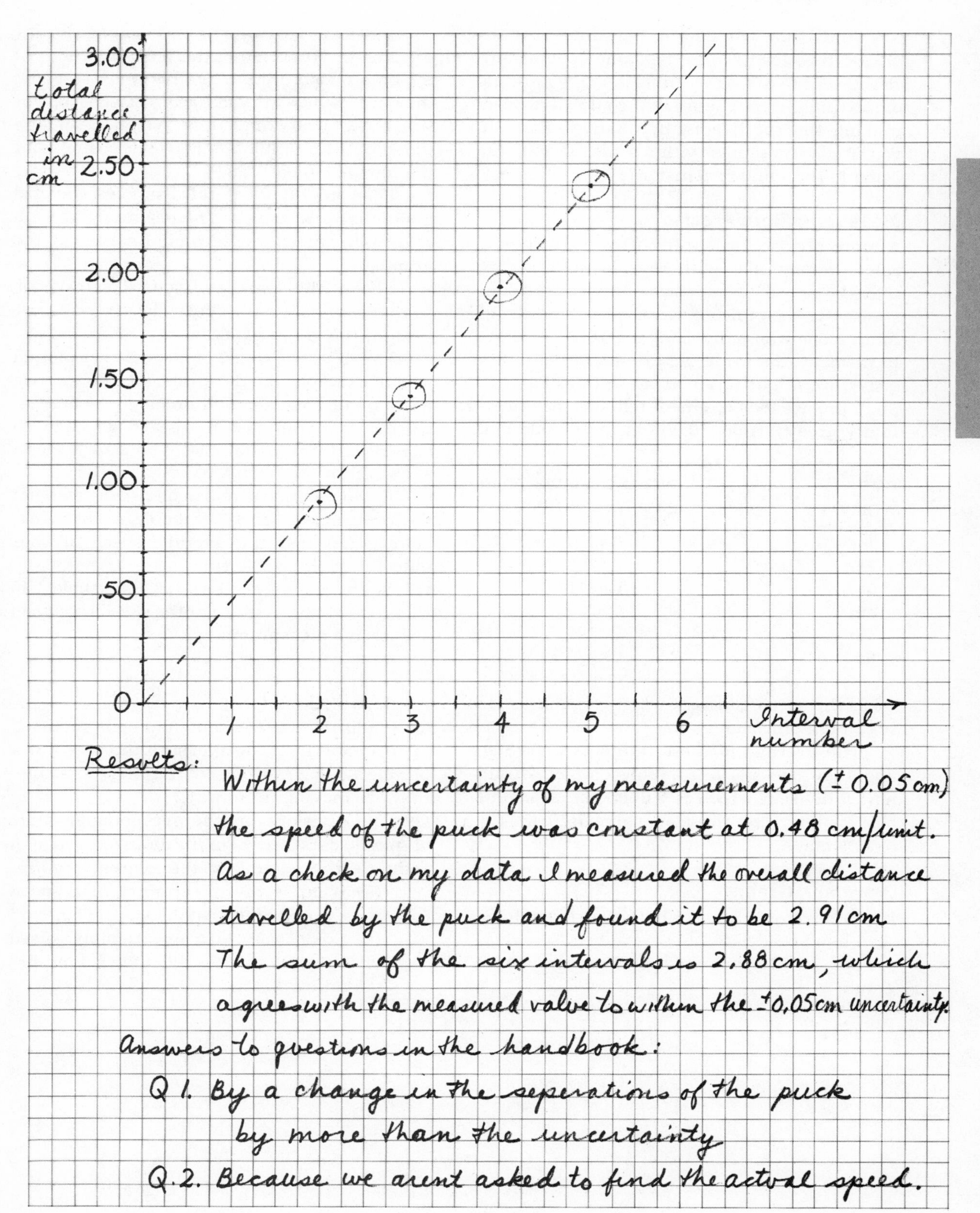

Results:

Within the uncertainty of my measurements (± 0.05 cm) the speed of the puck was constant at 0.48 cm/unit. As a check on my data I measured the overall distance travelled by the puck and found it to be 2.91 cm. The sum of the six intervals is 2.88 cm, which agrees with the measured value to within the ±0.05 cm uncertainty.

Answers to questions in the handbook:

Q 1. By a change in the seperations of the puck by more than the uncertainty

Q.2. Because we arent asked to find the actual speed.

Study your own data in the same way.

Q4 Do they lead you to the same conclusion? If your data vary as in Table 2, can you think of anything in your setup that could have been making the speed actually change? Even if you used a magnifier with a scale, do you still come to the same conclusion?

Measuring More Precisely

A more precise measuring instrument than a ruler or magnifier with a scale might show that the speed in our example was *not* constant. For example, if we used a measuring microscope whose divisions are 0.001 cm apart to measure the same picture again more precisely, we might arrive at the data in Table 3. Such precise measurement reduces the uncertainty greatly from ±0.05 cm to ±0.0005 cm.

TABLE 3

TIME INTERVAL	DISTANCE TRAVELED IN EACH TIME INTERVAL
1st	0.4826 cm
2nd	0.4593
3rd	0.4911
4th	0.5032
5th	0.4684
6th	0.4779

Q5 Is the speed constant when we measure to such high precision as this?

The average of these numbers is 0.4804, and they are all presumably correct within half a division which is 0.0005 cm. Thus our best estimate of the true value is 0.4804 ± 0.0005 cm.

Drawing a Graph

If you have read Sec. 1.5 in the *Text*, you have seen how speed data can be graphed. Your data provide an easy example to use in drawing a graph.

Just as in the example on *Text* page 19, lay off time intervals along the horizontal axis of the graph. Your units are probably not seconds; they are "blinks" if you used a stroboscope or simply "arbitrary time units" which mean here the equal time intervals between positions of the moving object.

Then lay off the *total* distances traveled along the vertical axis. The beginning of each scale is in the lower left-hand corner of the graph.

Choose the spacing of your scale division so that your data will, if possible, spread across most of the graph paper.

The data of Table 2 on page 23 are plotted as an example on the graph of the sample write up of Experiment 4 on pages 24 and 25.

Q6 In what way does the graph on page 25 shown uniform motion? Does your own graph show uniform motion too?

If the motion in your experiment was not uniform, review Sec. 1.7 of the *Text*. Then from your graph find the average speed of your object over the whole trip.

Q7 Is the average speed for the whole trip the same as the average of the speeds between successive measurements?

Additional Questions

Q8 Could you use the same methods you used in this experiment to measure the speed of a bicycle? a car? a person running? (Assume they are moving uniformly.)

Q9 The divisions on the speedometer scale of many cars are 5 mi/hr in size. You can estimate the reading to the nearest 1 mi/hr.

(a) What is the uncertainty in a speed measurement by such a speedometer?
(b) Could you measure reliably speed changes as small as 2 mi/hr? 1 mi/hr? 0.5 mi/hr? 0.3 mi/hr?

ACTIVITIES

USING THE ELECTRONIC STROBOSCOPE

Examine some moving objects illuminated by an electronic stroboscope. Put a piece of tape on a fan blade or mark it with chalk and watch the pattern as you turn the fan on and off. How can you tell when there is exactly one flash of light for each rotation of the fan blade?

Observe a stream of water from a faucet, objects tossed into the air, or the needle of a running sewing machine. If you can darken the room completely, try catching a thrown ball lighted only by a stroboscope. How many flashes do you need during the flight of the ball to be able to catch it reliably?

MAKING FRICTIONLESS PUCKS

Method 1. Use a flat piece of dry ice on a very smooth surface, like glass or Formica. When you push the piece of dry ice (frozen carbon dioxide), it moves in a frictionless manner because as the carbon dioxide changes to a vapor it creates a layer of CO_2 gas between the solid and the glass. (CAUTION: Don't touch dry ice with your bare hands; it can give you a severe frost bite!)

Method 2. Make a balloon puck if your lab does not have a supply. First cut a 4-inch diameter disk of 1-inch-thick Masonite. Drill a $\frac{1}{2}''$ diameter hole part way through the center of the disk so it will hold a rubber stopper. Then drill a $\frac{1}{32}''$ diameter hole on the same center the rest of the way through the disk. Drill a $\frac{1}{16}''$ hole through the center of a stopper in the hole in the masonite disk. Place the disk on glass or Formica.

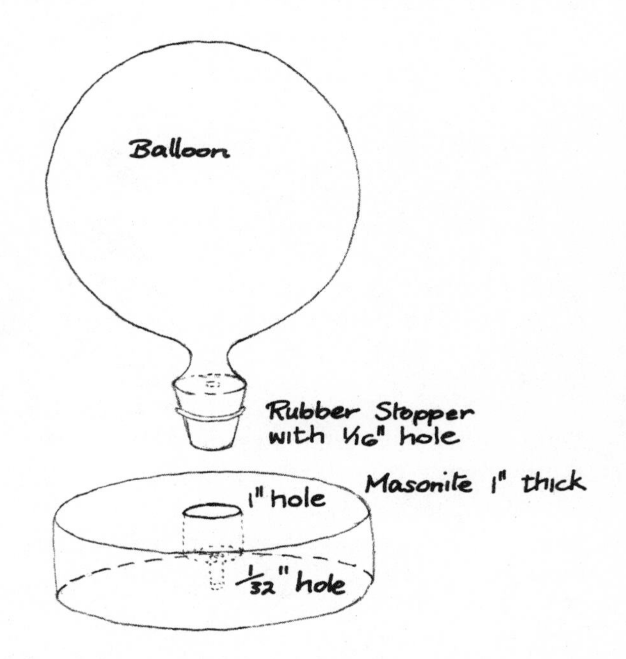

Method 3. Make a pressure pump puck. Make a disk as described in Method 2. Instead of using a balloon, attach a piece of flexible tubing, attached at the other end to the exhaust of a vacuum pump as shown in the diagram. Run the tubing over an overhead support so it does not interfere with the motion of the puck.

Method 4. Drill a $\frac{1}{32}''$ hole in the bottom of a smooth-bottomed cylindrical can, such as one for a typewriter ribbon. Break up dry ice (DON'T touch it with bare hands) and place the pieces inside the can. Seal the can with tape, and place it on a very smooth surface.

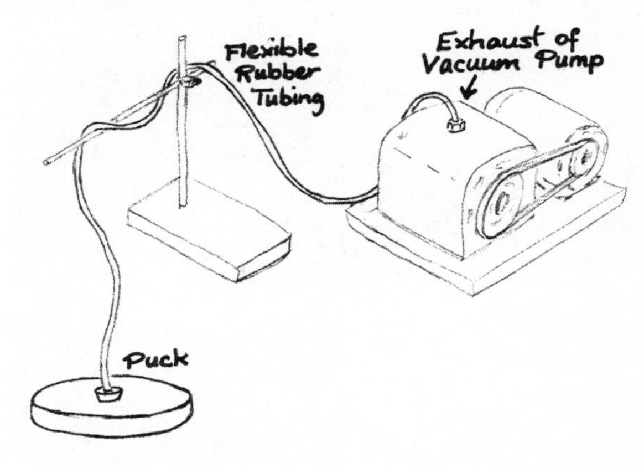

Chapter 2 Free Fall—Galileo Describes Motion

Accelerated motion goes on all around you every day. You experience many accelerations yourself, although not always as exciting as those shown in the photographs. What accelerations have you experienced today?

When you get up from a chair, or start to walk from a standstill, hundreds of sensations are gathered from all over your body in your brain, and you are aware of these normal accelerations. Taking off in a jet or riding on an express elevator, you experience much sharper accelerations. Often this feeling is in the pit of your stomach. These are very complex motions.

Note how stripped down and simple the accelerations are in the following experiments, film loops, activities. As you do these, you will learn to measure accelerations in a variety of ways, both old and new, and become more familiar with the fundamentals of acceleration.

If you do either of the first two experiments of this chapter, that is, numbers 5 and 6, you will try to find, as Galileo did, whether d/t^2 is a constant for motion down an inclined plane. The remaining experiments are measurements of the value of the acceleration due to gravity which is represented by the symbol a_g.

EXPERIMENT 5 A SEVENTEENTH-CENTURY EXPERIMENT

This experiment is similar to the one discussed by Galileo in the *Two New Sciences*. It will give you firsthand experience in working with tools similar to those of a seventeenth-century scientist. You will make quantitative measurements of the motion of a ball rolling down an incline, as described by Galileo.

From these measurements you should be able to decide for yourself whether Galileo's definition of acceleration was appropriate or not. Then you should be able to tell whether it was Aristotle or Galileo who was correct about his thinking concerning the acceleration of objects of different sizes.

Reasoning Behind the Experiment

You have read in Sec. 2.6 of the *Text* how Galileo expressed his belief that the speed of free-falling objects increases in proportion to the time of fall—in other words, that they accelerate uniformly. But since free fall was much too rapid to measure, he assumed that the speed of a ball rolling down an incline increased in the same way as an object in free fall did, only more slowly.

But even a ball rolling down a low incline still moved too fast to measure the speed for different parts of the descent accurately. So Galileo worked on the relationship $d \propto t^2$ (or $d/t^2 = \text{constant}$), an expression in which speed differences have been replaced by the *total time t* and *total distance d* rolled by the ball. Both these quantities can be measured.

Be sure to study *Text* Sec. 2.7 in which the derivation of this relationship is described. If Galileo's original assumptions were true, this relationship would hold for both freely falling objects and rolling balls. Since total distance and total time are not difficult to measure, seventeenth-century scientists now had a secondary hypothesis they could test by experiment. And so have you. Sec. 2.8 of the *Text* discusses much of this.

Apparatus

The apparatus that you will use is shown in Fig. 2-1 below. It is similar to that described by Galileo.

You will let a ball roll various distances down a channel about six feet long and time the motion with a water clock.

You use a water clock to time this experi-

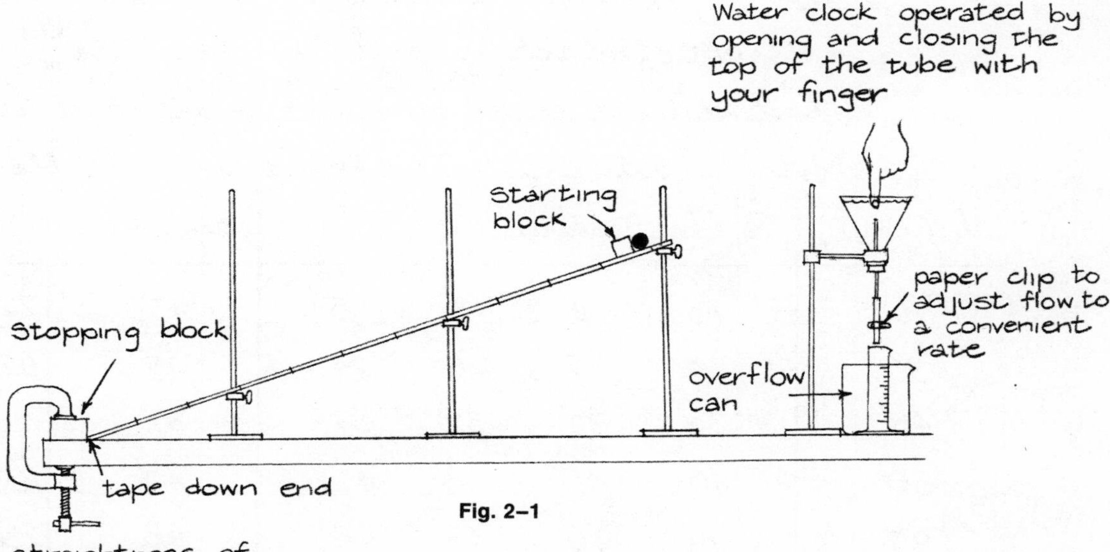

Fig. 2-1

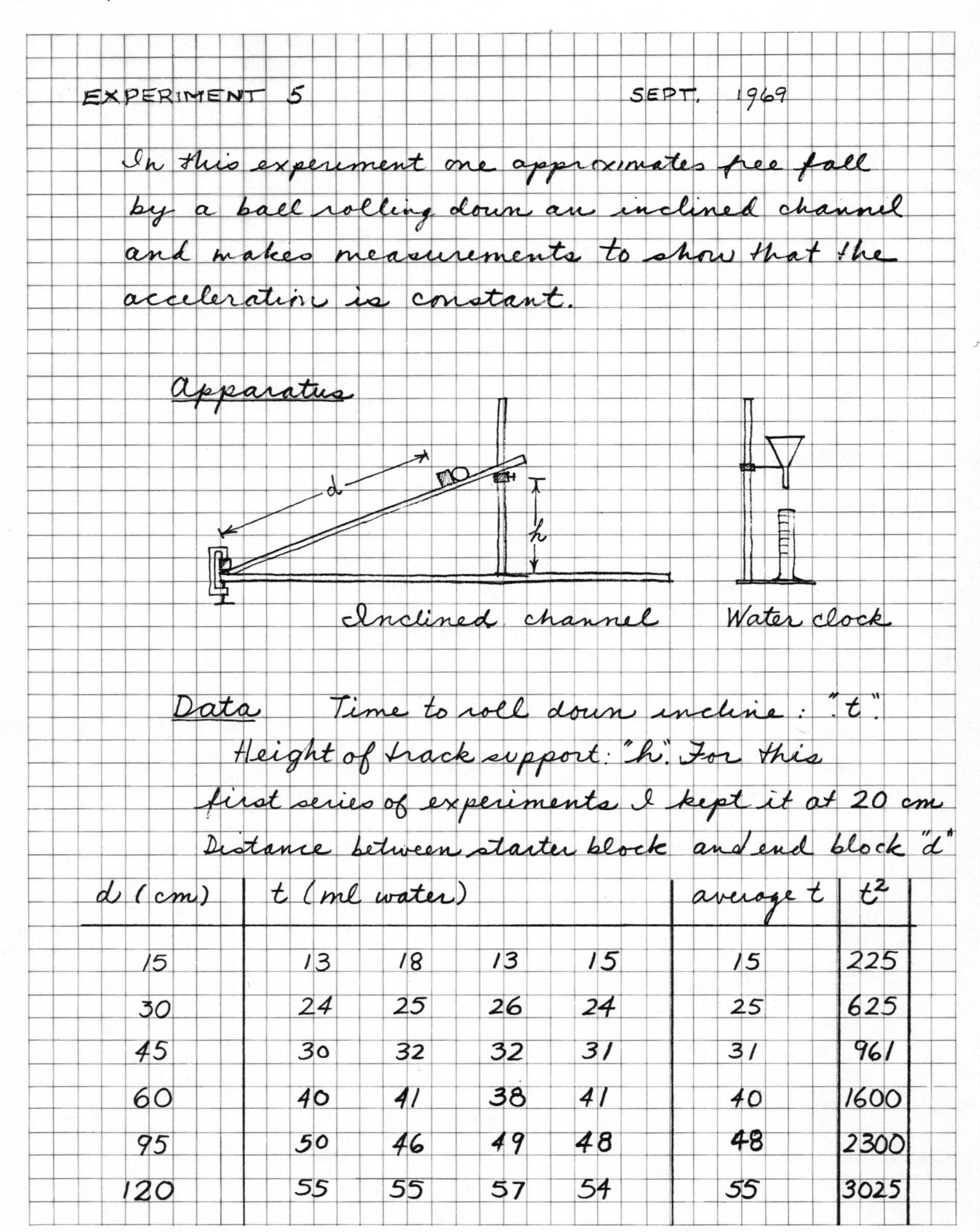

EXPERIMENT 5 SEPT. 1969

In this experiment one approximates free fall by a ball rolling down an inclined channel and makes measurements to show that the acceleration is constant.

Apparatus

d

h

Inclined channel Water clock

Data Time to roll down incline: "t".
Height of track support: "h". For this first series of experiments I kept it at 20 cm
Distance between starter block and end block "d"

d (cm)	t (ml water)				average t	t^2
15	13	18	13	15	15	225
30	24	25	26	24	25	625
45	30	32	32	31	31	961
60	40	41	38	41	40	1600
95	50	46	49	48	48	2300
120	55	55	57	54	55	3025

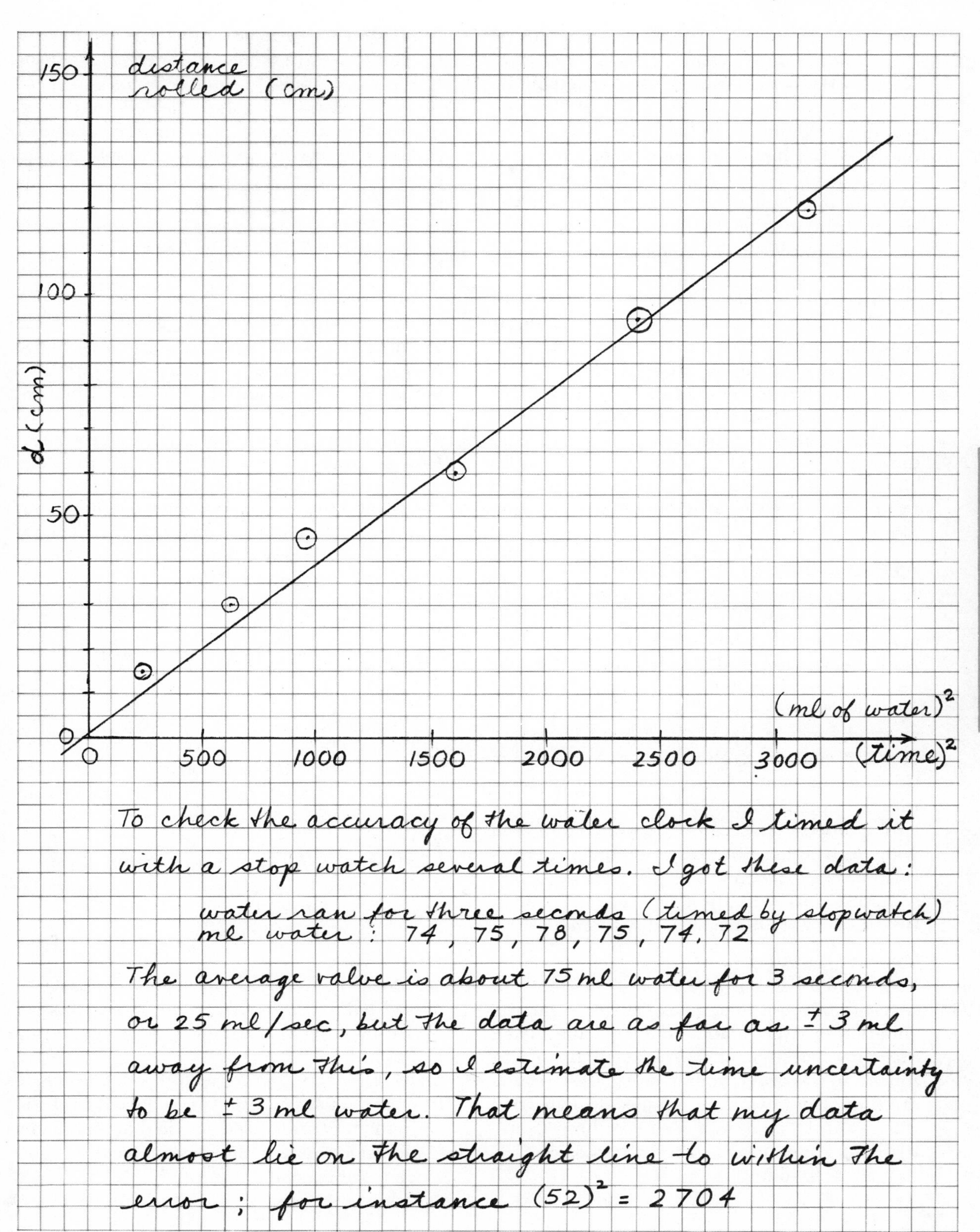

To check the accuracy of the water clock I timed it with a stop watch several times. I got these data:

water ran for three seconds (timed by stopwatch)
ml water : 74, 75, 78, 75, 74, 72

The average value is about 75 ml water for 3 seconds, or 25 ml/sec, but the data are as far as ± 3 ml away from this, so I estimate the time uncertainty to be ± 3 ml water. That means that my data almost lie on the straight line to within the error; for instance $(52)^2 = 2704$

ment because that was the best timing device available in Galileo's time. The way your own clock works is very simple. Since the volume of water is proportional to the time of flow, you can measure time in milliliters of water. Start and stop the flow with your fingers over the upper end of the tube inside the funnel. Whenever you refill the clock, let a little water run through the tube to clear out the bubbles.

Compare your water clock with a stop watch when the clock is full and when it is nearly empty to determine how accurate it is.
Q1 Does the clock's timing change? If so, by how much?

It is almost impossible to release the ball with your fingers without giving it a slight push or pull. Therefore, dam the ball up, with a ruler or pencil, and release it by quickly moving this dam away from it down the inclined plane. The end of the run is best marked by the sound of the ball hitting the stopping block.

Brief Comment on Recording Data

A good example of a way to record your data appears on page 30. We should emphasize again the need for neat, orderly work. Orderly work looks better and is more pleasing to you and everyone else. It may also save you from extra work and confusion. If you have an organized table of data, you can easily record and find your data. This will leave you free to think about your experiment or calculations rather than having to worry about which of two numbers on a scrap of paper is the one you want, or whether you made a certain measurement or not. A few minutes' preparation before you start work will often save you an hour or two of checking in books and with friends.

Operating Suggestions

You should measure times of descent for several different distances, keeping the inclination of the plane constant and using the same ball. Repeat each descent about four times, and average your results. Best results are found for very small angles of inclination (the top of the channel raised less than 30 cm). At greater inclinations, the ball tends to slide as well as to roll.

From Data to Calculations

Galileo's definition of uniform acceleration (*Text,* page 49) was "equal increases in speed in equal times." Galileo showed that if an object actually moved in this way, the total distance of travel should be directly proportional to the square of the total time of fall, or $d \propto t^2$.
Q2 Show how this follows from Galileo's definition. (See Sec. 2.7 in the *Text* if you cannot answer this.)

If two quantities are proportional, a graph of one plotted against the other will be a straight line. Thus, making a graph is a good way to check whether two quantities are proportional. Make a graph of d plotted against t^2.
Q3 Does your graph support the hypothesis? How accurate is the water clock you have been using to time this experiment?

If you have not already done so, check your water clock against a stopwatch or, better yet, repeat several trials of your experiment using a stopwatch for timing.
Q4 How many seconds is one milliliter of time for your water clock? Can the inaccuracy of your water clock explain the conclusion you arrived at in Q2 above?

Going Further

1. In Sec. 2.7 of the *Text* you learned that $a = 2d/t^2$. Use this relation to calculate the actual acceleration of the ball in one of your runs.
2. If you have time, go on to see whether Galileo or Aristotle was right about the acceleration of objects of various sizes. Measure d/t^2 for several different sizes of balls, all rolling the same distance down a plane of the same inclination.

Q5 Does the acceleration depend on the size of the ball? In what way does your answer refute or support Aristotle's ideas on falling bodies.
Q6 Galileo claimed his results were accurate to $\frac{1}{10}$ of a pulse beat. Do you believe his results were that accurate? Did you do that well? How could you improve the design of the water clock to increase its accuracy?

EXPERIMENT 6 TWENTIETH-CENTURY VERSION OF GALILEO'S EXPERIMENT

Galileo's seventeenth-century experiment had its limitations, as you read in the *Text*, Sec. 2.9. The measurement of time with a water clock was imprecise and the extrapolation from acceleration at a small angle of inclination to that at a verticle angle (90°) was extreme.

With more modern equipment you can verify Galileo's conclusions; further, you can get an actual value for acceleration in free fall (near the earth's surface). But remember that the idea behind the improved experiment is still Galileo's. More precise measurements do not always lead to more significant conclusions.

Determine a_g as carefully as you can. This is a fundamental measured value in modern science. It is used in many ways—from the determination of the shape of the earth and the location of oil fields deep in the earth's crust to the calculation of the orbits of earth satellites and spacecrafts in today's important space research programs.

Apparatus and Procedure

For an inclined plane use the air track. For timing the air track glider use a stopwatch instead of the water clock. Otherwise the procedure is the same as that used in Experiment 5. As you go to higher inclinations you should stop the glider by hand before it is damaged by hitting the stopping block.

Instead of a stopwatch, you may wish to use the Polaroid camera to make a strobe photo of the glider as it descends. A piece of white tape on the glider will show up well in the photograph. Or you can attach a small light source to the glider. You can use a magnifier with a scale attached to measure the glider's motion recorded on the photograph. Here the values of d will be millimeters on the photograph and t will be measured in an arbitrary unit, the "blink" of the stroboscope, or the "slot" of the strobe disk.

Plot your data as before on a graph of d vs. t^2. Compare your plotted lines with graphs of the preceding cruder seventeenth-century experiment, if they are available. Explain the differences between them.

Q1 Is d/t^2 constant for an air track glider?

Q2 What is the significance of your answer to the question above?

As further challenge, if time permits, try to predict the value of a_g, which the glider approaches as the air track becomes vertical. To do this, of course, you must express d and t in familiar units such as meters or feet, and seconds. The accepted value of a_g is 9.8 m/sec² or 32 ft/sec² near the earth's surface.

Q3 What is the percentage error in your calculated value? That is, what percent is your error of the accepted value?

Percentage error

$$= \frac{\text{accepted value} - \text{calculated value}}{\text{accepted value}} \times 100$$

so that if your value of a_g is 30 ft/sec² your percentage error

$$= \frac{32\ \text{ft/sec}^2 - 30\ \text{ft/sec}^2}{32\ \text{ft/sec}^2} \times 100$$

$$= \tfrac{2}{32} \times 100 = 6\%$$

Notice that you *cannot* carry this 6% out to 6.25% because you only know the 2 in the fraction $\frac{2}{32}$ to one digit. Hence, you can only know one digit in the answer, 6%. A calculated value like this is said to have one significant digit. You cannot know the second digit in the answer until you know the digit following the 2. To be significant, this digit would require a third digit in the calculated values of 30 and 32.

Q4 What are some of the sources of your error?

EXPERIMENT 7 MEASURING THE ACCELERATION OF GRAVITY a_g

Aristotle's idea that falling bodies on earth are seeking out their natural places sounds strange to us today. After all, we know the answer: It's gravity that makes things fall.

But just what is gravity? Newton tried to give operational meaning to the idea of gravity by seeking out the laws according to which it acts. Bodies near the earth fall toward it with a certain acceleration due to the gravitational "attraction" of the earth. But how can the earth make a body at a distance fall toward it? How is the gravitational force transmitted? Has the acceleration due to gravity always remained the same? These and many other questions about gravity have yet to be answered satisfactorily.

Whether you do one or several parts of this experiment, you will become more familiar with the effects of gravity—you find the acceleration of bodies in free fall yourself—and you will learn more about gravity in later chapters.

*Part A: a_g by Direct Fall**

In this experiment you measure the acceleration of a falling object. Since the distance and hence the speed of fall is too small for air resistance to become important, and since other sources of friction are very small, the acceleration of the falling weight is very nearly a_g.

Doing the Experiment

The falling object is an ordinary laboratory hooked weight of at least 200 g mass. (The drag on the paper strip has too great an effect on the fall of lighter weights.) The weight is suspended from about a meter of paper tape as shown in the photograph. Reinforce the tape by doubling a strip of masking tape over one end and punch a hole in the reinforcement one centimeter from the end. With careful handling, this can support at least a kilogram weight.

*Adapted from R. F. Brinckerhoff and D. S. Taft, *Modern Laboratory Experiments in Physics*, by permission of Science Electronics, Inc., Nashua, New Hampshire.

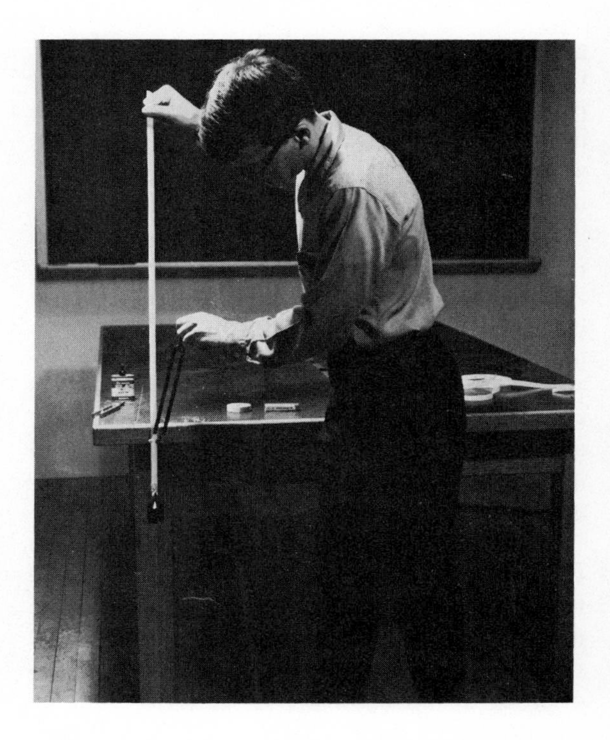

When the suspended weight is allowed to fall, a vibrating tuning fork will mark equal time intervals on the tape pulled down after the weight.

The tuning fork must have a frequency between about 100 vibrations/sec and about 400 vibrations/sec. In order to mark the tape, the fork must have a tiny felt cone (cut from a marking pen tip) glued to the side of one of its prongs close to the end. Such a small mass affects the fork frequency by much less than 1 vibration/sec. Saturate this felt tip with a drop or two of marking pen ink, set the fork in vibration, and hold the tip very gently against the tape. The falling tape is most conveniently guided in its fall by two thumbtacks in the edge of the table. The easiest procedure is to have an assistant hold the weighted tape straight up until you have touched the vibrating tip against it and said "Go." After a few practice runs, you will become expert enough to mark several feet of tape with a wavy line as the tape is accelerated past the stationary vibrating fork.

Instead of using the inked cone, you may press a corner of the vibrating tuning fork gently against a 1-inch square of carbon paper which the thumbtacks hold ink surface inwards over the falling tape. With some practice, this method can be made to yield a series of dots on the tape without seriously retarding its fall.

Analyzing Your Tapes

Label with an **A** one of the first wave crests (or dots) that is clearly formed near the beginning of the pattern. Count 10 intervals between wave crests (or dots), and mark the end of the tenth space with a **B**. Continue marking every tenth crest with a letter throughout the length of the record, which ought to be at least 40 waves long.

At **A,** the tape already had a speed of v_0. From this point to **B,** the tape moved in a time t, a distance we shall call d_1. The distance d_1 is described by the equation of free fall:

$$d_1 = v_0t + \frac{a_gt^2}{2}$$

In covering the distance from **A** to **C,** the tape took a time exactly twice as long, $2t$, and fell a distance d_2 described (on substituting $2t$ for t and simplifying) by the equation:

$$d_2 = 2v_0t + \frac{4a_gt^2}{2}$$

In the same way the distances **AB, AE,** etc., are described by the equations:

$$d_3 = 3v_0t + \frac{9a_gt^2}{2}$$

$$d_4 = 4v_0t + \frac{16a_gt^2}{2}$$

and so on.

All of these distances are measured from A, the arbitrary starting point. To find the distances fallen in each 10-crest interval, you must subtract each equation from the one before it, getting:

$$\mathbf{AB} = v_0t + \frac{a_gt^2}{2}$$

$$\mathbf{BC} = v_0t + \frac{3a_gt^2}{2}$$

$$\mathbf{CD} = v_0t + \frac{5a_gt^2}{2}$$

and

$$\mathbf{DE} = v_0t + \frac{7a_gt^2}{2}$$

From these equations you can see that the weight falls farther during each time interval. Moreover, when you subtract each of these distances, **AB, BC, CD,** . . . from the subsequent distance, you find that the *increase* in distance fallen is a constant. That is, each difference $\mathbf{BC} - \mathbf{AB} = \mathbf{CD} - \mathbf{BC} = \mathbf{DE} - \mathbf{CD} = a_gt^2$. This quantity is the increase in the distance fallen in each successive 10-wave interval and hence is an acceleration. Our formula shows that a body falls with a constant acceleration.

From your measurements of **AB, AC, AD,** etc., make a column of **AB, BC, CD, ED,** etc., and in the next column record the resulting values of a_gt^2. The values of a_gt^2 should all be equal (within the accuracy of your measurements). Why? Make all your measurements as precisely as you can with the equipment you are using.

Find the average of all your values of a_gt^2, the acceleration in centimeters/(10-wave interval)^2. You want to find the acceleration in cm/sec^2. If you call the frequency of the tuning fork n per second, then the length of the time interval t is $10/n$ seconds. Replacing t of 10 waves by $10/n$ seconds gives you the acceleration, a_g in cm/sec^2.

The ideal value of a_g is close to 9.8 m/sec^2, but a small force of friction impeding a falling object is sufficient to reduce the observed value by several percent.

Q1 What errors would be introduced by using a tuning fork whose vibrations are slower than about 100 vibrations per second? higher than about 400 vibrations per second?

Part B: a_g from a Pendulum

You can easily measure the acceleration due to gravity by timing the swinging of a pendulum.

Of course the pendulum is not falling straight down, but the time it takes for a round-trip swing still depends on a_g. The time T it takes for a round-trip swing is

$$T = 2\pi\sqrt{\frac{l}{a_g}}$$

In this formula l is the length of the pendulum. If you measure l with a ruler and T with a clock, you should be able to solve for a_g.

You may learn in a later physics course how to derive the formula. Scientists often use formulas they have not derived themselves, as long as they are confident of their validity.

Making the Measurements

The formula is derived for a pendulum with all the mass concentrated in the weight at the bottom, called the bob. Hence the best pendulum to use is one whose bob is a metal sphere hung by a fine thread. In this case you can be sure that almost all the mass is in the bob. The pendulum's length, l, is the distance from the point of suspension to the *center* of the bob.

Your suspension thread can have any convenient length. Measure l as accurately as possible, either in feet or meters.

Set the pendulum swinging with *small* swings. The formula doesn't work well for large swings, as you can test for yourself later.

Time at least 20 complete round trips, preferably more. By timing many round trips instead of just one you make the error in starting and stopping the clock a smaller fraction of the total time being measured. (When you divide by 20 to get the time for a single round trip, the error in the calculated value for one will be only $\frac{1}{20}$ as large as if you had measured only one.)

Divide the total time by the number of swings to find the time T of one swing.

Repeat the measurement at least once as a check.

Finally, substitute your measured quantities into the formula and solve it for a_g.

If you measured l in meters, the accepted value of a_g is 9.80 meters/sec².

If you measured l in feet, the accepted value of a_g is 32.1 ft/sec².

Finding Errors

You probably did not get the accepted value. Find your *percentage error* by dividing your error by the accepted value and multiplying by 100:

Percentage error

$$= \frac{\text{accepted value} - \text{your value}}{\text{accepted value}} \times 100$$

$$= \frac{\text{your error}}{\text{accepted value}} \times 100$$

With care, your value of a_g should agree within about 1%.

Which of your measurements do you think was the least accurate?

If you believe it was your measurement of length and you think you might be off by as much as 0.5 cm, change your value of l by 0.5 cm and calculate once more the value of a_g. Has a_g changed enough to account for your error? (If a_g went up and your value of a_g was already too high, then you should have altered your measured l in the opposite direction. Try again!)

If your possible error in measuring is not enough to explain your difference in a_g try changing your *total* time by a few tenths of a second—a possible error in timing. Then you must recalculate T and hence a_g.

If neither of these attempts work (nor both taken together in the appropriate direction) then you almost certainly have made an error in arithmetic or in reading your measuring instruments. It is most unlikely that a_g in your school differs from the accepted value by more than one unit in the third digit.

Q2 How does the length of the pendulum affect your value of T? of a_g?

Q3 How long is a pendulum for which $T = 2$ seconds? This is a useful timekeeper.

Part C: a_g with Slow-Motion Photography (Film Loop)

With a high speed movie camera you could photograph an object falling along the edge of a vertical measuring stick. Then you could determine a_g by projecting the film at standard speed and measuring the time for the object to fall specified distance intervals.

A somewhat similar method is used in *Film Loops 1* and 2. Detailed directions are given for their use in the Film Loop notes on pages 40–41.

Part D: a_g from Falling Water Drops

You can measure the acceleration due to gravity a_g simply with drops of water falling on a pie plate.

Put the pie plate or a metal dish or tray on the floor and set up a glass tube with a stopcock, valve, or spigot so that drops of water from the valve will fall at least a meter to the plate. Support the plate on three or four pencils to hear each drop distinctly, like a drum beat.

Adjust the valve carefully until one drop strikes the plate at the same instant the next drop from the valve begins to fall. You can do this most easily by *watching* the drops on the valve while listening for the drops hitting the plate. When you have exactly set the valve, the time it takes a drop to fall to the plate is equal to the time interval between one drop and the next.

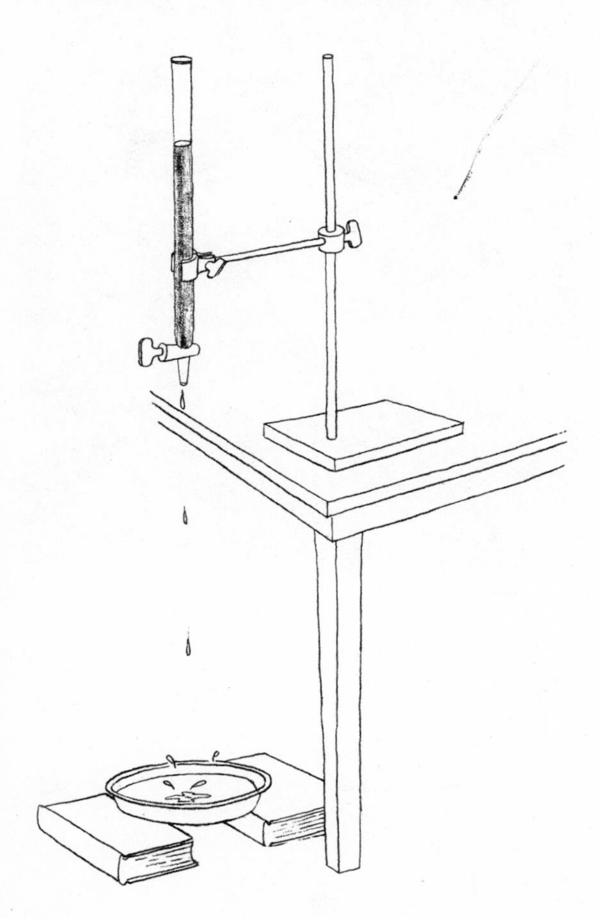

With the drip rate adjusted, now find the time interval t between drops. For greater accuracy, you may want to count the number of drops that fall in half a minute or a minute, or to time the number of seconds for 50 to 100 drops to fall.

Your results are likely to be more accurate if you run a number of trials, adjusting drip rate each time, and average your counts of drops or seconds. The average of several trials should be closer to actual drip rate, drop count, and time intervals than one trial would be.

Now you have all the data you need. You know the time t it takes a drop to fall a distance d from rest. From these you can calculate a_g, since you know that $d = \frac{1}{2}a_g t^2$ for objects falling from rest. Rewrite this relationship in the form $a_g = \ldots$.

Q4 When you have calculated a_g by this method, what is your percentage error? How does this compare with your percentage error by any other methods you have used? What do you think led to your error? Could it be leaking connections, allowing more water to escape sometimes? How would this affect your answer?

Distance of fall lessened by a puddle forming in the plate: How would this change your results?

Less pressure of water in the tube after a period of dripping: Would this increase or decrease the rate of dripping? Do you get the same counts when you refill the tube after each trial?

Would the starting and stopping of your counting against the watch or clock affect your answer? What other things may have shown up in your error?

Can you adapt this method of measuring the acceleration of gravity so that you can do it at home? Would it work in the kitchen sink?

or if the water fell a greater distance, such as down a stairwell?

Part E: a_g with Falling Ball and Turntable

You can measure a_g with a record-player turntable, a ring stand and clamp, carbon paper, two balls with holes in them, and thin thread.

Ball X and ball Y are draped across the prongs of the clamp. Line up the balls along a radius of the turntable, and make the lower ball hang just above the paper.

With the table turning, the thread is burned and each ball, as it hits the carbon paper, will leave a mark on the paper under it.

Measure the vertical distance between the balls and the angular distance between the marks. With these measurements and the speed of the turntable, determine the free-fall time. Calculate your percentage error and suggest its probable source.

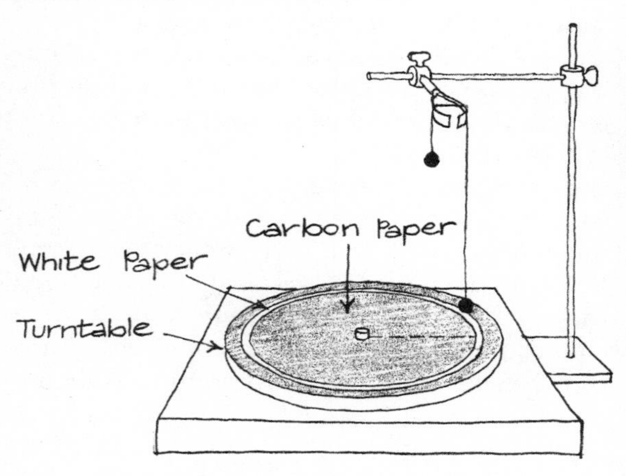

Part F: a_g with Strobe Photography

Photographing a falling light source with the Polaroid Land camera provides a record that can be graphed and analyzed to give an average value of a_g. The 12-slot strobe disk gives a very accurate 60 slots per second. (Or, a neon bulb can be connected to the ac line outlet in such a way that it will flash a precise 60 times per second, as determined by the line frequency. Your teacher has a description of the approximate circuit for doing this.)

ACTIVITIES

WHEN IS AIR RESISTANCE IMPORTANT?

By taking strobe photos of various falling objects, you can find when air resistance begins to play an important role. You can find the actual value of the terminal speed for less dense objects such as a Ping-Pong or styrofoam ball by dropping them from greater and greater heights until the measured speeds do not change with further increases in height. (A Ping-Pong ball achieves terminal speed within 2 m.) Similarly, ball bearings and marbles can be dropped in containers of liquid shampoo or cooking oil to determine factors affecting terminal speed in a liquid as shown in the adjoining photograph.

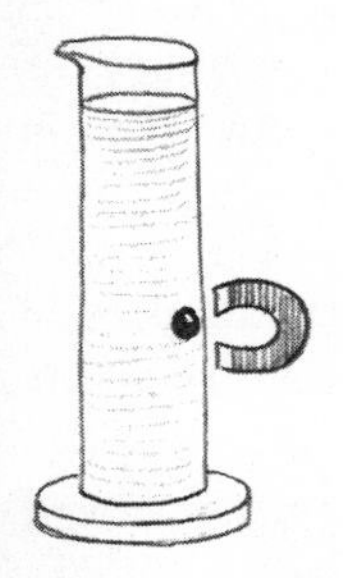

A magnet is a handy aid in raising the steel ball to the top of the container.

MEASURING YOUR REACTION TIME

Your knowledge of physics can help you calculate your reaction time. Have someone hold the top of a wooden ruler while you space your thumb and forefinger around the bottom (zero) end of the ruler. As soon as the other person releases the ruler, you catch it. You can compute your reaction time from the relation

$$d = \frac{1}{2}a_g t^2$$

by solving for t. Compare your reaction time with that of other people, both older and younger than yourself. Also try it under different conditions—lighting, state of fatigue, distracting noise, etc. Time can be saved by computing d for $\frac{1}{10}$ sec or shorter intervals, and then taping reaction-time marks on the ruler.

A challenge is to try this with a one-dollar bill, telling the other person that he can have the dollar if he can catch it.

FALLING WEIGHTS

This demonstration shows that the time it takes a body to fall is proportional to the square root of the vertical distance ($d \propto t^2$). Suspend a string, down a stairwell or out of a window, on which metal weights are attached at the following heights above the ground: 3″, 1′, 2′3″, 4′, 6′3″, 9′, 12′3″, 16′. Place a metal tray or ashcan cover under the string and then drop or cut the string at the point of suspension. The weights will strike the tray at equal intervals of time—about $\frac{1}{8}$ second.

Compare this result with that obtained using a string on which the weights are suspended at equal distance intervals.

EXTRAPOLATION

Many arguments regarding private and public policies depend on how people choose to extrapolate from data they have gathered. From magazines, make a report on the problems of extrapolating in various cases. For example:

1. The population explosion
2. The number of students in your high school ten years from now
3. The number of people who will die in traffic accidents over next holiday weekend
4. The number of lung cancer cases that will occur next year among cigarette smokers
5. How many gallons of punch you should order for your school's Junior prom

To become more proficient in making statistics support your pet theory—and more cautious about common mistakes—read *How to Lie with Statistics* by Darrell Huff, published by W. W. Norton and Company.

FILM LOOPS

FILM LOOP 1 ACCELERATION DUE TO GRAVITY – I

A bowling ball in free fall was filmed in real time and in slow motion. Using the slow-motion sequence, you can measure the acceleration of the ball due to gravity. This film was exposed at 3900 frames/sec and is projected at about 18 frames/sec; therefore, the slow-motion factor is 3900/18, or about 217. However, your projector may not run at exactly 18 frames/sec. To calibrate your projector, time the length of the entire film which contains 3331 frames. (Use the yellow circle as the zero frame.)

To find the acceleration of the falling body using the definition

$$\text{acceleration} = \frac{\text{change in speed}}{\text{time interval}}$$

you need to know the instantaneous speed at two different times. You cannot directly measure instantaneous speed from the film, but you can determine the average speed during small intervals. Suppose the speed increases steadily, as it does for freely falling bodies. During the first half of any time interval, the instantaneous speed is less than the average speed; during the second half of the interval, the speed is greater than average. Therefore, for uniformly accelerated motion, the average speed v_{av} for the interval is the same as the instantaneous speed at the mid-time of the interval.

If you find the instantaneous speed at the midtimes of each of two intervals, you can calculate the acceleration a from

$$a = \frac{v_2 - v_1}{t_2 - t_1}$$

where v_1 and v_2 are the average speeds during the two intervals, and where t_1 and t_2 are the midtimes of these intervals.

Two intervals 0.5 meter in length are shown in the film. The ball falls 1 meter before reaching the first marked interval, so it has some initial speed when it crosses the first line. Using a watch with a sweep second hand, time the ball's motion and record the times at which the ball crosses each of the four lines. You can make measurements using either the bottom edge of the ball or the top edge. With this information, you can determine the time (in apparent seconds) between the midtimes of the two intervals and the time required for the ball to move through each $\frac{1}{2}$-meter interval. Repeat these measurements at least once and then find the average times. Use the slow-motion factor to convert these times to real seconds; then, calculate the two values of v_{av}. Finally, calculate the acceleration a.

This film was made in Montreal, Canada, where the acceleration due to gravity, rounded off to ± 1%, is 9.8 m/sec². Try to decide from the internal consistency of your data (the repeatability of your time measurements) how precisely you should write your result.

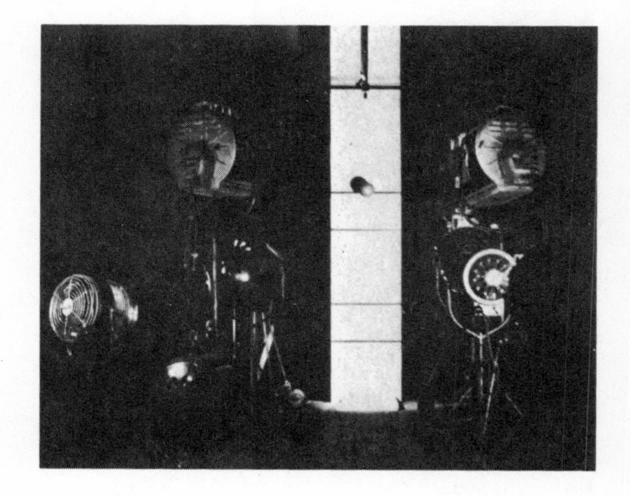

FILM LOOP 2 ACCELERATION DUE TO GRAVITY –II

A bowling ball in free fall was filmed in slow motion. The film was exposed at 3415 frames/sec and it is projected at about 18 frames/sec. You can calibrate your projector by timing the length of the entire film, 3753 frames. (Use the yellow circle as a reference mark.)

If the ball starts from rest and steadily acquires a speed v after falling through a distance d, the change in speed Δv is $v - 0$, or v, and the average speed is $v_{av} = \frac{0+v}{2} = \frac{1}{2}v$. The time required to fall this distance is given by

$$\Delta t = \frac{d}{v_{av}} = \frac{d}{\frac{1}{2}v} = \frac{2d}{v}$$

The acceleration a is given by

$$a = \frac{\text{change of speed}}{\text{time interval}} = \frac{\Delta v}{\Delta t} = \frac{v}{2d/v} = \frac{v^2}{2d}$$

Thus, if you know the instantaneous speed v of the falling body at a distance d below the starting point, you can find the acceleration. Of course you cannot directly measure the instantaneous speed but only average speed over the interval. For a small interval, however, you can make the approximation that the average speed is the instantaneous speed at the midpoint of the interval. (The average speed is the instantaneous speed at the mid*time,* not the mid*point;* but the error is small if you use a short enough interval.)

In the film, small intervals of 20 cm are centered on positions 1m, 2m, 3m, and 4m below the starting point. Determine four average speeds by timing the ball's motion across the 20 cm intervals. Repeat the measurements several times and average out errors of measurement. Convert your measured times into real times using the slow-motion factor. Compute the speeds, in m/sec, and then compute the value of $v^2/2d$ for each value of d.

Make a table of calculated values of a, in order of increasing values of d. Is there any evidence for a systematic trend in the values? Would you expect any such trend? State the results by giving an average value of the acceleration and an estimate of the possible error. This error estimate is a matter of judgment based on the consistency of your four measured values of the acceleration.

The Wizard of Id by Parker and Hart

Chapter 3 The Birth of Dynamics—Newton Explains Motion

EXPERIMENT 8 NEWTON'S SECOND LAW

Newton's second law of motion is one of the most important and useful laws of physics. Review *Text* Sec. 3.7 on Newton's second law to make sure you are familiar with it.

Newton's second law is part of a much larger body of theory than can be proved by any simple set of laboratory experiments. Our experiment on the second law has two purposes.

First, because the law *is* so important, it is useful to get a feeling for the behavior of objects in terms of force (F), mass (m), and acceleration (a). You do this in the first part of the experiment.

Second, the experiment permits you to consider the uncertainties of your measurements. This is the purpose of the latter part of the experiment.

You will apply different forces to carts of different masses and measure the acceleration.

Fig. 3–1

How the Apparatus Works

You are about to find the mass of a loaded cart on which you then exert a measurable force. From Newton's second law you can predict the resulting acceleration of the loaded cart.

Arrange the apparatus as shown in Fig. 3-1. A spring scale is firmly taped to a dynamics cart. The cart, carrying a blinky, is pulled along by a cord attached to the hook of the spring scale. The scale therefore measures the force exerted on the cart.

The cord runs over a pulley at the edge of the lab table and from its end hangs a weight.

Fig. 3–2

(Fig. 3-2.) The hanging weight can be changed so as to produce various tensions in the cord and hence various accelerating forces on the cart.

Now You Are Ready to Go

Measure the total mass of the cart, the blinky, the spring scale, and any other weights you may want to include with it to vary the mass. This is the mass m being accelerated.

Release the cart and allow it to accelerate. Repeat the motion several times while watching the spring-scale pointer. You may notice that the pointer has a range of positions. The midpoint of this range is a fairly good measurement of the average force F_{av} producing the acceleration.

Record F_{av} in newtons.

Our faith in Newton's law is such that we assume the acceleration is the same and is constant every time this particular F_{av} acts on the mass m.

Use Newton's law to predict what the average acceleration a_{av} was during the run.

Then find a directly to see how accurate your prediction was.

To measure the average acceleration a_{av} take a Polaroid photograph through a rotating disk stroboscope of a light source mounted on the cart. As alternatives you might use a liquid surface accelerometer described in detail on page 46, or a blinky. Analyze your results just as in the experiments on uniform and accelerated motion 4, 5, and 6 to find a_{av}.

This time, however, you must know the distance traveled in meters and the time interval in seconds, not just in blinks, flashes or other arbitrary time units.

Q1 Does F_{av} (as measured) equal ma_{av} (as computed from measured values)?

You may wish to observe the following effects without actually making numerical measurements.

1. Keep the mass of the cart constant and observe how various forces affect the acceleration.
2. Keep the force constant and observe how various masses of the cart affect the acceleration.

Q2 Do your observations support Newton's second law? Explain.

Experimental Errors

It is unlikely that your values of F_{av} and ma_{av} were equal.

Does this mean that you have done a poor job of taking data? Not necessarily. As you think about it, you will see that there are at least two other possible reasons for the inequality. One may be that you have not yet measured everything necessary in order to get an accurate value for each of your three quantities.

In particular, the force used in the calculation ought to be the net, or resultant, force on the cart—not just the towing force that you measured. Friction force also acts on your cart, opposing the accelerating force. You can measure it by reading the spring scale as you tow the cart by hand at *constant speed.* Do it several times and take an average, F_f. Since F_f acts in a direction opposite to the towing force F_T,

$$F_{net} = F_T - F_f.$$

If F_f is too small to measure, then $F_{net} = F_T$, which is simply the towing force that you wrote as F_{av} in the beginning of the experiment.

Another reason for the inequality of F_{av} and m_{av} may be that your value for each of these quantities is based on *measurements* and every measurement is uncertain to some extent.

You need to estimate the uncertainty of each of your measurements.

Uncertainty in average force F_{av} Your uncertainty in the measurement of F_{av} is the amount by which your reading of your spring scale varied above and below the average force, F_{av}. Thus if your scale reading ranged from 1.0 to 1.4N the average is 1.2N, and the range of uncertainty is 0.2N. The value of F_{av} would be reported as 1.2 ± 0.2N.

Q3 What is *your* value of F_{av} and its uncertainty?

Uncertainty in mass m Your uncertainty in m is roughly half the smallest scale reading of the balance with which you measured it. The mass consisted of a cart, a blinky, and a spring scale (and possibly an additional mass). If the smallest scale reading is 0.1 kg, your record of the mass of each of these in kilograms might be as follows:

$$m_{cart} = 0.90 \pm 0.05 \text{ kg}$$
$$m_{blinky} = 0.30 \pm 0.05 \text{ kg}$$
$$m_{scale} = 0.10 \pm 0.05 \text{ kg}$$

The total mass being accelerated is the sum of these masses. The uncertainty in the total mass is the sum of the three uncertainties. Thus, in our example, $m = 1.30 \pm 0.15$ kg.

Q4 What is *your* value of m and its uncertainty?

Uncertainty in average acceleration a_{av} Finally, consider a_{av}. You found this by measuring $\Delta d/\Delta t$ for each of the intervals between the points on your blinky photograph.

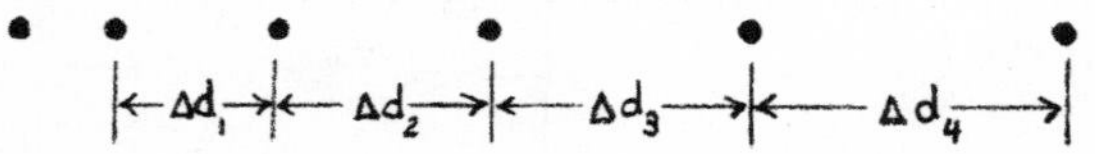

Fig. 3–3

Suppose the points in Fig. 3-3 represent images of a light source photographed through a single slot—giving 5 images per second. Calculate $\Delta d/\Delta t$ for several intervals.

If you assume the time between blinks to have been measured very accurately, the uncertainty in each value of $\Delta d/\Delta t$ is due primarily to the fact that the photographic images are a bit fuzzy. Suppose that the uncertainty in locating the distance between the centers of the dots is 0.1 cm as shown in the first column of the Table below.

Average speeds	Average accelerations
$\Delta d_1/\Delta t = 2.5 \pm 0.1$ cm/sec	$\Delta v_1/\Delta t = 0.9 \pm 0.2$ cm/sec²
$\Delta d_2/\Delta t = 3.4 \pm 0.1$ cm/sec	$\Delta v_2/\Delta t = 0.6 \pm 0.2$ cm/sec²
$\Delta d_3/\Delta t = 4.0 \pm 0.1$ cm/sec	$\Delta v_3/\Delta t = 0.8 \pm 0.2$ cm/sec²
$\Delta d_4/\Delta t = 4.8 \pm 0.1$ cm/sec	Average $= 0.8 \pm 0.2$ cm/sec²

When you take the differences between successive values of the speeds, $\Delta d/\Delta t$, you get the accelerations, $\Delta v/\Delta t$, which are recorded in the second column. When a difference in two measurements is involved, you find the uncertainty of the differences (in this case, $\Delta v/\Delta t$) by *adding* the uncertainties of the two measurements. This results in an uncertainty in acceleration of $(\pm 0.1) + (\pm 0.1)$ or ± 0.2 cm/sec² as recorded in the table.

Q5 What is *your* value of a_{av} and its uncertainty?

Comparing Your Results

You now have values of F_{av}, m and a_{av}, their uncertainties, and you consider the uncertainty of ma_{av}. When you have a value for the uncertainty of this *product* of two quantities, you will then compare the value of ma_{av} with the value of F_{av} and draw your final conclusions. For convenience, we have dropped the "av" from the symbols in the equations in the following discussion. When two quantities are multiplied, the *percentage* uncertainty in the product never exceeds the sum of the percentage uncertainties in each of the factors. In our example, $m \times a = 1.30$ kg $\times$ 0.8 cm/sec² $= 1.04$ newtons. The uncertainty in a (0.8 $\pm$ 0.2 cm/sec²) is 25% (since 0.2 is 25% of 0.8). The uncertainty in m is 11%. Thus the uncertainty in ma is 25% + 11% = 36% and we can write our product as $ma = 1.04$ N $\pm 0.36\%$ which is, to two significant figures,

$$ma = 1.04 \pm 0.36 \text{ N}$$

(The error is so large here that it really isn't appropriate to use the two decimal places; we could round off to 1.0 $\pm$ 0.4 N.) In our example we found from direct measurement that $F_{net} = 1.2 \pm 0.2$ N. Are these the same quantity?

Although 1.0 does not equal 1.2, the range of 1.0 $\pm$ 0.4 overlaps the range of 1.2 + 0.2, so we can say that "the two numbers agree within the range of uncertainty of measurement."

An example of lack of agreement would be 1.0 $\pm$ 0.2 and 1.4 $\pm$ 0.1. These are presumably not the same quantity since there is no overlap of expected uncertainties.

In a similar way, work out your own values of F_{net} and ma_{av}.

Q6 Do your own values agree within the range of uncertainty of your measurement?

Q7 Is the relationship $F_{net} = ma_{av}$ consistent with your observations?

CHAPTER 3

EXPERIMENT 9 MASS AND WEIGHT

You know from your own experience that an object that is pulled strongly toward the earth (like, say, an automobile) is difficult to accelerate by pushing. In other words, objects with great weight also have great inertia. But is there some simple, exact relationship between the masses of objects and the gravitational forces acting on them? For example, if one object has twice the mass of another, does it also weigh twice as much?

Measuring Mass

The masses of two objects can be compared by observing the accelerations they each experience when acted on by the same force. Accelerating an object in one direction with a constant force for long enough to take measurements is often not practical in the laboratory. Fortunately there is an easier way. If you rig up a puck and springs between two rigid supports as shown in the diagram, you

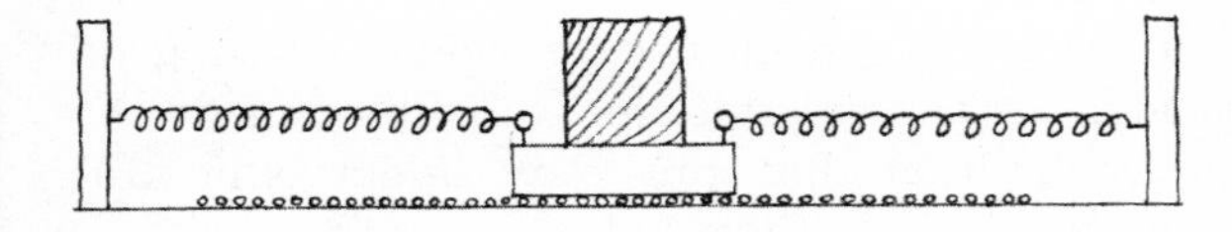

can attach objects to the puck and have the springs accelerate the object back and forth. The greater the mass of the object, the less the magnitude of acceleration will be, and the longer it will take to oscillate back and forth.

To "calibrate" your oscillator, first time the oscillations. The time required for 5 complete round trips is a convenient measure. Tape pucks on top of the first one, and time the period for each new mass. (The units of mass are not essential here, for we will be interested only in the ratio of masses.) Then plot a graph of mass against the oscillation period, drawing a smooth curve through your experimental plot points. Do not leave the pucks stuck together.

Q1 Does there seem to be a simple relationship between mass and period? Could you write an algebraic expression for the relationship?

Weight

To compare the gravitational forces on two objects, they can be hung on a spring scale. In this investigation the units on the scale are not important, because we are interested only in the ratio of the weights.

Comparing Mass and Weight

Use the puck and spring oscillator and calibration graph to find the masses of two objects (say, a dry cell and a stapler). Find the gravitational pulls on these two objects by hanging each from a spring scale.

Q2 How does the ratio of the gravitational forces compare to the ratio of the masses?

Q3 Describe a similar experiment that would compare the masses of two iron objects to the magnetic forces exerted on them by a large magnet.

You probably will not be surprised to find that, to within your uncertainty of measurement, the ratio of gravitational forces is the same as the ratio of masses. Is this really worth doing an experiment to find out, or is the answer obvious to begin with? Newton didn't think it was obvious. He did a series of very precise experiments using many different substances to find out whether gravitational force was always proportional to inertial mass. To the limits of his precision, he found the proportionality to hold exactly. (Newton's results have been confirmed to a precision of $\pm 0.000000001\%$, and extended to gravitational attraction to bodies other than the earth).

Newton could offer no explanation from his physics as to why the attraction of the earth for an object should increase in exact proportion to the object's inertia. No other forces bear such a simple relation to inertia, and this remained a complete puzzle for two centuries until Einstein related inertia and gravitation theoretically. (See "Outside and Inside the Elevator" in the Unit 5 *Reader*.) Even before Einstein, Ernst Mach made the ingenious suggestion that inertia is not the property of an object by itself, but is the result of the gravitational forces exerted on an object by everything else in the universe.

ACTIVITIES

CHECKER SNAPPING

Stack several checkers. Put another checker on the table and snap it into the stack. On the basis of Newton's first law, can you explain what happened?

BEAKER AND HAMMER

One teacher suggests placing a glass beaker half full of water on top of a pile of three wooden blocks. Three quick back-and-forth swipes (NOT FOUR!) of a hammer leave the beaker sitting on the table.

PULLS AND JERKS

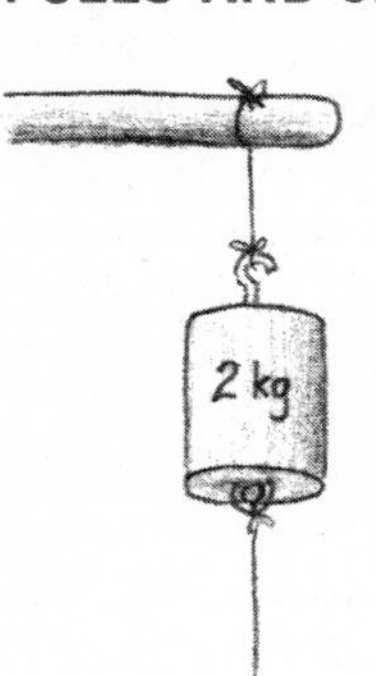

Hang a weight (such as a heavy wooden block) by a string that just barely supports it, and tie another identical string below the weight. A slow, steady pull on the string below the weight breaks the string *above* the weight. A quick jerk breaks it *below* the weight. *Why*?

EXPERIENCING NEWTON'S SECOND LAW

One way for you to get the feel of Newton's second law is actually to pull an object with a constant force. Load a cart with a mass of several kilograms. Attach one end of a long rubber band to the cart and, pulling on the other end, move along at such a speed that the rubber band is maintained at a constant length —say 70 cm. Holding a meter stick above the band with its 0-cm end in your hand will help you to keep the length constant.

The acceleration will be very apparent to the person applying the force. Vary the mass on the cart and the number of rubber bands (in parallel) to investigate the relationship between F, m, and a.

MAKE ONE OF THESE ACCELEROMETERS

An accelerometer is a device that measures acceleration. Actually, anything that has mass could be used for an accelerometer. Because you have mass, you were acting as an accelerometer the last time you lurched forward in the seat of your car as the brakes were applied. With a knowledge of Newton's laws and certain information about you, anybody who measured how far you leaned forward and how tense your muscles were would get a good idea of the magnitude and direction of the acceleration that you were undergoing. But it would be complicated.

Here are four accelerometers of a much simpler kind. With a little practice, you can learn to read accelerations from them directly, without making any difficult calculations.

A. The Liquid-Surface Accelerometer

This device is a hollow, flat plastic container

By permission of John Hart and Field Enterprises, Inc.

partly filled with a colored liquid. When it is not being accelerated, the liquid surface is horizontal, as shown by the dotted line in Fig. 3–4. But when it is accelerated toward the left (as shown) with a uniform acceleration a, the surface becomes tilted, with the level of the liquid rising a distance h above its normal position at one end of the accelerometer and falling the same distance at the other end. The greater the acceleration, the more steeply the surface of the liquid is slanted. This means that the slope of the surface is a measure of the magnitude of the acceleration a.

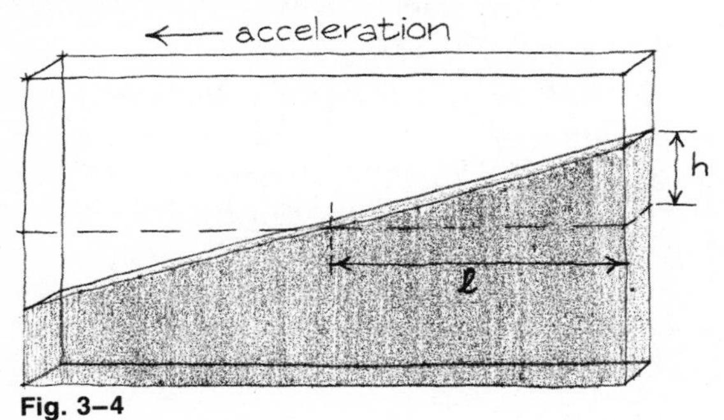

Fig. 3–4

The length of the accelerometer is $2l$, as shown in Fig. 3–4 above. So the slope of the surface may be found by

$$\text{slope} = \frac{\text{vertical distance}}{\text{horizontal distance}} = \frac{2h}{2l} = \frac{h}{l}$$

Theory gives you a very simple relationship between this slope and the acceleration a:

$$\text{slope} = \frac{h}{l} = \frac{a}{a_g}$$

Notice what this equation tells you. It says that if the instrument is accelerating in the direction shown with just a_g (one common way to say this is that it has a "one-G acceleration"), the acceleration of gravity, then the slope of the surface is just 1; that is, $h = l$ and the surface makes a 45° angle with its normal, horizontal direction. If it is accelerating with $\frac{1}{2}$ a_g, then the slope will be $\frac{1}{2}$; that is $h = \frac{1}{2}\,l$. In the same way, if $h = \frac{1}{4}\,l$, then $a = \frac{1}{4}\,a_g$, and so on with any acceleration you care to measure.

To measure h, stick a piece of centimeter tape on the front surface of the accelerometer as shown in Fig. 3–5 below. Then stick a piece of white paper or tape to the back of the instrument to make it easier to read the level of the liquid. Solving the equation above for a gives

$$a = a_g \times \frac{h}{l}$$

white paper on back of cell
Unaccelerated liquid level
cm scale on front of cell
$l = 9.8$ cm
Accelerated liquid surface

Fig. 3–5

B.C. by John Hart

This shows that if you place a scale 10 scale units away from the center you can read accelarations directly in $\frac{1}{10}$th's of "G's." Since a_g is very close to 9.8m/sec^2 at the earth's surface if you place the scale 9.8 scale units from the center you can read accelerations directly in m/sec^2. For example, if you stick a centimeter tape just 9.8 cm from the center of the liquid surface, one cm on the scale is equivalent to an acceleration of one m/sec^2.

Calibration of the Accelerometer

You do not have to trust blindly the theory mentioned above. You can test it for yourself. Does the accelerometer really measure accelerations directly in m/sec^2? Stroboscopic methods give you an independent check on the correctness of the theoretical prediction.

Set the accelerometer on a dynamics cart and arrange strings, pulleys, and masses as you did in Experiment 9 to give the cart a uniform acceleration on a long tabletop. Don't forget to put a block of wood at the end of the cart's path to stop it. Make sure that the accelerometer is fastened firmly enough so that it will not fly off the cart when it stops suddenly. Make the string as long as you can, so that you use the entire length of the table.

Give the cart a wide range of accelerations by hanging different weights from the string. Use a stroboscope to record each motion. To measure the accelerations from your strobe records, plot t^2 against d, as you did in Experiment 5. (What relationship did Galileo discover between d/t^2 and the acceleration?) Or use the method of analysis you need in Experiment 9.

Compare your stroboscopic measurements with the readings on the accelerometer during each motion. It takes some cleverness to read the accelerometer accurately, particularly near the end of a high-acceleration run. One way is to have several students along the table observe the reading as the cart goes by; use the average of their reports. If you are using a xenon strobe, of course, the readings on the accelerometer will be visible in the photograph; this is probably the most accurate method.

Plot the accelerometer readings against the stroboscopically measured accelerations. This graph is called a "calibration curve." If the two methods agree perfectly, the graph will be a straight line through the origin at a 45° angle to each axis. If your curve turns out to have some other shape, you can use it to convert "accelerometer readings" to "accelerations"—if you are willing to assume that your strobe measurements are more accurate than the accelerometer. (If you are not willing, what can you do?)

B. Automobile Accelerometer–I

With a liquid-surface accelerometer mounted on the front-back line of a car, you can measure the magnitude of acceleration along its path. Here is a modification of the liquid-surface design that you can build for yourself. Bend a small glass tube (about 30 cm long) into a U-shape, as shown in Fig. 3–6 below.

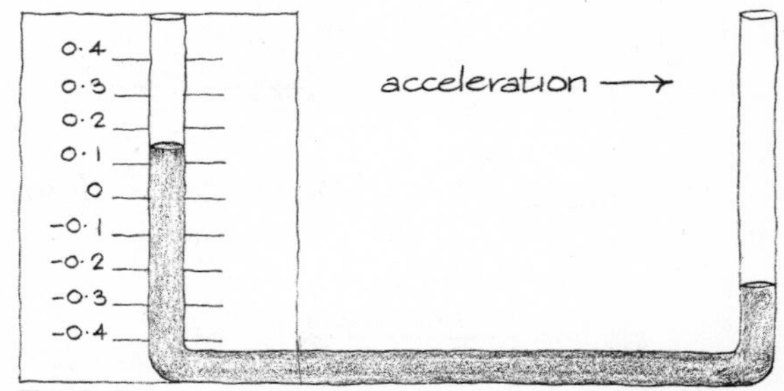

Fig. 3–6

Calibration is easiest if you make the long horizontal section of the tube just 10 cm long; then each 5 mm on a vertical arm represents an acceleration of $\frac{1}{10}$ g = (about) 1 m/sec^2, by the same reasoning as before. The two vertical arms should be at least three-fourths as long as the horizontal arm (to avoid splashing out the liquid during a quick stop). Attach a scale to one of the vertical arms, as shown. Holding the long arm horizontal, pour colored water into the tube until the water level in the arm comes up to the zero mark. How can you be sure the long arm is horizontal?

To mount your accelerometer in a car, fasten the tube with staples (carefully) to a piece of plywood or cardboard a little bigger than the U-tube. To reduce the hazard from broken glass while you do this, cover all but the scale (and the arm by it) with cloth or cardboard, but leave both ends open. It is essential that the accelerometer be horizontal if its readings are to be accurate. When you are measuring acceleration in a car, be sure the road is level. Otherwise, you will be reading the tilt of the car as well as its acceleration. When a car accelerates—in any direction—it tends to tilt on the suspension. This will introduce error in the accelerometer readings. Can you think of a way to avoid this kind of error?

C. Automobile Accelerometer–II

An accelerometer that is more directly related to $F = ma$ can be made from a 1-kg cart and a spring scale marked in newtons. The spring scale is attached between a wood frame and the cart as in the sketch below. If the frame is kept level, the acceleration of the system can

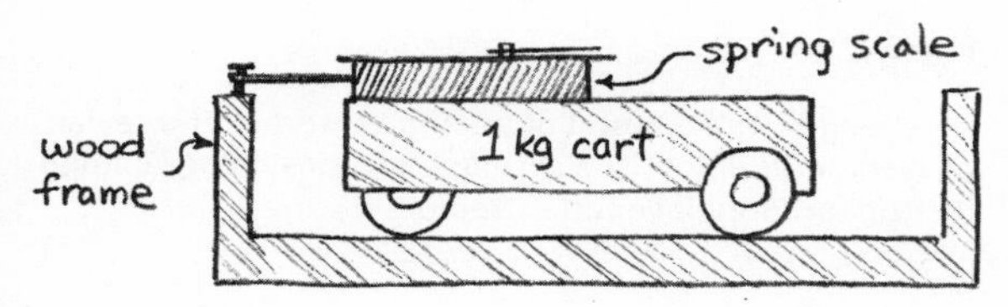

be read directly from the spring scale, since one newton of force on the 1-kg mass indicates an acceleration of one m/sec^2. (Instead of a cart, any 1-kg object can be used on a layer of low-friction plastic beads.)

D. Damped-Pendulum Accelerometer

One advantage of liquid-surface accelerometers is that it is easy to put a scale on them and read accelerations directly from the instrument. They have a drawback, though; they give only the component of acceleration that is parallel to their horizontal side. If you accelerate one at right angles to its axis, it doesn't register any acceleration at all. And if you don't know the direction of the acceleration, you have to use trial-and-error methods to find it with the accelerometers we have discussed up to this point.

A damped-pendulum accelerometer, on the other hand, indicates the direction of any horizontal acceleration; it also gives the magnitude, although less directly than the previous instruments do.

Hang a small metal pendulum bob by a short string fastened to the middle of the lid of a one-quart mason jar as shown on the left hand side of the sketch at the bottom of the page. Fill the jar with water and screw the lid on tight. For any position of the pendulum, the angle that it makes with the vertical depends upon your position. What would you see, for example, if the bottle were accelerating straight toward you? Away from you? Along a table with you standing at the side? (Careful: this last question is trickier than it looks.

To make a fascinating variation on the damped-pendulum accelerometer, simply replace the pendulum bob with a cork and turn the bottle upside down as shown on the right hand side of the sketch at the bottom of the page. If you have punched a hole in the bottle lid to fasten the string, you can prevent leakage with the use of sealing wax, parafin, or tape.

This accelerometer will do just the opposite from what you would expect. The explanation of this odd behavior is a little beyond the scope of this course: it is thoroughly explained in *The Physics Teacher*, vol. 2, no. 4 (April 1964) page 176.

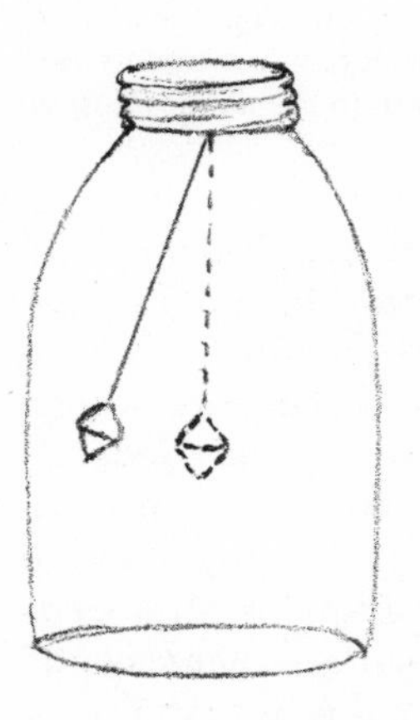

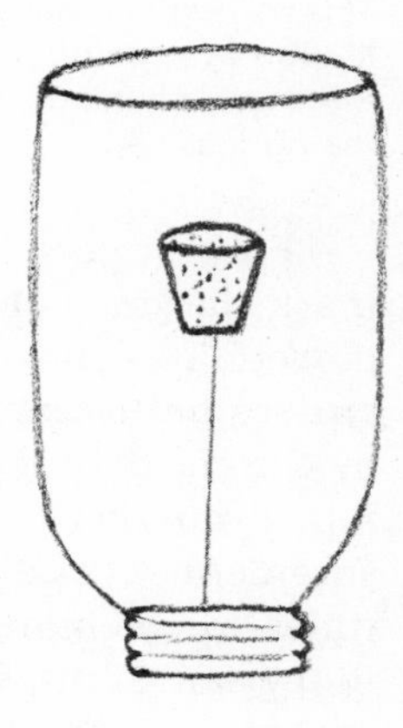

FILM LOOP

FILM LOOP 3 VECTOR ADDITION— VELOCITY OF A BOAT

A motorboat was photographed from a bridge in this film. The boat heads upstream, then downstream, then directly across stream, and at an angle across the stream. The operator of the boat tried to keep the throttle at a constant setting to maintain a steady speed relative to the water. The task before you is to find out if he succeeded.

This photograph was taken from one bank of the stream. It shows the motorboat heading across the stream and the camera filming this loop fixed on the scaffolding on the bridge.

First project the film on graph paper and mark the lines along which the boat's image moves. You may need to use the reference crosses on the markers. Then measure speeds by timing the motion through some predetermined number of squares. Repeat each measurement several times, and use the average times to calculate speeds. Express all speeds in the same unit, such as "squares per second" (or "squares per cm" where cm refers to measured separations between marks on the moving paper of a dragstrip recorder). Why is there no need to convert the speeds to meters per second? Why is it a good idea to use a large distance between the timing marks on the graph paper?

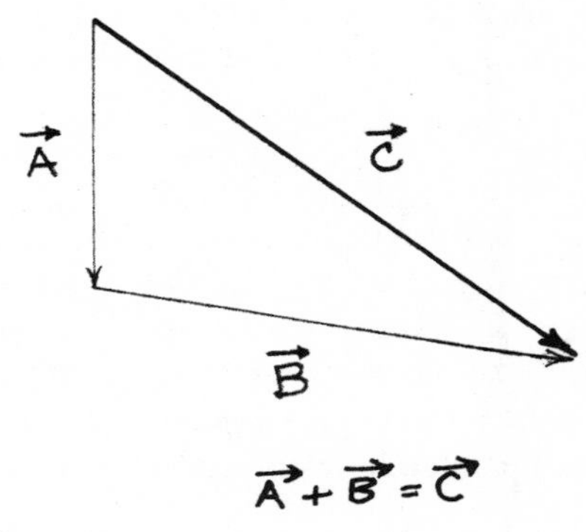

Fig. 3–7

The head-to-tail method of adding vectors. For a review of vector addition see Project Physics Programmed instruction Booklet entitled *Vectors II*.

The head-to-tail method of adding vectors is illustrated in Fig. 3-7. Since velocity is a vector with both magnitude and direction, you can study vector addition by using velocity vectors. An easy way of keeping track of the velocity vectors is by using subscripts:

$\vec{v}_{BE}$ velocity of boat relative to earth

$\vec{v}_{BW}$ velocity of boat relative to water

$\vec{v}_{WE}$ velocity of water relative to earth

Then

$$\vec{v}_{BE} = \vec{v}_{BW} + \vec{v}_{WE}$$

For each heading of the boat, a vector diagram can be drawn by laying off the velocities to scale. A suggested procedure is to record data (direction and speed) for each of the five scenes in the film, and then draw the vector diagram for each.

Scene 1: Two blocks of wood are dropped overboard. Time the blocks. Find the speed of the river, the magnitude of v_{WE}.

Scene 2: The boat heads upstream. Measure $\vec{v}_{BE}$, then find $\vec{v}_{BW}$ using a vector diagram similar to Fig. 3-8.

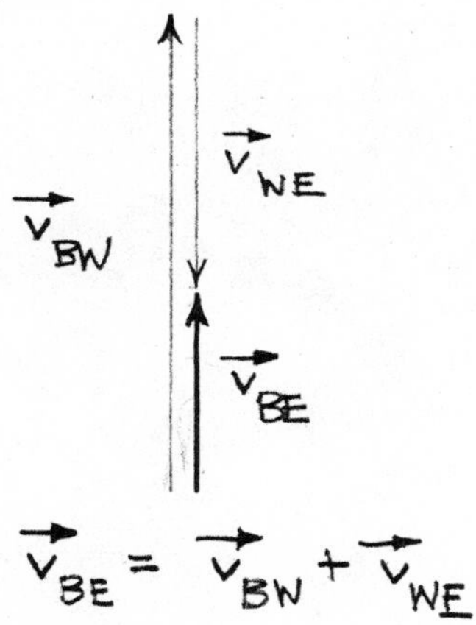

Fig. 3–8

Scene 3: The boat heads downstream. Measure $\vec{v}_{BE}$, then find $\vec{v}_{BW}$ using a vector diagram.

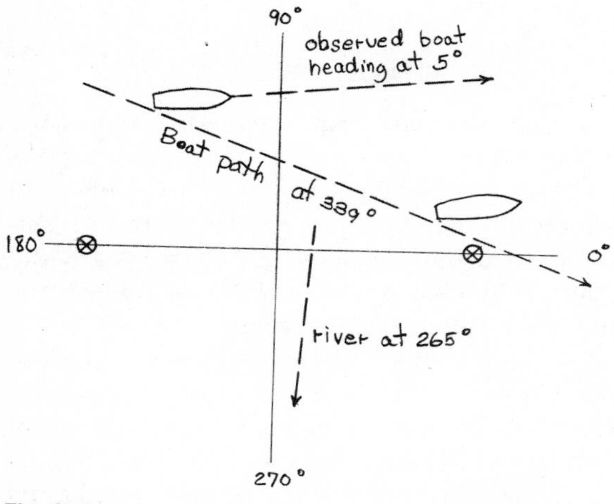

Fig. 3–9

Scene 4: The boat heads across stream and drifts downstream. Measure the speed of the boat and the direction of its path to find $\vec{v}_{BE}$. Also measure the *direction* of $\vec{v}_{BW}$, the direction the boat points. One way to record data is to use a set of axes with the 0° - 180° axis passing through the markers anchored in the river. A diagram, such as Fig. 3-9, will help you record and analyze your measurements. (Note that the numbers in the diagram are deliberately not correct.) Your vector diagram should be something like Fig. 3-10.

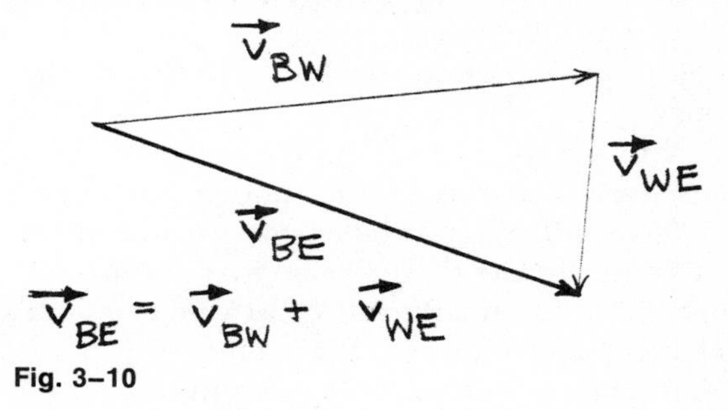

Fig. 3–10

Scene 5: The boat heads upstream at an angle, but moves directly across stream. Again find a value for $\vec{v}_{BW}$.

Checking your work: (a) How well do the four values of the magnitude of $\vec{v}_{BW}$ agree with each other? Can you suggest reasons for any discrepancies? (b) From scene 4, you can *calculate* the heading of the boat. How well does this angle agree with the *observed* boat heading? (c) In scene 5, you determine a direction for $\vec{v}_{BW}$. Does this angle agree with the observed boat heading?

Chapter 4 Understanding Motion

EXPERIMENT 10 CURVES OF TRAJECTORIES

Imagine you are a ski-jumper. You lean forward at the top of the slide, grasp the railing on each side, and yank yourself out into the track. You streak down the trestle, crouch and give a mighty leap at the takeoff lip, and soar up and out, looking down at tiny fields far below. The hill flashes into view and you thump on its steep incline, bobbing to absorb the impact.

This exciting experience involves a more complex set of forces and motions than you can deal with in the laboratory at one time. Let's concentrate therefore on just one aspect: your flight through the air. What kind of a path, or trajectory, would your flight follow?

At the moment of projection into the air a skier has a certain velocity (that is, a certain speed in a given direction), and throughout his flight he must experience the downward acceleration due to gravity. These are circumstances that we can duplicate in the laboratory. To be sure, the flight path of an actual ski-jumper is probably affected by other factors, such as air velocity and friction; but we now know that it usually pays to begin experiments with a simplified approximation that allows us to study the effects of a few factors at a time. Thus, in this experiment you will launch a steel ball from a ramp into the air and try to determine the path it follows.

How to Use the Equipment

If you are assembling the equipment for this experiment for the first time, follow the manufacturer's instructions.

The apparatus being used by the students in the photograph on page 53 consists of two ramps down which you can roll a steel ball. Adjust one of the ramps (perhaps with the help of a level) so that the ball leaves it horizontally.

Tape a piece of squared graph paper to the plotting board with its left-hand edge behind the end of the launching ramp.

To find a path that extends all across the graph paper, release the ball from various points up the ramp until you find one from which the ball falls close to the bottom right-hand corner of the plotting board. Mark the point of release on the ramp and release the ball each time from this point.

Attach a piece of carbon paper to the impact board, with the carbon side facing the ramp. Then tape a piece of thin onionskin paper over the carbon paper.

Now when you put the impact board in its way, the ball hits it and leaves a mark that you can see through the onionskin paper, automatically recording the point of impact between ball and board. (Make sure that the impact board doesn't move when the ball hits it; steady the board with your hand if necessary.) Transfer the point to the plotting board by making a mark on it just next to the point on the impact board.

Do not hold the ball in your fingers to release it—it is impossible to let go of it in the same way every time. Instead, dam it up with

a ruler held at a mark on the ramp and release the ball by moving the ruler quickly away from it down the ramp.

Try releasing the ball several times (always from the same point) for the same setting of the impact board. Do all the impact points exactly coincide?

Repeat this for several positions of the impact board to record a number of points on the ball's path. Move the board *equal distances* every time and always release the ball from the same spot on the ramp. Continue until the ball does not hit the impact board any longer.

Now remove the impact board, release the ball once more, and watch carefully to see that the ball moves along the points marked on the plotting board.

The curve traced out by your plotted points represents the *trajectory* of the ball. By observing the path the ball follows, you have completed the first phase of the experiment.

If you have time, you will find it worth while to go further and explore some of the properties of your trajectory.

Analyzing Your Data

To help you analyze the trajectory, draw a horizontal line on the paper at the level of the end of the launching ramp. Then remove the paper from the plotting board and draw a smooth continuous curve through the points as shown in the figure at the bottom of the page.

You already know that a moving object on which there is no net force acting will move at constant speed. There is no appreciable horizontal force acting on the ball during its fall, so we can make an *assumption* that its horizontal progress is at a constant speed. Then equally spaced lines will indicate equal time intervals.

Draw vertical lines through the points on your graph. Make the first line coincide with the end of the launching ramp. *Because of your plotting procedure* these lines should be equally spaced. If the horizontal speed of the ball is uniform, these vertical lines are drawn through positions of the ball separated by equal time intervals.

Now consider the vertical distances fallen in each time interval. Measure down from your horizontal line the vertical fall to each of your

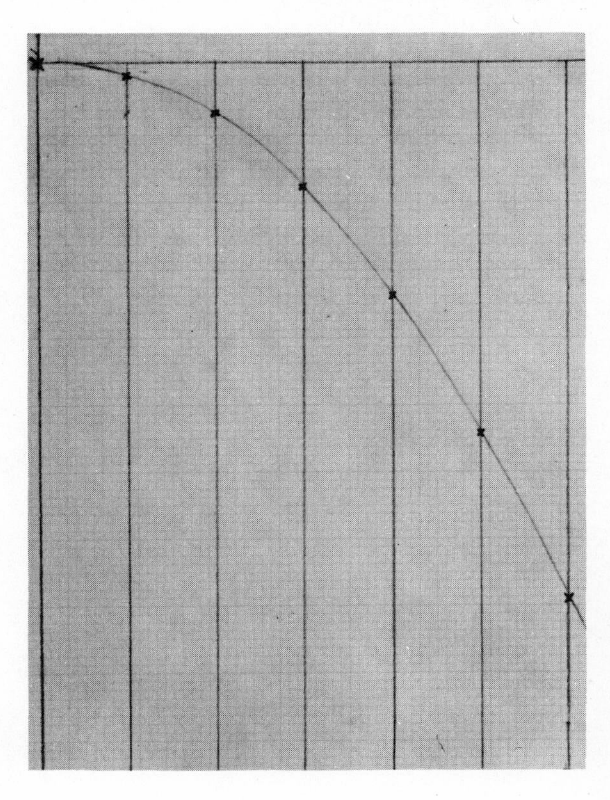

plotted points. Record your measurements in a column. Alongside them record the corresponding horizontal distances measured from the first vertical line. A sample of results as recorded in a student notebook is shown on the right.

Q1 What would a graph look like on which you plot horizontal distance against time?

Earlier in your work with accelerated motion you learned how to recognize uniform acceleration (see Secs. 2.5–2.8 in the *Text* and Experiment 5). Use the data you have just collected to decide whether the vertical motion of the ball was uniformly accelerated motion.

Q2 What do you find?

Q3 Do the horizontal and the vertical motions affect each other in any way?

Q4 Write an equation that describes the horizontal motion in terms of horizontal speed v, the horizontal distance, Δx, and the time of travel, Δt.

Q5 What is the equation that describes the vertical motion in terms of the distance fallen vertically, Δy, the vertical acceleration, a_g, and the time of travel, Δt?

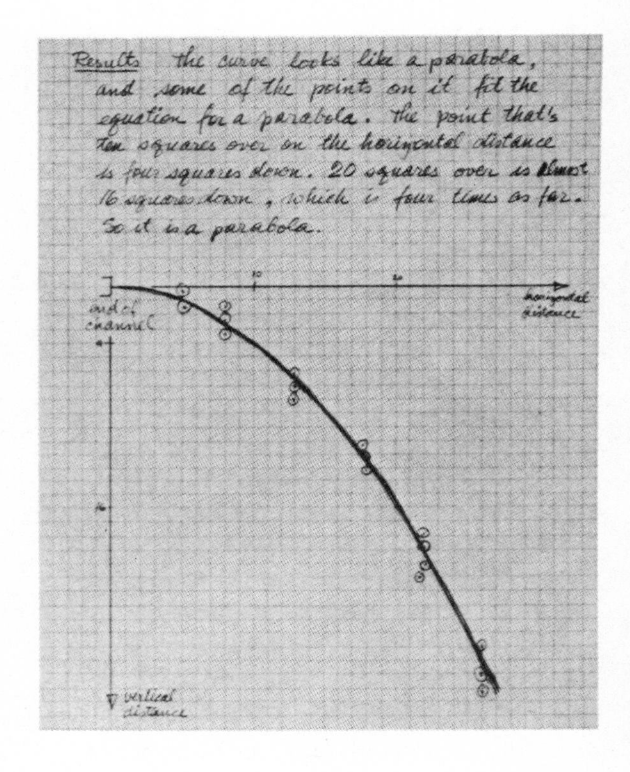

Results the curve looks like a parabola, and some of the points on it fit the equation for a parabola. the point that's ten squares over on the horizontal distance is four squares down. 20 squares over is almost 16 squares down, which is four times as far. So it is a parabola.
10
20
end of channel
horizontal distance
vertical distance

Try These Yourself

There are many other things you can do with this apparatus. Some of them are suggested by the following questions.

Q6 What do you expect would happen if you repeated the experiment with a glass marble of the same size instead of a steel ball?

Q7 What will happen if you next try to repeat the experiment starting the ball from a different point on the ramp?

Q8 What do you expect if you use a smaller or larger ball starting always from the same reference point on the ramp?

Q9 Plot the trajectory that results when you use a ramp that launches the ball at an angle to the horizontal. In what way is this curve similar to your first trajectory?

EXPERIMENT 11 PREDICTION OF TRAJECTORIES

You can predict the landing point of a ball launched horizontally from a tabletop at any speed. If you know the speed v of the ball as it leaves the table, the height of the table above the floor and a_g, you can then use the equation for projectile motion to predict where on the floor the ball will land.

You know an equation for horizontal motion:

$$\Delta x = v\,\Delta t$$

and you know an equation for free-fall from rest:

$$\Delta y = \tfrac{1}{2} a_g\,(\Delta t)^2$$

The time interval is difficult to measure. Besides, in talking about the *shape* of the path, all we really need to know is how Δy relates to Δx. Since, as you found in the previous experiment, these two equations still work when an object is moving horizontally and falling at the same time, we can combine them to get an equation relating Δy and Δx, without Δt appearing at all. We can rewrite the equation for horizontal motion as:

$$\Delta t = \frac{\Delta x}{v}$$

Then we can substitute this expression for t into the equation for fall:

$$\Delta y = \tfrac{1}{2} a_g \frac{(\Delta x)^2}{v^2}$$

Thus the equation we have derived should describe how Δy changes with Δx—that is, it should give us the shape of the trajectory. If we want to know how far out from the edge of the table the ball will land (Δx), we can calculate if from the height of the table (Δy), a_g, and the ball's speed v along the table.

Doing the Experiment

Find v by measuring with a stopwatch the time t that the ball takes to roll a distance d along the tabletop. (See Fig. 4–1 below.) Be sure to have the ball caught as it comes off the end of the table. Repeat the measurement a few times, always releasing the ball from the same place on the ramp, and take the average value of v.

Measure Δy and then use equation for Δy to calculate Δx. Place a target, a paper cup, perhaps, on the floor at your predicted landing spot as shown on the next page. How confident are you of your prediction? Since it is based on *measurement*, some uncertainty is involved. Mark an area around the spot to indicate your uncertainty.

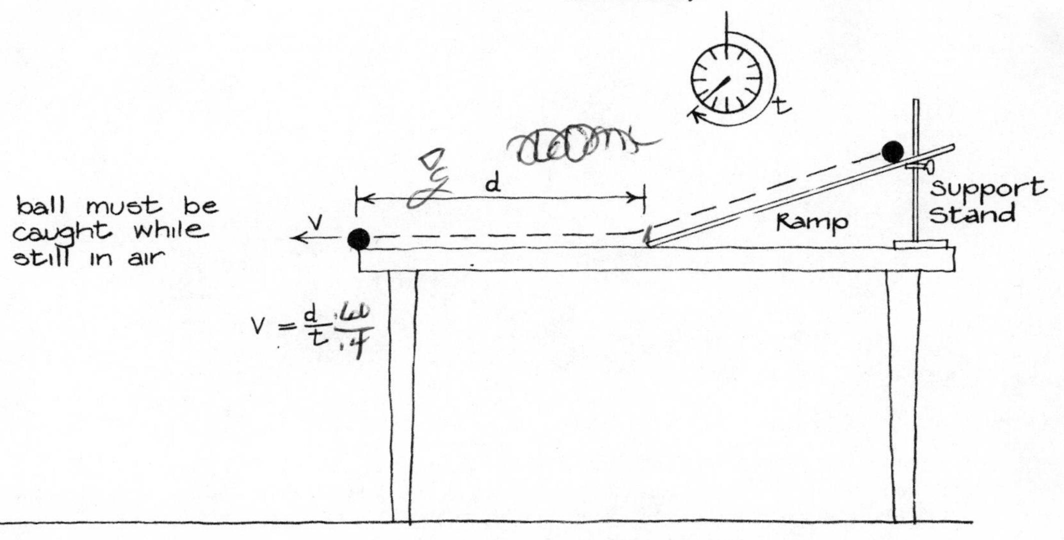

Fig. 4–1

Measuring Δ x

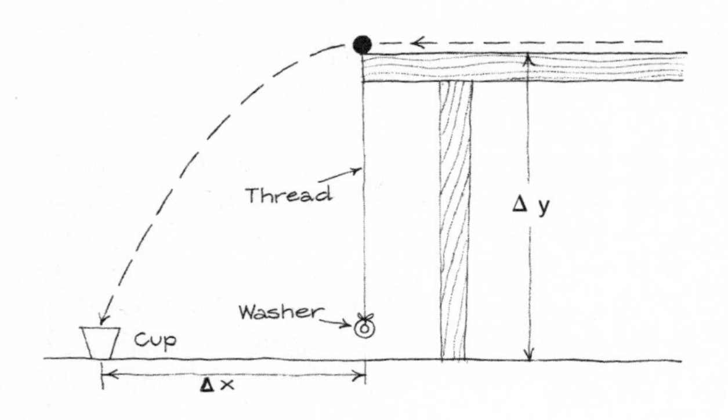

Now release the ball once more. This time, let it roll off the table and land, hopefully, on the target as shown in the figure above.

If the ball actually does fall within the range of values of Δx you have estimated, then you have supported the assumption on which your calculation was based, that vertical and horizontal motion are not affected by each other.

Q1 How could you determine the range of a ball launched horizontally by a slingshot?

Q2 Assume you can throw a baseball 40 meters on the earth's surface. How far could you throw that same ball on the surface of the moon, where the acceleration of gravity is one-sixth what it is at the surface of the earth?

Q3 Will the assumptions made in the equations $\Delta x = v\Delta t$ and $\Delta y = \frac{1}{2}ag(\Delta t)^2$ hold for a Ping-Pong ball? If the table were 1000 meters above the floor, could you still use these equations? Why or why not?

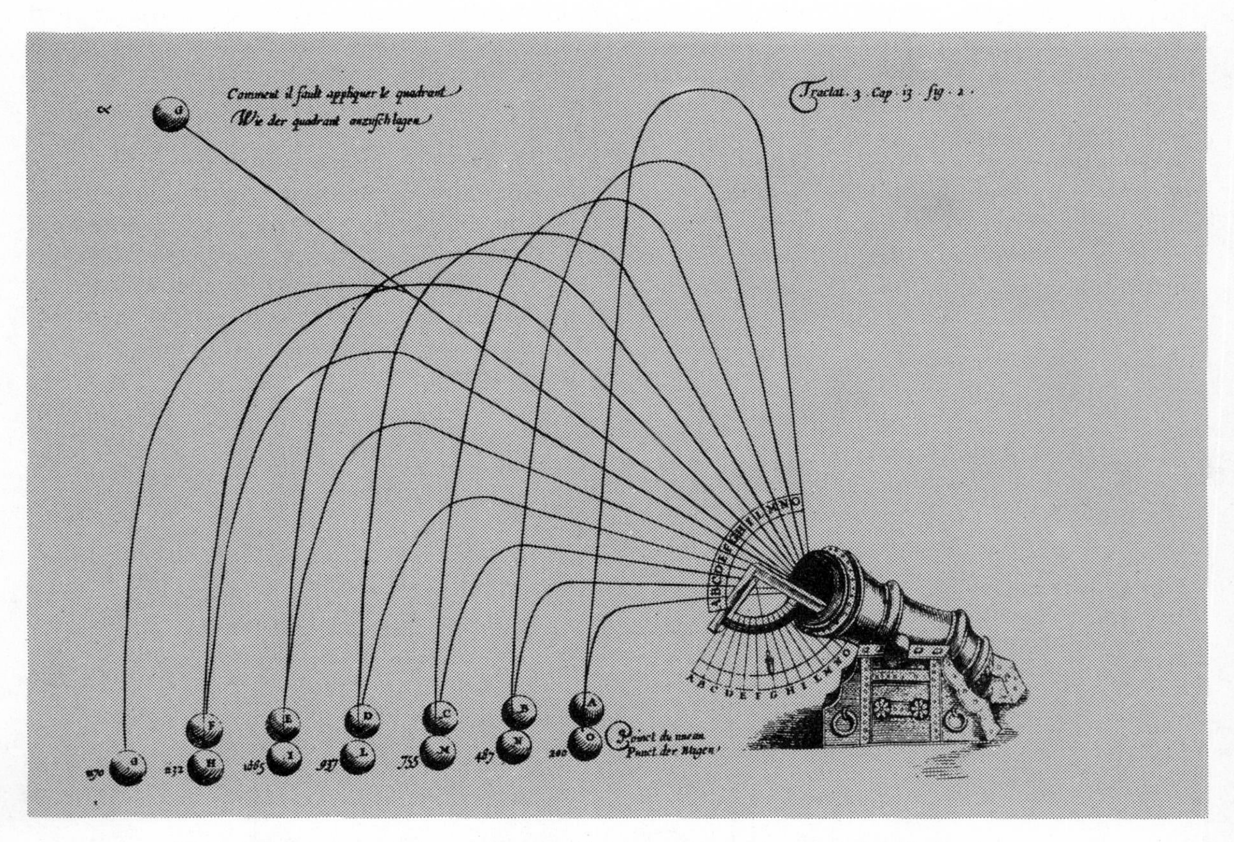

The path taken by a cannon ball according to a drawing by Ufano (1621). He shows that the same horizontal distance can be obtained by two different firing angles. Gunners had previously found this by experience. What angles give the maximum range? What is wrong with the way Ufano has drawn the trajectories?

EXPERIMENT 12 CENTRIPETAL FORCE

The motion of an earth satellite and of a weight swung around your head on the end of a string are described by the same laws of motion. Both are accelerating toward the center of their orbit due to the action of an unbalanced force.

In the following experiment you can discover for yourself how this centripetal force depends on the mass of the satellite and on its speed and distance from the center.

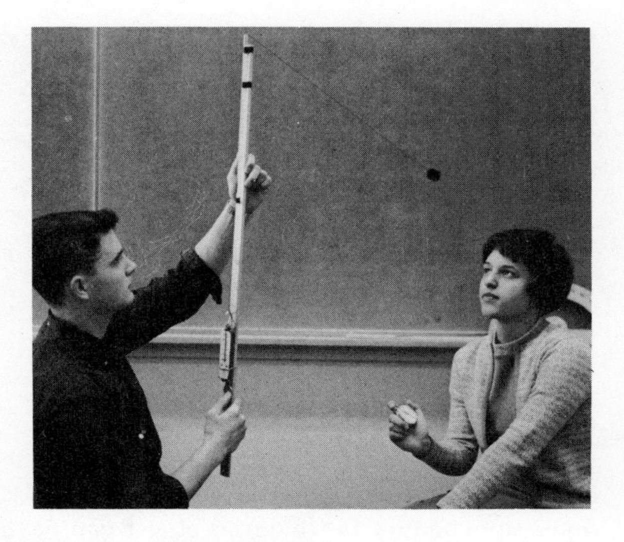

How the Apparatus Works

Your "satellite" is one or more rubber stoppers. When you hold the apparatus in both hands, as shown in the photo above, and swing the stopper around your head, you can measure the centripetal force on it with a spring scale at the base of the stick. The scale should read in newtons or else its readings should be converted to newtons.

You can change the length of the string so as to vary the radius R of the circular orbit, and you can tie on more stoppers to vary the satellite mass m.

The best way to set the frequency f is to swing the apparatus in time with some periodic sound from a metronome or an earphone attachment to a blinky. You keep the rate constant by adjusting the swinging until you see the stopper cross the same point in the room at every tick.

Hold the stick vertically and have as little motion at the top as possible, since this would change the radius. Because the stretch of the spring scale also alters the radius, it is helpful to have a marker (knot or piece of tape) on the string. You can move the spring scale up or down slightly to keep the marker in the same place.

Doing the Experiment

The object of the experiment is to find out how the force F read on the spring scale varies with m, with f, and with R.

You should only change *one* of these three quantities at a time so that you can investigate the effect of each quantity independently of the others. It's easiest to either double or triple m, f, and R (or halve them, and so on, if you started with large values).

Two or three different values should be enough in each case. Make a table and clearly record your numbers in it.

Q1 How do changes in m affect F when R and f are kept constant? Write a formula that states this relationship.

Q2 How do changes in f affect F when m and R are kept constant? Write a formula to express this too.

Q3 What is the effect of R on F?

Q4 Can you put m, f, and R all together in a single formula for centripetal force, R?

How does your formula compare with the expression derived in Sec. 4.7 of the *Text*.

EXPERIMENT 13 CENTRIPETAL FORCE ON A TURNTABLE

You may have had the experience of spinning around on an amusement park contraption known as the Whirling Platter. The riders seat themselves at various places on a large flat polished wooden turntable about 40 feet in diameter. The turntable gradually rotates faster and faster until everyone (except for the person at the center of the table) has slid off. The people at the edge are the first to go. Why do the people slide off?

Unfortunately you probably do not have a Whirling Platter in your classroom, but you do have a Masonite disk that fits on a turntable. The object of this experiment is to predict the maximum radius at which an object can be placed on the rotating turntable without sliding off.

If you do this under a variety of conditions, you will see for yourself how forces act in circular motion.

Before you begin, be sure you have studied Sec. 4.6 in your *Text* where you learned that the centripetal force needed to hold a rider in a circular path is given by $F = mv^2/R$.

Studying Centripetal Force

For these experiments it is more convenient to write the formula $F = mv^2/R$ in terms of the frequency f. This is because f can be measured more easily than v. We can rewrite the formula as follows:

$$v = \frac{\text{distance traveled}}{\text{in one revolution}} \times \frac{\text{number of revolu-}}{\text{tions per sec}}$$

$$= 2\pi R \times f$$

Substituting this expression for v in the formula gives:

$$F = \frac{m \times (2\pi R f)^2}{R}$$

$$= \frac{4\pi^2 m R^2 f^2}{R}$$

$$= 4\pi^2 m R f^2$$

You can measure all the quantities in this equation.

Friction on a Rotating Disk

For objects on a rotating disk, the centripetal force is provided by friction. On a frictionless disk there could be no such centripetal force. As you can see from the equation we have just derived, the centripetal acceleration is proportional to R and to f^2. Since the frequency f is the same for any object moving around with a turntable, the centripetal acceleration is directly proportional to R, the distance from the center. The further an object is from the center of the turntable, therefore, the greater the centripetal force must be to keep it in a circular path.

You can measure the maximum force F_{max} that friction can provide on the object, measure the mass of the object, and then calculate the maximum distance from the center R_{max} that the object can be without sliding off. Solving the centripetal force equation for R gives

$$R_{max} = \frac{F_{max}}{4\pi^2 m f^2}$$

Use a spring scale to measure the force needed to make some object (of mass m from 0.2 to 1.0 kg) start to slide across the motionless disk. This will be a measure of the maximum friction force that the disk can exert on the object.

Then make a chalk mark on the turntable and time it (say, for 100 sec)—or accept the marked value of rpm—and calculate the frequency in rev/sec.

Make your predictions of R_{max} for turntable frequencies of 33 revolutions per minute (rpm), 45 rpm, and 78 rpm.

Then try it!

Q1 How great is the percentage difference between prediction and experiment for each turntable frequency? Is this reasonable agreement?

Q2 What effect would decreasing the mass have on the predicted value of *R*? Careful! Decreasing the mass has an effect on *F* also. Check your answer by doing an experiment.

Q3 What is the smallest radius in which you can turn a car if you are moving 60 miles an hour and the friction force between tires and road is one-third the weight of the car? (Careful! Remember that weight is equal to $a_g \times m$.)

By permission of John Hart and Field Enterprises, Inc.

ACTIVITIES

PROJECTILE MOTION DEMONSTRATION

Here is a simple way to demonstrate projectile motion. Place one coin near the edge of a table. Place an identical coin on the table and snap it with your finger so that it flies off the table, just ticking the first coin enough that it falls almost straight down from the edge of the table. The fact that you hear only a single ring as both coins hit shows that both coins took the same time to fall to the floor from the table. Incidentally, do the coins *have* to be identical? Try different ones.

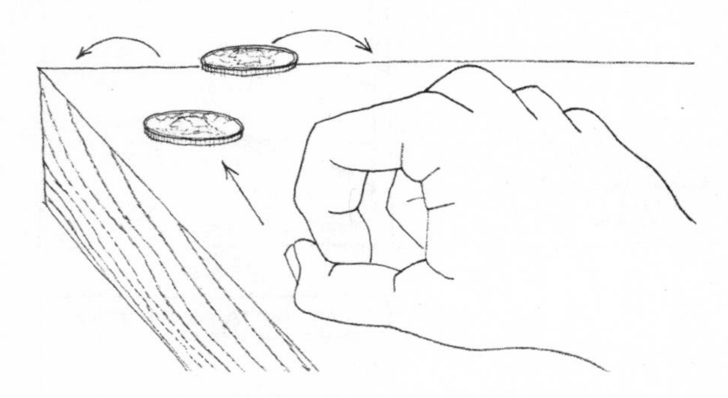

SPEED OF A STREAM OF WATER

You can use the principles of projectile motion to calculate the speed of a stream of water issuing from a horizontal nozzle. Measure the vertical distance Δy from the nozzle to the ground, and the horizontal distance Δx from the nozzle to where the water hits the ground.

Use the equation relating Δx and Δy that was derived in Experiment 11, solving it for v:

$$y = \frac{1}{2}a_g\frac{(\Delta x)^2}{v^2}$$

so

$$v^2 = \frac{1}{2}a_g\frac{(\Delta x)^2}{y}$$

and

$$v = \Delta x\sqrt{\frac{a_g}{2\Delta y}}$$

The quantities on the right can all be measured and used to compute v.

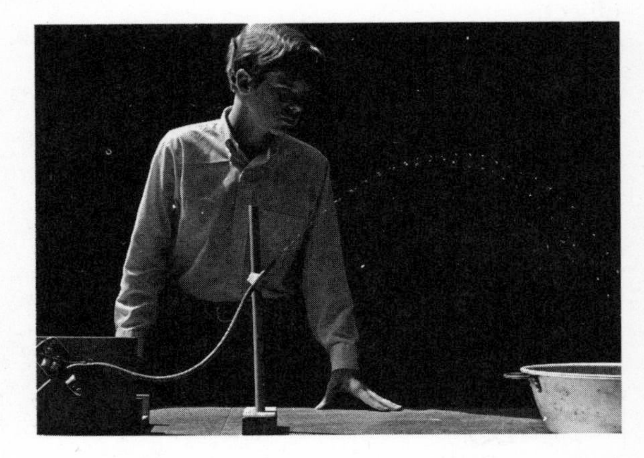

PHOTOGRAPHING A WATERDROP PARABOLA

Using an electronic strobe light, a doorbell timer, and water from a faucet, you can photograph a water drop parabola. The principle of independence of vertical and horizontal motions will be clearly evident in your picture.

Remove the wooden block from the timer. Fit an "eye dropper" barrel in one end of some tubing and fit the other end of the tubing onto a water faucet. (Instead of the timer you can use a doorbell without the bell.) Place the tube through which the water runs under the clapper so that the tube is given a steady series of sharp taps. This has the effect of breaking the stream of water into separate, equally spaced drops (see photo on the next page).

To get more striking power, run the vibrator from a variable transformer (Variac) connected to the 110 volt a.c., gradually increasing the Variac from zero just to the place where the striker vibrates against the tubing. Adjust the water flow through the tube and eye dropper nozzle. By viewing the drops with the xenon strobe set at the same frequency as the timer, a parabola of motionless drops is seen. A spot-light and disk strobe can be used instead of the electronic strobe light, but it is more difficult to match the frequencies of vibrator and strobe. The best photos are made by lighting the parabola from the side (that is, putting the light source in the plane of the parabola). The photo above was made in that

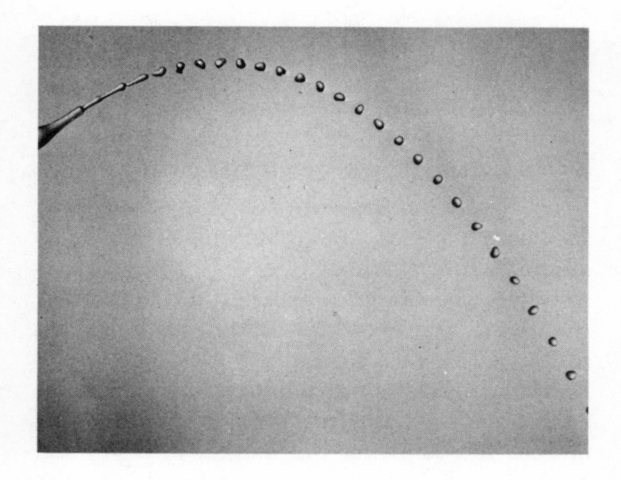

way. With front lighting, the shadow of the parabola can be projected onto graph paper for more precise measurement.

Some heating of the doorbell coil results, so the striker should not be run continuously for long periods of time.

BALLISTIC CART PROJECTILES

Fire a projectile straight up from a cart or toy locomotive as shown in the photo below that is rolling across the floor with nearly uniform velocity. You can use a commercial device called a ballistic cart or make one yourself. A spring-loaded piston fires a steel ball when you pull a string attached to a trigger pin. Use the electronic strobe to photograph the path of the ball.

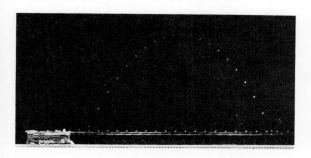

Of course projectile trajectories can be photographed of any object thrown into the air using the electronic strobe and Polaroid Land camera. By fastening the camera (securely!) to a pair of carts, you can photograph the action from a moving frame of reference.

MOTION IN A ROTATING REFERENCE FRAME

Here are three ways you can show how a moving object would appear in a rotating reference frame.

Method I Attach a piece of paper to a phonograph turntable. Draw a line across the paper as a turntable is turning (see Fig. 4–2 below), using as a guide a meter stick supported on books at either side of the turntable. The line should be drawn at a constant speed.

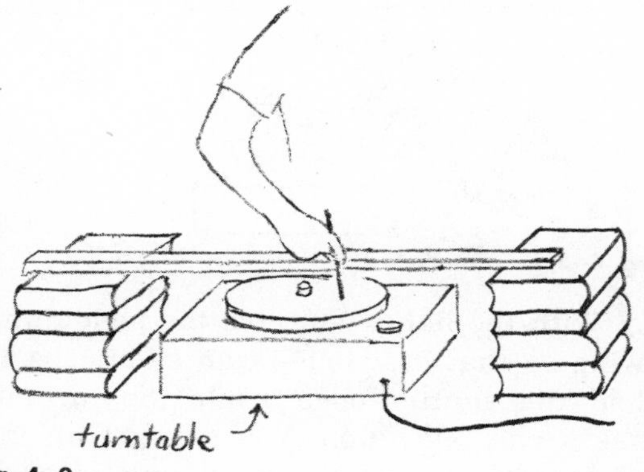

Fig. 4–2

Method II Place a Polaroid camera on the turntable on the floor and let a tractor run along the edge of a table, with a flashlight bulb on a pencil taped to the tractor so that it sticks out over the edge of the table.

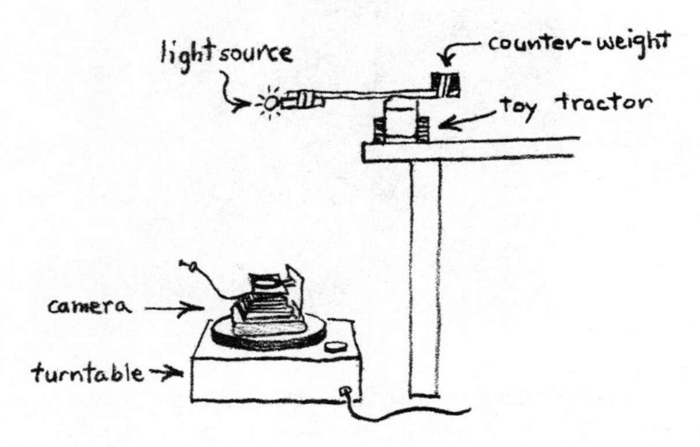

Method III How would an elliptical path appear if you were to view it from a rotating reference system? You can find out by placing a Polaroid camera on a turntable on the floor, with the camera aimed upwards. (See Fig. 4–3 below.) For a pendulum, hang a flashlight bulb and an AA dry cell. Make the pendulum long enough so that the light is about 4 feet from the camera lens.

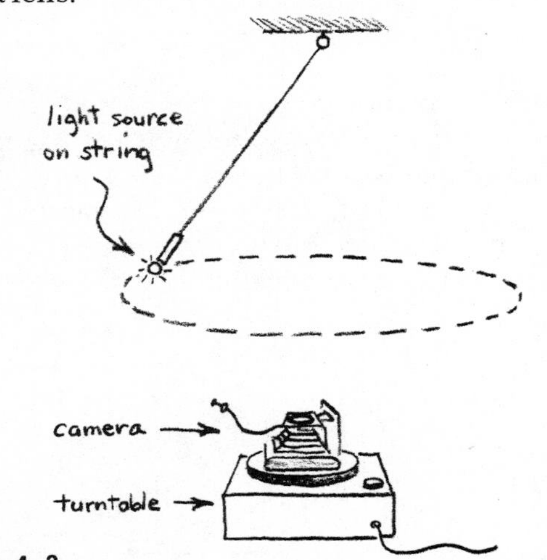

Fig. 4–3

With the lights out, give the pendulum a swing so that it swings in an elliptical path. Hold the shutter open while the turntable makes one revolution. You can get an indication of how fast the pendulum moves at different points in its swing by using a motor strobe in front of the camera, or by hanging a blinky.

PENNY AND COAT HANGER

Bend a coat hanger into the shape shown in the sketch below. Bend the end of the hook slightly with a pair of pliers so that it points to where the finger supports the hanger. File the end of the hook flat. Balance a penny on the hook. Move your finger back and forth so that the hanger (and balanced penny) starts swinging like a pendulum. Some practice will enable you to swing the hanger in a vertical circle, or around your head and still keep the penny on the hook. The centripetal force provided by the hanger keeps the penny from flying off on a straight-line path. Some people have done this demonstration successfully with a pile of as many as five pennies at once.

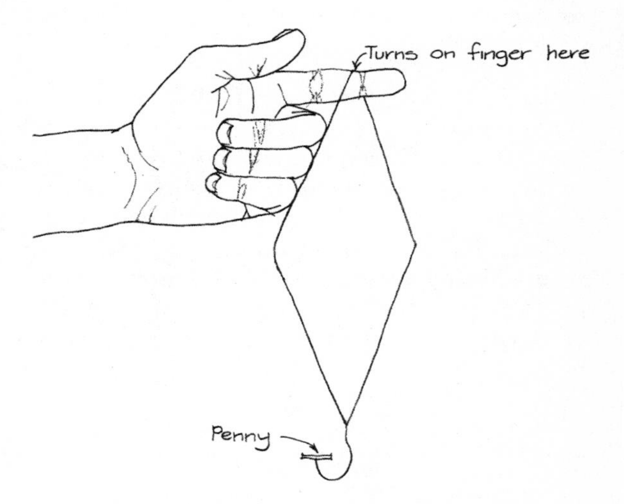

MEASURING UNKNOWN FREQUENCIES

Use a calibrated electronic stroboscope or a hand-stroboscope and stopwatch to measure the frequencies of various motions. Look for such examples as an electric fan, a doorbell clapper, and a banjo string.

On page 108 of the *Text* you will find tables of frequencies of rotating objects. Notice the enormous range of frequencies listed, from the electron in the hydrogen atom to the rotation of our Milky Way galaxy.

FILM LOOPS

FILM LOOP 4 A MATTER OF RELATIVE MOTION

Two carts of equal mass collide in this film. Three sequences labeled Event A, Event B, and Event C are shown. Stop the projector after *each* event and describe these events in words, as they appear to you. View the loop now, before reading further.

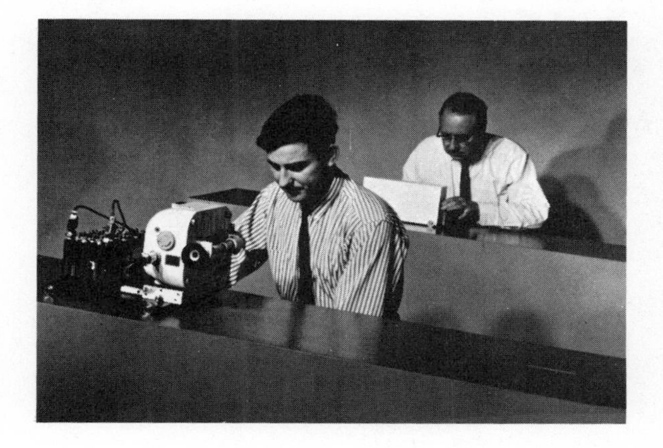

Even though Events A, B, and C are visibly different to the observer, in each the carts interact similarly. The laws of motion apply for each case. Thus, these events could be the *same* event observed from different reference frames. They *are* closely similar events photographed from different frames of reference, as you see after the initial sequence of the film.

The three events are photographed by a camera on a cart which is on a second ramp parallel to the one on which the colliding carts move. The camera is your frame of reference, your coordinate system. This frame of reference may or may not be in motion with respect to the ramp. As photographed, the three events appear to be quite different. Do such concepts as position and velocity have a meaning independently of a frame of reference, or do they take on a precise meaning only when a frame of reference is specified? Are these three events really similar events, viewed from different frames of reference?

You might think that the question of which cart is in motion is resolved by sequences at the end of the film in which an experimenter, Franklin Miller of Kenyon College, stands near the ramp to provide a reference object. Other visual clues may already have provided this information. The events may appear different when this reference object is present. But is this *fixed* frame of reference any more fundamental than one of the moving frames of reference? fixed relative to what? Or is there a "completely" fixed frame of reference?

If you have studied the concept of momentum, you can also consider each of these three events from the standpoint of momentum conservation. Does the total momentum depend on the frame of reference? Does it seem reasonable to assume that the carts would have the same *mass* in all the frames of reference used in the film?

B.C. by John Hart

CHAPTER 4

FILM LOOP 5 GALILEAN RELATIVITY— BALL DROPPED FROM MAST OF SHIP

This film is a partial actualization of an experiment described by Sagredo in Galileo's *Two New Sciences:*

> If it be true that the impetus with which the ship moves remains indelibly impressed in the stone after it is let fall from the mast; and if it be further true that this motion brings no impediment or retardment to the motion directly downwards natural to the stone, then there ought to ensue an effect of a very wondrous nature. Suppose a ship stands still, and the time of the falling of a stone from the mast's round top to the deck is two beats of the pulse. Then afterwards have the ship under sail and let the same stone depart from the same place. According to what has been premised, it shall take up the time of two pulses in its fall, in which time the ship will have gone, say, twenty yards. The true motion of the stone will then be a transverse line (i.e., a curved line in the vertical plane), considerably longer than the first straight and perpendicular line, the height of the mast, and yet nevertheless the stone will have passed it in the same time. Increase the ship's velocity as much as you will, the falling stone shall describe its transverse lines still longer and longer and yet shall pass them all in those selfsame two pulses.

In the film a ball is dropped three times:

Scene 1: The ball is dropped from the mast. As in Galileo's discussion, the ball continues to move horizontally with the boat's velocity, and also it falls vertically relative to the mast.

Scene 2: The ball is tipped off a stationary support as the boat goes by. It has no forward velocity, and it falls vertically relative to the water surface.

Scene 3: The ball is picked up and held briefly before being released.

The ship and earth are frames of reference in constant relative motion. Each of the three events can be described as viewed in either frame of reference. The laws of motion apply for all six descriptions. The fact that the laws of motion work for *both* frames of reference, one moving at constant velocity with respect to the other, is what is meant by "Galilean relativity." (The positions and velocities are relative to the frame of reference, but the *laws of motion* are not. A "relativity" principle also states what is *not* relative.)

Scene 1 can be described from the boat frame as follows: "A ball, initially at rest, is released. It accelerates downward at 9.8 m/sec^2 and strikes a point directly beneath the starting point." Scene 1 described differently from the earth frame is: "A ball is projected horizontally toward the left; its path is a parabola and it strikes a point below and to the left of the starting point."

To test your understanding of Galilean relativity, you should describe the following: Scene 2 from the boat frame; Scene 2 in earth frame; Scene 3 from the boat frame; Scene 3 from the earth frame.

FILM LOOP 6 GALILEAN RELATIVITY—OBJECT DROPPED FROM AIRCRAFT

A Cessna 150 aircraft 23 feet long is moving about 100 ft/sec at an altitude of about 200 feet. The action is filmed from the ground as a flare is dropped from the aircraft. Scene 1 shows part of the flare's motion; Scene 2, shot from a greater distance, shows several flares dropping into a lake; Scene 3 shows the vertical motion viewed head-on. Certain frames of the film are "frozen" to allow measurements. The time interval between freeze frames is always the same.

Seen from the earth's frame of reference, the motion is that of a projectile whose original velocity is the plane's velocity. If gravity is the only force acting on the flare, its motion should be a parabola. (Can you check this?) Relative to the airplane, the motion is that of a body falling freely from rest. In the frame of reference of the plane, the motion is vertically downward.

The plane is flying approximately at uniform speed in a straight line, but its path is not necessarily a horizontal line. The flare starts with the plane's velocity, in both magnitude and in direction. Since it also falls freely under the action of gravity, you expect the flare's downward displacement below the plane to be $d = \frac{1}{2}at^2$. But the trouble is that you cannot be sure that the first freeze frame occurs at the very instant the flare is dropped. However, there is a way of getting around this difficulty. Suppose a time B has elapsed between the release of the flare and the first freeze frame. This time must be added to each of the freeze frame times (conveniently measured from the first freeze frame) and so you would have

$$d = \tfrac{1}{2}a(t + B)^2$$

To see if the flare follows an equation such as this, take the square root of each side:

$$\sqrt{d} = (\text{constant})\ (t + B)$$

Now if we plot $\sqrt{d}$ against t, we expect a straight line. Moreover, if $B = 0$, this straight line will also pass through the origin.

Suggested Measurements

(a) **Vertical motion.** Project Scene 1 on paper. At each freeze frame, when the motion on the screen is stopped briefly, mark the positions of the flare and of the aircraft cockpit. Measure the displacement d of the flare below the plane. Use any convenient units. The times can be taken as integers, $t = 0, 1, 2, \ldots$, designating successive freeze frames. Plot $\sqrt{d}$ versus t. Is the graph a straight line? What would be the effect of air resistance, and how would this show up in your graph? Can you detect any signs of this? Does the graph pass through the origin?

(b) **Analyze Scene 2 in the same way.**

(c) **Horizontal motion.** Use another piece of graph paper with time (in intervals) plotted horizontally and displacements (in squares) plotted vertically. Using measurements from your record of the flare's path, make a graph of the two motions in Scene 2. What are the effects of air resistance in the horizontal motion? the vertical motion? Explain your findings between the effect of air friction on the horizontal and vertical motions.

(d) **Acceleration due to gravity.** The "constant" in your equation, d = (constant) $(t + B)$, is $\frac{1}{2}a$; this is the slope of the straight-line graph obtained in part (a). The square of the slope gives $\frac{1}{2}a$ so the acceleration is twice the square of the slope. In this way you can obtain the acceleration in squares/(interval)^2. To convert your acceleration into ft/sec^2 or m/sec^2, you can estimate the size of a "square" from the fact that the length of the plane is 23 ft (7 m). The time interval in seconds between freeze frames can be found from the slow-motion factor.

FILM LOOP 7 GALILEAN RELATIVITY—PROJECTILE FIRED VERTICALLY

A rocket tube is mounted on bearings that leave the tube free to turn in any direction. When the tube is hauled along the snow-covered surface of a frozen lake by a "ski-doo," the bearings allow the tube to remain pointing vertically upward in spite of some roughness of path. Equally spaced lamps along the path allow you to judge whether the ski-doo has constant velocity or whether it is accelerating. A preliminary run shows the entire scene; the setting is in the Laurentian Mountains in the Province of Quebec at dusk.

Four scenes are photographed. In each case a rocket flare is fired vertically upward. With care you can trace a record of the trajectories.

Scene 1: The ski-doo is stationary relative to the earth. How does the flare move?

Scene 2: The ski-doo moves at uniform velocity relative to the earth. Describe the motion of the flare relative to the earth; describe the motion of the flare relative to the ski-doo.

Scenes 3 and 4: The ski-doo's speed changes after the shot is fired. In each case describe the motion of the ski-doo and describe the flare's motion relative to the earth and relative to the ski-doo. In which cases are the motions a parabola?

How do the events shown in this film illustrate the principle of Galilean relativity? In which frames of reference does the rocket flare behave the way you would expect it to behave in all four scenes knowing that the force is constant, and assuming Newton's laws of motion? In which systems do Newton's laws fail to predict the correct motion in some of the scenes?

FILM LOOP 8 ANALYSIS OF A HURDLE RACE–I

The initial scenes in this film show a regulation hurdle race, with 1-meter-high hurdles spaced 9 meters apart. (Judging from the number of hurdles knocked over, the competitors were of something less than Olympic caliber!) Next, a runner, Frank White, a 75-kg student at McGill University, is shown in medium slow-motion (slow-motion factor 3) during a 50-meter run. His time was 8.1 seconds. Finally, the beginning of the run is shown in extreme slow motion (slow-motion factor of 80). "Analysis of a Hurdle Race II" has two more extreme slow-motion sequences.

To study the runner's motion, measure the average speed for each of the 1-meter intervals in the slow-motion scene. A "drag-strip" chart recorder is particularly convenient for recording the data on a single viewing of the loop. Whatever method you use for measuring time, the small but significant variations in speed will be lost in experimental uncertainty unless you work very carefully. Repeat each measurement several times.

The extreme slow-motion sequence shows the runner from 0 m to 6 m. The seat of the runner's white shorts might serve as a reference mark. (What are other reference points on the runner that could be used? Are all ref-

erence points equally useful?) Measure the time to cover each of the distances, 0-1, 1-2, 2-3, 3-4, 4-5, and 5-6 m. Repeat the measurements several times, viewing the film over again, and average your results for each interval. Your accuracy might be improved by forming a grand average that combines your average with others in the class. (Should you use *all* the measurements in the class?) Calculate the average speed for each interval, and plot a graph of speed versus displacement. Draw a smooth graph through the points. Discuss any interesting features of the graph.

You might assume that the runner's legs push between the time when a foot is directly beneath his hip and the time when that foot is off the ground. Is there any relationship between your graph of speed and the way the runner's feet push on the track?

The initial acceleration of the runner can be estimated from the time to move from the starting point to the 1-meter mark. You can use a watch with a sweep second hand. Calculate the average acceleration, in m/sec², during this initial interval. How does this forward acceleration compare with the magnitude of the acceleration of a falling body? How much force was required to give the runner this acceleration? What was the origin of this force?

FILM LOOP 9 ANALYSIS OF A HURDLE RACE–II

This film loop, which is a continuation of "Analysis of a Hurdle Race I," shows two scenes of a hurdle race which was photographed at a slow-motion factor of 80.

In Scene 1 the hurdler moves from 20 m to 26 m, clearing a hurdle at 23 m. (See photograph.) In Scene 2 the runner moves from 40 m to 50 m, clearing a hurdle at 41 m and sprinting to the finish line at 50 m. Plot graphs of these motions, and discuss any interesting features. The seat of the runner's pants furnishes a convenient reference point for measurements. (See the film-notes about the "Analysis of a Hurdle Race I" for further details.)

No measurement is entirely precise; measurement error is always present, and it cannot be ignored. Thus it may be difficult to tell if the small changes in the runner's speed are significant, or are only the result of measurement uncertainties. You are in the best tradition of experimental science when you pay close attention to errors.

It is often useful to display the experimental uncertainty graphically, along with the measured or computed values.

For example, say that the dragstrip timer was used to make three different measurements of the time required for the first meter of the run: 13.7 units, 12.9 units, and 13.5 units, which give an average time of 13.28

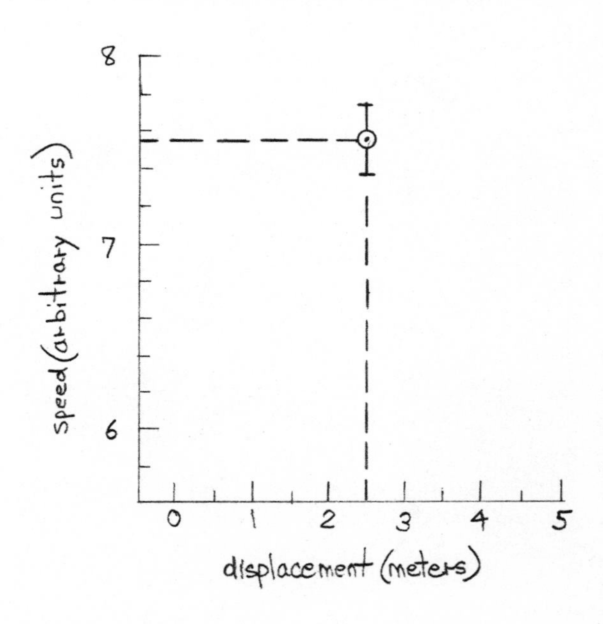

units. (If you wish to convert the dragstrip units to seconds, it will be easier to wait until the graph has been plotted using just units, and then add a seconds scale to the graph.) The lowest and highest values are about 0.4 units on either side of the average, so we could report the time as 13.3 + 0.4 units. The uncertainty 0.4 is about 3% of 13.3, therefore the percentage uncertainty in the time is 3%. If we assume that the distance was exactly one meter, so that all the uncertainty is in the time, then the percentage uncertainty in the speed will be the same as for the time—3%. The slow-motion speed is 100 cm/13.3 time units, which equals 7.53 cm/unit. Since 3% of 7.53 is 0.23, the speed can be reported as 7.53 + 0.23 cm/unit. In graphing this speed value, you plot a point at 7.53 and draw an "error bar" extending 0.23 above and below the point. Now estimate the limit of error for a typical point on your graph and add error bars showing the range to each plotted point.

Your graph for this experiment may well look like some commonly obtained in scientific research. For example, in the figure at the right a research team has plotted its experimental data; they published their results in spite of the considerable scattering of plotted points and even though some of the plotted points have errors as large as 5%.

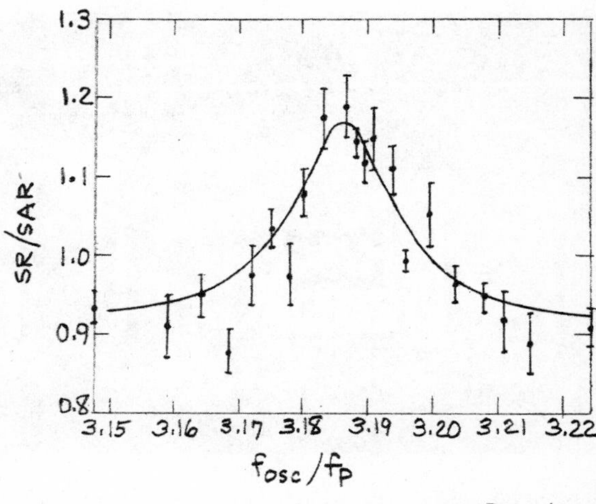

Stroboscopic Coincidence Positive Muons in Bromoform
f_{osc} = 48.63 Mc/Sec

How would you represent the uncertainty in measuring *distance*, if there were significant errors here also?

The Project Physics Course

Handbook 2

Motion in the Heavens

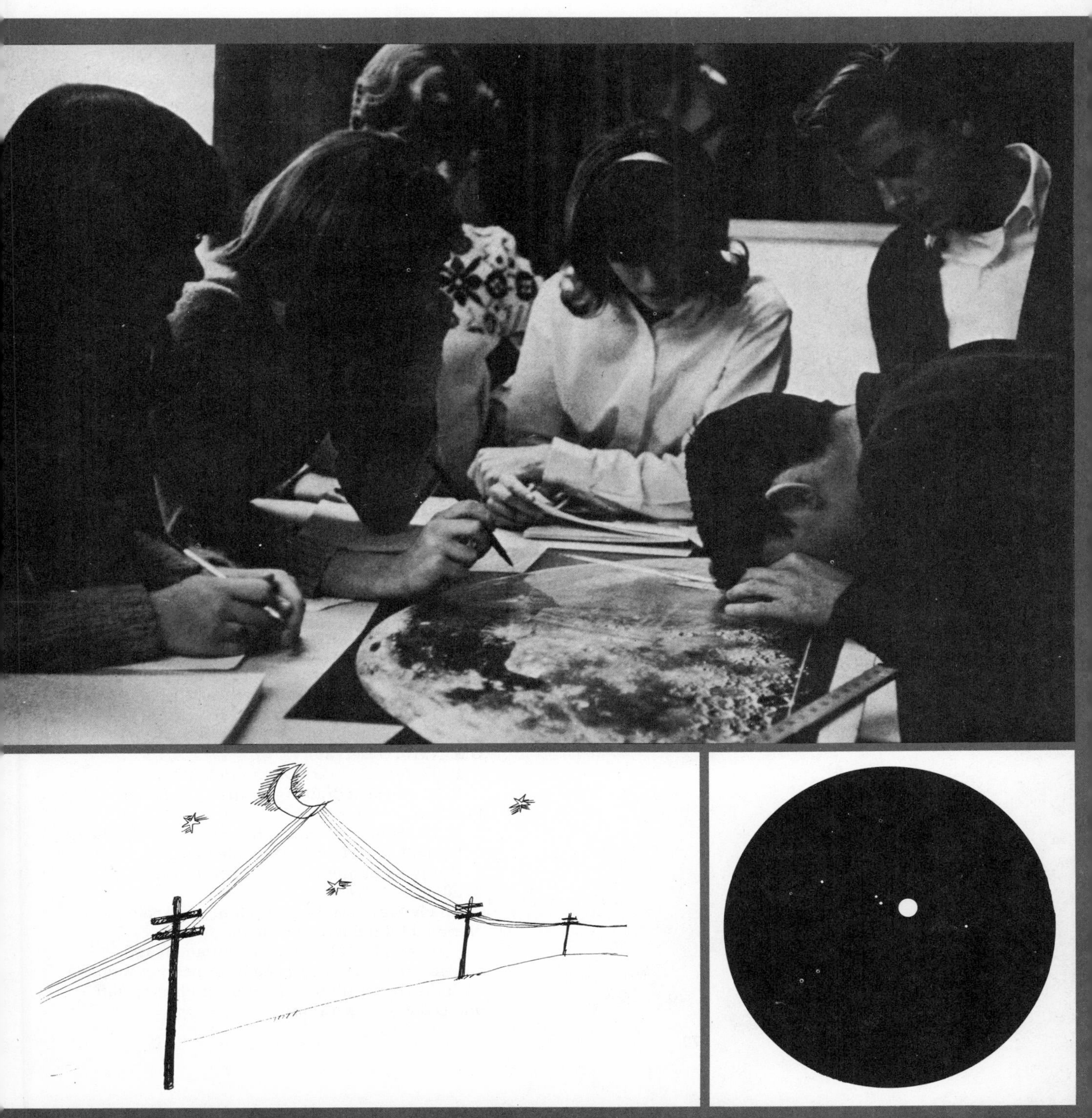

Picture Credits, Unit 2

Cover: (lower left) Cartoon by Charles Gary Solin and reproduced by his permission only.

Pp. 73, 97, 99, 104, 113, 114, 119, 123, 135 Mount Wilson and Palomar Observatories photographs.

Pp. 74, 76, 78, 83, 86, 92, 98, 108 (cartoons). By permission of Johnny Hart and Field Enterprises, Inc.

Pp. 76, 81, 105 National Aeronautics and Space Administration photograph.

P. 79 Lick Observatory photograph.

P. 84 Fig. 5–13. Photograph courtesy of Damon Corporation, Educational Division.

P. 89 Map courtesy of Rand McNally and Co.

P. 92 Sun film strip photographs courtesy of the U.S. Naval Observatory.

P. 97 Venus, Yerkes Observatory photograph.

P. 99 Sun spot drawings courtesy of Robert A. Pinkham.

P. 100 Nubbin Cartoon, King Features Syndicate.

P. 106 Lowell Observatory photograph.

P. 127 By permission of United Features Syndicate.

All photographs used with film loops courtesy of National Film Board of Canada.

Photographs of laboratory equipment and of students using laboratory equipment were supplied with the cooperation of the Project Physics staff and Damon Corporation.

Contents HANDBOOK, Unit 2

Chapter 5 Where is the Earth?—The Greek's Answers

EXPERIMENT 14 NAKED-EYE ASTRONOMY (Continued from Unit 1, Experiment 1)

Weather permitting, you have been watching events in the day and night sky since this course started. Perhaps you have followed the sun's path, or viewed the moon, planets, or stars.

From observations much like your own, scientists in the past developed a remarkable sequence of theories. The more aware you are of the motions in the sky and the more you interpret them yourself, the more easily you can follow the development of these theories. If you do not have your own data, you can use the results provided in the following sections.

A. One Day of Sun Observations

One student made the following observations of the sun's position on September 23.

Eastern Daylight Time (EDT)	Sun's Altitude	Sun's Azimuth
7:00 A.M.	------	------
8:00	08°	097°
9:00	19	107
10:00	29	119
11:00	38	133
12:00	45	150
1:00 P.M.	49	172
2:00	48	197
3:00	42	217
4:00	35	232
5:00	25	246
6:00	14	257
7:00	03	267

If you plot altitude (vertically) against azimuth (horizontally) on a graph and mark the hours for each point, it will help you to answer these questions.

1. What was the sun's greatest altitude during the day?
2. What was the latitude of the observer?
3. At what time (EDT) was the sun highest?
4. When during the day was the sun's direction (azimuth) changing fastest?

5. When during the day was the sun's altitude changing fastest?
6. At what time of day did the sun reach its greatest altitude? How do you explain the fact that it is not exactly at 12:00? (Remember that daylight time is an hour ahead.)

B. A Year of Sun Observations

One student made the following monthly observations of the sun through a full year. (He had remarkably clear weather!)

Dates	Sun's Noon Altitude	Sunset Azimuth	Time Between Noon and Sunset
Jan 1	20°	238°	$4_h 25_m$*
Feb 1	26	245	4 50
Mar 1	35	259	5 27
Apr 1	47	276	6 15
May 1	58	291	6 55
Jun 1	65	300	7 30
Jul 1	66	303	7 40
Aug 1	61	295	7 13
Sep 1	52	282	6 35
Oct 1	40	267	5 50
Nov 1	31	250	5 00
Dec 1	21	239	4 30

*h = hours, m = minutes.

Use these data to make three plots (different colors or marks on the same sheet of graph

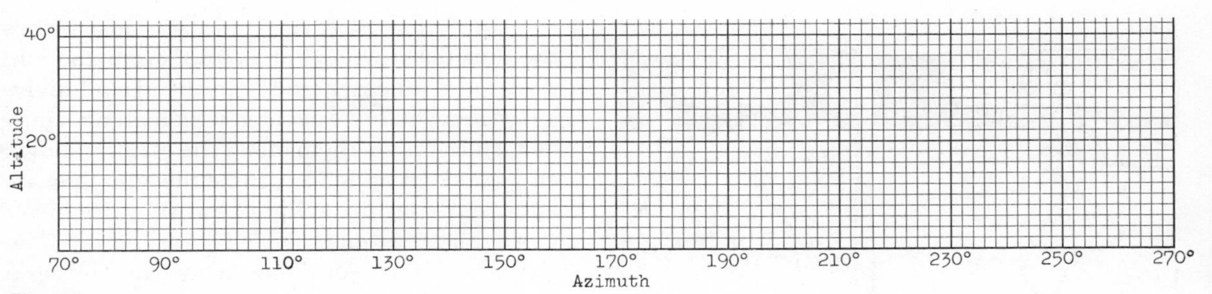

Fig. 5-1

paper) of the sun's noon altitude, direction at sunset, and time of sunset after noon. Place these data on the vertical axis and the dates on the horizontal axis.

1. What was the sun's noon altitude at the equinoxes (March 21 and September 23)?
2. What was the observer's latitude?
3. If the observer's longitude was 71°W, what city was he near?
4. Through what range (in degrees) did his sunset point change during the year?
5. By how much did the observer's time of sunset change during the year?
6. If the time from sunrise to noon was always the same as the time between noon and sunset, how long was the sun above the horizon on the shortest day? on the longest day?

C. Moon Observations

During October 1966 a student in Las Vegas, Nevada made the following observations of the moon at sunset when the sun had an azimuth of about 255°.

Date	Angle from Sun to Moon	Moon Altitude	Moon Azimuth
Oct.			
16	032°	17°	230°
18	057	25	205
20	081	28	180
22	104	30	157
24	126	25	130
26	147	16	106
28	169	05	083

•1. Plot these positions of the moon on a chart such as in Fig. 5-1.

•2. From the data and your plot, estimate the dates of new moon, first quarter moon, and full moon.

•3. For each of the points you plotted, sketch the shape of the lighted area of the moon.

Phases of the moon: (1) 23 days, (2) 26 days, (3) 17 days, (4) 5 days, (5) 3 days after new moon.

B.C. By John Hart

D. Locating the Planets

Table 1, Planetary Longitudes lists the position of each major planet along the ecliptic. The positions are given, accurate to the nearest degree, for every ten-day interval. By interpolation you can find a planet's position on any given day.

The column headed "J.D." shows the corresponding Julian Day calendar date for each entry. This calendar is simply a consecutive numbering of days that have passed since an arbitrary "Julian Day 1" in 4713 B.C.: September 30, 1970, for example, is the same as J.D. 2,440,860.

Julian dates are used by astronomers for convenience. For example, the number of days between March 8 and September 26 of this year is troublesome to figure out, but it is easy to find by simple subtraction if the Julian dates are used instead.

Look up the sun's present longitude in the table. Locate the sun on your SC-1 Constellation Chart: The sun's path, the ecliptic, is the curved line marked off in 360 degrees of longitude.

A planet that is just to the west of the sun's position (to the right on the chart) is "ahead of the sun," that is, it rises and sets just before the sun does. One that is 180° from the sun rises near sundown and is in the sky all night.

When you have decided which planets may be visible, locate them along the ecliptic shown on your sky map SC-1. Unlike the sun, they are not exactly on the ecliptic, but they are never more than 8° from it. Once located on the Constellation Chart you know where to look for a planet among the fixed stars.

E. Graphing the Position of the Planets

Here is a useful way to display the information in Table I, Planetary Longitudes. On ordinary graph paper, plot the sun's longitude versus time. Use Julian Day numbers along the horizontal axis, beginning close to the present date. The plotted points should fall on a nearly straight line, sloping up toward the right until they reach 360° and then starting again at zero.

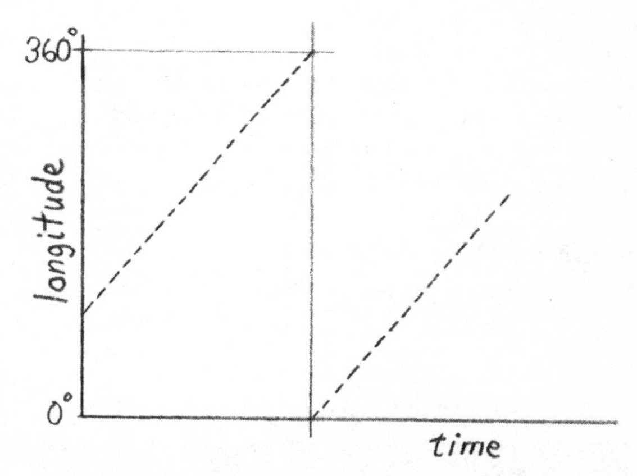

How long will it be before the sun again has the same longitude it has today? Would the answer to that question be the same if it were asked three months from now? What is the sun's average angular speed (in degrees per day) over a whole year? When is its angular speed greatest?

Plot Mercury's longitudes on the same graph (use a different color or shape for the points). According to your plot, how far (in longitude) does Mercury get from the sun? (This is Mercury's "maximum elongation.") At what time interval does Mercury pass the sun?

Table 1 Planet Longitudes at 10-Day Intervals

Date	J.D.	Sun	Mer	Ven	Mars	Jup	Sat
Nov 24	0550	242	246	227	314	206	32
Dec 4	0560	252	262	240	322	207	31
Dec 14	0570	262	278	252	329	209	31
Dec 24	0580	272	292	265	337	211	31
Jan 3	0590	283	300	277	344	212	31
Jan 13	0600	293	293	290	352	214	32
Jan 23	0610	303	284	303	359	215	32
Feb 2	0620	313	288	315	6	215	32
Feb 12	0630	323	298	327	13	216	33
Feb 22	0640	333	312	340	21	216	34
Mar 4	0650	343	327	353	28	216	35
Mar 14	0660	353	345	5	35	215	36
Mar 24	0670	3	4	17	42	214	37
Apr 3	0680	13	25	30	49	213	38
Apr 13	0690	23	42	42	56	212	39
Apr 23	0700	33	52	55	63	211	41
May 3	0710	42	52	67	70	209	42
May 13	0720	52	45	80	77	208	43
May 23	0730	62	42	92	83	207	45
Jun 2	0740	71	47	104	90	207	46
Jun 12	0750	81	58	115	97	206	47
Jun 22	0760	90	74	127	103	206	48
Jul 2	0770	100	94	139	109	206	49
Jul 12	0780	109	116	150	116	206	50
Jul 22	0790	119	135	162	122	207	50
Aug 1	0800	128	152	173	129	208	51
Aug 11	0810	138	165	184	136	209	52
Aug 21	0820	148	175	194	142	211	53
Aug 31	0830	157	178	204	149	212	53
Sep 10	0840	167	172	213	155	214	53
Sep 20	0850	177	164	222	161	216	53
Sep 30	0860	187	168	229	167	217	52
Oct 10	0870	197	184	234	174	219	52
Oct 20	0880	206	201	236	180	221	51
Oct 30	0890	216	218	234	186	224	50
Nov 9	0900	227	234	230	192	226	49
Nov 19	0910	237	250	224	199	228	48
Nov 29	0920	247	265	220	205	231	47
Dec 9	0930	257	278	222	212	233	46
Dec 19	0940	267	285	226	218	235	45
Dec 29	0950	277	276	233	224	237	44
Jan 8	0960	288	268	242	231	238	44
Jan 18	0970	298	273	251	237	240	44
Jan 28	0980	308	285	261	243	242	44
Feb 7	0990	318	299	272	249	243	45
Feb 17	1000	328	315	283	256	244	46
Feb 27	1010	338	332	294	262	245	46
Mar 9	1020	348	351	306	268	246	48
Mar 19	1030	358	10	318	274	246	49
Mar 29	1040	8	27	330	280	247	50
Apr 8	1050	18	34	341	286	246	51
Apr 18	1060	28	31	353	292	246	52
Apr 28	1070	37	24	5	297	245	53
May 8	1080	47	23	17	303	244	55
May 18	1090	57	31	29	308	242	56
May 28	1100	66	43	41	314	241	58
Jun 7	1110	76	60	53	318	240	59
Jun 17	1120	85	80	66	320	239	60
Jun 27	1130	95	103	78	322	238	61
Jul 7	1140	104	122	90	323	237	62
Jul 17	1150	114	138	103	324	237	63
Jul 27	1160	123	151	115	323	236	64
Aug 6	1170	133	159	127	320	237	65
Aug 16	1180	143	160	140	315	237	65
Aug 26	1190	152	153	152	313	238	66
Sep 5	1200	162	147	164	312	239	66
Sep 15	1210	172	154	177	312	240	66
Sep 25	1220	182	170	189	314	242	67
Oct 5	1230	191	189	202	315	243	67
Oct 15	1240	201	206	214	318	245	66
Oct 25	1250	211	222	227	322	247	66
Nov 4	1260	221	238	239	328	249	65
Nov 14	1270	231	252	252	333	251	64
Nov 24	1280	241	264	264	340	253	63
Dec 4	1290	252	269	277	346	256	61
Dec 14	1300	262	259	289	352	258	60
Dec 24	1310	272	252	302	358	261	59
Jan 3	1320	282	259	314	5	263	59
Jan 13	1330	292	272	326	12	265	59

Yr.	Date	J.D.	Sun	Mer	Ven	Mars	Jup	Sat
1972	Jan 23	1340	303	286	338	18	267	58
1972	Feb 2	1350	313	302	350	24	269	58
1972	Feb 12	1360	323	319	2	31	271	59
1972	Feb 22	1370	333	337	14	38	273	59
1972	Mar 3	1380	343	356	26	44	274	60
1972	Mar 13	1390	353	12	37	51	276	61
1972	Mar 23	1400	3	16	48	57	277	61
1972	Apr 2	1410	13	10	59	64	278	62
1972	Apr 12	1420	23	4	69	70	278	64
1972	Apr 22	1430	32	6	78	77	279	65
1972	May 2	1440	42	15	86	84	279	66
1972	May 12	1450	52	29	91	90	278	67
1972	May 22	1460	61	46	95	97	278	69
1972	Jun 1	1470	71	66	94	103	277	70
1972	Jun 11	1480	80	89	89	109	276	72
1972	Jun 21	1490	90	108	83	115	274	73
1972	Jul 1	1500	99	124	77	122	273	74
1972	Jul 11	1510	109	136	76	128	271	75
1972	Jul 21	1520	119	142	80	134	270	76
1972	Jul 31	1530	128	140	85	141	269	77
1972	Aug 10	1540	138	133	93	147	269	78
1972	Aug 20	1550	147	130	101	153	268	79
1972	Aug 30	1560	157	139	111	160	268	79
1972	Sep 9	1570	167	157	121	166	269	80
1972	Sep 19	1580	176	176	132	173	269	80
1972	Sep 29	1590	186	194	143	179	270	80
1972	Oct 9	1600	196	210	154	186	271	80
1972	Oct 19	1610	206	225	166	192	273	80
1972	Oct 29	1620	216	239	178	198	274	81
1972	Nov 8	1630	226	249	190	205	276	80
1972	Nov 18	1640	236	253	202	212	278	79
1972	Nov 28	1650	246	241	214	218	280	78
1972	Dec 8	1660	257	237	227	225	282	77
1972	Dec 18	1670	267	246	239	232	284	76
1972	Dec 28	1680	277	259	252	238	287	75
1973	Jan 7	1690	287	274	264	245	289	74
1973	Jan 17	1700	297	290	277	252	291	73
1973	Jan 27	1710	307	306	289	259	294	73
1973	Feb 6	1720	317	324	302	266	296	73
1973	Feb 16	1730	328	342	315	273	298	73
1973	Feb 26	1740	338	356	327	280	300	73
1973	Mar 8	1750	348	358	340	287	303	74
1973	Mar 18	1760	358	349	352	294	305	74
1973	Mar 28	1770	8	345	4	301	306	75
1973	Apr 7	1780	17	350	17	308	308	76
1973	Apr 17	1790	27	0	29	315	309	77
1973	Apr 27	1800	37	14	41	322	310	78
1973	May 7	1810	47	32	54	329	311	79
1973	May 17	1820	56	52	66	336	312	81
1973	May 27	1830	66	75	79	343	312	82
1973	Jun 6	1840	75	94	91	350	312	83
1973	Jun 16	1850	85	109	103	357	312	85
1973	Jun 26	1860	94	119	116	4	312	86
1973	Jul 6	1870	104	123	128	10	311	87
1973	Jul 16	1880	114	120	140	16	310	89
1973	Jul 26	1890	123	113	152	22	308	89
1973	Aug 5	1900	133	114	164	28	306	90
1973	Aug 15	1910	142	125	176	33	305	91
1973	Aug 25	1920	152	143	188	37	304	92
1973	Sep 4	1930	162	163	200	39	303	93
1973	Sep 14	1940	171	181	211	40	302	94
1973	Sep 24	1950	181	198	223	40	302	94
1973	Oct 4	1960	191	213	235	88	302	94
1973	Oct 14	1970	201	226	246	36	302	95
1973	Oct 24	1980	211	235	257	32	303	94
1973	Nov 3	1990	221	236	268	27	304	94
1973	Nov 13	2000	231	224	278	25	305	94
1973	Nov 23	2010	241	222	288	24	306	93
1973	Dec 3	2020	251	232	297	25	308	93
1973	Dec 13	2030	261	246	304	26	310	92
1973	Dec 23	2040	272	262	309	28	312	91
1974	Jan 2	2050	282	278	313	32	314	90
1974	Jan 12	2060	292	294	311	37	317	89
1974	Jan 22	2070	302	311	306	41	319	88
1974	Feb 1	2080	312	328	300	46	321	87
1974	Feb 11	2090	322	341	296	51	324	87
1974	Feb 21	2100	332	340	297	57	326	87
1974	Mar 3	2110	342	330	301	63	329	87
1974	Mar 13	2120	352	327	308	68	331	88

Yr.	Date	J.D.	Sun	Mer	Ven	Mars	Jup	Sat
1974	Mar 23	2130	2	335	316	73	333	88
1974	Apr 2	2140	12	346	325	79	335	89
1974	Apr 12	2150	22	1	335	85	337	89
1974	Apr 22	2160	32	18	346	91	340	90
1974	May 2	2170	42	38	357	97	342	91
1974	May 12	2180	51	61	8	103	343	93
1974	May 22	2190	61	80	19	109	345	94
1974	Jun 1	2200	70	94	31	115	346	95
1974	Jun 11	2210	80	102	42	121	347	96
1974	Jun 21	2220	89	103	54	128	347	97
1974	Jul 1	2230	99	97	66	134	348	99
1974	Jul 11	2240	109	94	77	140	348	100
1974	Jul 21	2250	118	97	89	146	347	102
1974	Jul 31	2260	128	110	101	152	348	103
1974	Aug 10	2270	137	129	114	159	347	104
1974	Aug 20	2280	147	150	126	165	345	105
1974	Aug 30	2290	157	169	139	171	344	106
1974	Sep 9	2300	166	185	151	177	342	106
1974	Sep 19	2310	176	200	163	184	341	107
1974	Sep 29	2320	186	212	176	191	339	108
1974	Oct 9	2330	196	220	188	197	338	108
1974	Oct 19	2340	206	219	201	204	338	109
1974	Oct 29	2350	216	207	213	211	337	109
1974	Nov 8	2360	226	207	226	218	338	109
1974	Nov 18	2370	236	218	238	224	338	109
1974	Nov 28	2380	246	234	251	231	339	109
1974	Dec 8	2390	256	250	264	238	340	108
1974	Dec 18	2400	266	266	276	245	341	108
1974	Dec 28	2410	276	281	289	253	342	107
1975	Jan 7	2420	287	298	301	260	344	105
1975	Jan 17	2430	297	314	314	267	346	104
1975	Jan 27	2440	307	325	326	274	348	103
1975	Feb 6	2450	317	322	339	281	350	103
1975	Feb 16	2460	327	312	351	289	353	102
1975	Feb 26	2470	337	312	4	296	355	102
1975	Mar 8	2480	347	320	16	304	357	102
1975	Mar 18	2490	357	332	28	312	0	102
1975	Mar 28	2500	7	347	41	319	2	102
1975	Apr 7	2510	17	5	53	327	5	103
1975	Apr 17	2520	27	25	65	334	7	103
1975	Apr 27	2530	36	46	76	342	9	104
1975	May 7	2540	46	65	88	349	12	105
1975	May 17	2550	56	78	99	357	14	106
1975	May 27	2560	65	83	110	4	17	107
1975	Jun 6	2570	75	81	120	12	18	108
1975	Jun 16	2580	84	75	131	19	20	110
1975	Jun 26	2590	94	74	140	27	21	111
1975	Jul 6	2600	103	82	148	33	22	112
1975	Jul 16	2610	113	96	155	40	24	114
1975	Jul 26	2620	123	115	160	47	24	115
1975	Aug 5	2630	132	137	162	54	24	116
1975	Aug 15	2640	142	156	160	60	25	117
1975	Aug 25	2650	151	172	154	66	25	118
1975	Sep 4	2660	161	187	149	72	24	119
1975	Sep 14	2670	171	198	145	77	23	120
1975	Sep 24	2680	181	205	146	82	22	121
1975	Oct 4	2690	190	202	150	86	21	122
1975	Oct 14	2700	200	190	156	90	19	122
1975	Oct 24	2710	210	192	164	92	18	123
1975	Nov 3	2720	220	205	174	93	16	123
1975	Nov 13	2730	230	221	184	92	15	123
1975	Nov 23	2740	240	237	194	89	14	123

Courtesy of William D. Stahlman and Owen Gingerich, Solar and Planetary Longitudes for Years −2500 to +2500 by 10-day Intervals, the University of Wisconsin Press, Madison.

B.C. By John Hart

By permission of John Hart and Field Enterprises, Inc.

Plot the positions of the other planets using a different color for each one. The data on the resulting chart is much like the data that puzzled the ancients. In fact, the table of longitudes is just an updated version of the tables that Ptolemy, Copernicus, and Tycho had made.

The graph contains a good deal of useful information. For example, when will Mercury and Venus next be close enough to each other so that you can use bright Venus to help you find Mercury? Where are the planets, relative to the sun, when they go through their retrograde motions?

A "full earth" photograph from 22,300 miles in space.

EXPERIMENT 15 SIZE OF THE EARTH

People have been telling you for many years that the earth has a diameter of about 8000 miles and a circumference of about 25,000 miles. You've believed what they told you. But suppose someone challenged you to prove it? How would you go about it?

The first recorded calculation of the size of the earth was made a long time ago—in the third century B.C., by Eratosthenes. He compared the lengths of shadows cast by the sun at two different points in Egypt. The points were rather far apart, but nearly on a north-south line on the earth's surface. The experiment you do here uses a similar method. Instead of measuring the length of a shadow, you will measure the angle between the vertical and the sight line to a star or to the sun.

You will need a colleague at least 200 miles away, due north or south of your position, to take simultaneous measurements. The two of you will need to agree in advance on the star, the date, and the time for your observations. See how close you can come to calculating the actual size of the earth.

Assumptions and Theory of the Experiment

The experiment is based on the assumptions that

1. the earth is a perfect sphere,
2. a plumb line points towards the center of the earth, and
3. the distance from the earth to the stars and sun is very great compared with the earth's diameter.

The two observers must be located at points nearly north and south of each other. Suppose they are at points A and B, separated by a distance s, as shown in Fig. 5-2. The observer at A and the observer at B both sight on the same star at the prearranged time, when the star is on or near their meridian, and measure the angle between the vertical of the plumb line and the sight line to the star.

Light rays from the star reaching locations A and B are parallel (this is implied by assumption 3).

You can therefore relate the angle θ_A at A to the angle θ_B at B, and to the angle ϕ between the two radii, as shown in Fig. 5-3.

In the triangle A′BO

$$\phi = (\theta_A - \theta_B) \qquad (1)$$

If C is the circumference of the earth, and s is an arc of the meridian, then you can make the proportion

$$\frac{s}{C} = \frac{\phi}{360°} \qquad (2)$$

Combining equations (1) and (2), you have

$$C = \frac{360°}{\theta_A - \theta_B} s,$$

where θ_A and θ_B are measured in degrees.

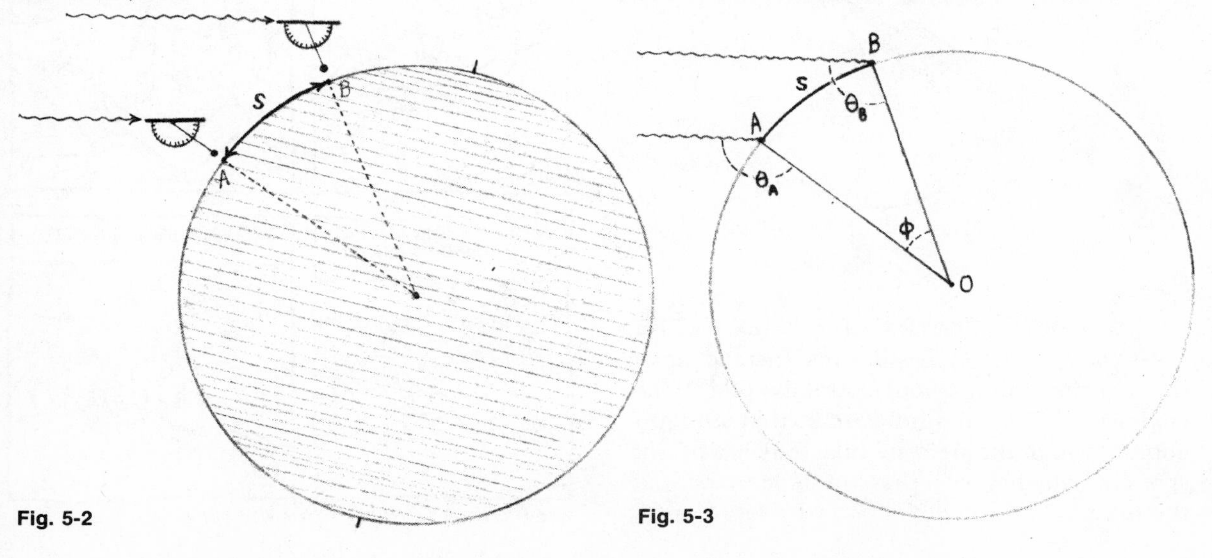

Fig. 5-2

Fig. 5-3

Doing the Experiment

For best results, the two locations A and B should be directly north and south of each other, and the observations should be made when the star is near its highest point in the sky.

You will need some kind of instrument to measure the angle θ. Such an instrument is called an astrolabe. If your teacher does not have an astrolabe, you can make one fairly easily from a protractor, a small sighting tube, and a weighted string assembled according to the design in Fig. 5-4.

Align your astrolabe along the north-south line and measure the angle from the vertical to the star as it crosses the north-south line.

If the astrolabe is not aligned along the north-south line or meridian, the star will be observed before or after it is highest in the sky. An error of a few minutes from the time of crossing the meridian will make little difference in the angle measured.

An alternative method would be to measure the angle to the sun at local noon. This means the time when the sun is highest in the sky, and not necessarily 12 o'clock by standard time. (Remember that the sun, seen from the earth, is itself $\frac{1}{2}°$ wide.)

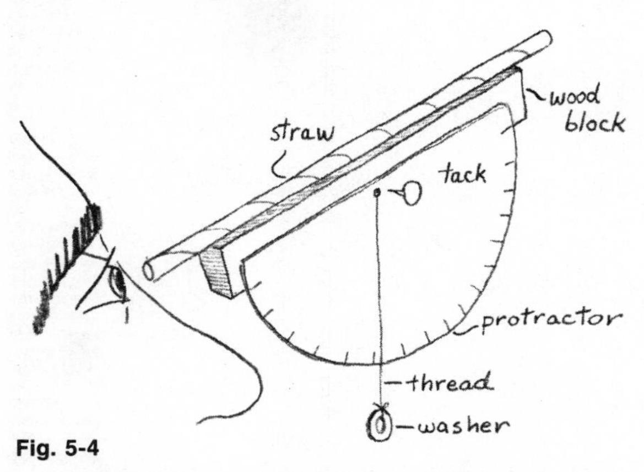

Fig. 5-4

DO NOT TRY SIGHTING DIRECTLY AT THE SUN. You may damage your eyes. Instead, have the sighting tube of your astrolabe pierce the center of a sheet of cardboard so that sunlight going through the sighting tube makes a bright spot on a shaded card that you hold up behind the tube.

An estimate of the uncertainty in your measurement of θ is important. Take several measurements on the same star (at the same time) and take the average value of θ. Use the spread in values of θ to estimate the uncertainty of your observations and of your result.

Your value for the earth's circumference depends also in knowing the over-the-earth distance between the two points of observation. You should get this distance from a map, using its scale of miles. For a description of what earth measurements over the years have disclosed about the earth's shape, see: "The Shape of the Earth," *Scientific American*, October, 1967, page 67.

Q1 How does the uncertainty of the over-the-earth distance compare with the uncertainty in your value for θ?

Q2 What is your calculated value for the circumference of the earth and what is the uncertainty of your value?

Q3 Astronomers have found that the average circumference of the earth is about 24,900 miles (40,000 km). What is the percentage error of your result?

Q4 Is this acceptable, in terms of the uncertainty of your measurement?

EXPERIMENT 16 THE HEIGHT OF PITON, A MOUNTAIN ON THE MOON

Closeup photographs of the moon's surface have been radioed back to earth from Lunar Orbiter spacecraft (Fig. 5-10) and from Surveyor vehicles that have made "soft landings" on the moon (Fig. 5–11, p. 81), and carried back by the Apollo astronauts. Scientists are discovering a great deal about the moon from such photographs, as well as from the landings made by astronauts in Apollo spacecraft.

But long before the space age, indeed since Galileo's time, astronomers have been learning about the moon's surface without even leaving the earth. In this experiment, you will use a photograph (Fig. 5-5) taken with a 36-inch telescope in California to estimate the height of a mountain on the moon. You will use a method similar in principle to one used by Galileo, although you should be able to get a more accurate value than he could working with his small telescope (and without photographs!).

The photograph of the moon in Fig. 5-5 was taken at the Lick Observatory very near the time of the third quarter. The photograph does not show the moon as you see it in the sky at third quarter because an astronomical telescope gives an inverted image—reversing top-and-bottom and left-and-right. (Thus north is at the bottom.) Fig. 5-6 is a 10X enlargement of the area within the white rectangle in Fig. 5-5.

Why Choose Piton?

Piton, a mountain in the moon's Northern Hemisphere, is fairly easy to measure because it is a slab-like pinnacle in an otherwise fairly flat area. When the photograph was made, with the moon near third quarter phase, Piton was quite close to the line separating the lighted portion from the darkened portion of the moon. (This line is called the *terminator*.)

You will find Piton to the south and west of the large, dark-floored crater, Plato (numbered 230 on your moon map) which is located at a longitude of −10° and a latitude of +50°.

Fig. 5-5

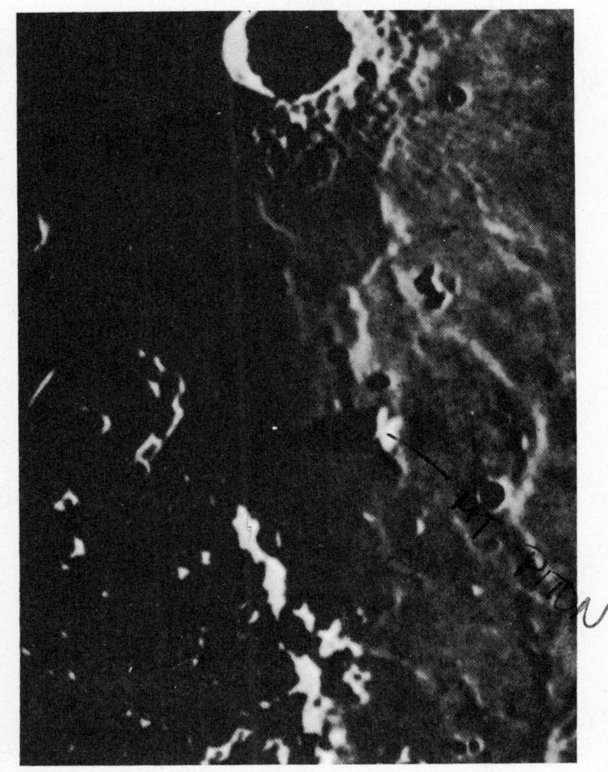

Fig. 5-6

Assumptions and Relations

Fig. 5-7 represents the third-quarter moon of radius r, with Piton P, its shadow of length l, at a distance d from the terminator.

The rays of light from the sun can be considered to be parallel because the moon is a great distance from the sun. Therefore, the

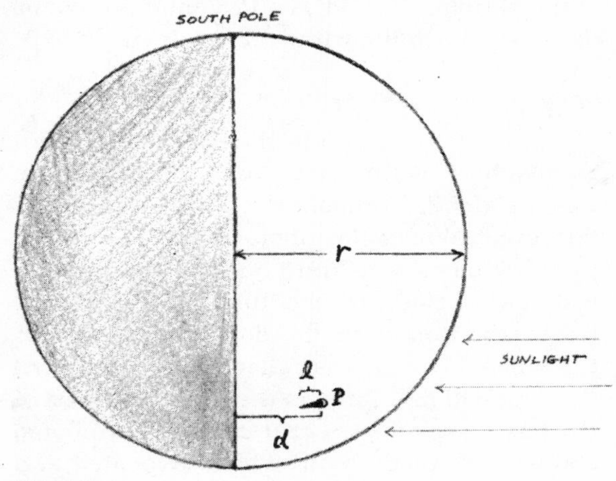

Fig. 5-7

angle at which the sun's rays strike Piton will not change if, in imagination, we rotate the moon on an axis that points toward the sun. In Fig. 5-8, the moon has been rotated just enough to put Piton on the lower edge. In this position it is easier to work out the geometry of the shadow.

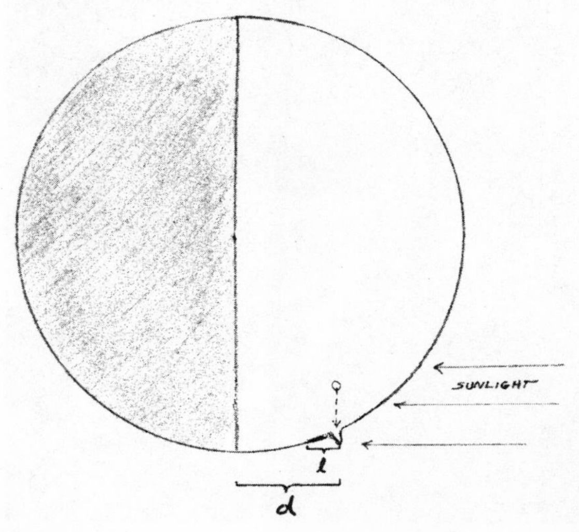

Fig. 5-8

Fig. 5-9 shows how the height of Piton can be found from similar triangles. h represents the height of the mountain, l is the apparent length of its shadow, d is the distance of the mountain from the terminator; r is a radius of the moon (drawn from Piton at P to the center of the moon's outline at O.)

It can be proven geometrically (and you can see from the drawing) that the small triangle BPA is similar to the large triangle PCO. The corresponding sides of similar triangles are proportional, so we can write

$$\frac{h}{l} = \frac{d}{r}$$

$$\text{and } h = \frac{l \times d}{r}$$

All of the quantities on the right can be measured from the photograph.

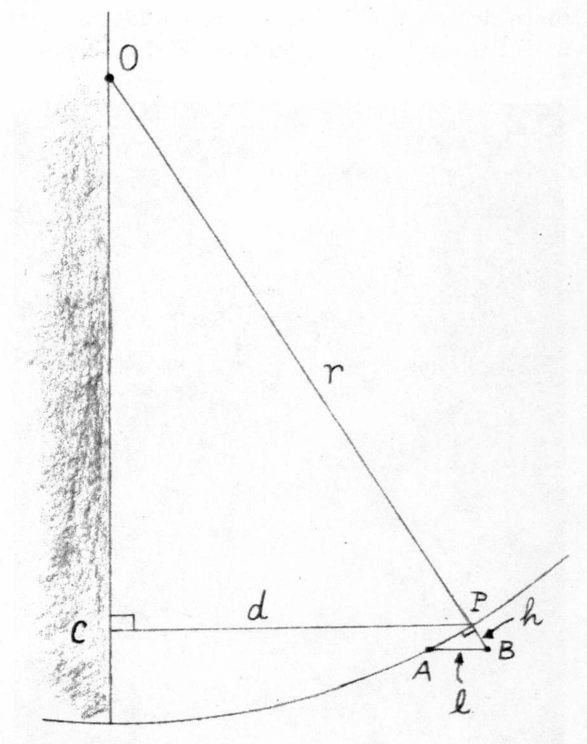

Fig. 5-9

The curvature of the moon's surface introduces some error into the calculations, but as long as the height and shadow are very small compared to the size of the moon, the error is not too great.

Measurements and Calculations

Unless specifically instructed by your teacher, you should work on a tracing of the moon picture rather than in the book itself. Trace the outline of the moon and the location of Piton. If the photograph was made when the moon was exactly at third quarter phase, then the moon was divided exactly in half by the terminator. The terminator appears ragged because highlands cast shadows across the lighted side and peaks stick up out of the shadow side. Estimate the best overall straight line for the terminator and add it to your tracing. Use a cm ruler to measure the length of Piton's shadow and the distance from the terminator to Piton's peak.

It probably will be easiest for you to do all the calculations in the scale of the photograph, find the height of Piton in cm, and then finally change to the real scale of the moon.

Q1 How high is Piton in cm on the photograph scale?

Q2 The diameter of the moon is 3,476 km. What is the scale of the photograph?

Q3 What value do you get for the actual height of Piton?

Q4 Which of your measurements is the least certain? What is your estimate of the uncertainty of your height for Piton?

Q5 Astronomers, using more complicated methods than you used, find Piton to be about 2.3 km high (and about 22 km across at its base). Does your value differ from the accepted value by more than your experimental uncertainty? If so, can you suggest why?

Fig. 5-10 A fifty square mile area of the moon's surface near the large crater, Goclenius. An unusual feature of this crater is the prominent rille that crosses the crater rim.

Fig. 5-11 A four-inch rock photographed on the lunar surface by Surveyor VII in 1968.

ACTIVITIES

MAKING ANGULAR MEASUREMENTS

For astronomical observations, and often for other purposes, you need to estimate the angle between two objects. You have several instant measuring devices handy once you calibrate them. Held out at arm's length in front of you, they include:

1. Your thumb,
2. Your fist not including thumb knuckles,
3. Two of your knuckles, and
4. The span of your hand from thumb-tip to tip of little finger when your hand is opened wide.

For a first approximation, your fist is about 8° and thumb-tip to little finger is between 15° and 20°.

However, since the length of people's arms and the size of their hands vary, you can calibrate yours using the following method.

To find the angular size of your thumb, fist, and hand span at your arm's length, you make use of one simple relationship. An object viewed from a distance that is 57.4 times its diameter covers an angle of 1°. For example, a one-inch coin viewed from 57.4 inches away has an angular size of 1°.

Set a 1″ ruler on the blackboard chalk tray. Stand with your eye at a distance of $57\frac{1}{2}$″ from the scale. From there observe how many inches of the scale are covered by your thumb, etc. Make sure that you have your arm straight out in front of your nose. Each inch covered corresponds to 1°. Find some convenient measuring dimensions on your hand.

A Mechanical Aid

You can use a 3″ × 5″ file card and a meter stick or yard stick to make a simple instrument for measuring angles. Remember that when an object with a given diameter is placed at a distance from your eye equal to 57.4 times its diameter, it cuts off an angle of 1°. This means that a one inch object placed at a distance of 57.4 inches from your eye would cut off an angle of 1°. An instrument 57.4 inches long would be a bit cumbersome, but we can scale down the measurements to a convenient size. A $\frac{1}{2}$ inch diameter object placed at a distance of 27.7 inches (call it 28 inches) from your eye, would cut off an angle of 1°. At this same distance, 28 inches, a 1 inch diameter object would cut off 2° and a $2\frac{1}{2}$ inch object 5°.

Now you can make a simple device that you can use to estimate angles of a few degrees.

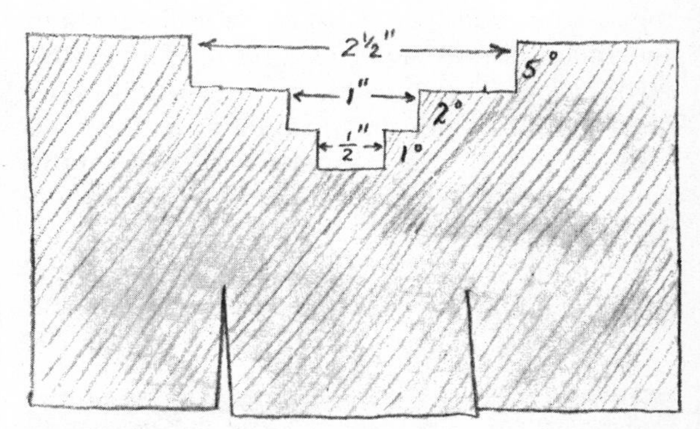

Fig. 5-12

Cut a series of step-wise slots as shown in Fig. 5-12, in the file card. Make the slots $\frac{1}{2}$ inch for 1°, one inch for 2°, and $2\frac{1}{2}$ inches for 5°. Mount the card vertically at the 28 inch mark on a yard stick. (If you use a meter stick, put the card at 57 cm and make the slots 1 cm wide for 1°, 2 cm for 2°, etc.) Cut flaps in the bottom of the card, fold them to fit along the stick and tape the card to the stick. Hold the zero end of the stick against your upper lip—and observe. (Keep a stiff upper lip!)

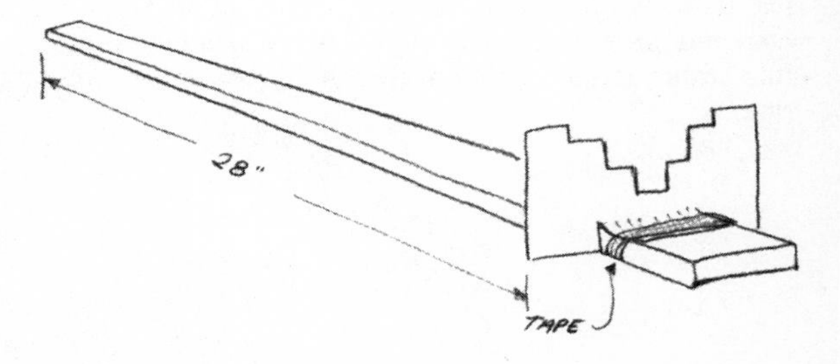

Things to Observe

1. What is the visual angle between the pointers of the Big Dipper?
2. What is the angular length of Orion's Belt?

B.C. By John Hart

By permission of John Hart and Field Enterprises, Inc.

3. How many degrees away from the sun is the moon? Observe on several nights at sunset.
4. What is the angular diameter of the moon? Does it change between the time the moon rises and the time when it is highest in the sky on a given day? To most people, the moon seems larger when near the horizon. Is it? See: "The Moon Illusion," *Scientific American*, July 1962, p. 120.

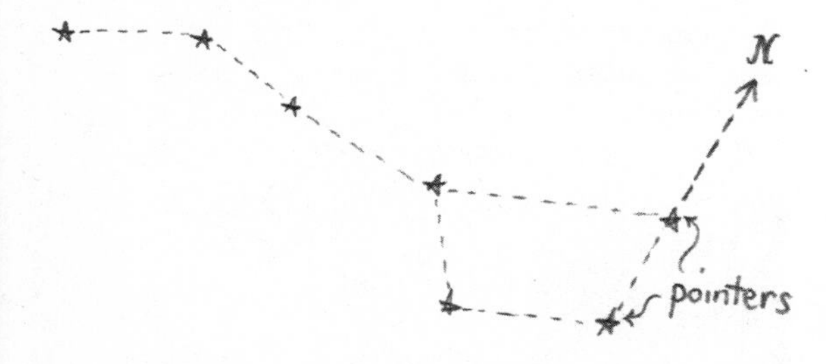

EPICYCLES AND RETROGRADE MOTION

The hand-operated epicycle machine allows you to explore the motion produced by two circular motions. You can vary both the ratio of the turning rates and the ratio of the radii to find the forms of the different curves that may be traced out.

The epicycle machine has three possible gear ratios: 2 to 1 (producing two loops per revolution), 1 to 1 (one loop per revolution) and 1 to 2 (one loop per two revolutions). To change the ratio, simply slip the drive band to another set of pulleys. The belt should be twisted in a figure 8 so the deferent arm (the long arm) and the epicycle arm (the short arm) rotate in the same direction.

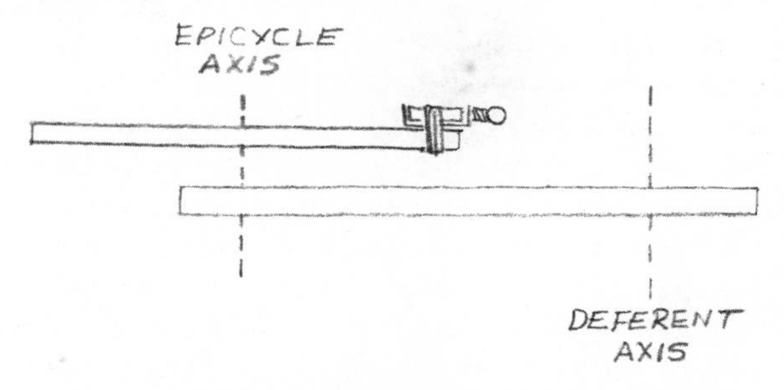

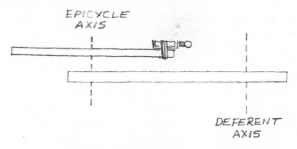

Tape a light source (pen-light cell, holder and bulb) securely to one end of the short, epicycle arm and counter-weight the other end of the arm with, say, another (unlit) light source. If you use a fairly high rate of rotation in a darkened room, you and other observers should be able to see the light source move in an epicycle.

The form of the curve traced out depends not only on the gear ratio but also on the relative lengths of the arms. As the light is moved closer to the center of the epicycle arm, the epicycle loop decreases in size until it becomes

Fig. 5-13

a cusp. When the light is very close to the center of the epicycle arm, as it would be for the motion of the moon around the earth, the curve will be a slightly distorted circle. (Fig. 5-14).

To relate this machine to the Ptolemaic model, in which planets move in epicycles around the earth as a center, you should really stand at the center of the deferent arm (earth) and view the lamp against a distant fixed background. The size of the machine, however, does not allow you to do this, so you must view the motion from the side. (Or, you can glue a spherical glass Christmas-tree ornament at the center of the machine; the reflection you see in the bulb is just what you would see if you were at the center.) The lamp then goes into retrograde motion each time an observer in front of the machine sees a loop. The retrograde motion is most pronounced with the light source far from the center of the epicycle axis.

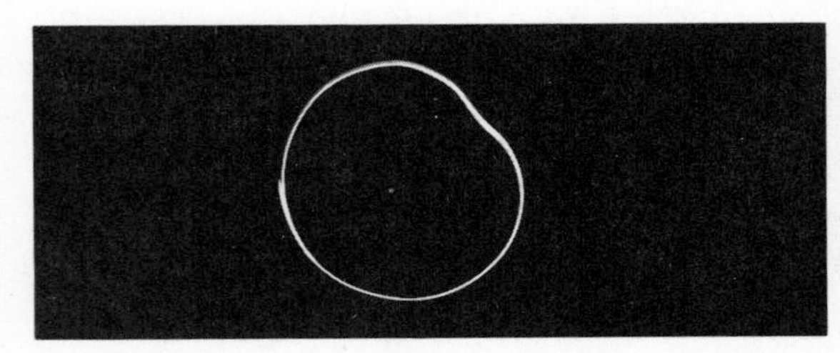

Fig. 5-14

Photographing Epicycles

The motion of the light source can be photographed by mounting the epicycle machine on a turntable and holding the center pulley stationary with a clamp (Fig. 5-15). Alternatively, the machine can be held in a burette clamp on a ringstand and turned by hand.

Running the turntable for more than one revolutión may show that the traces do not exactly overlap (Fig. 5-13). (This probably occurs because the drive band is not of uniform thickness, particularly at its joint, or because the pulley diameters are not in exact proportion.) As the joining seam in the band runs over either pulley, the ratio of speeds changes momentarily and a slight displacement of the axes takes place. By letting the turntable rotate for some time, the pattern will eventually begin to overlap.

A time photograph of this motion can reveal very interesting geometric patterns. You might enjoy taking such pictures as an afterclass activity. Figures 5-16, a through d, show four examples of the many different patterns that can be produced.

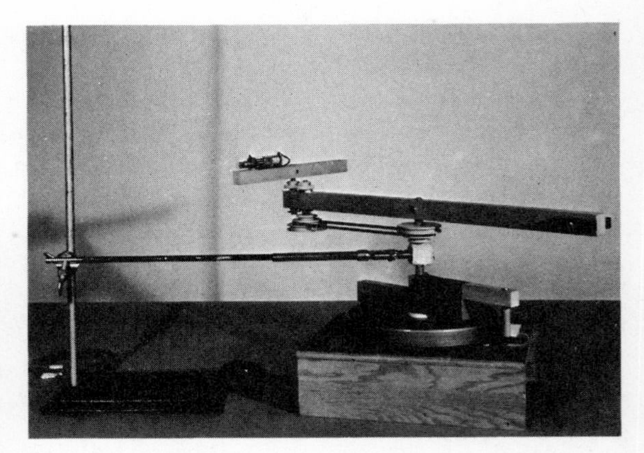

Fig. 5-15 An epicycle demonstrator connected to a turntable.

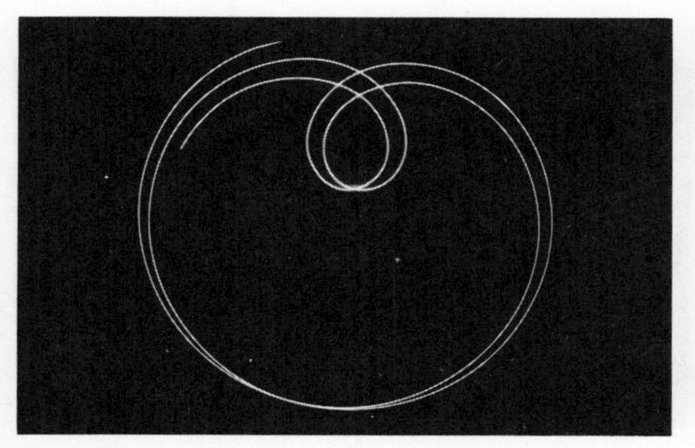

Fig. 5-16a

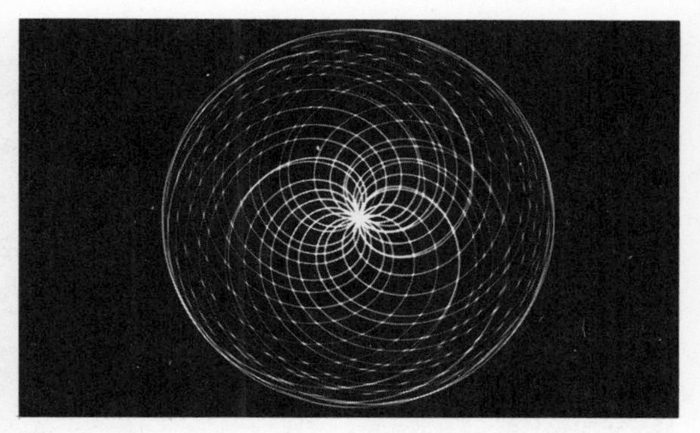

Fig. 5-16b

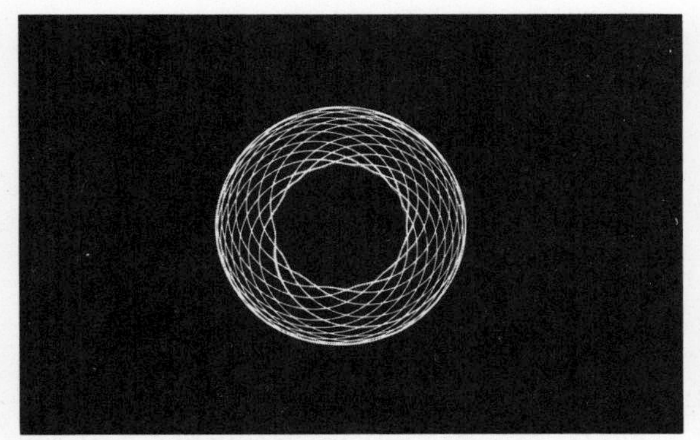

Fig. 5-16c

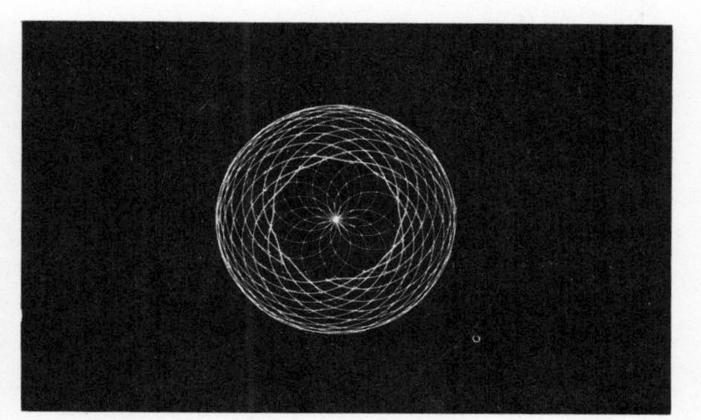

Fig. 5-16d

CELESTIAL SPHERE MODEL*

You can make a model of the celestial sphere with a round-bottom flask. With it, you can see how the appearance of the sky changes as you go northward or southward and how the stars appear to rise and set.

To make this model, you will need, in addition to the round-bottom flask, a one-hole rubber stopper to fit its neck, a piece of glass tubing, paint, a fine brush (or grease pencil), a star map or a table of star positions, and considerable patience.

On the bottom of the flask, locate the point opposite the center of the neck. Mark this point and label it "N" for north celestial pole. With a string or tape, determine the circumference of the flask—the greatest distance around it. This will be 360° in your model. Then, starting at the north celestial pole, mark points that are $\frac{1}{4}$ of the circumference, or 90°, from the North Pole point. These points lie around the flask on a line that is the celestial equator. You can mark the equator with a grease pencil (china marking pencil), or with paint.

*Adapted from *You and Science*, by Paul F. Brandwein, et al., copyright 1960 by Harcourt, Brace and World, Inc.

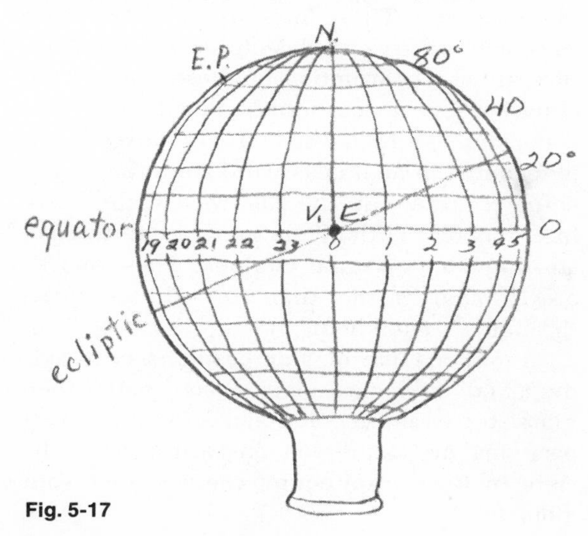

Fig. 5-17

B.C. By John Hart

YOU CERTAINLY
PICKED A LOUSY
NIGHT TO NAME
THE CONSTELLATIONS.

To locate the stars accurately on your "globe of the sky," you will need a coordinate system. If you do not wish to have the coordinate system marked permanently on your model, put on the lines with a grease pencil.

Mark a point $23\frac{1}{2}°$ from the North Pole (about $\frac{1}{4}$ of 90°). This will be the *pole of the ecliptic* marked E.P. in Fig. 5-17. The ecliptic (path of the sun) will be a great circle 90° from the ecliptic pole. The point where the ecliptic crosses the equator from south to north is called the *vernal equinox*, the position of the sun on March 21. All positions in the sky, are located eastward from this point, and north or south from the equator.

To set up the north-south scale, measure off eight circles, about 10° apart, that run east and west in the northern hemisphere parallel to the equator. These lines are like altitude on the earth but are called *declination* in the sky. Repeat the construction of these lines of declination for the southern hemisphere.

A star's position, called its *right ascension*, is recorded in hours eastward from the vernal equinox. To set up the east-west scale, mark intervals of 1/24th of the total circumference starting at the vernal equinox. These marks are 15° apart (rather than 10°)—the sky turns through 15° each hour.

From a table of star positions or a star map, you can locate a star's coordinates, then mark the star on your globe. All east-west positions are expressed eastward, or to the right of the vernal equinox as you face your globe.

To finish the model, put the glass tube into the stopper so that it almost reaches across the flask and points to your North Pole point. Then put enough inky water in the flask so that, when you hold the neck straight down, the water just comes up to the line of the equator. For safety, wrap wire around the neck of the flask and over the stopper so it will not fall out (Fig. 5-18).

Now, as you tip the flask you have a model of the sky as you would see it from different latitudes in the Northern Hemisphere. If you were at the earth's North Pole, the north celestial pole would be directly overhead and you would see only the stars in the northern half of the sky. If you were at latitude 45°N, the north

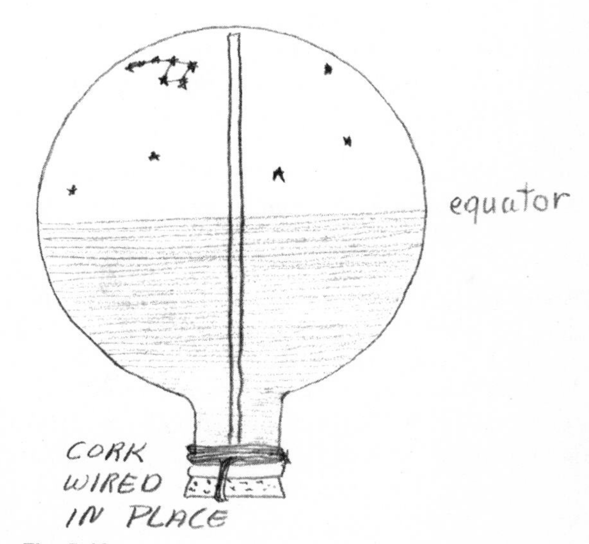
equator
CORK
WIRED
IN PLACE

Fig. 5-18

celestial pole would be halfway between the horizon and the point directly overhead. You can simulate the appearance of the sky at 45°N by tipping the axis of your globe to 45° and rotating it. If you hold your globe with the axis horizontal, you would be able to see how the sky would appear if you were at the equator.

HOW LONG IS A SIDEREAL DAY?

A sidereal day is the time interval it takes a star to travel completely around the sky. To measure a sidereal day you need an electric clock and a screw eye.

Choose a neighboring roof, fence, etc., towards the west. Then fix a screw-eye as an eye-piece in some rigid support such as a post or a tree so that a bright star, when viewed through the screw-eye will be a little above the roof (Fig. 5-19).

Record the time when the star viewed through the screw-eye just disappears behind the roof, and again on the next night. How long did it take to go around? What is the uncertainty in your measurement? If you doubt your result, you can record times for several nights in a row and average the time intervals; this should give you a very accurate measure of a sidereal day. (If your result is not exactly 24 hours, calculate how many days would be needed for the error to add up to 24 hours.)

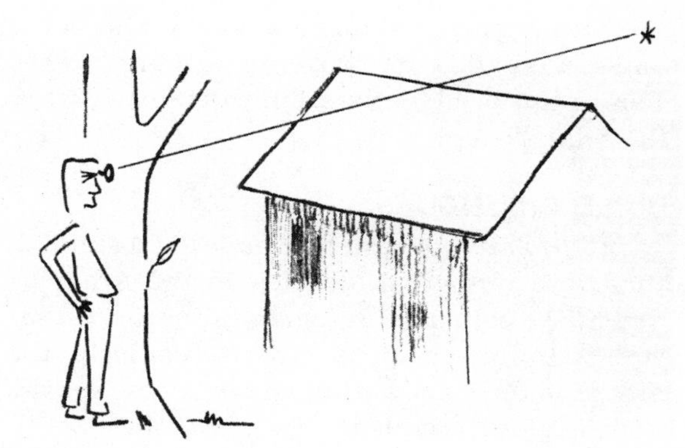

Fig. 5-19

SCALE MODEL OF THE SOLAR SYSTEM

Most drawings of the solar system are badly out of scale, because it is impossible to show both the sizes of the sun and planets and their relative distances on an ordinary-sized piece of paper. Constructing a simple scale model will help you develop a better picture of the real dimensions of the solar system.

Let a three-inch tennis ball represent the sun. The distance of the earth from the sun is 107 times the sun's diameter, or for this model, about 27 feet. (You can confirm this easily. In the sky the sun has a diameter of half a degree – about half the width of your thumb when held upright at arm's length in front of your nose. Check this, if you wish, by comparing your thumb to the angular diameter of the moon, which is nearly equal to that of the sun; both have diameters of $\frac{1}{2}°$. Now hold your thumb in the same upright position and walk away from the tennis ball until its diameter is about half the width of your thumb. You will be between 25 and 30 feet from the ball!) Since the diameter of the sun is about 1,400,000 kilometers (870,000 miles), in the model one inch represents about 464,000 kilometers. From this scale, the proper scaled distances and sizes of all the other planets can be derived.

The moon has an average distance of 384,000 kilometers from the earth and has a diameter of 3,476 kilometers. Where is it on the scale model? How large is it? Completion of the column for the scale-model distances in the table to the left will yield some surprising results.

A scale model of the solar system

Object	Solar Distance		Diameter		Sample Object
	AU	Model (ft)	km (approx.)	Model (inches)	
Sun	------	------	1,400,000	3	tennis ball
Mercury	0.39		4,600		
Venus	0.72		12,000		
Earth	1.00	27	13,000		pinhead
Mars	1.52		6.600		
Jupiter	5.20		140,000		
Saturn	9.45		120,000		
Uranus	19.2		48,000		
Neptune	30.0		45,000		
Pluto	39.5		6,000		
Nearest star	2.7×10^5				

The average distance between the earth and sun is called the "astronomical unit" (AU). This unit is used for describing distance within the solar system.

BUILD A SUNDIAL

If you are interested in building a sundial, there are numerous articles in the Amateur Scientist section of *Scientific American* that you will find helpful. See particularly the article in the issue of August 1959. Also see the issues of September 1953, October 1954, October 1959, or March 1964. The book *Sundials* by Mayall and Mayall (Charles T. Branford, Co., publishers, Boston) gives theory and building instructions for a wide variety of sundials. Encyclopedias also have helpful articles.

PLOT AN ANALEMMA

Have you seen an analemma? Examine a globe of the earth, and you will usually find a graduated scale in the shape of a figure 8, with dates on it. This figure is called an analemma. It is used to summarize the changing positions of the sun during the year.

You can plot your own analemma. Place a small square mirror on a horizontal surface so that the reflection of the sun at noon falls on a south-facing wall. Make observations each day at exactly the same time, such as noon, and mark the position of the reflection on a sheet of paper fastened to the wall. If you remove the paper each day, you must be sure to replace it in exactly the same position. Record the date beside the point. The north-south motion is most evident during September–October and March–April. You can find more about the east–west migration of the marks in astronomy texts and encyclopedias under the subject Equation of Time.

STONEHENGE

Stonehenge (pages 1 and 2 of your Unit 2 *Text*) has been a mystery for centuries. Some scientists have thought that it was a pagan temple, others that it was a monument to slaughtered chieftains. Legends invoked the power of Merlin to explain how the stones were brought to their present location. Recent studies indicate that Stonehenge may have been an astronomical observatory and eclipse computer.

Read "Stonehenge Physics," in the April, 1966 issue of *Physics Today: Stonehenge Decoded,* by G. S. Hawkins and J. B. White; or see *Scientific American,* June, 1953. Make a report and/or a model of Stonehenge for your class.

MOON CRATER NAMES

Prepare a report about how some of the moon craters were named. See Isaac Asimov's *Biographical Encyclopedia of Science and Technology* for material about some of the scientists whose names were used for craters.

LITERATURE

The astronomical models that you read about in Chapters 5 and 6, Unit 2, of the *Text* strongly influenced the Elizabethan view of the world and the universe. In spite of the ideas of Galileo and Copernicus, writers, philosophers, and theologians continued to use Aristotelian and Ptolemaic ideas in their works. In fact, there are many references to the crystal-sphere model of the universe in the writings of Shakespeare, Donne, and Milton. The references often are subtle because the ideas were commonly accepted by the people for whom the works were written.

For a quick overview of this idea, with reference to many authors of the period, read the paperbacks *The Elizabethan World Picture,* by E. M. W. Tillyard, Vintage Press, or Basil Willey, *Seventeenth Century Background,* Doubleday. See also the articles by H. Butterfield and B. Willey in Project Physics *Reader 1.*

An interesting specific example of the prevailing view, as expressed in literature, is found in Christopher Marlowe's *Doctor Faustus,* when Faustus sells his soul in return for the secrets of the universe. Speaking to the devil, Faustus says:

". . . Come, Mephistophilis, let us
dispute again
And argue of divine astrology.
Tell me, are there many heavens
above the moon?
Are all celestial bodies but one
globe.
As is the substance of this centric
earth? . . .

THE SIZE OF THE EARTH—SIMPLIFIED VERSION

Perhaps, for lack of a distant colleague, you were unable to determine the size of the earth as described in Experiment 13. You may still do so if you measure the maximum altitude of one of the objects on the following list and then use the attached data as described below.

In Santiago, Chile, Miss Maritza Campusano Reyes made the following observations of the maximum altitude of stars and of the sun: (all were observed *north* of her zenith)

Antares (Alpha Scorpio)	83.0°
Vega (Alpha Lyra)	17.5
Deneb (Alpha Cygnus)	11.5
Altair (Alpha Aquila)	47.5
Fomalhaut (Alpha Pisces Austr.)	86.5
Sun: October 1	59.4°
15	64.8°
November 1	70.7°
15	74.8°

Since Miss Reyes made her observations when the objects were highest in the sky, the values depend only upon her latitude and not upon her longitude or the time at which the observations were made.

From the map below, find how far north you are from Santiago. Next, measure the maximum altitude of one or more of these objects at your location. Then using the reasoning in Experiment 13, calculate a value for the circumference of the earth.

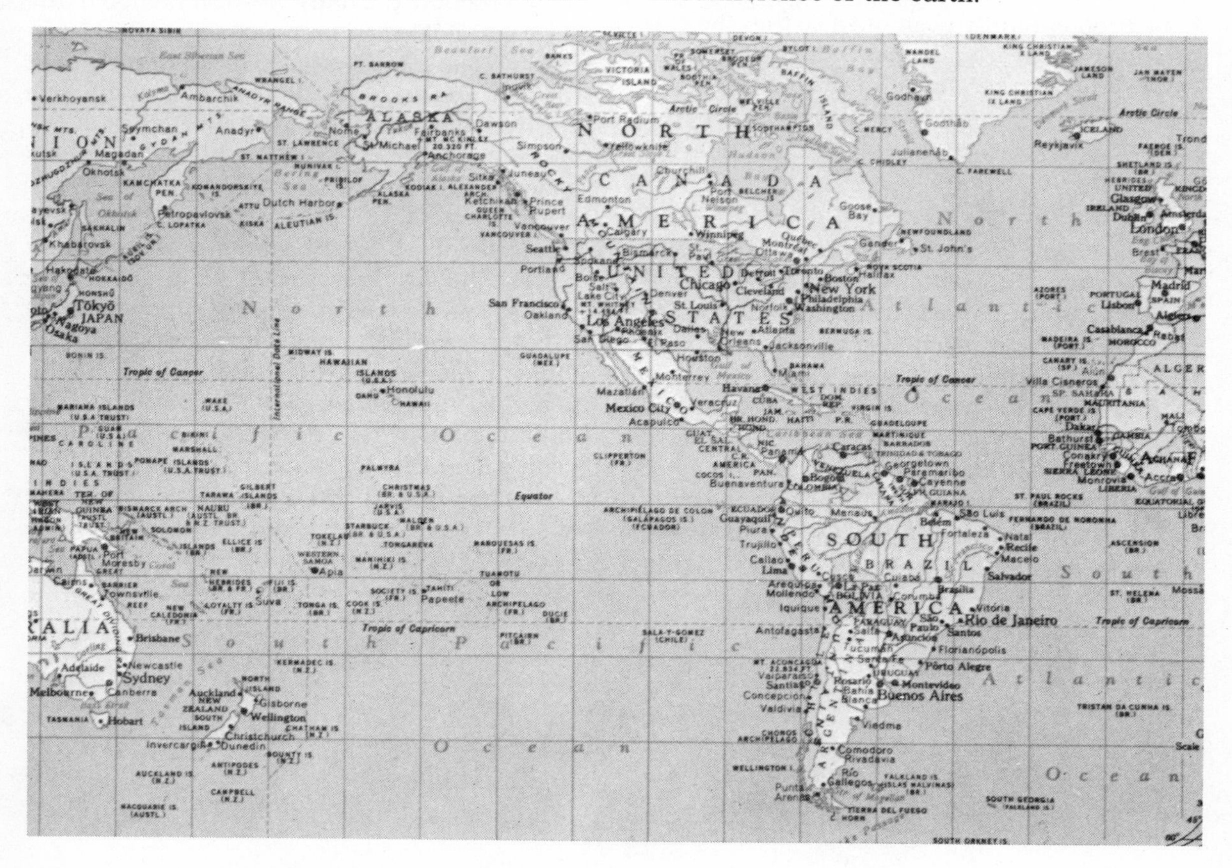

FILM STRIP RETROGRADE MOTION OF MARS

Photographs of the positions of Mars, from the files of the Harvard College Observatory, are shown for three oppositions of Mars, in 1941, 1943, and 1946. The first series of twelve frames shows the positions of Mars before and after the opposition of October 10, 1941. The series begins with a photograph on August 3, 1941 and ends with one on December 6, 1941. The second series shows positions of Mars before and after the opposition of December 5, 1943. This second series of seven photographs begins on October 28, 1943 and ends on February 19, 1944.

The third set of eleven pictures, which shows Mars during 1945-46, around the opposition of January 14, 1946, begins with October 16, 1945 and ends with February 23, 1946.

The film strip is used in the following way:

1. The star fields for each series of frames have been carefully positioned so that the star positions are nearly identical. If the frames of each series can be shown in rapid succession, the stars will be seen as stationary on the screen, while the motion of Mars among the stars is quite apparent. This would be like viewing a flip-book.

2. The frames can be projected on a paper screen where the positions of various stars and of Mars can be marked. If the star pattern for each frame is adjusted to match that plotted from the first frame of that series, the positions of Mars can be marked accurately for the various dates. A continuous line through these points will be a track for Mars. The dates of the turning points (when Mars begins and ends its retrograde motion) can be estimated. From these dates, the duration of the retrograde motion can be found. By use of the scale (10°) shown on one frame, the angular size of the retrograde loop can also be derived.

During 1943-44 and again in 1945-46, Mars and Jupiter came to opposition at approximately the same time. As a result, Jupiter appears in the frames and also shows its retrograde motion. Jupiter's oppositions were: January 11, 1943; February 11, 1944; March 13, 1945; and April 13, 1946. Jupiter's position can also be tracked, and the duration and size of its retrograde loop derived. The durations and angular displacements can be compared to the average values listed in Table 5.1 of the Unit 2 *Text*. This is the type of observational information which Ptolemy, Copernicus and Kepler attempted to explain by their theories.

The photographs were taken by the routine Harvard Sky Patrol with a camera of 6-inch focal length and a field of 55°. During each exposure, the camera was driven by a clockwork to follow the daily western motion of the stars and hold their images fixed on the photographic plate. Mars was never in the center of the field and was sometimes almost at the edge because the photographs were not made especially to show Mars. The planet just happened to be in the star fields being photographed.

The images of the stars and planets are not of equal brightness on all pictures because the sky was less clear on some nights and the exposures varied somewhat in duration. Also, the star images show distortions from limitations of the camera's lens. Despite these limitations, however, the pictures are adequate for the uses described above.

From a purely artistic point of view, some of the frames show beautiful pictures of the Milky Way in Taurus (1943) and Gemini (1945).

FILM LOOPS

FILM LOOP 10A RETROGRADE MOTION OF MARS AND MERCURY

To illustrate the retrograde motions of all the planets, the retrograde motions of Mercury and Mars are shown. The changing positions of each planet against the background of stars are shown during several months by animated drawings. Stars are represented by small disks whose sizes are proportional to the brightness of the stars.

Mercury first moves eastward, stops, and moves westward in retrograde motion. During the retrograde motion, Mercury passes between the earth and the sun, which has been moving steadily eastward. Mercury stops its westward motion and resumes its eastward motion following the sun. Time flashes appear for each five days. The star field includes portions of the constellations of Aries and Taurus; the familiar cluster of stars known as the Pleiades is in the upper left part of the field.

Mars similarly moves eastward across the star field, stops, moves westward in retrograde motion, stops, and resumes its eastward motion. Time flashes appear for each ten days. The star field included parts of the constellations of Leo and Cancer. The open star cluster at the upper right is Praesepe (the Beehive) which is faintly visible on a moonless night (and beautiful in a small telescope).

An angular scale (10°) allows the magnitude of the retrograde motions to be measured, while the time flashes permit a determination of the duration of those motions. The disks representing the planets change in brightness in the same manner as observed for the planets in the sky.

FILM LOOP 10 RETROGRADE MOTION —GEOCENTRIC MODEL

The film illustrates the motion of a planet such as Mars, as seen from the earth. It was made using a large "epicycle machine," as a model of the Ptolemaic system.

First, from above, you see the characteristic retrograde motion during the "loop" when the planet is closest to the earth. Then the studio lights go up and you see that the motion is due to the combination of two circular motions. One arm of the model rotates at the end of the other.

The earth, at the center of the model, is then replaced by a camera that points in a fixed direction in space. The camera views the motion of the planet relative to the fixed stars (so the rotation of the earth on its axis is being ignored). This is the same as if you were looking at the stars and planets from the earth toward one constellation of the zodiac, such as Sagittarius.

The planet, represented by a white globe, is seen along the plane of motion. The direct motion of the planet, relative to the fixed stars, is eastward, toward the left (as it would be if you were facing south). A planet's retrograde motion does not always occur at the same place in the sky, so some retrograde motions are not visible in the chosen direction of observation. To simulate observations of planets better, an additional three retrograde loops were photographed using smaller bulbs and slower speeds.

Note the changes in apparent brightness and angular size of the globe as it sweeps close to the camera. Actual planets appear only as points of light to the eye, but a marked change in brightness can be observed. This was not considered in the Ptolemaic system, which focused only on positions in the sky.

Another film loop, described on page 101, Chapter 6 of this *Handbook*, shows a similar model based on a heliocentric theory.

CHAPTER 6

Chapter 6 Does the Earth Move?—The Work of Copernicus and Tycho

EXPERIMENT 17 THE SHAPE OF THE EARTH'S ORBIT

Ptolemy and most of the Greeks thought that the sun revolved around the earth. But after the time of Copernicus the idea gradually became accepted that the earth and other planets revolve around the sun. Although you probably believe the Copernican model, the evidence of your senses gives you no reason to prefer one model over the other.

With your unaided eyes you see the sun going around the sky each day in what appears to be a circle. This apparent motion of the sun is easily accounted for by imagining that it is the *earth* which rotates once a day. But the sun also has a *yearly* motion with respect to the stars. Even if we argue that the daily motion of objects in the sky is due to the turning of the earth, it is still possible to think of the earth as being at the center of the universe, and to imagine the sun moving in a year-long orbit around the earth. Simple measurements show that the sun's angular size increases and decreases slightly during the year as if it were alternately changing its distance from the earth. An interpretation that fits these observations is that the sun travels around the earth in a slightly off-center circle.

The purpose of this laboratory exercise is to plot the sun's apparent orbit with as much accuracy as possible.

Plotting the Orbit

You know the sun's direction on each date that the sun is observed. From its observed diameter on that date you can find its relative distance from the earth. So, date by date, you can plot the sun's direction and relative distance. When you connect your plotted points by a smooth curve, you will have drawn the sun's apparent orbit.

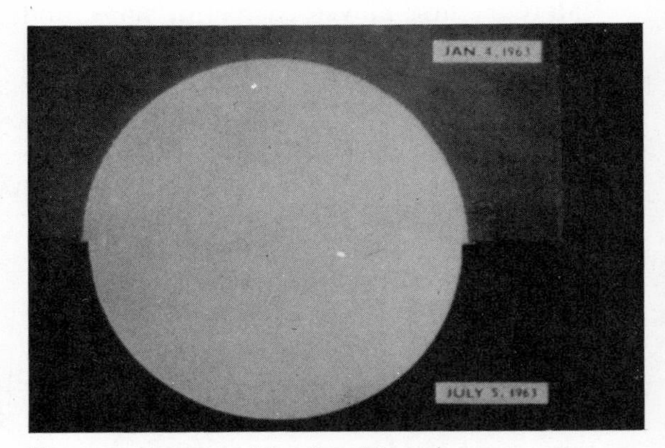

Fig. 6-1 Frame 4 of the Sun Filmstrip.

For observations you will use a series of sun photographs taken by the U.S. Naval Observatory at approximately one-month intervals and printed on a film strip. Frame 4, in which the images of the sun in January and in July are placed adjacent to each other, has been reproduced in Fig. 6-1 so you can see how

B.C. By John Hart

much the apparent size of the sun changes during the year. Then note in Fig. 6-2 how the apparent size of an object is related to its distance from you.

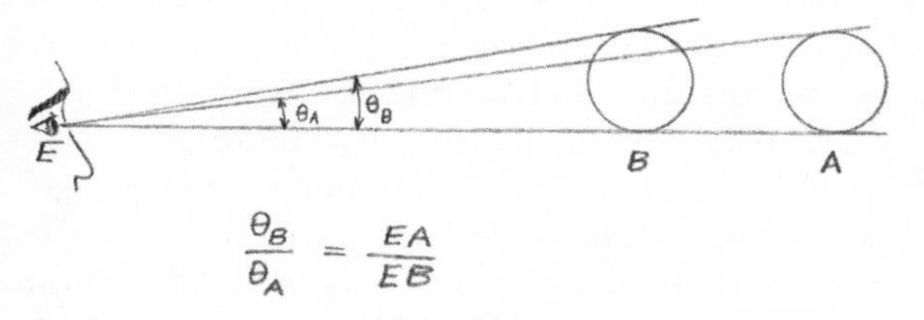

Fig. 6-2 When an object is closer to your eye, it looks bigger; it fills a larger angle as seen by your eye. In fact, the angles θ_A and θ_B are inversely proportional to the distances EA and EB:

$$\frac{\theta_B}{\theta_A} = \frac{EA}{EB}$$

In this drawing EB $= \frac{3}{4}$ EA, so angle $\theta_B = \frac{4}{3}$ angle θ_A.

Procedure

On a large sheet of graph paper (16″ × 20″, or four $8\frac{1}{2}$″ × 11″ pieces taped together) make a dot at the center to represent the earth. It is particularly important that the graph paper be this large if you are going on to plot the orbit of Mars (Experiments 17 and 19) which uses the results of the present experiment.

Take the 0° direction (toward a reference point among the stars) to be along the graph-paper lines toward the right. This will be the direction of the sun as seen from the earth on March 21. (Fig. 6-3) The dates of all the photographs and the directions to the sun, measured counterclockwise from this zero direction, are given in the table below. Use a protractor in order to draw accurately a fan of lines radiating from the earth in these different directions.

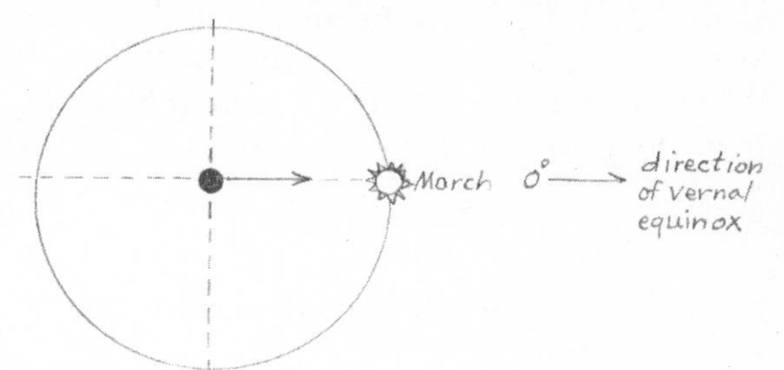

Fig. 6-3

Date	Direction from earth to sun	Date	Direction from earth to sun
March 21	000°	Oct. 4	191°
April 6	015	Nov. 3	220
May 6	045	Dec. 4	250
June 5	074	Jan. 4	283
July 5	102	Feb. 4	315
Aug. 5	132	March 7	346
Sept. 4	162		

Measure carefully the diameter of the projected image on each of the frames of the film strip. The apparent diameter of the sun depends inversely on how far away it is. You can get a set of relative distances to the sun by choosing some constant and then dividing it by the apparent diameters. An orbit with a radius of about 10 cm will be a particularly convenient size for later use. If you measure the sun's diameter to be about 50 cm, a convenient constant to choose would be 500, since $\frac{500}{50} = 10$. A larger image 51.0 cm in diameter leads to a smaller earth-sun distance:

$$\frac{500}{51.0} = 9.8 \text{ cm.}$$

Make a table of the relative distances for each of the thirteen dates.

Along each of the direction lines you have drawn, measure off the relative distance to the sun for that date. Through the points located in this way draw a smooth curve. This is the apparent orbit of the sun relative to the earth. (Since the distances are only relative, you cannot find the actual distance in miles from the earth to the sun from this plot.)

Q1 Is the orbit a circle? If so, where is the center of the circle? If the orbit is not a circle, what shape is it?

Q2 Locate the major axis of the orbit through the points where the sun passes closest to and farthest from the earth. What are the approximate dates of closest approach and greatest distance? What is the ratio of the largest distance to the smallest distance?

A Heliocentric System

Copernicus and his followers adopted the sun-centered model because they believed that the solar system could be described more simply that way. They had no new data that could not be accounted for by the old model.

Therefore, you should be able to use the same data to turn things around and plot the earth's orbit around the sun. Clearly there's going to be some similarity between the two plots.

You already have a table of the relative distances between the sun and the earth. The dates of largest and smallest distances from the earth won't change, and your table of relative distances is still valid because it wasn't based on which body was moving, only on the distance between them. Only the directions used in your plotting will change.

To figure out how the angles will change, remember that when the earth was at the center of the plot the sun was in the direction 0° (to the right) on March 21.

Q3 This being so, what is the direction of the earth as seen from the sun on that date? See if you can't answer this question for yourself before studying Fig. 6-4. Be sure you understand it before going on.

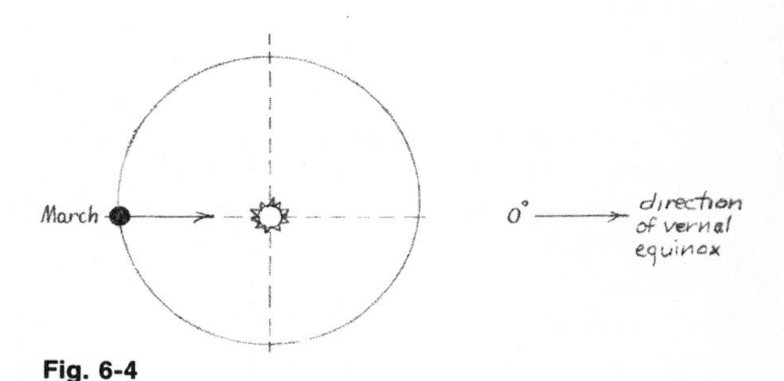

Fig. 6-4

At this stage the end is in sight. Perhaps you can see it already without doing any more plotting. But if not, here is what you can do.

If the sun is in the 0° direction from the earth, then from the sun the earth will appear to be in just the opposite direction, 180° away from 0°. You could make a new table of data giving the earth's apparent direction from the sun on the 12 dates, just by changing all the directions 180° and then making a new sun-centered plot. An easier way is to rotate your plot until top and bottom are reversed; this will change all of the directions by 180°. Relabel the 0° direction; since it is toward a reference point among the distant stars, it will still be toward the right. You can now label the center as the sun, and the orbit as the earth's.

EXPERIMENT 18 USING LENSES TO MAKE A TELESCOPE

In this experiment, you will first examine some of the properties of single lenses. Then, you will combine these lenses to form a telescope, which you can use to observe the moon, the planets, and other heavenly (as well as earth-bound) objects.

The Simple Magnifier

You certainly know something about lenses already—for instance, that the best way to use a magnifier is to hold it immediately in front of the eye and then move the object you want to examine until its image appears in sharp focus.

Examine some objects through several different lenses. Try lenses of various shapes and sizes. Separate the lenses that magnify from those that don't. Describe the difference between lenses that magnify and those that do not.

Q1 Arrange the lenses in order of their magnifying powers. Which lens has the highest magnifying power?

Q2 What physical feature of a lens seems to determine its power or ability to magnify—is it diameter, thickness, shape, the curvature of its surface? To vary the diameter, simply put pieces of paper or cardboard with various sizes of holes in them over the lens.

Sketch side views of a high-power lens, of a low-power lens, and of the highest-power and lowest-power lenses you can imagine.

Real Images

With one of the lenses you have used, project an image of a ceiling light or an outdoor scene on a sheet of paper. Describe all the properties of the image that you can observe. An image that can be projected is called a *real image*.

Q3 Do all your lenses form real images?

Q4 How does the size of the image depend on the lens?

Q5 If you want to look at a real image without using the paper, where do you have to put your eye?

Q6 The image (or an interesting part of it) may be quite small. How can you use a second lens to inspect it more closely? Try it.

Q7 Try using other combinations of lenses. Which combination gives the greatest magnification?

Making a Telescope

With two lenses properly arranged, you can magnify distant objects. Figure 6-5 shows a simple assembly of two lenses to form a telescope. It consists of a large lens (called the *objective*) through which light enters and either of two interchangeable lenses for eyepieces.

The following notes will help you assemble your telescope.

1. If you lay the objective down on a flat clean surface, you will see that one surface is more curved than the other. The more curved surface should face the front of the telescope.
2. Clean dust, etc., off the lenses (using lens tissue or clean handkerchief) before assembling and try to keep fingerprints off it during assembly.
3. Wrap rubber bands around the slotted end of the main tube to give a convenient amount

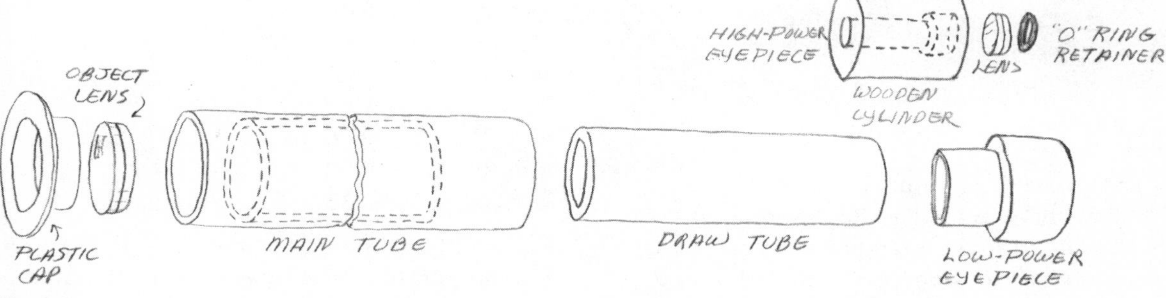

Fig. 6-5

of friction with the draw-tube—tight enough so as not to move once adjusted, but loose enough to adjust without sticking. Focus by sliding the draw tube with a rotating motion, not by moving the eyepiece in the tube.

4. To use high power satisfactorily, a steady support (a tripod) is essential.

5. Be sure that the lens lies flat in the high-power eyepiece.

Use your telescope to observe objects inside and outside the lab. Low power gives about 12× magnification. High power gives about 30× magnification.

Mounting the Telescope

If no tripod mount is available, the telescope can be held in your hands for low-power observations. Grasp the telescope as far forward and as far back as possible (Fig. 6-6) and brace both arms firmly against a car roof, telephone pole, or other rigid support.

With the higher power you must use a mounting. If a swivel-head camera tripod is available, the telescope can be held in a wooden saddle by rubber bands, and the saddle attached to the tripod head by the head's standard mounting screw. Because camera tripods are usually too short for comfortable viewing from a standing position, it is strongly recommended that you be seated in a reasonably comfortable chair.

Fig. 6-6

Aiming and Focusing

You may have trouble finding objects, especially with the high-power eyepiece. One technique is to sight over the tube, aiming slightly below the object, and then to tilt the tube up slowly while looking through it and sweeping left and right. To do this well, you will need some practice.

Focusing by pulling or pushing the sliding tube tends to move the whole telescope. To avoid this, rotate the sliding tube while moving it as if it were a screw.

Eyeglasses will keep your eye farther from the eyepiece than the best distance. Far-sighted or near-sighted observers are generally able to view more satisfactorily by removing their glasses and refocusing. Observers with astigmatism have to decide whether or not the distorted image (without glasses) is more annoying than the reduced field of view (with glasses).

Many observers find that they can keep their eye in line with the telescope while aiming and focusing if the brow and cheek rest lightly against the forefinger and thumb. (Fig. 6-6) When using a tripod mounting, remove your hands from the telescope while actually viewing to minimize shaking the instrument.

Limitations of Your Telescope

You can get some idea of how much fine detail to expect when observing the planets by comparing the angular sizes of the planets with the resolving power of the telescope. For a telescope with a 1 inch diameter object lens, to distinguish between two details, they must be at least 0.001° apart as seen from the location of the telescope. The low-power Project Physics eyepiece may not quite show this much detail, but the high power will be more than sufficient.

The angular sizes of the planets as viewed from the earth are:

Venus:	0.003°	(minimum)
	0.016	(maximum)
Mars:	0.002	(minimum)
	0.005	(maximum)
Jupiter:	0.012	(average)
Saturn:	0.005	(average)
Uranus:	0.001	(average)

Galileo's first telescope gave 3× magnification, and his "best" gave about 30× magnification. (But, he used a different kind of eyepiece that gave a much smaller field of view.) You should find it challenging to see whether you can observe all the phenomena he saw which are mentioned in Sec. 7.7 of the *Text*.

Observations You Can Make

The following group of suggested objects have been chosen because they are (1) fairly easy to find, (2) representative of what is to be seen in the sky, and (3) very interesting. You should observe all objects with the low power first and then the high power. For additional information on current objects to observe, see the paperback *New Handbook of the Heavens*, or the last few pages of each monthly issue of the magazines *Sky and Telescope*, *Natural History*, or *Science News*.

Venus: No features will be visible on this planet, but you can observe its phases, as shown in the photographs below (enlarged to equal sizes) and on page 72 of the Unit 2 *Text*. When Venus is very bright you may need to reduce the amount of light coming through the telescope in order to tell the true shape of the image. A paper lens cap with a round hole in the center will reduce the amount of light (and the resolution of detail!) You might also try using sunglasses as a filter.

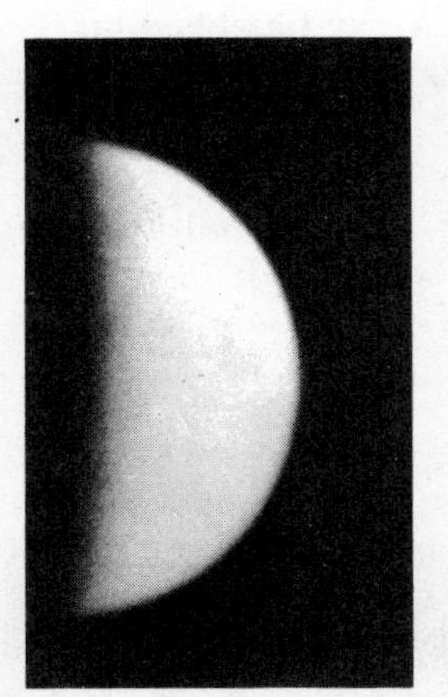

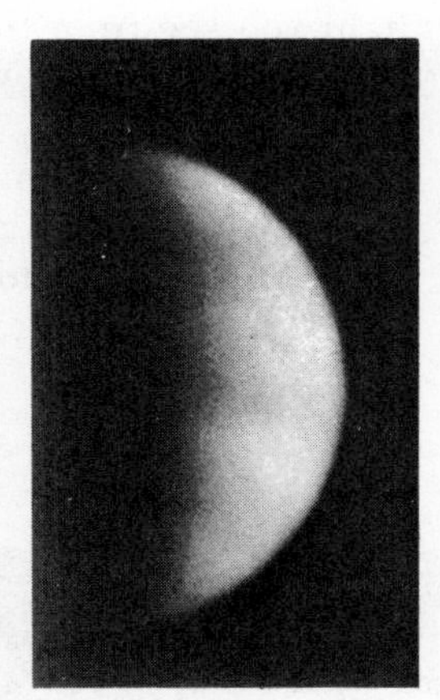

Venus, photographed at Yerkes Observatory with the 82-inch reflector telescope.

Saturn: The planet is so large that you can resolve the projection of the rings beyond the disk, but you probably can't see the gap between the rings and the disk with your 30× telescope. Compare your observations to the sketches on page 73 of the *Text*.

Saturn photographed with the 100-inch telescope at Mount Wilson.

Jupiter: Observe the four satellites that Galileo discovered. Observe them several times, a few hours or a day apart, to see changes in their positions. By keeping detailed data over several months time, you can determine the period for each of the moons, the radii of their orbits, and then the mass of Jupiter. (See the notes for the Film Loop, "Jupiter Satellite Orbit," in Chapter 8 of this *Handbook* for directions on how to analyze your data.)

Jupiter is so large that some of the detail on its disk—like a broad, dark, equatorial cloud belt—can be detected (especially if you know it should be there!)

Jupiter photographed with the 200-inch telescope at Mount Palomar.

Moon: Moon features stand out mostly because of shadows. Best observations are made about the time of half-moon, that is, around the first and last quarter. Make sketches of your observations, and compare them to Galileo's sketch on page 66 of your *Text*. Look carefully for walls, mountains in the centers of craters, bright peaks on the dark side beyond the terminator, and craters in other craters.

B.C. By John Hart

By permission of John Hart and Field Enterprises, Inc.

The Pleiades: A beautiful little star cluster, this is located on the right shoulder of the bull in the constellation Taurus. These stars are almost directly overhead in the evening sky in December. The Pleiades were among the objects Galileo studied with his first telescope. He counted 36 stars, which the poet Tennyson described as "a swarm of fireflies tangled in a silver braid."

The Hyades: This group of stars is also in Taurus, near the star Aldebaran, which forms the bull's eye. Mainly, the Hyades look like a "v." The high-power may show that several stars are double.

The Great Nebula in Orion: Look about halfway down the row of stars that form the sword of Orion. It is in the southeastern sky during December and January. Use low power.

Algol: This famous variable star is in the constellation Perseus, south of Cassiopeia. Algol is high in the eastern sky in December, and nearly overhead during January. Generally it is a second-magnitude star, like the Pole Star. After remaining bright for almost $2\frac{1}{2}$ days, Algol fades for 5 hours and becomes a fourth-magnitude star, like the faint stars of the Little Dipper. Then, the variable star brightens during 5 hours to its normal brightness. From one minimum to the next, the period is 2 days, 20 hours, 49 minutes.

Great Nebula in Andromeda: Look high in the western sky in the early evening in December for this nebula, for by January it is low on the horizon. It will appear like a fuzzy patch of light, and is best viewed with *low* power. The light you see from this galaxy has been on its way for two million years.

The Milky Way: This is particularly rich in Cassiopeia and Cygnus (if air pollution in your area allows it to be seen at all).

B.C. By John Hart

By permission of John Hart and Field Enterprises, Inc.

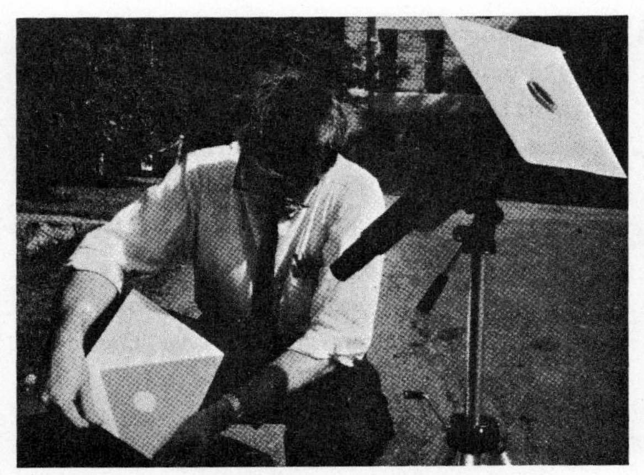

Fig. 6-7 Observing sunspots with a telescope.

Observing sunspots: DO NOT LOOK AT THE SUN THROUGH THE TELESCOPE. THE SUNLIGHT WILL INJURE YOUR EYES. Figure 6-7 shows an arrangement of a tripod, the low-power telescope, and a sheet of paper for projecting sunspots. Cut a hole in a piece of cardboard so it fits snugly over the object end of the telescope. This acts as a shield so there is a shadow area where you can view the sunspots. First focus the telescope, using the high-power eyepiece, on some distant object. Then, project the image of the sun on a piece of white paper a couple of feet behind the eyepiece. Focus the image by moving the draw-tube slightly further out. When the image is in focus, you may see some small dark spots on the paper. To tell marks on the paper from sunspots, jiggle the paper back and forth. How can you tell that the spots aren't on the lenses? By focusing the image farther from the telescope, you can make the image larger and not so bright. It may be easier to get the best focus by moving the paper rather than the eyepiece tube.

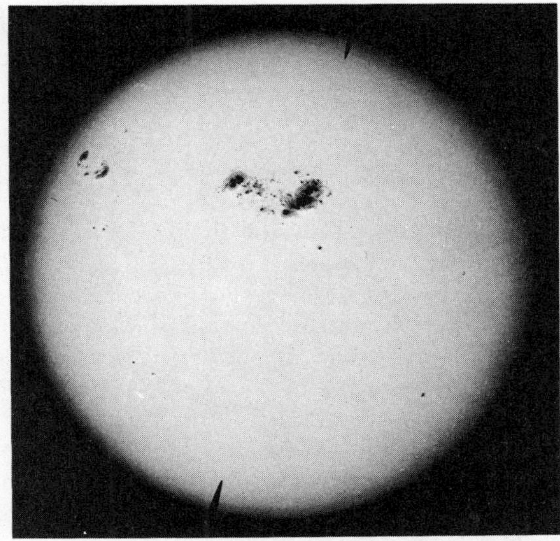

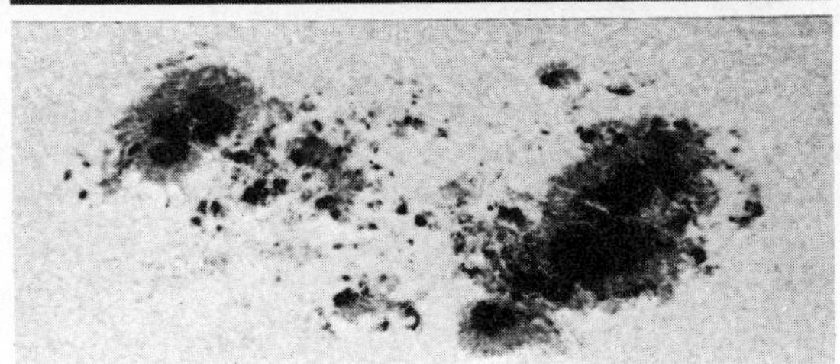

The sunspots of April 7, 1947.

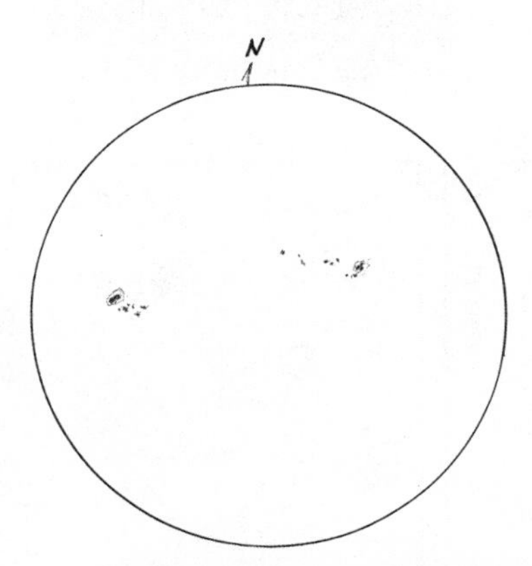

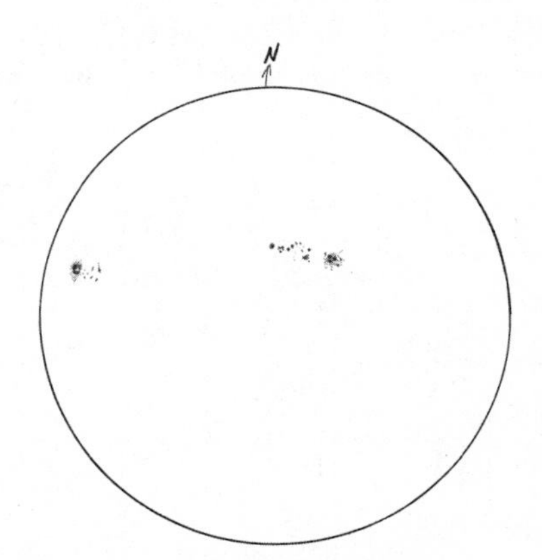

Drawings of the projected image of the sun on Aug. 26 and Aug. 27, 1966, drawn by an amateur astronomer in Walpole, Mass.

ACTIVITIES

TWO ACTIVITIES ON FRAMES OF REFERENCE

1. You and a classmate take hold of opposite ends of a meter stick or a piece of string a meter or two long. If you rotate about on one fixed spot so that you are always facing him while he walks around you in a circle, you will see him moving around you against a background of walls and furniture. But, how do you appear to him? Ask him to describe what he sees when he looks at you against the background of walls and furniture. How do your reports compare? In what direction did you see him move—toward your left or your right? In which direction did he see you move—toward his left or his right?

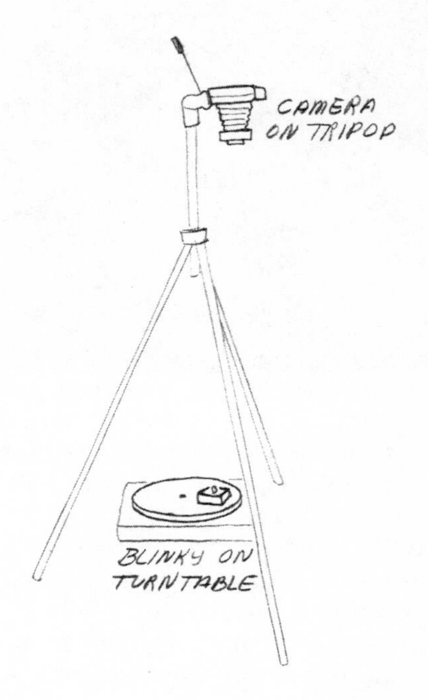

2. The second demonstration involves a camera, tripod, blinky, and turntable. Mount the camera on the tripod (using motor-strobe bracket if camera has no tripod connection) and put the blinky on a turntable. Aim the camera straight down.

Take a time exposure with the camera at rest and the blinky moving one revolution in a circle. If you do not use the turntable, move the blinky by hand around a circle drawn faintly on the background. Then take a second print, with the blinky at rest and the camera on time exposure moved steadily by hand about the axis of the tripod. Try to move the camera at the same rotational speed as the blinky moved in the first photo.

Can you tell, just by looking at the photos whether the camera or the blinky was moving?

Nubbin

FILM LOOP

FILM LOOP 11 RETROGRADE MOTION —HELIOCENTRIC MODEL

This film is based on a large heliocentric mechanical model. Globes represent the earth and a planet moving in concentric circles around the sun (represented by a yellow globe). The earth (represented by a light blue globe) passes inside a slower moving outer planet such as Mars (represented by an orange globe).

Then the earth is replaced by a camera having a 25° field of view. The camera points in a fixed direction in space, indicated by an arrow, thus ignoring the daily rotation of the earth and concentrating on the motion of the earth relative to the sun.

The view from the moving earth is shown for more than 1 year. First the sun is seen in direct motion, then Mars comes to opposition and undergoes a retrograde motion loop, and finally you see the sun again in direct motion.

Scenes are viewed from above and along the plane of motion. Retrograde motion occurs whenever Mars is in opposition, that is, whenever Mars is opposite the sun as viewed from the earth. But not all these oppositions take place when Mars is in the sector the camera sees. The time between oppositions averages about 2.1 years. The film shows that the earth moves about 2.1 times around its orbit between oppositions.

You can calculate this value. The earth makes one cycle around the sun per year and Mars makes one cycle around the sun every 1.88 years. So the frequencies of orbital motion are:

$$f_{earth} = 1 \text{ cyc/yr and } f_{mars} = 1 \text{ cyc/1.88 yr} = 0.532 \text{ cyc/yr}$$

The frequency of the earth relative to Mars is $f_{earth} - f_{mars}$:

$$f_{earth} - f_{mars} = 1.00 \text{ cyc/yr} - 0.532 \text{ cyc/yr} = 0.468 \text{ cyc/yr}$$

That is, the earth catches up with and passes Mars once every

$$\frac{1}{0.468} = 2.14 \text{ years.}$$

Note the increase in apparent size and brightness of the globe representing Mars when it is nearest the earth. Viewed with the naked eye, Mars shows a large variation in brightness (ratio of about 50:1) but always appears to be only a point of light. With the telescope we can see that the angular size also varies as predicted by the model.

The heliocentric model is in some ways simpler than the geocentric model of Ptolemy, and gives the general features observed for the planets: angular position, retrograde motion, and variation in brightness. However, detailed numerical agreement between theory and observation cannot be obtained using circular orbits.

A film of a similar model for the geocentric theory of Ptolemy is described on page 91, Chapter 5 of this *Handbook*.

Chapter 7 A New Universe Appears—the Work of Kepler and Galileo

EXPERIMENT 19 THE ORBIT OF MARS

In this laboratory activity you will derive an orbit for Mars around the sun by the same method that Kepler used in discovering that planetary orbits are elliptical. Since the observations are made from the earth, you will need the orbit of the earth that you developed in Experiment 17, "The Shape of the Earth's Orbit." Make sure that the plot you use for this experiment represents the orbit of the earth around the sun, not the sun around the earth.

If you did not do the earth-orbit experiment, you may use, for an approximate orbit, a circle of 10 cm radius drawn in the center of a large sheet of graph paper (16″ × 20″ or four $8\frac{1}{2}''$ × 11″ joined). Because the eccentricity of the earth's orbit is very small (0.017) you can place the sun at the center of the orbit without introducing a significant error in this experiment.

From the sun (at the center), draw a line to the right, parallel to the grid of the graph paper (Fig. 7-1). Label the line 0°. This line is directed toward a point on the celestial sphere called the vernal equinox and is the reference direction from which angles in the plane of the earth's orbit (the ecliptic plane) are measured. The earth crosses this line on September 23. When the earth is on the other side of its orbit on March 21, the sun is between the earth and the vernal equinox.

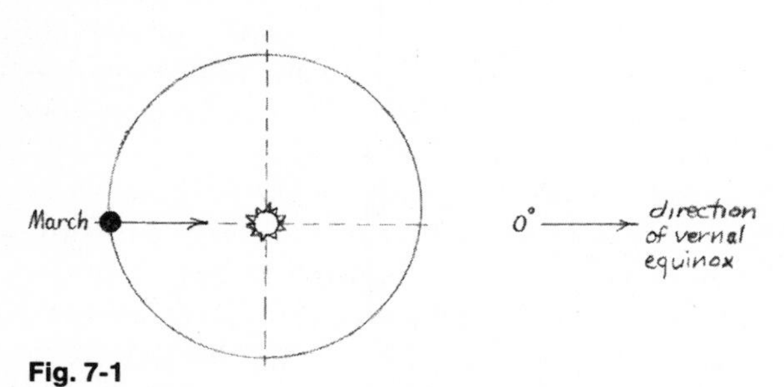

Fig. 7-1

Photographic Observations of Mars

You will use a booklet containing sixteen enlarged sections of photographs of the sky showing Mars among the stars at various dates between 1931 and 1950. All were made with the same small camera used for the Harvard Observatory Sky Patrol. In some of the photographs Mars was near the center of the field. In many other photographs Mars was near the edge of the field where the star images are distorted by the camera lens. Despite these distortions the photographs can be used to provide positions of Mars that are satisfactory for this study. Photograph P is a double exposure, but it is still quite satisfactory.

Changes in the positions of the stars relative to each other are extremely slow. Only a few stars near the sun have motions large enough to be detected after many years observations with the largest telescopes. Thus you can consider the pattern of stars as fixed.

Finding Mars' Location

Mars is continually moving among the stars but is always near the ecliptic. From several hundred thousand photographs at the Harvard Observatory sixteen were selected, with the aid of a computer, to provide pairs of photographs separated by 687 days – the period of Mars around the sun as determined by Copernicus. Thus, each pair of photographs shows Mars at one place in its orbit.

During these 687 days, the earth makes nearly two full cycles of its orbit, but the interval is short of two full years by 43 days. Therefore, the position of the earth, from which we can observe Mars, will not be the same for the two observations of each pair. If you can determine the direction from the earth towards Mars for each of the pairs of observations, the two sight lines must cross at a point on the orbit of Mars. (See Fig. 7-2.)

Coordinate System Used

When you look into the sky you see no coordinate system. Coordinate systems are created for various purposes. The one used here centers on the ecliptic. Remember that the ecliptic is the imaginary line on the celestial sphere along which the sun appears to move.

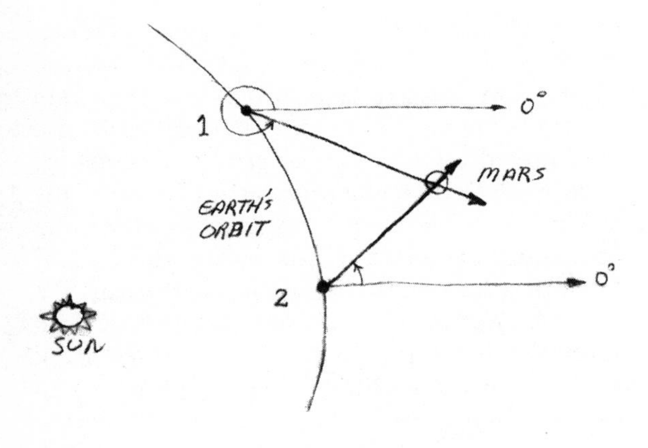

Fig. 7-2 Point 2 is the position of the earth 687 days after leaving point 1. In 687 days, Mars has made exactly one revolution and so has returned to the same point on the orbit. The intersection of the sight lines from the earth determines that point on Mars' orbit.

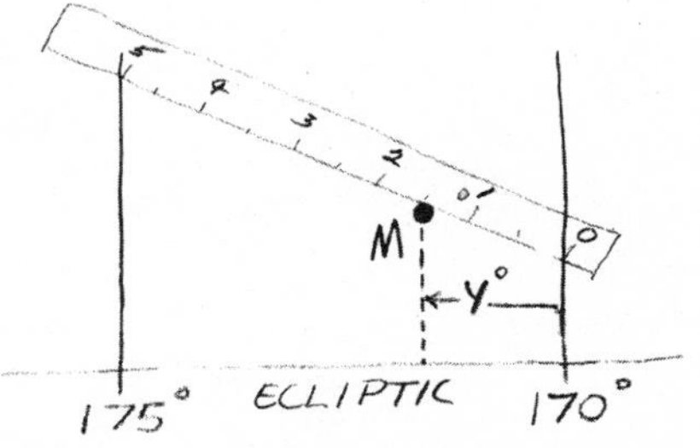

Fig. 7-3 Interpolation between coordinate lines. In the sketch, Mars (M), is at a distance y° from the 170° line. Take a piece of paper or card at least 10 cm long. Make a scale divided into 10 equal parts and label alternate marks), 1, 2, 3, 4, 5. This gives a scale in $\frac{1}{2}$° steps. Notice that the numbering goes from right to left on this scale. Place the scale so that the edge passes through the position of Mars. Now tilt the scale so that the 0 and 5 marks each fall on a grid line. Read off the value of y from the scale. In the sketch, $y = 1\frac{1}{2}$°, so that the longitude of M is $171\frac{1}{2}$°.

Along the ecliptic, *longitudes* are always measured eastward from the 0° point (the vernal equinox). This is toward the left on star maps. *Latitudes* are measured perpendicular to the ecliptic north or south to 90°. (The small movement of Mars above and below the ecliptic is considered in the Activity, "The Inclination of Mars' Orbit.")

To find the coordinates of a star or of Mars you must project the coordinate system upon the sky. To do this you are provided with transparent overlays that show the coordinate system of the ecliptic for each frame, A to P. The positions of various stars are circled. Adjust the overlay until it fits the star positions. Then you can read off the longitude and latitude of the position of Mars. Figure 7-3 shows how you can interpolate between marked coordinate lines. Because you are interested in only a small section of the sky on each photograph, you can draw each small section of the ecliptic as a straight line. For plotting, an accuracy of $\frac{1}{2}$° is satisfactory.

In a chart like the one shown in Figure 7-4, record the longitude and latitude of Mars for each photograph. For a simple plot of Mars' orbit around the sun you will use only the first column—the longitude of Mars. You will use the columns for latitude, Mars' distance from the sun, and the sun-centered coordinates if you do the Activity on the inclination, or tilt, of Mars' orbit on page 109.

Finding Mars' Orbit

When your chart is completed for all eight pairs of observations, you are ready to locate points on the orbit of Mars.

1. On the plot of the earth's orbit, locate the position of the earth for each date given in the

Fig. 7-4 Observed Positions of Mars

Frame	Date	Geocentric Long.	Geocentric Lat.	Mars to Earth Distance	Mars to Sun Distance	Heliocentric Long.	Heliocentric Lat.
A	Mar. 21, 1931						
B	Feb. 5, 1933						
C	Apr. 20, 1933						
D	Mar. 8, 1935						
E	May 26, 1935						
F	Apr. 12, 1937						
G	Sept.16, 1939						
H	Aug. 4, 1941						
I	Nov. 22, 1941						
J	Oct. 11, 1943						
K	Jan. 21, 1944						
L	Dec. 9, 1945						
M	Mar. 19, 1946						
N	Feb. 3, 1948						
O	Apr. 4, 1948						
P	Feb. 21, 1950						

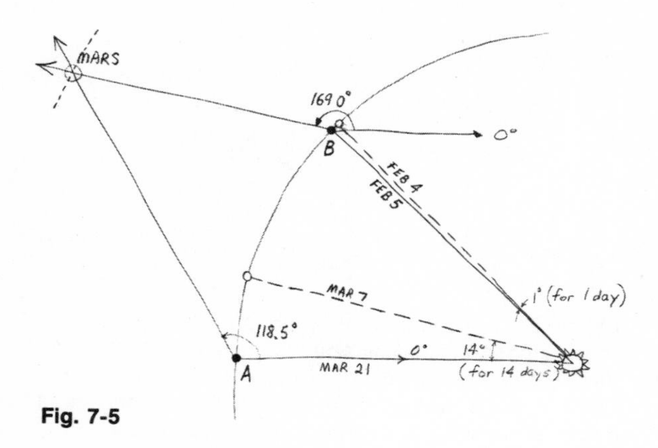

Fig. 7-5

16 photographs. You may do this by interpolating between the dates given for the earth's orbit experiment. Since the earth moves through 360° in about 365 days, you may use ±1° for each day ahead or behind the date given in the previous experiment. For example, frame A is dated March 21. The earth was at 166° on March 7: fourteen days later on March 21, the earth will have moved 14° from 166° to 180°. Always work from the earth-position date nearest the date of the Mars photograph.

2. Through each earth-position point draw a "0° line" parallel to the line you drew from the sun toward the vernal equinox (the grid on the graph paper is helpful). Use a protractor and a sharp pencil to mark the angle between the 0° line and the direction to Mars on that date as seen from the earth (longitude of Mars). The two lines drawn from the earth's positions for each pair of dates will intersect at a point. This is a point on Mars' orbit. Figure 7-5 shows one point on Mars' orbit obtained from the data of the first pair of photographs. By drawing the intersecting lines from the eight pairs of positions, you establish eight points on Mars' orbit.

3. You will notice that there are no points in one section of the orbit. You can fill in the missing part because the orbit is symmetrical about its major axis. Use a compass and, by trial and error, find a circle that best fits the plotted points. Perhaps you can borrow a French curve or long spline from the mechanical drawing or mathematics department.

Now that you have plotted the orbit, you have achieved what you set out to do: you have used Kepler's method to determine the path of Mars around the sun.

If you have time to go on, it is worthwhile to see how well your plot agrees with Kepler's generalization about planetary orbits.

Kepler's Laws from Your Plot

Q1 Does your plot agree with Kepler's conclusion that the orbit is an ellipse?

Photographs of Mars made with a 60 inch reflecting telescope (Mount Wilson and Palomar Observatories) during closest approach to the earth in 1956. Left: August 10; right: Sept. 11. Note the shrinking of the polar cap.

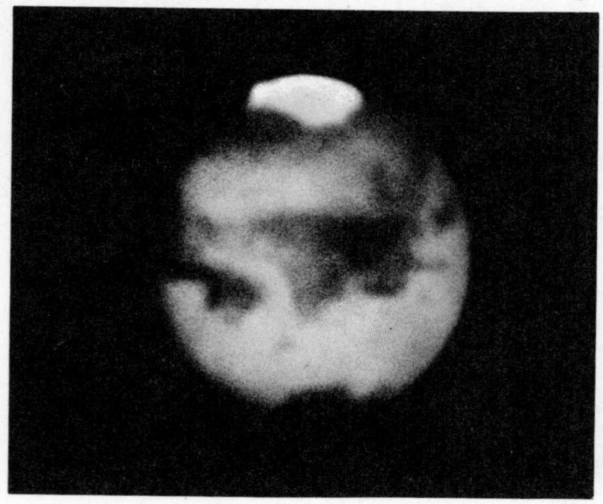

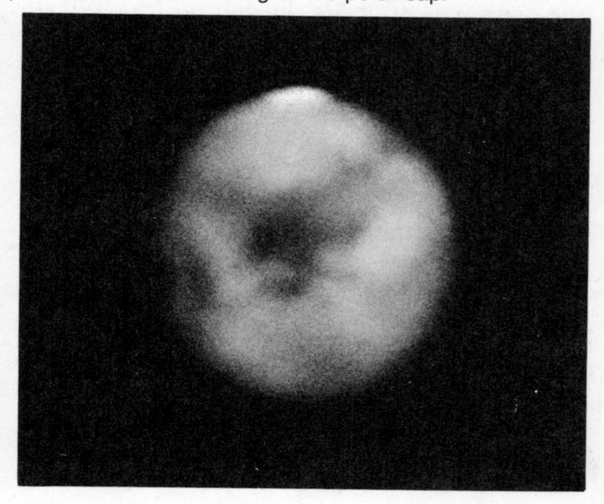

Q2 What is the average sun-to-Mars distance in AU?

Q3 As seen from the sun, what is the direction (longitude) of Mars' nearest and farthest positions?

Q4 During what month is the earth closest to the orbit of Mars? What would be the minimum separation between the earth and Mars?

Q5 What is the eccentricity of the orbit of Mars?

Q6 Does your plot of Mars' orbit agree with Kepler's law of areas, which states that a line drawn from the sun to the planet sweeps out areas proportional to the time intervals? From your orbit, you see that Mars was at point B′ on February 5, 1933, and at point C′ on April 20, 1933, as shown in Fig. 7-6. There are eight such pairs of dates in your data. The time intervals are different for each pair.

Connect these pairs of positions with a line to the *sun* (Fig. 7-6). Find the areas of squares on the graph paper (count a square when more than half of it lies within the area). Divide the area (in squares) by the number of days in the interval to find an "area per day" value. Are these values nearly the same?

Q7 How much (by what percentage) do they vary?

Q8 What is the uncertainty in your area measurements?

Q9 Is the uncertainty the same for large areas as for small?

Q10 Do your results bear out Kepler's law of areas?

This is by no means all that you can do with the photographs you used to make the plot of Mars' orbit. If you want to do more, look at the Activity, "The Inclination of Mars' Orbit."

Fig. 7-6 In this example, the time interval is 74 days.

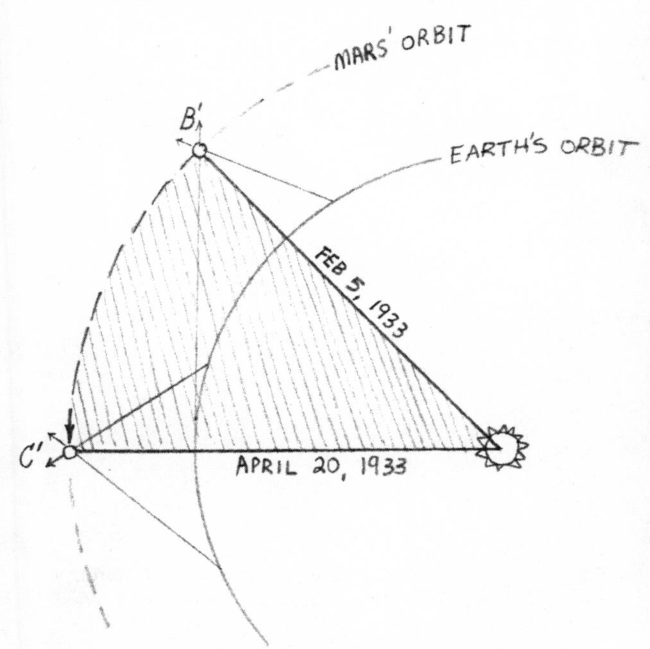

Television picture of a 40 × 50 mile area just below Mars' equator, radioed from the Mariner 6 Mars probe during its 1969 fly-by.

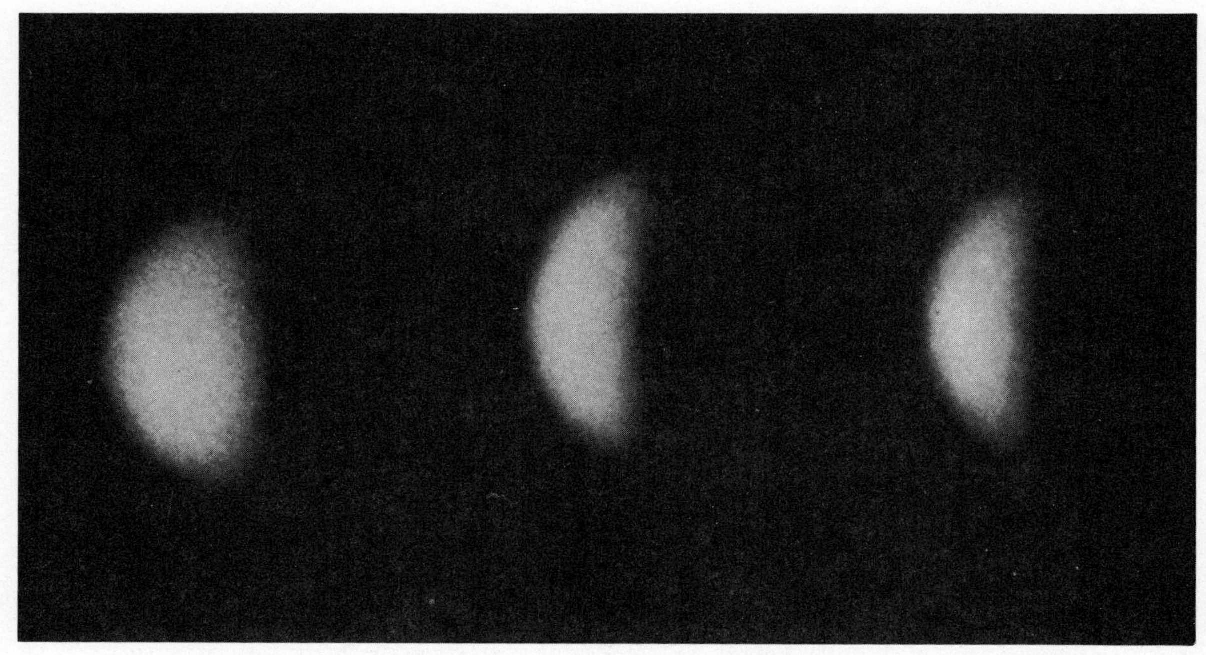

Fig. 7-7 Mercury, first quarter phase, taken June 7, 1934 at the Lowell Observatory, Flagstaff, Ariz.

EXPERIMENT 20 THE ORBIT OF MERCURY

Mercury, the innermost planet, is never very far from the sun in the sky. It can be seen only close to the horizon, just before sunrise or just after sunset, and viewing is made difficult by the glare of the sun. (Fig. 7-7.)

Except for Pluto, which differs in several respects from the other planets, Mercury has the most eccentric planetary orbit in our solar system ($e = 0.206$). The large eccentricity of Mercury's orbit has been of particular importance, since it has led to one of the tests for Einstein's General Theory of Relativity. For a planet with an orbit inside the earth's, there is a simpler way to plot the orbit than by the paired observations you used for Mars. In this experiment you will use this simpler method to get the approximate shape of Mercury's orbit.

Mercury's Elongations

Let us assume a heliocentric model for the solar system. Mercury's orbit can be found from Mercury's maximum angles of elongation east and west from the sun as seen from the earth on various known dates.

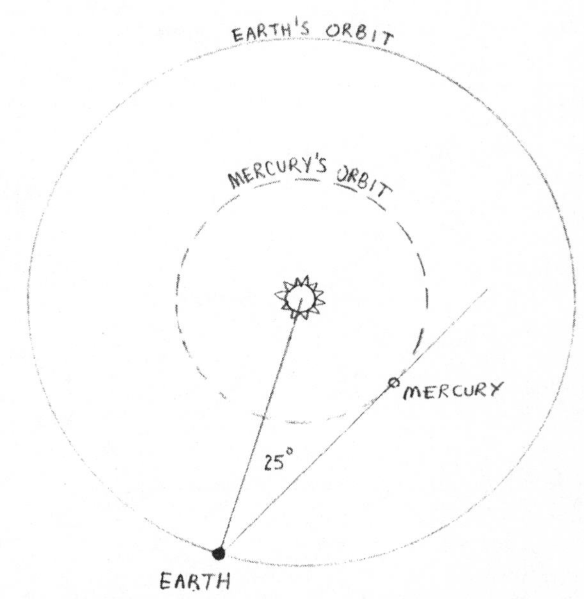

Fig. 7-8 The greatest western elongation of Mercury, May 25, 1964. The elongation had a value of 25° West.

The angle (Fig. 7-8), between the sun and Mercury as seen from the earth, is called the "elongation." Note that when the elongation reaches its maximum value, the sight lines from the earth are tangent to Mercury's orbit.

Since the orbits of Mercury and the earth are both elliptical, the greatest value of the elongation varies from revolution to revolution. The 28° elongation given for Mercury on page 14 of the *Text* refers to the maximum value. Table 1 gives the angles of a number of these greatest elongations.

TABLE 1 SOME DATES AND ANGLES OF GREATEST ELONGATION FOR MERCURY (from the American Ephemeris and Nautical Almanac)

Date	Elongation
Jan. 4, 1963	19° E
Feb. 14	26 W
Apr. 26	20 E
June 13	23 W
Aug. 24	27 E
Oct. 6	18 W
Dec. 18	20 E
Jan. 27, 1964	25 W
Apr. 8	19 E
May 25	25 W

Plotting the Orbit

You can work from the plot of the earth's orbit that you established in Experiment 17. Make sure that the plot you use for this experiment represents the orbit of the earth around the sun, not of the sun around the earth.

If you did not do the earth's orbit experiment, you may use, for an approximate earth orbit, a circle of 10 cm radius drawn in the center of a sheet of graph paper. Because the eccentricity of the earth's orbit is very small (0.017) you can place the sun at the center of the orbit without introducing a significant error in the experiment.

Draw a reference line horizontally from the center of the circle to the right. Label the line 0°. This line points toward the vernal equinox and is the reference from which the earth's position in its orbit on different dates can be established. The point where 0° line from the sun crosses the earth's orbit is the earth's position in its orbit on September 23.

The earth takes about 365 days to move once around its orbit (360°). Use the rate of approximately 1° per day, or 30° per month, to establish the position of the earth on each of the dates given in Table 1. Remember that the earth moves around this orbit in a *counterclockwise* direction, as viewed from the north celestial pole. Draw radial lines from the sun to each of the earth positions you have located.

Now draw sight lines from the earth's orbit for the elongation angles. Be sure to note, from Fig. 7-8, that for an *eastern* elongation, Mercury is to the *left* of the sun as seen from the earth. For a *western* elongation, Mercury is to the right of the sun.

You know that on a date of greatest elongation Mercury is somewhere along the sight line, but you don't know exactly where on the line to place the planet. You also know that the sight line is tangent to the orbit. A reasonable assumption is to put Mercury at the point along the sight line closest to the sun.

You can now find the orbit of Mercury by drawing a smooth curve through, or close to, these points. Remember that the orbit must *touch* each sight line without crossing any of them.

Finding R_{av}

The average distance of a planet in an elliptical orbit is equal to one half the long diameter of the ellipse, the "semi-major axis."

To find the size of the semi-major axis *a* of Mercury's orbit, relative to the earth's semi-major axis, you must first find the aphelion and perihelion points of the orbit. You can use a drawing compass to find these points on the orbit farthest from and closest to the sun.

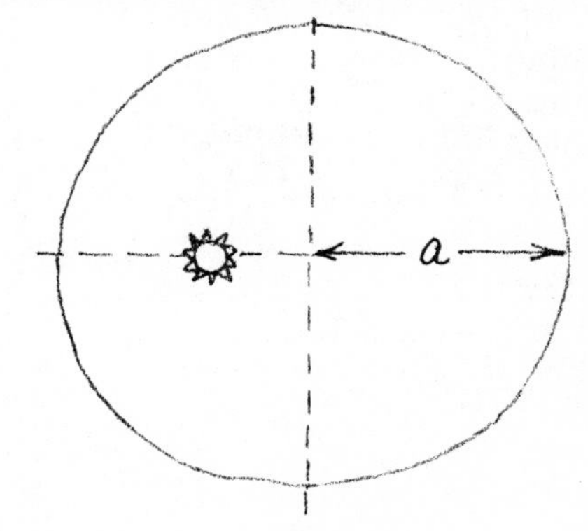

Measure the greatest diameter of the orbit along the line perihelion-sun-aphelion. Since 10.0 cm corresponds to one AU (the semi-major axis of the earth's orbit) you can now obtain the semi-major axis of Mercury's orbit in AU's.

Calculating Orbital Eccentricity

Eccentricity is defined as $e = c/a$ (Fig. 7-9). Since c, the distance from the center of Mercury's ellipse to the sun, is small on your plot, you lose accuracy if you try to measure it directly.

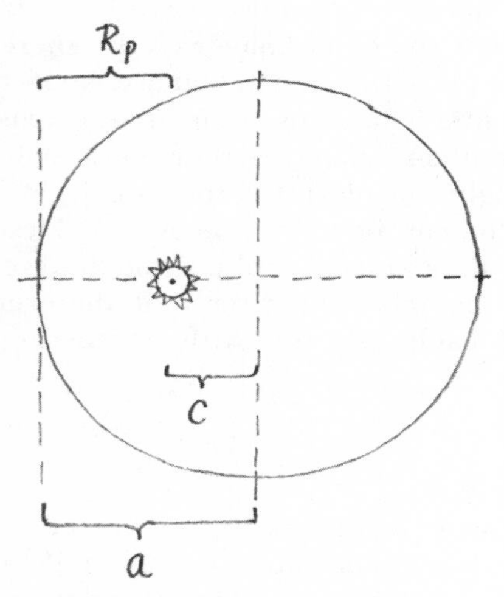

Fig. 7-9

From Fig. 7-9, you can see that c is the difference between Mercury's perihelion distance R_p and the semi-major axis a. That is:

$$c = a - R_p$$

So

$$e = \frac{c}{a} = \frac{a - R_p}{a} = 1 - \frac{R_p}{a}$$

You can measure R_p and a with reasonable accuracy from your plotted orbit. Compute e, and compare your value with the accepted value, $e \approx 0.206$.

Kepler's Second Law

You can test Kepler's equal-area law on your Mercury orbit in the same way as that described in Experiment 19, The Orbit of Mars. By counting squares you can find the area swept out by the radial line from the sun to Mercury between successive dates of observation, such as January 4 to February 14, and June 13 to August 24. Divide the area by the number of days in the interval to get the "area per day." This should be constant, if Kepler's law holds for your plot. Is it constant?

CHAPTER 7

ACTIVITIES

THREE-DIMENSIONAL MODEL OF TWO ORBITS

You can make a three-dimensional model of two orbits quickly with two small pieces of cardboard (or 3″ × 5″ cards). On each card draw a circle or ellipse, but have one larger than the other. Mark clearly the position of the focus (sun) on each card. Make a straight cut *to the sun*, on one card from the left, on the other from the right. Slip the cards together until the sun-points coincide. (Fig. 7-10) Tilt the two cards (orbit planes) at various angles.

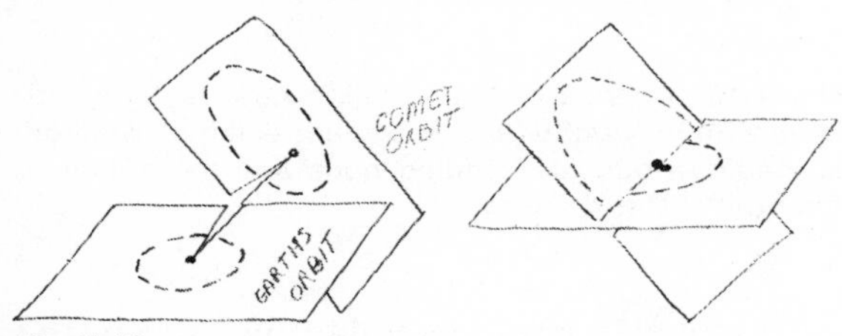

Fig. 7-10

INCLINATION OF MARS' ORBIT

When you plotted the orbit of Mars in Experiment 17, you ignored the slight movement of the planet above and below the ecliptic. This movement of Mars north and south of the ecliptic shows that the plane of its orbit is slightly inclined to the plane of the earth's orbit. In this activity, you may use the table of values for Mars latitude (which you made in Experiment 17) to determine the inclination of Mar's orbit.

Do the activity, "Three-dimensional model of two orbits," just before this activity, to see exactly what is meant by the inclination of orbits.

Theory

From each of the photographs in the set of 16 that you used in Experiment 17, you can find the observed latitude (angle from the ecliptic) of Mars at a particular point in its orbital plane. Each of these angles is measured on a photograph taken *from the earth.* As you can see from Fig. 7-11, however, it is the *sun*, not the earth, which is at the center

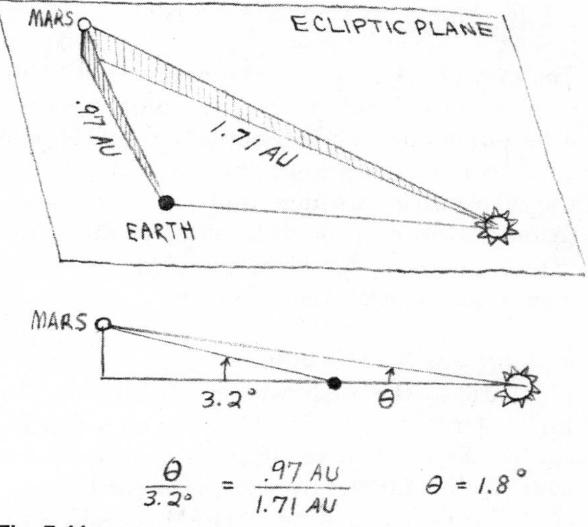

Fig. 7-11

of the orbit. The inclination of Mars' orbit must, therefore, be an angle measured *at the sun.* It is this angle (the heliocentric latitude) that you wish to find.

Figure 7-11 shows that Mars can be represented by the head of a pin whose point is stuck into the ecliptic plane. We see Mars from the earth to be north or south of the ecliptic, but we want the N-S angle of Mars as seen from the sun. The following example shows how you can derive the angles as if you were seeing them from the sun.

In Plate A (March 21, 1933), Mars was about 3.2° north of the ecliptic *as seen from the earth.* But the earth was considerably closer to Mars on this date than the sun was. Can you see how the angular elevation of Mars above the ecliptic plane as seen from the sun will therefore be considerably less than 3.2°?

For very small angles, the apparent angular sizes are inversely proportional to the distances. For example, if the sun were twice as far from Mars as the earth was, the angle at the sun would be $\frac{1}{2}$ the angle at the earth.

Measurement on the plot of Mars' orbit (Experiment 17) gives the earth-Mars distance as 9.7 cm (0.97 AU) and the distance sun-Mars as 17.1 cm (1.71 AU) on the date of the photograph. The heliocentric latitude of Mars is therefore

$$\frac{9.7}{17.1} \times 3.2°\text{N} = 1.8°\text{N}$$

You can check this value by finding the heliocentric latitude of this same point in Mars' orbit on photograph B (February 5, 1933). The earth was in a different place on this date so the geocentric latitude and the earth-Mars distance will both be different, but the heliocentric latitude should be the same to within your experimental uncertainty.

Making the Measurements

Turn to the table you made that is like Fig. 7-4 in Experiment 17, on which you recorded the geocentric latitudes λ_g of Mars. On your Mars' orbit plot from Experiment 17, measure the corresponding earth-Mars and sun-Mars distances and note them in the same table.

From these two sets of values, calculate the heliocentric latitudes as explained above. The values of heliocentric latitude calculated from the two plates in each pair (A and B, C and D, etc.) should agree within the limits of your experimental procedure.

On the plot of Mars' orbit, measure the *heliocentric longitude* λ_h for each of the eight Mars positions. Heliocentric longitude is measured from the sun, counterclockwise from the 0° direction (direction toward vernal equinox), as shown in Fig. 7-12.

Complete the table given in Fig. 7-4, Experiment 17, by entering the earth-to-Mars and sun-to-Mars distances, the geocentric and heliocentric latitudes, and the geocentric and heliocentric longitudes for all sixteen plates.

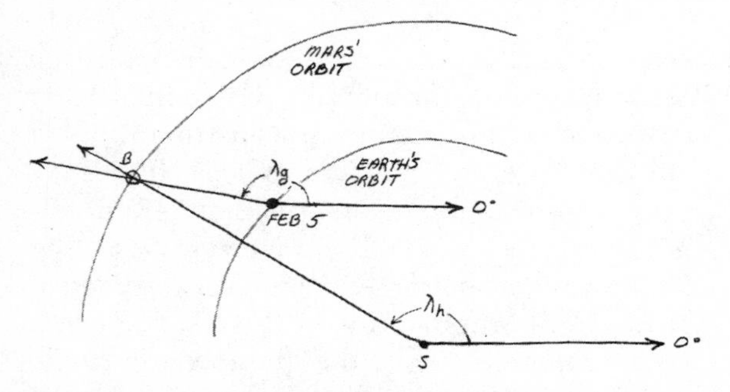

Fig. 7-12 On February 5, the heliocentric longitude (λ_h) of Point B on Mars' orbit is 150°; the geocentric longitude (λ_g) measured from the earth's position is 169°.

Make a graph, like Fig. 7-13, that shows how the heliocentric latitude of Mars changes with its heliocentric longitude.

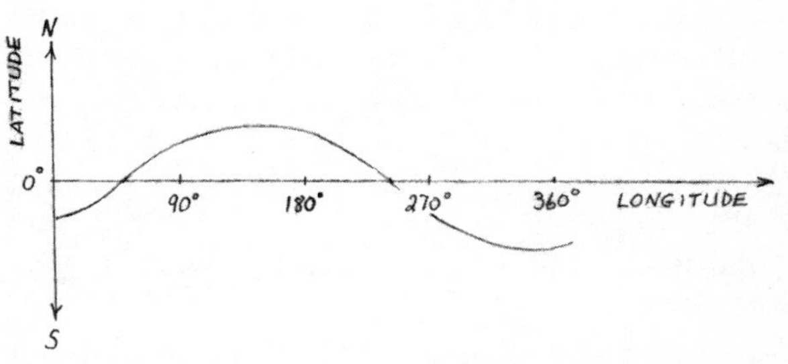

Fig. 7-13 Change of Mars' heliocentric latitude with heliocentric longitude. Label the ecliptic, latitude, ascending node, descending node and inclination of the orbit in this drawing.

From this graph, you can find two of the elements that describe the orbit of Mars with respect to the ecliptic. The point at which Mars crosses the ecliptic from south to north is called the ascending node. (The descending node, on the other side of the orbit, is the point at which Mars crosses the ecliptic from north to south.)

The angle between the plane of the earth's orbit and the plane of Mars' orbit is the inclination of Mars' orbit, i. When Mars reaches its maximum latitude above the ecliptic, which occurs at 90° beyond the ascending node, the planet's maximum latitude equals the inclination of the orbit, i.

Elements of an Orbit

Two angles, the longitude of the ascending node, Ω, and the inclination, i, locate the plane of Mars' orbit with respect to the plane of the ecliptic. One more angle is needed to orient the orbit of Mars in its orbital plane. This is the "argument of perihelion" ω, shown in Fig. 7-14 which is the angle in the *orbit plane* between the ascending node and perihelion point. On your plot of Mars' orbit measure the angle from the ascending node Ω to the direction of peri-

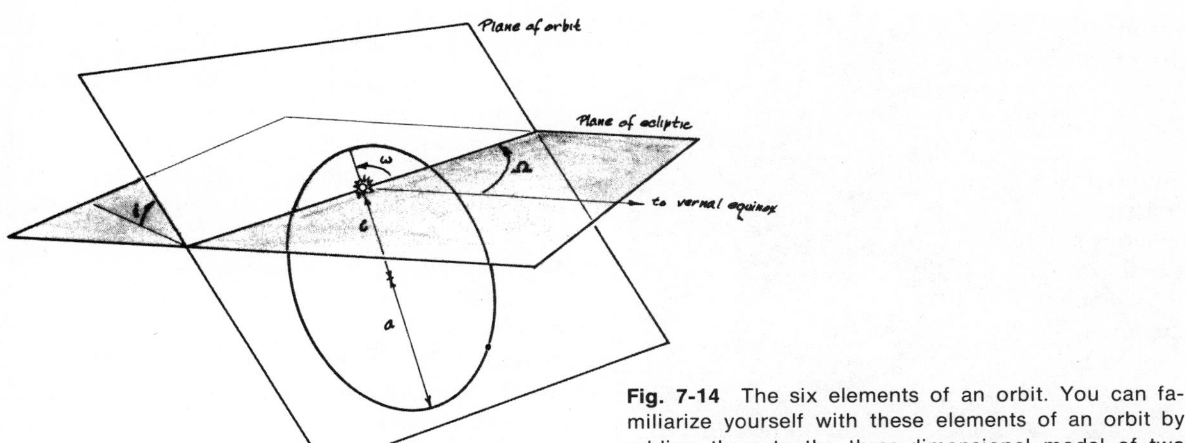

Fig. 7-14

Fig. 7-14 The six elements of an orbit. You can familiarize yourself with these elements of an orbit by adding them to the three-dimensional model of two orbits, assuming that the earth's orbit is in the plane of the ecliptic.

helion to obtain the argument of the perihelion ω.

If you have worked along this far, you have determined five of the six elements that define *any* orbit:

a – semi-major axis, or average distance (which determines the period)

e – eccentricity (shape of orbit as given by e/a in Fig. 7-14)

i – inclination (tilt of orbital plane)

Ω – longitude of ascending node (where orbital plane crosses ecliptic)

ω – argument of perihelion (which orients the orbit in its plane)

These five elements (shown in Fig. 7-14) fix the orbital plane of any planet or comet in space, tell the size and shape of the orbit, and also give its orientation within the orbital plane. To compute a complete timetable, or ephemeris, for the body, you need only to know T, a zero date when the body was at a particular place in the orbit. Generally, T is given as the date of a perihelion passage. Photograph G was made on September 16, 1933. From this you can estimate a date of perihelion passage for Mars.

DEMONSTRATING SATELLITE ORBITS

A piece of thin plastic or a rubber sheet can be stretched tight and clamped in an embroidery hoop about 22″ in diameter. Place the hoop on some books and put a heavy ball, for example, a 2″-diameter steel ball bearing, in the middle of the plastic. The plastic will sag so that there is a greater force toward the center on an object when it is closer to the center than when it is far away. (See Fig. 7-15.)

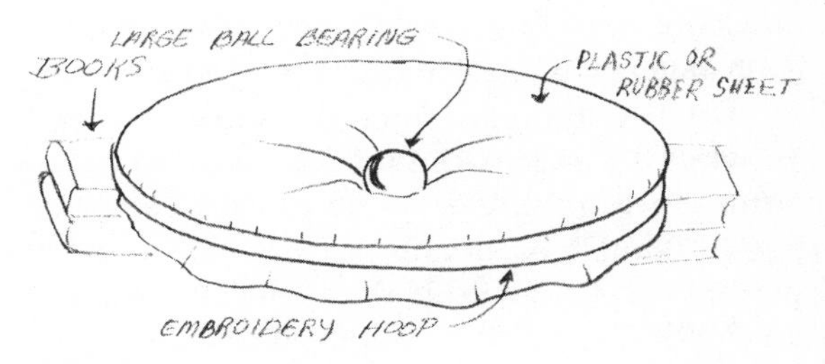

Fig. 7-15

You can use a smaller hoop (about 14″) on the stage of an overhead projector. Use small ball bearings, marbles or beads as "satellites." Then you will have a shadow projection of the large central mass, with the small satellites racing around it. Be careful not to drop the ball through the glass.

If you take strobe photos of the motion, you can check whether Kepler's three laws are satisfied; you can see where satellites travel fastest in their orbits, and how the orbits themselves turn in space. To take the picture, set

up the hoop on the floor with black paper under it.

You can use either the electronic strobe light or the slotted disc stroboscope to take the pictures. In either case, place the camera directly over the hoop and the light source at the side, slightly enough above the plane of the hoop so that the floor under the hoop is not well lighted. A ball bearing or marble will make the best pictures.

Here are some questions to think about:

1. Does your model give a true representation of the gravitational force around the earth? In what ways does the model fail (other than suffering from fingernail holes in the plastic)?
2. Is it much harder to put a satellite into a perfectly circular orbit than into an elliptical one? What conditions must be satisfied for a circular orbit?
3. Are Kepler's three laws really verified? Should they be?

For additional detail and ideas see "Satellite Orbit Simulator," *Scientific American*, October, 1958.

GALILEO

Read Bertoldt Brecht's play, *Galileo*, and present a part of it for the class. There is some controversy about whether the play truly reflects what historians believe were Galileo's feelings. For comparison, you could read *The Crime of Galileo*, by Giorgio de Santillana; *Galileo and the Scientific Revolution*, by Laura Fermi; The Galileo Quadricentennial Supplement in *Sky and Telescope*, February, 1964; or articles in the April, 1966 issue of *The Physics Teacher*, "Galileo: Antagonist," and "Galileo Galilei: An Outline of His Life."

CONIC-SECTIONS MODELS

Obtain from a mathematics teacher a demonstration cone that has been cut along several different planes so that when it is taken apart the planes form the four conic sections.

If such a cone is not available, tape a cone of paper to the front of a small light source, like a flashlight bulb, as shown in Fig. 7-17. Shine the light on the wall and tilt the cone at different angles with respect to the wall. You can make all the conic sections shown in section 7.3 of *Text*.

If you have a wall lamp with a circular shade, the shadows cast on the wall above and below the lamp are usually hyperbolas. You can check this by tracing the curve on a large piece of paper, and seeing whether the points satisfy the definition of an hyperbola.

CHALLENGING PROBLEM: FINDING EARTH-SUN DISTANCE FROM VENUS PHOTOS

Here's a teaser: assume that Venus has the same diameter as the earth. Also assume that the scale of the pictures on page 72 of the Unit 2 *Text* is 1.5 seconds of arc per millimeter.

Determine the distance from the earth to the sun in miles.

MEASURING IRREGULAR AREAS

Are you tired of counting squares to measure the area of irregular figures? A device called a planimeter can save you much drudgery. There are several styles, ranging from a simple pocket knife to a complex arrangement of worm gears and pivoted arms. See the Amateur Scientist section of *Scientific American*, August, 1958 and February, 1959.

An interstellar cloud of gas obscures bright nebulosity in the constellation *Monoceros*, photographed by the 200-inch telescope at Mount Palomar.

Chapter 8 Unity of Earth and Sky—the Work of Newton

EXPERIMENT 21 STEPWISE APPROXIMATION TO AN ORBIT*

Photograph of the comet Cunningham made at Mount Wilson and Palomar Observatories December 21, 1940. Can you explain why the stars leave trails and the comet does not?

You have seen in the *Text* how Newton analyzed the motions of bodies in orbit, using the concept of a centrally directed force. On the basis of the discussion in *Text* Sec. 8.4, you are now ready to apply Newton's method to develop an approximate orbit of a satellite or a comet around the sun. You can also, from your orbit, check Kepler's law of areas and other relationships discussed in the *Text*.*

Imagine a ball rolling over a smooth level surface such as a piece of plate glass.

Q1 What would you predict for the path of the ball, based on your knowledge from Unit 1 of Newton's laws of motion?

Q2 Suppose you were to strike the ball from one side. Would the path direction change?

Q3 Would the speed change? Suppose you gave the ball a series of "sideways" blows of equal force as it moves along, what do you predict its path might be?

Reread Sec. 8.4 of the *Text* if you have difficulties answering these questions.

*This experiment is based on a similar one developed by Dr. Leo Lavatelli, University of Illinois, *American Journal of Physics*, vol. 33, p. 605.

Your Assumptions

A planet or satellite in orbit has a continuous force acting on it. But as the body moves, the magnitude and direction of this force change. To predict exactly the orbit under the application of this continually changing force requires advanced mathematics. However, you can get a reasonable approximation to the orbit by breaking the continuous attraction into many small steps, in which the force acts as a sharp "blow" toward the sun once every sixty days. (See Fig. 8-1.)

The application of repeated steps is known as "iteration." It is a powerful technique for solving problems. Modern high-speed digital computers use repeated steps to solve complex problems, such as the best path (or paths) for a Mariner probe to follow between earth and Mars.

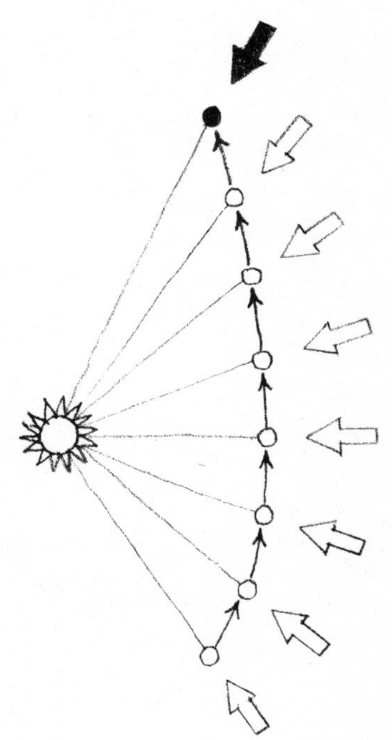

Fig. 8-1 A body, such as a comet, moving in the vicinity of the sun will be deflected from its straight-line path by a gravitational force. The force acts continuously but Newton has shown that we can think about the orbit as though it were produced by a series of intermittent sharp blows.

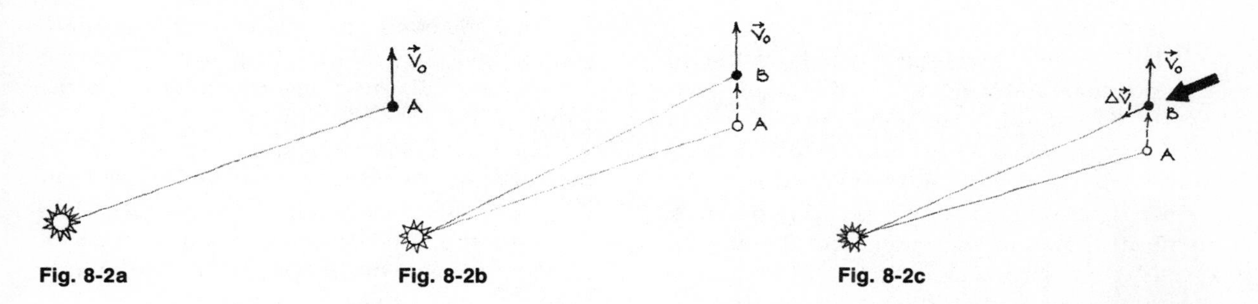

Fig. 8-2a

Fig. 8-2b

Fig. 8-2c

You can now proceed to plot an approximate comet orbit if you will make these additional assumptions:

1. The force on the comet is an attraction toward the sun.
2. The force of the blow varies inversely with the square of the comet's distance from the sun.
3. The blows occur regularly at equal time intervals, in this case, 60 days. The magnitude of each brief blow is assumed to equal the total effect of the continuous attraction of the sun throughout a 60-day interval.

Effect of the Central Force

From Newton's second law you know that the gravitational force will cause the comet to accelerate toward the sun. If a force $\vec{F}$ acts for a time interval Δt on a body of mass m, you know that

$$F = m\vec{a} = m\frac{\Delta\vec{v}}{\Delta t} \text{ and therefore}$$

$$\Delta\vec{v} = \frac{\vec{F}}{m}\Delta t$$

This equation relates the change in the body's velocity to its mass, the force, and the time for which it acts. The mass m is constant. So is Δt (assumption 3 above). The change in velocity is therefore proportional to the force, $\Delta\vec{v} \propto \vec{F}$. But remember that the force is *not* constant in magnitude; it varies inversely with the square of the distance from comet to sun.

Q4 Is the force of a blow given to the comet when it is near the sun greater or smaller than one given when the comet is far from the sun?

Q5 Which blow causes the biggest velocity change?

In Fig. 8-2a the vector $\vec{v}_0$ represents the comet's velocity at the point A. During the first 60 days, the comet moves from A to B (Fig. 8-2b). At B a blow causes a velocity change $\Delta\vec{v}_1$ (Fig. 8-2c). The new velocity after the blow is $\vec{v}_1 = \vec{v}_0 + \Delta\vec{v}_1$, and is found by completing the vector triangle (Fig. 8-2d).

The comet therefore leaves point B with velocity $\vec{v}_1$ and continues to move with this velocity for another 60-day interval. Because the time intervals between blows are always the same (60 days), the displacement along the path is proportional to the velocity, $\vec{v}$. You therefore use a length proportional to the comet's velocity to represent its displacement during each time interval. (Fig. 8-2e.)

Each new velocity is found, as above, by adding to the previous velocity the $\Delta\vec{v}$ given by the blow. In this way, step by step, the comet's orbit is built up.

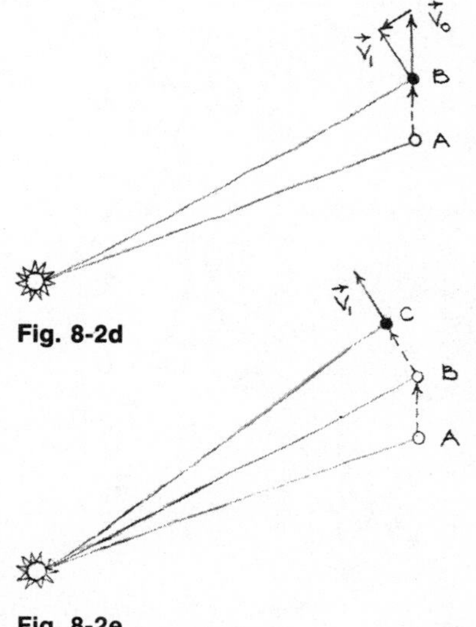

Fig. 8-2d

Fig. 8-2e

Scale of the Plot

The shape of the orbit depends on the initial position and velocity, and on the force acting. Assume that the comet is first spotted at a distance of 4 AU from the sun. Also assume that the comet's velocity at this point is $v = 2$ AU per year (about 20,000 miles per hour) at right angles to the sun-comet distance R.

The following scale factors will reduce the orbit to a scale that fits conveniently on a 16″ x 20″ piece of graph paper. (Make this up from four $8\frac{1}{2}$″ x 11″ pieces if necessary.)

1. Let 1 AU be scaled to 2.5 inches (or 6.5 cm) so that 4 AU becomes 10 inches (or about 25 cm).
2. Since the comet is hit every 60 days, it is convenient to express the velocity in AU per 60 days. Suppose you adopt a scale factor in which a velocity vector of 1 AU/60 days is represented by an arrow 2.5 inches (or 6.5 cm) long.

The comet's initial velocity of 2 AU per year can be given as 2/365 AU per day, or 2/365 × 60 = 0.33 AU per 60 days. This scales to an arrow 0.83 inches (or 2.11 cm) long. This is the *displacement* of the comet in the first 60 days.

Computing Δv

On the scale and with the 60-day iteration interval that has been chosen, the force field of the sun is such that the Δv given by a blow when the comet is 1 AU from the sun is 1 AU/60 days.

To avoid computing Δv for each value of R, you can plot Δv against R on a graph. Then for any value of R you can immediately find the value of Δv.

Table 1 gives values of R in AU and in inches and in centimeters to fit the scale of your orbit plot. The table also gives for each value of R the corresponding value of Δv in AU/60 days and in inches and in centimeters to fit the scale of your orbit plot.

Table 1 Scales for R and Δv

Distance from sun, R			Change in speed, Δv		
AU	inches	cm	AU/60 days	inches	cm
0.75	1.87	4.75	1.76	4.44	11.3
0.8	2.00	5.08	1.57	3.92	9.97
0.9	2.25	5.72	1.23	3.07	7.80
1.0	2.50	6.35	1.00	2.50	6.35
1.2	3.0	7.62	0.69	1.74	4.42
1.5	3.75	9.52	0.44	1.11	2.82
2.0	5.0	12.7	0.25	0.62	1.57
2.5	6.25	15.9	0.16	0.40	1.02
3.0	7.50	19.1	0.11	0.28	0.71
3.5	8.75	22.2	0.08	0.20	0.51
4.0	10.00	25.4	0.06	0.16	0.41

Graph these values on a separate sheet of paper at least 10 inches long, as illustrated in Fig. 8-3, and carefully connect the points with a smooth curve.

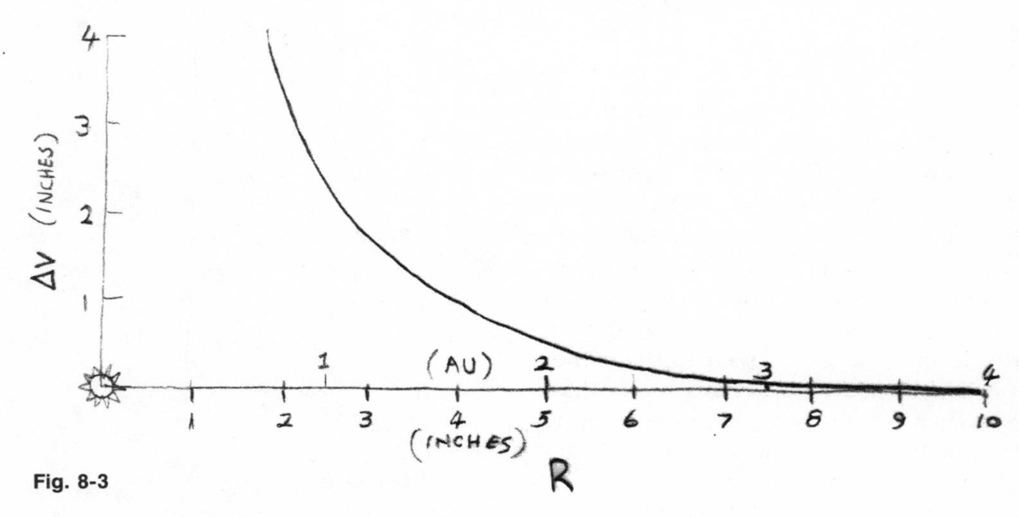

Fig. 8-3

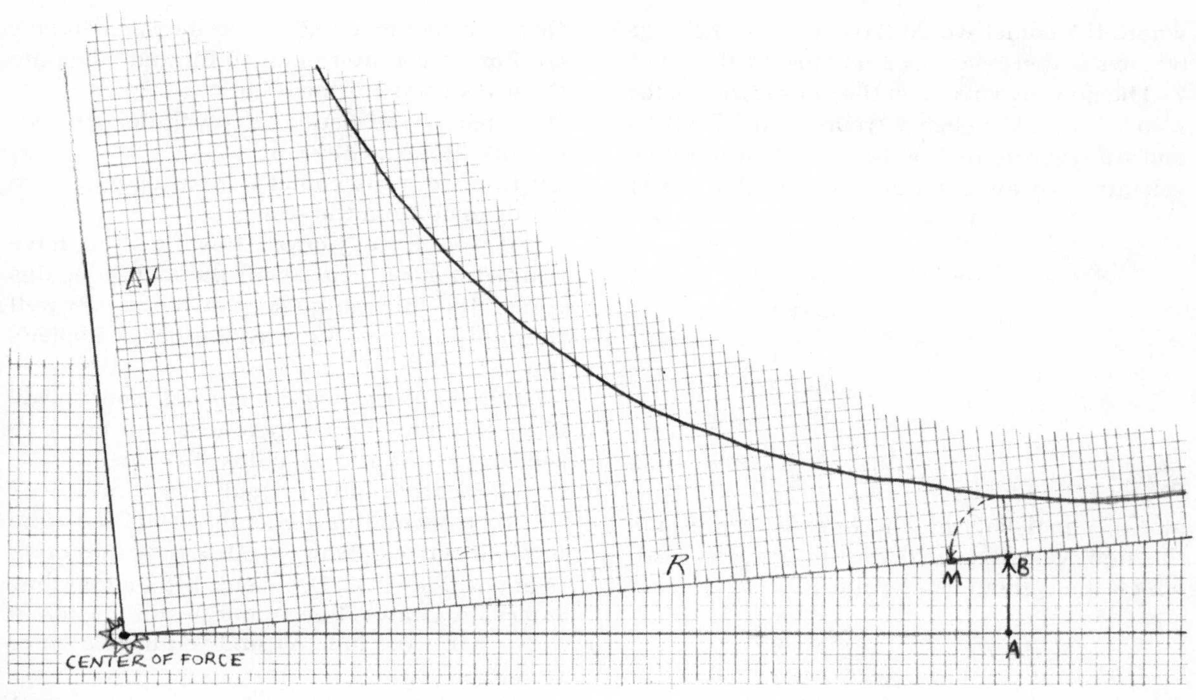

Fig. 8-4

You can use this curve as a simple graphical computer. Cut off the bottom margin of the graph paper, or fold it under along the *R* axis. Lay this edge on the orbit plot and measure the distance from the sun to a blow point (such as *B* in Fig. 8-4). With dividers or a drawing compass pick off the value of Δv corresponding to this *R* and lay off this distance along the radius line toward the sun (see Fig. 8-4).

Making the Plot

1. Mark the position of the sun S halfway up the large graph paper (held horizontally) and 12 inches (or 30 cm) from the right edge.
2. Locate a point 10 inches (or 25 cm), 4 AU, that is, to the right from the sun S. This is point A where you first find the comet.

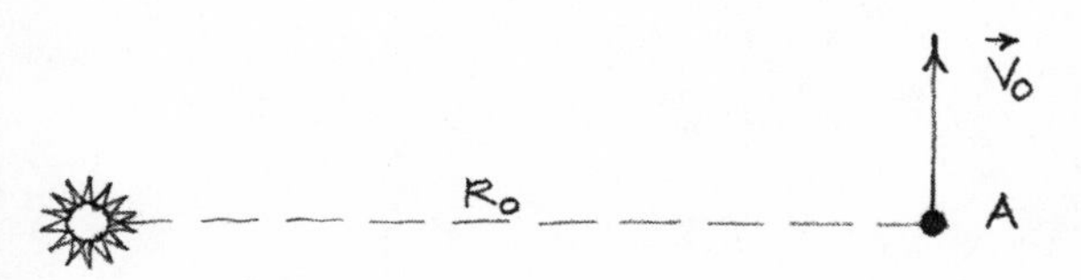

3. To represent the comet's initial velocity draw vector AB perpendicular to SA. B is the comet's position at the end of the first 60-day interval. At B a blow is struck which causes a change in velocity Δv_1.
4. Use your Δv graph to measure the distance of B from the sun at S, and to find Δv_1 for this distance (Fig. 8-4).
5. The force, and therefore the change in velocity, is always directed toward the sun. From B lay off $\Delta \vec{v}_1$ toward S. Call the end of this short line M.

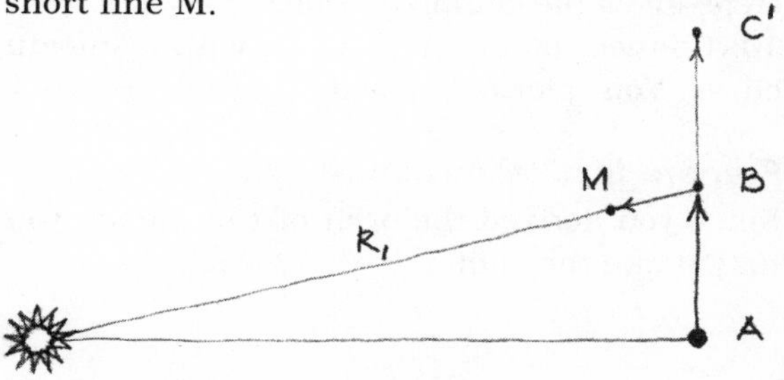

6. Draw the line BC′, which is a continuation of AB and has the same length as AB. That is

where the comet would have gone in the next 60 days if there had been no blow at B.

7. The new velocity after the blow is the vector sum of the old velocity (represented by BC′) and $\Delta\vec{v}$ (represented by BM). To find the new velocity $\vec{v}_1$ draw the line C′C parallel to BM

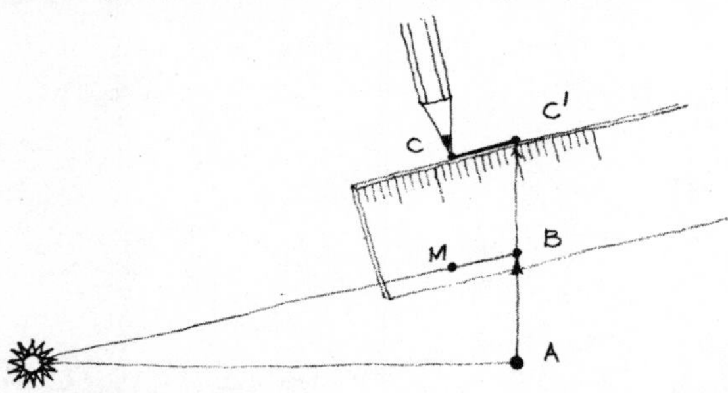

and of equal length. The line BC represents the new velocity vector $\vec{v}_1$, the velocity with which the comet leaves point B.

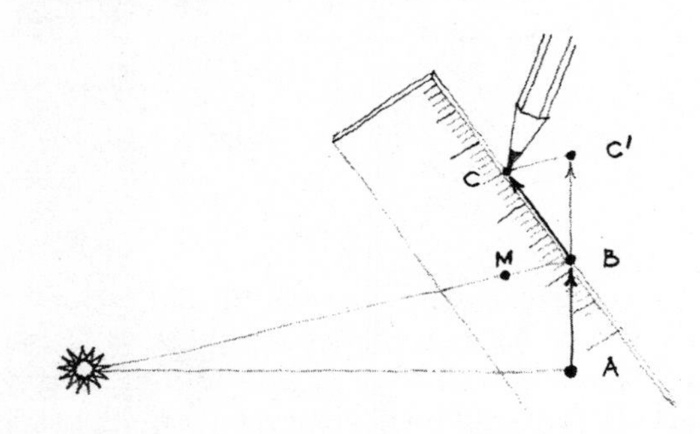

8. Again the comet moves with uniform velocity for 60 days, arriving at point C. Its displacement in that time is $\Delta\vec{d}_1 = \vec{v}_1 \times 60$ days; and because of the scale factor chosen, the displacement is represented by the line BC.

9. Repeat steps 1 through 8 to establish point D and so forth, for at least 14 or 15 steps (25 steps gives the complete orbit).

10. Connect points A, B, C . . . with a smooth curve. Your plot is finished.

Prepare for Discussion

Since you derived the orbit of this comet, you may name the comet.

Q6 From your plot, find the perihelion distance.

Q7 Find the center of the orbit and calculate the eccentricity of the orbit.

Q8 What is the period of revolution of your comet? (Refer to *Text*, Sec. 7.3.)

Q9 How does the comet's speed change with its distance from the sun?

If you have worked this far, you have learned a great deal about the motion of this comet. It is interesting to go on to see how well the orbit obtained by iteration obeys Kepler's laws.

Q10 Is Kepler's law of ellipses confirmed? (Can you think of a way to test your curve to see how nearly it is an ellipse?)

Q11 Is Kepler's law of equal areas confirmed?

To answer this remember that the time interval between blows is 60 days, so the comet is at positions B, C, D . . . , etc., after equal time intervals. Draw a line from the sun to each of these points (include A), and you have a set of triangles.

Find the area of each triangle. The area A of a triangle is given by $A = \frac{1}{2}ab$ where a and b are altitude and base, respectively. Or you can count squares to find the areas.

More Things to Do

1. The graphical technique you have practiced can be used for many problems. You can use it to find out what happens if different initial speeds and/or directions are used. You may wish to use the $1/R^2$ graph, or you may construct a new graph. To do this, use a different law (e.g., force proportional to $1/R^3$, or to $1/R$ or to R) to produce different paths; actual gravitational forces are *not* represented by such force laws.

2. If you use the same force graph but reverse the direction of the force to make it a repulsion, you can examine how bodies move under such a force. Do you know of the existence of any such repulsive force?

Spiral nebula in the constellation *Leo*, photographed by the 200-inch telescope at Mount Palomar.

ACTIVITIES

MODEL OF THE ORBIT OF HALLEY'S COMET

Halley's comet is referred to several times in your *Text*. You will find that its orbit has a number of interesting features if you construct a model of it.

Since the orbit of the earth around the sun lies in one plane and the orbit of Halley's comet lies in another plane intersecting it, you will need two large pieces of stiff cardboard for planes, on which to plot these orbits.

The Earth's Orbit

Make the earth's orbit first. In the center of one piece of cardboard, draw a circle with a radius of 5 cm (1 AU) for the orbit of the earth. On the same piece of cardboard, also draw approximate (circular) orbits for Mercury (radius 0.4 AU) and Venus (radius 0.7 AU). For this plot, you can consider that all of these planets lie roughly in the one plane. Draw a line from the sun at the center and mark this line as 0° longitude.

The table on page 93 of this *Handbook* lists the apparent position of the sun in the sky on thirteen dates. By adding 180° to each of the tabled values, you can get the position of the earth in its orbit on those dates. Mark these positions on your drawing of the earth's orbit. (If you wish to mark more than those thirteen positions, you can do so by using the technique described on page 104.)

The Comet's Orbit

Figure 8-9 shows the positions of Halley's comet near the sun in its orbit, which is very nearly a parabola. You will construct your own orbit of Halley's comet by tracing Fig. 8-9 and mounting the tracing on stiff cardboard.

Combining the Two Orbits

Now you have the two orbits, the comet's and the earth's in their planes, each of which contains the sun. You need only to fit the two together in accordance with the elements of orbits shown in Fig. 7-1 that you may have used in the activity on the "Inclination of Mars Orbit" in Chapter 7.

The line along which the comet's orbital plane cuts the ecliptic plane is called the "line of nodes." Since you have the major axis drawn, you can locate the ascending node, in the orbital plane, by measuring ω, the angle from perihelion in a direction *opposite* to the comet's motion (see Fig. 8-9).

To fit the two orbits together, cut a narrow slit in the ecliptic plane (earth's orbit) along the line of the *ascending* node in as far as the sun. The longitude of the comet's ascending node Ω was at 57° as shown in Fig. 8-5. Then slit the comet's orbital plane on the side of the *descending* node in as far as the sun (see Fig. 8-6). Slip one plane into the other along the cuts until the sun-points on the two planes come together.

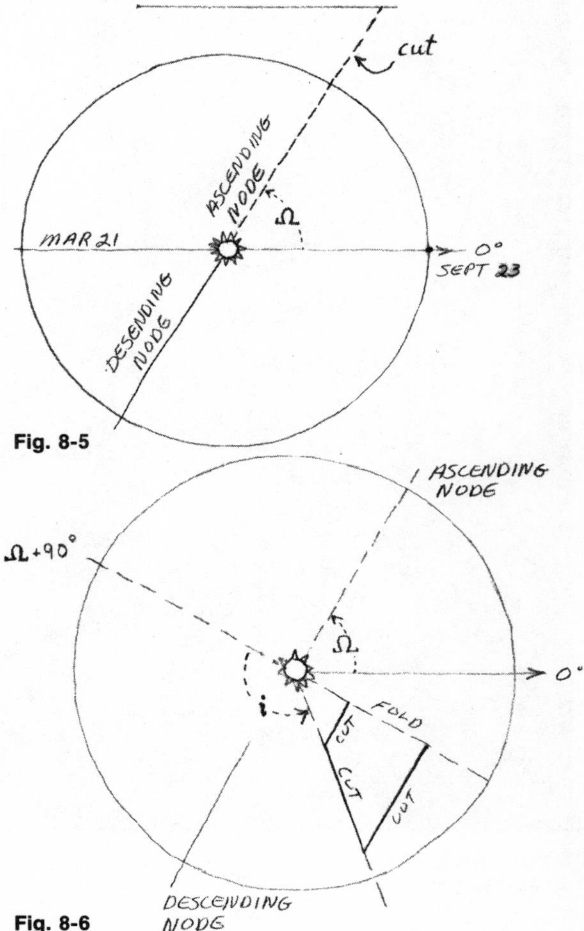

Fig. 8-5

Fig. 8-6

To establish the model in three dimensions you must now fit the two planes together at the correct angle. Remember that the inclination i, 162°, is measured upward (northward) from the ecliptic in the direction of $\Omega + 90°$ (see Fig. 8-7). When you fit the two planes together you will find that the comet's orbit is on the underside of the cardboard. The simplest way to transfer the orbit to the top of the cardboard is to prick through with a pin at enough points so that you can draw a smooth curve through them. Also, you can construct a small tab to support the orbital plane in the correct position.

Halley's comet moves in the opposite sense to the earth and other planets. Whereas the earth and planets move counterclockwise when viewed from above (north of) the ecliptic, Halley's comet moves clockwise.

If you have persevered this far, and your model is a fairly accurate one, it should be easy to explain the comet's motion through the sky shown in Fig. 8-8. The dotted line in the figure is the ecliptic.

With your model of the comet orbit you can now answer some very puzzling questions about the behavior of Halley's comet in 1910.

1. Why did the comet appear to move westward for many months?
2. How could the comet hold nearly a stationary place in the sky during the month of April 1910?

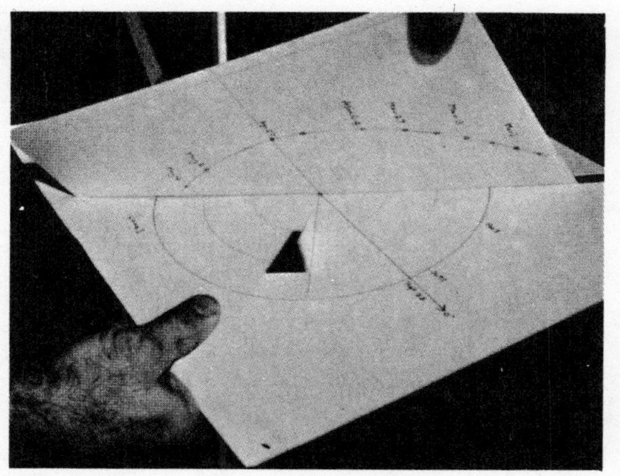

Fig. 8-7

3. After remaining nearly stationary for a month, how did the comet move nearly halfway across the sky during the month of May 1910?
4. What was the position of the comet in space relative to the earth on May 19th?
5. If the comet's tail was many millions of miles long on May 19th, is it likely that the earth passed through part of the tail?
6. Were people worried about the effect a comet's tail might have on life on the earth? (See newspapers and magazines of 1910!)
7. Did anything unusual happen? How dense is the material in a comet's tail? Would you expect anything to have happened?

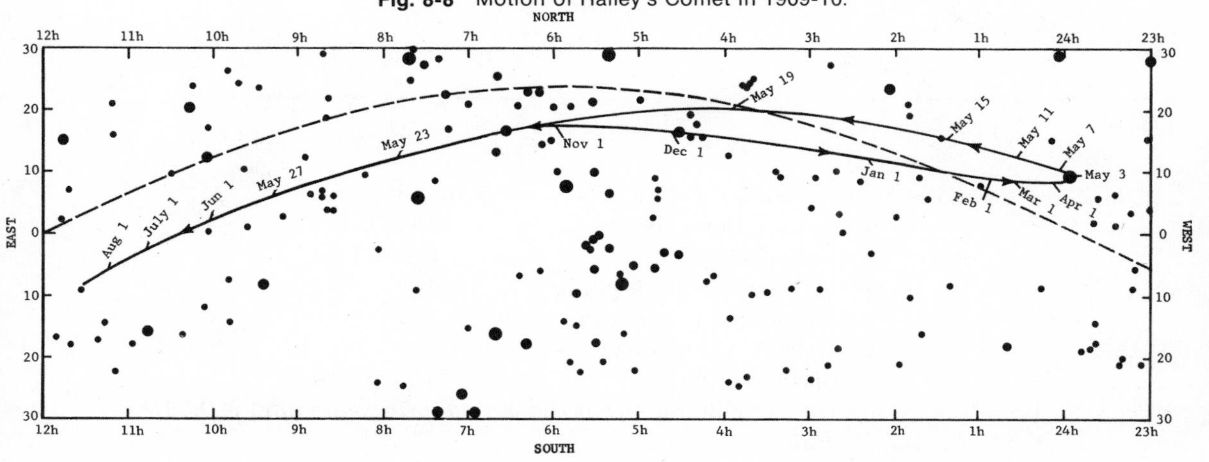

Fig. 8-8 Motion of Halley's Comet in 1909-10.

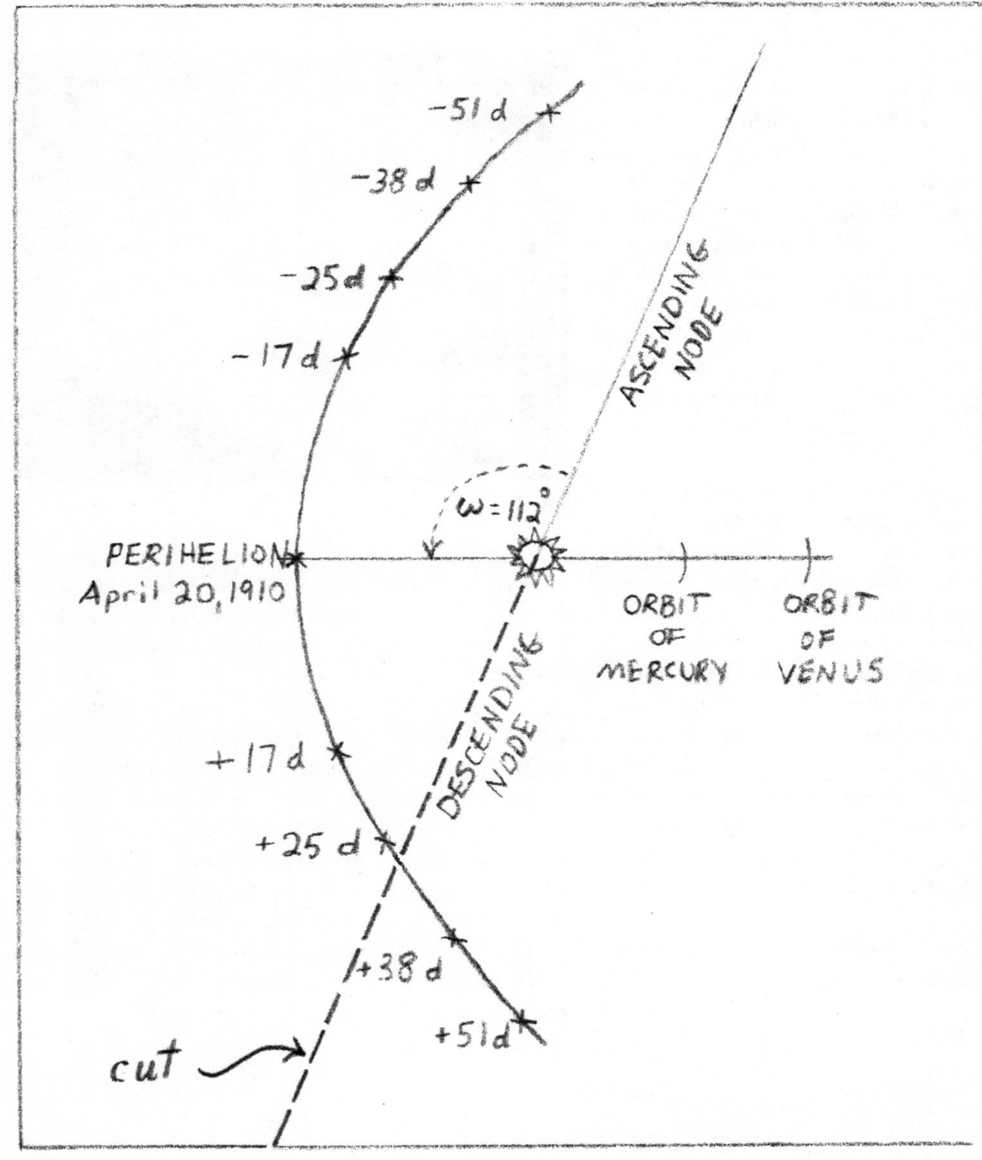

Fig. 8-9

The elements of Halley's comet are, approximately:

a (semi-major axis)	17.9 AU
e (eccentricity)	0.967
i (inclination Forbit plane)	162°
Ω (longitude of ascending node)	057°
ω (angle to perihelion)	112°
T (perihelion date)	April 20, 1910

From these data we can calculate that the period is 76 years, and is 0.59 AU the perihelion distance.

OTHER COMET ORBITS

If you enjoyed making a model of the orbit of Halley's comet, you may want to make models of some other comet orbits. Data are given below for several others of interest.

Encke's comet is interesting because it has the shortest period known for a comet, only 3.3 years. In many ways it is representative of all short-period comet orbits. All have orbits of low inclination and pass near the orbit of Jupiter, where they are often strongly deviated. The full ellipse can be drawn at the scale of 10 cm for 1 AU. The orbital elements for Encke's comet are:

$a = 2.22$ AU
$e = 0.85$
$i = 15°$
$\Omega = 335°$
$\omega = 185°$

From these data we can calculate that the perihelion distance R_r is 0.33 AU and the aphelion distance R_a is 4.11 AU.

The comet of 1680 is discussed extensively in Newton's *Principia,* where approximate orbital elements are given. The best parabolic orbital elements known are:

T = Dec. 18, 1680
$\omega = 350.7°$
$\Omega = 272.2°$
$i = 60.16°$
$R_p = 0.00626$ AU

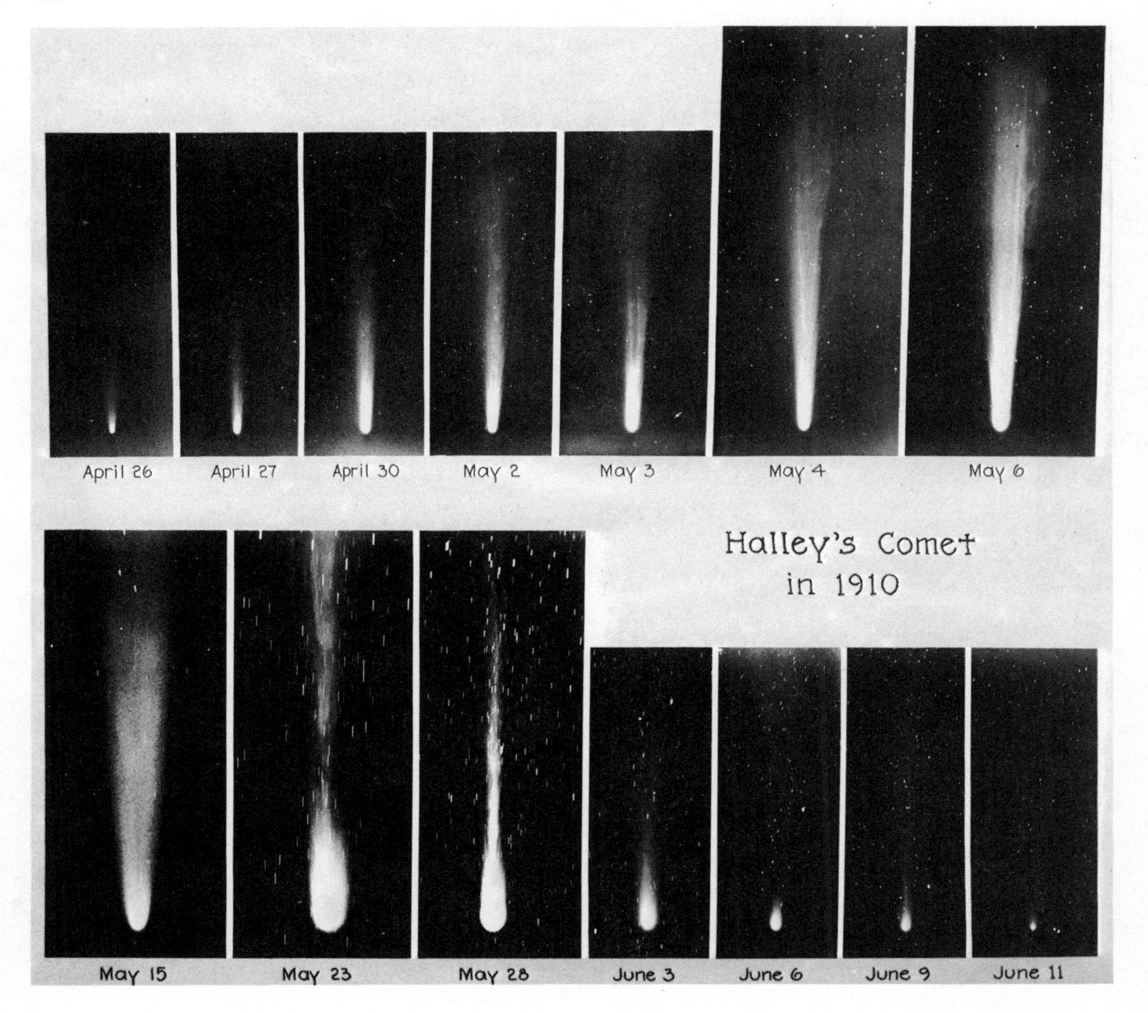

M. Babinet prévenu par sa portière de la visite de la comète. A lithograph by the French artist Honoré Daumier (1808-1879) Museum of Fine Arts, Boston.

Note that this comet passed very close to the sun. At perihelion it must have been exposed to intense destructive forces like the comet of 1965.

Comet Candy (1960N) had the following parabolic orbital elements:

T = Feb. 8, 1961
$\omega = 136.3°$
$\Omega = 176.6$
$i = 150.9$
$R_p = 1.06$ AU

FORCES ON A PENDULUM

If a pendulum is drawn aside and released with a small sideways push, it will move in an almost elliptical path. This looks vaguely like the motion of a planet about the sun, but there are some differences.

To investigate the shape of the pendulum orbit and see whether the motion follows the law of areas, you can make a strobe photo with the setup shown in Fig. 8-10. Use either an electronic strobe flashing from the side, or use a small light and AA battery cell on the pendulum and a motor strobe disk in front of the lens. If you put the tape over one slot of a 12-slot disk to make it half as wide as the rest, it will make every 12th dot fainter giving a handy time marker, as shown in Fig. 8-11. You can also set the camera on its back on the floor with the motor strobe above it, and suspend the pendulum overhead.

Are the motions and the forces similar for the pendulum and the planets? The center of force for planets is located at one focus of the ellipse. Where is the center of force for the

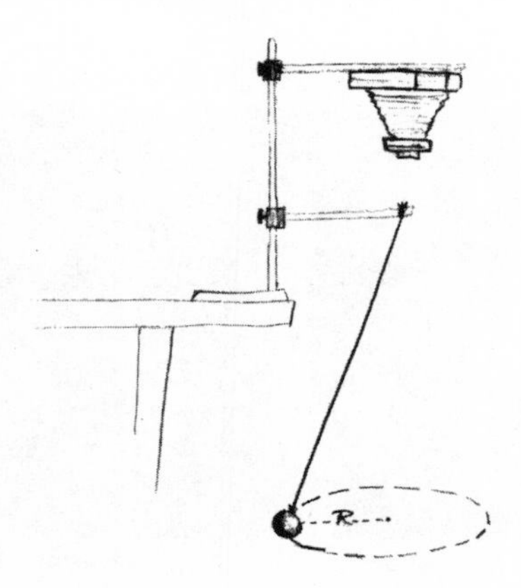

Fig. 8-10

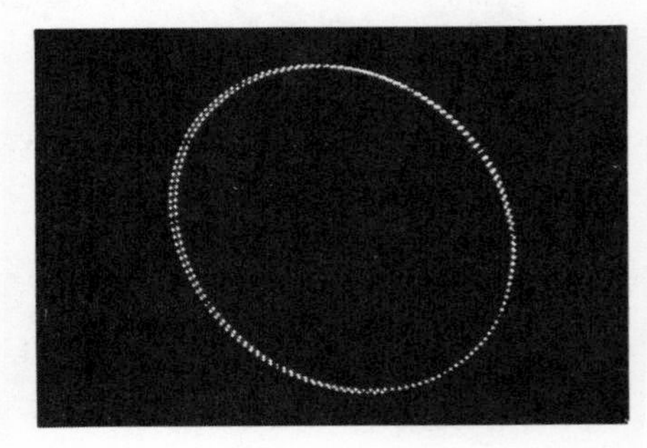

Fig. 8-11

pendulum? Measure your photos to determine whether the pendulum bob follows the law of areas for motion under a central force.

In the case of the planets, the force varies inversely with the square of the distance between the sun and the planet. From your photograph you can find how the restoring force on the pendulum changes with distance R from the rest point. Find Δv between strobe flashes for two sections of the orbit, one near and one far from the rest point. How do the accelerations as indicated by the Δv's compare with the distances R? Does the restoring force depend on distance in the same way as it does for a planet? If you have a copy of Newton's *Principia* available, read Proposition X.

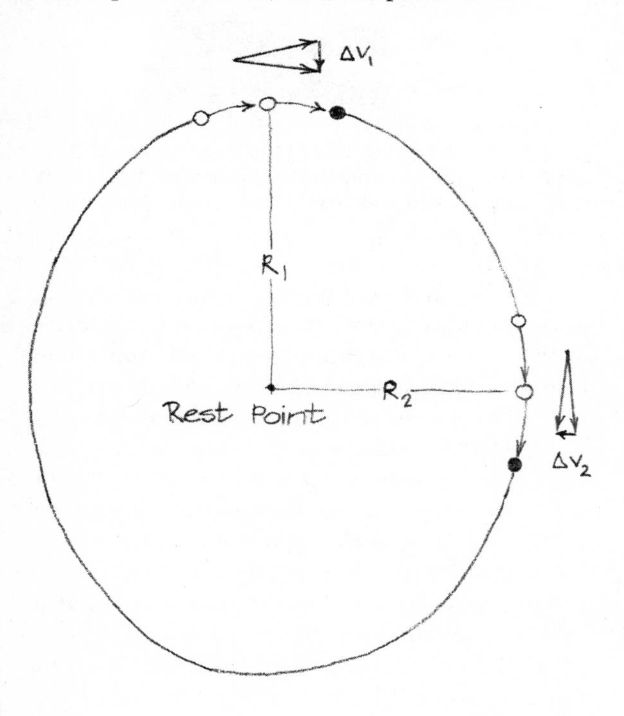

HAIKU

If you are of a literary turn of mind, try your hand at using Japanese haiku (hi-koo), a form of poetry, to summarize what you have learned so far in physics. The rules are quite simple: a haiku must have three lines, the first and third having five syllables and the second having seven syllables. No rhyming is necessary. Several student haiku (the plural is pronounced and spelled the same as the singular) are given below:

An epicycle
Those most complicated things
were rid by Kepler.

The orbit of Mars
Was a great task to achieve
From that of the earth

Kepler took star dates,
The shining stars in the sky,
Physics came this way.

TRIAL OF COPERNICUS

Hold a mock trial for Copernicus. Two groups of students represent the prosecution and the defense. If possible, have English, social studies, and language teachers serve as the jury for your trial.

DISCOVERY OF NEPTUNE AND PLUTO

The Project Physics supplementary unit, *Discoveries in Physics,* describes how Newton's law of universal gravitation was used to predict the existence of Neptune and Pluto before they were observed with a telescope. Read it, and decide whether you think it possible that we may yet discover another planet beyond Pluto. Other accounts appear in *The World of Mathematics,* "John Couch Adams and The Discovery of Neptune," and in Owen Gingerich's article "Solar System beyond Neptune" in the April 1959 *Scientific American.*

HOW TO FIND THE MASS OF A DOUBLE STAR*

To demonstrate the power of Newton's laws, let us study the motion of a double star. You can derive its mass from your own observations.

An interesting double star of short period, which can be seen as a double star with a six-inch telescope, is Kruger 60. The finding chart shows its location less than one degree

*Adapted from a paper by James F. Wanner of the Sproul Observatory, Swarthmore College, in *Sky and Telescope*, January 1967.

south of the variable star Delta Cephei in the northern sky.

The sequential photographs (right), spaced in proportion to their dates show the double star on the right. Another star, which just happens to be in the line of sight, is seen on the left. The photographs show the revolution within the double-star system, which has a period of about 45 years. As you can see, the components were farthest apart, about 3.4 seconds of arc, in the mid-1940's. The chart of the relative positions of the two components (page 127) shows that they will be closest together at 1.4 seconds of arc around 1971. The large cross marks the center of mass of the two-star system. If you measure the direction and distance of one star relative to the other at five-year intervals, you can make a plot on graph paper which shows the motion of one star relative to the other. Would you expect this to be an ellipse? Should Kepler's law of areas apply? Does it? Have you assumed that the orbital plane is perpendicular to the line of sight?

The sequential pictures show that the center of mass of Kruger 60 is drifting away from the star at the left. If you were to extend the lines back to earlier dates, you would find that in the 1860's Kruger 60 passed only 4 seconds of arc from that reference star.

The drift of Kruger 60 relative to the reference star shows that the stars do move relative to each other. For most stars, which are at great distances, this motion, called *proper motion*, is too small to be detected. Kruger 60 is, however, relatively nearby, only about 13 light years away. This is the distance light travels in 13 years at 3.0×10^8 meters per second. The distance to Kruger 60 is then 13 light years $\times\ 3.0 \times 10^8$ m/sec $\times\ 3.2 \times 10^7$ sec/yr, or 13×10^{16} m, or 8.7×10^5 A.U. (One year contains about 3.2×10^7 seconds. One A.U. is 1.5×10^{11} meters.)

From the sequential photographs and the scale given there we can derive the change in distance of Kruger 60 from the reference star between 1919 and 1965.

From the photographs our measurements give the distances as 55 seconds of arc in 1919 and 99 seconds of arc in 1965. (One second of arc is the angle subtended by a unit length, perpendicular to the line of sight, at a distance of 2.1×10^5 of the units.) Thus the proper motion was 44 seconds in 46 years, very nearly 1.0 second of arc per year. This angle is about $1/2.1 \times 10^5$ of the distance to the star. Then in one year the star moves 13×10^{16} m$/2.1 \times 10^5$, or 6.7×10^{11} m/yr. In one second the component of the star's velocity vector across the sky is 1.9×10^4m; hence its velocity perpendicular to the line of sight is 19 km/sec. Probably the star also has a component of motion along the line of sight, called the radial velocity, but this must be found from another type of observation.

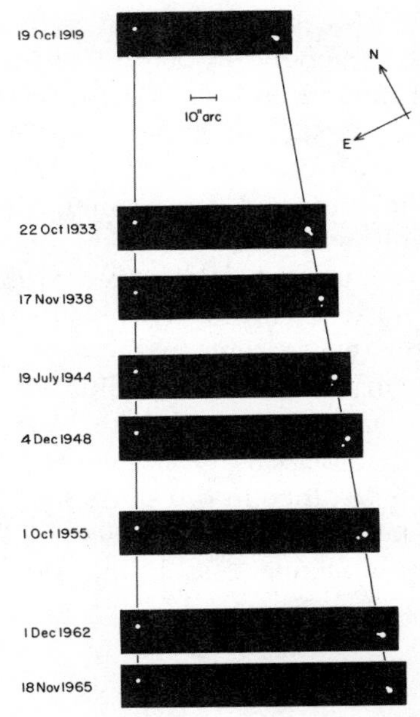

The orbital and linear motions of the visual binary, Kruger 60, are both shown in the chart above which is made up of photographs taken at Leander McCormick Observatory (1919 and 1933) and at Sproul Observatory (1938 to 1965).

The masses of the two stars of Kruger 60 can be found from the photographs shown above and the application of the equation below. When we developed this equation we assumed that the mass of one body of each pair (sun-planet, or planet-satellite) was negligible. In the equation the mass is actually the sum of the two; so for the double star we must write $(m_1 + m_2)$. Thus we have

$$\frac{(m_1 + m_2)_{\text{pair}}}{m_{\text{sun}}} = \left(\frac{T_E}{T_{\text{pair}}}\right)^2 \left(\frac{R_{\text{pair}}}{R_E}\right)^3$$

The arithmetic is greatly simplified if we take the periods in years and the distances in astronomical units (A.U.) which are both units for the earth. The period of Kruger 60 is about 45 years. The mean distance of the components can be found in seconds of arc from the diagram above. The mean separation is

$$\frac{\text{max} + \text{min}}{2} = \frac{3.4 \text{ seconds} + 1.4 \text{ seconds}}{2}$$

$$= \frac{4.8 \text{ seconds}}{2} = 2.4 \text{ seconds.}$$

Earlier we found that the distance from the sun to the pair is nearly 8.7×10^5 A.U. Then the mean angular separation of 2.4 seconds equals

$$\frac{2.4 \times 8.7 \times 10^5 \text{ A.U.}}{2.1 \times 10^5} = 10 \text{ A.U.}$$

or the stars are separated from each other by about the same distance as Saturn is from the sun.

Now, upon substituting the numbers into the equation we have

$$\frac{(m_1 + m_2)_{\text{pair}}}{\text{m}_{\text{sun}}} = \frac{1}{45}^2 \frac{10^3}{1}$$

$$= \frac{1000}{2025} = 0.50,$$

or, the two stars together have about half the mass of the sun.

We can even separate this mass into the two components. In the diagram of motions relative to the center of mass we see that one star has a smaller motion, and we conclude that it must be more massive. For the positions of 1970 (or those observed a cycle earlier in

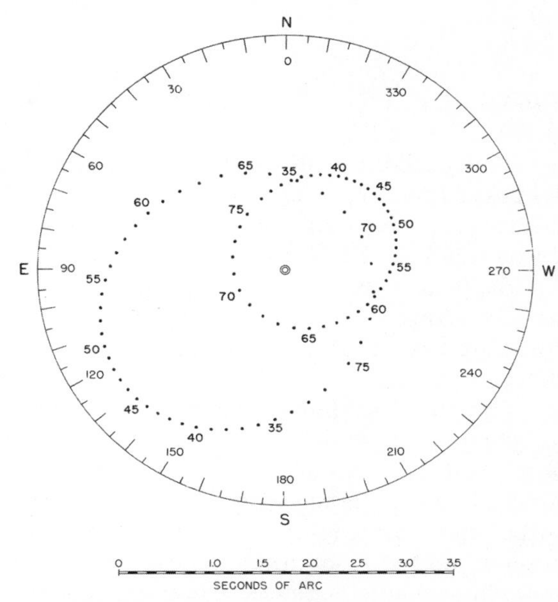

Kruger 60's components trace elliptical orbits, indicated by dots, around their center of mass, marked by a double circle. For the years 1932 to 1975, each dot is plotted on September 1. The outer circle is calibrated in degrees, so the position angle of the companion may be read directly, through the next decade. (Positions after 1965 by extrapolation from data for 1932 to 1965).

1925) the less massive star is 1.7 times farther than the other from the center of mass. So the masses of the two stars are in the ratio 1.7:1. Of the total mass of the pair, the less massive star has

$$\frac{1}{1 + 1.7} \times 0.5 = 0.18$$

the mass of the sun, while the other star has 0.32 the mass of the sun. The more massive star is more than four times brighter than the smaller star. Both stars are red dwarfs, less massive and considerably cooler than the sun.

Peanuts **By Charles M. Schulz**

FILM LOOPS

FILM LOOP 12 JUPITER SATELLITE ORBIT

This time-lapse study of the orbit of Jupiter's satellite, Io, was filmed at the Lowell Observatory in Flagstaff, Arizona, using a 24-inch refractor telescope.

Exposures were made at 1-minute intervals during seven nights in 1967. An almost complete orbit of Io is reconstructed using all these exposures.

The film first shows a segment of the orbit as photographed at the telescope; a clock shows the passage of time. Due to small errors in guiding the telescope, and atmospheric turbulence, the highly magnified images of Jupiter and its satellites dance about. To remove this unsteadiness, each image—over 2100 of them!—was optically centered in the frame. The stabilized images were joined to give a continuous record of the motion of Io. Some variation in brightness was caused by haze or cloudiness.

The four Galilean satellites are listed in Table 1. On Feb. 3, 1967, they had the configuration shown in Fig. 8-12. The satellites move nearly in a plane which we view almost edge-on; thus they seem to move back and forth along a line. The field of view is large enough to include the entire orbits of I and II, but III and IV are outside the camera field when they are farthest from Jupiter.

The position of Io in the last frame of the Jan. 29 segment matches the position in the first frame of the Feb. 7 segment. However, since these were photographed 9 days apart, the other three satellites had moved varying distances, so you see them pop in and out while the image of Io is continuous. Lines identify Io in each section. Fix your attention on the steady motion of Io and ignore the comings and goings of the other satellites.

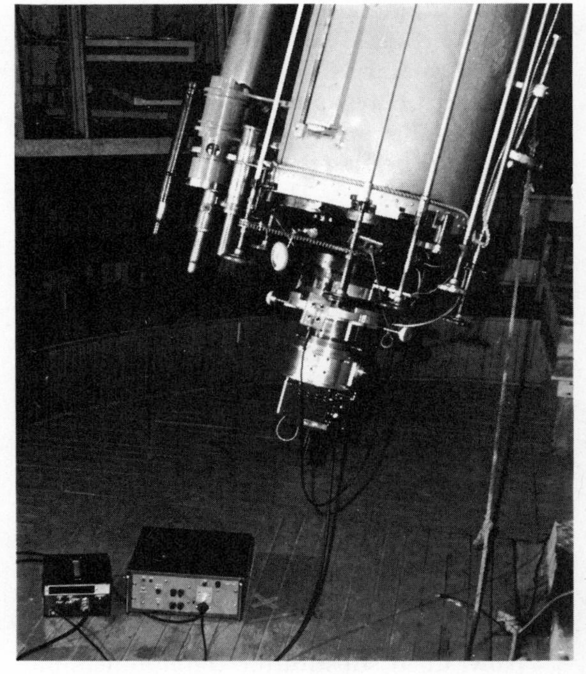

Business end of the 24-inch refractor at Lowell Observatory.

TABLE 1
SATELLITES OF JUPITER

	NAME	PERIOD	RADIUS OF ORBIT (miles)	ECCENTRICITY OF ORBIT	DIAMETER (miles)
I	Io	$1^d\ 18^h\ 28^m$	262,000	0.0000	2,000
II	Europa	$3^d\ 13^h\ 14^m$	417,000	0.0003	1,800
III	Ganymede	$7^d\ 3^h\ 43^m$	666,000	0.0015	3,100
IV	Callisto	$16^d\ 16^h\ 32^m$	1,171,000	0.0075	2,800

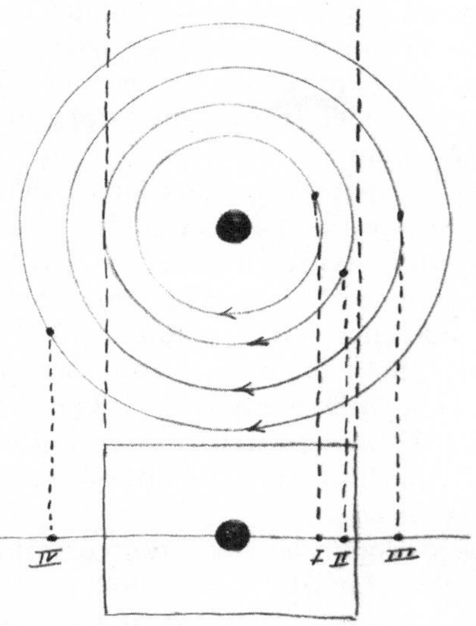

Fig. 8-12

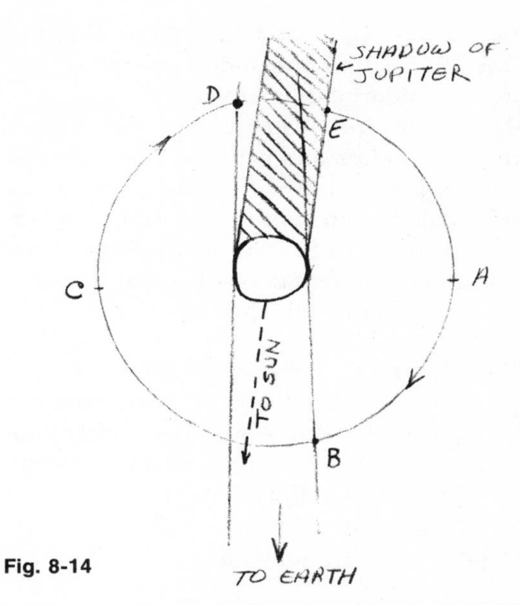

Fig. 8-14

Interesting Features of the Film

1. At the start Io appears almost stationary at the right, at its greatest elongation; another satellite is moving toward the left and overtakes it.

2. As Io moves toward the left (Fig. 8-13), it passes in front of Jupiter, a *transit*. Another satellite, *Ganymede*, has a transit at about the same time. Another satellite moves toward the right and disappears behind Jupiter, an *occultation*. It is a very active scene! If you look closely during the transit, you may see the shadow of Ganymede and perhaps that of Io, on the left part of Jupiter's surface.

Fig. 8-13 Still photograph from Film Loop 12 showing the positions of three satellites of Jupiter at the start of the transit and occultation sequence. Satellite IV is out of the picture, far to the right of Jupiter.

3. Near the end of the film, Io (moving toward the right) disappears; an occulation begins. Look for Io's reappearance—it emerges from an eclipse and appears to the right of Jupiter. Note that Io is out of sight part of the time because it is behind Jupiter as viewed from the earth and part of the time because it is in Jupiter's shadow. It cannot be seen as it moves from O to E in Fig. 8-14.

4. Jupiter is seen as a flattened circle because its rapid rotation period (9 h 55 m) has caused it to flatten at the poles and bulge at the equator. The effect is quite noticeable: the equatorial diameter 89,200 miles and the polar diameter is 83,400 miles.

Measurements

1. *Period of orbit.* Time the motion between transit and occulation (from B to D in Fig. 8-14), half a revolution, to find the period. The film is projected at about 18 frames/sec, so that the speed-up factor is 18 × 60, or 1080. How can you calibrate your projector more accurately? (There are 3969 frames in the loop.) How does your result for the period compare with the value given in the table?

2. *Radius of orbit.* Project on paper and mark the two extreme positions of the satellite,

farthest to the right (at A) and farthest to the left (at C). To find the radius in miles, use Jupiter's equatorial diameter for a scale.

3) *Mass of Jupiter*. You can use your values for the orbit radius and period to calculate the mass of Jupiter relative to that of the sun (a similar calculation based on the satellite Callisto is given in SG 8.9 of the *Text*). How does your experimental result compare with the accepted value, which is $m_j/m_s = 1/1048$?

FILM LOOP 13 PROGRAM ORBIT I

A student (right, Fig. 8-15) is plotting the orbit of a planet, using a stepwise approximation. His teacher (left) is preparing the computer program for the same problem. The computer and the student follow a similar procedure.

Fig. 8-15

The computer "language" used was FORTRAN. The FORTRAN program (on a stack of punched cards) consists of the "rules of the game": the laws of motion and of gravitation. These describe precisely how the calculation is to be done. The program is translated and stored in the computer's memory before it is executed.

The calculation begins with the choice of initial position and velocity of the planet. The initial position values of X and Y are selected and also the initial components of velocity XVEL and YVEL. (XVEL is the name of a single variable, not a product of four variables X, V, E, and L.)

Then the program instructs the computer to calculate the force on the planet from the sun from the inverse-square law of gravitation. Newton's laws of motion are used to calculate how far and in what direction the planet moves after each blow.

The computer's calculations can be displayed in several ways. A table of X and Y values can be typed or printed. An X-Y plotter can draw a graph from the values, similar to the hand-constructed graph made by the student. The computer results can also be shown on a cathode ray tube (CRT), similar to that in a television set, in the form of a visual trace. In this film, the X-Y plotter was the mode of display used.

The dialogue between the computer and the operator for trial 1 is as follows. The numerical values are entered at the computer typewriter by the operator after the computer types the messages requesting them.

Computer: GIVE ME INITIAL POSITION IN AU . . .
Operator: X = 4
Y = 0
Computer: GIVE ME INITIAL VELOCITY IN AU/YR . . .
Operator: XVEL = 0
YVEL = 2
Computer: GIVE ME CALCULATION STEP IN DAYS . . .
Operator: 60.
Computer: GIVE ME NUMBER OF STEPS FOR EACH POINT PLOTTED . . .
Operator: 1.
Computer: GIVE ME DISPLAY MODE . . .
Operator: X-Y PLOTTER.

You can see that the orbit displayed on the X-Y plotter, like the student's graph, does not close. This is surprising, as you know that the orbits of planets are closed. Both orbits fail to close exactly. Perhaps too much error is introduced by using such large steps in the step-by-step approximation. The blows may be too infrequent near perihelion, where the force is largest, to be a good approximation to a continuously acting force. In the Film Loop, "Program Orbit II," the calculations are based upon smaller steps, and you can see if this explanation is reasonable.

FILM LOOP 14 PROGRAM ORBIT II

In this continuation of the film "Program Orbit I," a computer is again used to plot a planetary orbit with a force inversely proportional to the square of the distance. The computer program adopts Newton's laws of motion. At equal intervals, blows act on the body. We guessed that the orbit calculated in the previous film failed to close because the blows were spaced too far apart. You could calculate the orbit using many more blows, but to do this by hand would require much more time and effort. In the computer calculation we need only specify a smaller time interval between the calculated points. The laws of motion are the same as before, so the same program is used.

A portion of the "dialogue" between the computer and the operator for trial 2 is as follows:

Computer: GIVE ME CALCULATION STEP IN DAYS . . .
Operator: 3.
Computer: GIVE ME NUMBER OF STEPS FOR EACH POINT PLOTTED . . .
Operator: 7.
Computer: GIVE ME DISPLAY MODE . . .
Operator: X-Y PLOTTER.

Points are now calculated every 3 days (20 times as many calculations as for trial 1 on the "Program Orbit I" film), but, to avoid a graph with too many points, only 1 out of 7 of the calculated points is plotted.

The computer output in this film can also be displayed on the face of a cathode ray tube (CRT). The CRT display has the advantage of speed and flexibility and we will use it in the other loops in this series, *Film loops* 15, 16 and 17. On the other hand, the permanent record produced by the X-Y plotter is sometimes very convenient.

Orbit Program

The computer program for orbits is written in FORTRAN II and includes "ACCEPT" (data) statements used on an IBM 1620 input typewriter. (Example at the right.)

```
      PROGRAM ORBIT
C
C        HARVARD PROJECT PHYSICS ORBIT PROGRAM.
C        EMPIRICAL VERIFICATION OF KEPLERS LAWS
C        FROM NEWTONS LAW OF UNIVERSAL GRAVITATION.
C
      G=40.
    4 CALL MARKF(0.,0.)
    6 PRINT 7
    7 FORMAT(9HGIVE ME Y )
      X=0.
      ACCEPT 5,Y
      PRINT 8
    8 FORMAT(12HGIVE ME XVEL)
    5 FORMAT(F10.6)
      ACCEPT 5,XVEL
      YVEL=0.
      PRINT 9
    9 FORMAT(49HGIVE ME DELTA IN DAYS,  AND NUMBER BETWEEN PRINTS)
      ACCEPT 5,DELTA
      DELTA=DELTA/365.25
      ACCEPT 5,PRINT
      IPRINT = PRINT
      INDEX = 0
      NFALLS = 0
   13 CALL MARKF(X,Y)
      PRINT 10,X,Y
   15 IF(SENSE SWITCH 3) 20,16
   20 PRINT 21
   10 FORMAT(2F7.3)
      NFALLS = NFALLS + IPRINT
   21 FORMAT(23HTURN OFF SENSE SWITCH 3 )
   22 CONTINUE
      IF(SENSE SWITCH 3) 22,4
   16 RADIUS = SQRTF(X*X + Y*Y)
      ACCEL = -G/(RADIUS*RADIUS)
      XACCEL = (X/RADIUS)*ACCEL
      YACCEL = (Y/RADIUS)*ACCEL
C  FIRST TIME THROUGH WE WANT TO GO ONLY 1/2 DELTA
      IF(INDEX) 17,17,18
   17 XVEL = XVEL + 0.5 * XACCEL * DELTA
      YVEL = YVEL + 0.5 * YACCEL * DELTA
      GO TO 19
C  DELTA V = ACCELERATION TIMES DELTA T
   18 XVEL = XVEL + XACCEL * DELTA
      YVEL = YVEL + YACCEL * DELTA
C  DELTA X = XVELOCITY TIMES DELTA T
   19 X = X + XVEL * DELTA
      Y = Y + YVEL * DELTA
      INDEX = INDEX + 1
      IF(INDEX - NFALLS) 15,15,13
      END
```

With slight modification it worked on a CDC 3100 and CDC 3200, as shown in the film loops 13 and 14, "Program Orbit I" and "Program Orbit II." With additional slight modifications (in statement 16 and the three succeeding statements) it can be used for other force laws. The method of computation is the scheme used in Project Physics *Reader 1* "Newton's Laws of Dynamics." A similar program is presented and explained in *FORTRAN for Physics* (Alfred M. Bork, Addison-Wesley, 1967).

Note that it is necessary to have a subroutine MARK. In our case we used it to plot the points on an X-Y plotter, but MARK could be replaced by a PRINT statement to print the X and Y coordinates.

FILM LOOP 15 CENTRAL FORCES—ITERATED BLOWS

In Chapter 8 and in Experiment 19 and Film Loop 13 on the stepwise approximation or orbits we find that Kepler's law of areas applied to objects acted on by a central force. The force in each case was attractive and was either constant or varied smoothly according to some pattern. But suppose the central force is repulsive; that is, directed *away* from the center? or sometimes attractive and sometimes repulsive? And what if the amount of force applied each time varies unsystematically? Under these circumstances would the law of areas still hold? You can use this film to find out.

The film was made by photographing the face of a cathode ray tube (CRT) which displayed the output of a computer. It is important to realize the role of the computer program in this film: it controlled the change in direction and change in speed of the "object" as a result of a "blow." This is how the computer program uses Newton's laws of motion to predict the result of applying a brief impulsive force, or blow. The program remained the same for all parts of the loop, just as Newton's laws remain the same during all experiments in a laboratory. However, at one place in the program, the operator had to specify how he wanted the force to vary.

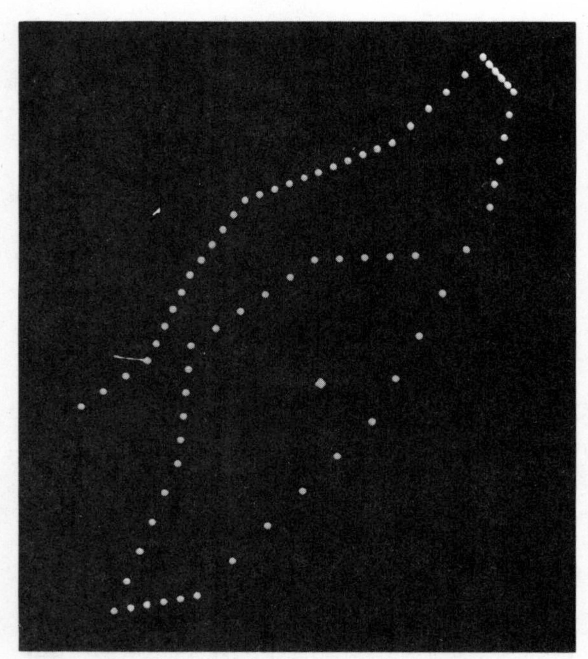

Fig. 8-16

Random Blows

The photograph (Fig. 8-16) shows part of the motion of the body as blows are repeatedly applied at equal time intervals. No one decided in advance how great each blow was to be. The computer was programmed to select a number at random to represent the magnitude of the blow. The directions toward or away from the center were also selected at random, although a slight preference for attractive blows was built in so the pattern would be likely to stay on the face of the CRT. The dots appear at equal time intervals. The intensity and direction of each blow is represented by the length of line at the point of the blow.

Study the photograph. How many blows were attractive? How many were repulsive? Were any blows so small as to be negligible?

You can see if the law of areas applies to this random motion. Project the film on a piece of paper, mark the center and mark the points where the blows were applied. Now measure the areas of the triangles. Does the moving body sweep over equal areas in equal time intervals?

Force Proportional to Distance

If a weight on a string is pulled back and released with a sideways shove, it moves in an elliptical orbit with the force center (lowest point) at the center of the ellipse. A similar path is traced on the CRT in this segment of the film. Notice how the force varies at different distances from the center. A smooth orbit is approximated by the computer by having the blows come at shorter time intervals. In 2(a), 4 blows are used for a full orbit; in 2(b) there are 9 blows, and in 2(c), 20 blows which give a good approximation to the ellipse that is observed with this force. Geometrically, how does this orbit differ from planetary orbits? How is it different physically?

Inverse-square Force

A similar program is used with two planets simultaneously, but with a force on each varying *inversely* as the *square* of the distance from a force center. Unlike the real situation, the program assumes that the planets do not exert forces on one another. For the resulting ellipses, the force center is at one *focus* (Kepler's first law), not at the center of the ellipse as in the previous case.

In this film, the computer has done thousands of times faster what you could do if you had enormous patience and time. With the computer you can change conditions easily, and thus investigate many different cases and display the results. And, once told what to do, the computer makes fewer calculation errors than a person!

FILM LOOP 16 KEPLER'S LAWS

A computer program similar to that used in the film "Central forces–iterated blows" causes the computer to display the motion of two planets. Blows directed toward a center (the sun), act on each planet in equal time intervals. The force exerted by the planets on one another is ignored in the program; each is attracted only by the sun, by a force which varies inversely as the square of the distance from the sun.

Initial positions and initial velocities for the planets were selected. The positions of the planets are shown as dots on the face of the cathode ray tube at regular intervals. (Many more points were calculated between those displayed.)

You can check Kepler's three laws by projecting on paper and marking successive positions of the planets. The law of areas can be verified by drawing triangles and measuring areas. Find the areas swept out in at least three places: near perihelion, near aphelion, and at a point approximately midway between perihelion and aphelion.

Kepler's third law holds that in any given planetary system the squares of the periods of the planets are proportional to the cubes of their average distances from the object around which they are orbiting. In symbols,

$$T^2 \propto R_{av}^3$$

where T is the period and R_{av} is the average distance. Thus in any one system, the value of T^2/R_{av}^3 ought to be the same for all planets.

We can use this film to check Kepler's law of periods by measuring T and for each of the two orbits shown, and then computing T^2/R_{av}^3 for each. To measure the periods of revolution, use a clock or watch with a sweep second hand. Another way is to count the number of plotted points in each orbit. To find R_{av} for each orbit, measure the perihelion and aphelion distances (R_p and R_a) and take their average (Fig. 8-17).

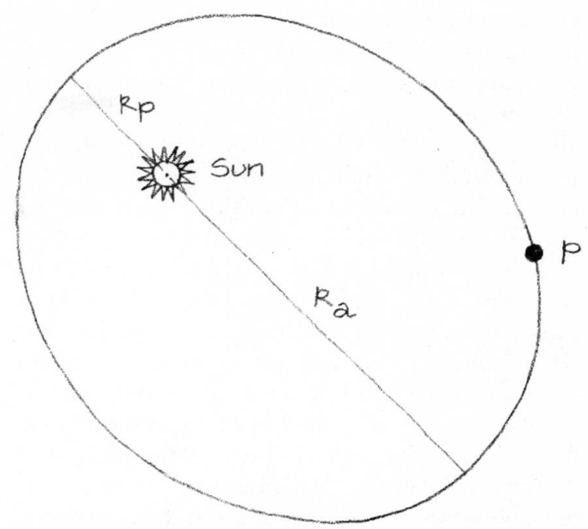

Fig. 8-17 The mean distance R_{av} of a planet P orbiting about the sun is $(R_p + R_a)/2$.

How close is the agreement between your two values of T^2/R_{av}^3? Which is the greater source of error, the measurement of T or of R_{av}?

To check Kepler's first law, see if the orbit is an ellipse with the sun at a focus. You can use string and thumbtacks to draw an ellipse. Locate the empty focus, symmetrical with respect to the sun's position. Place tacks in a

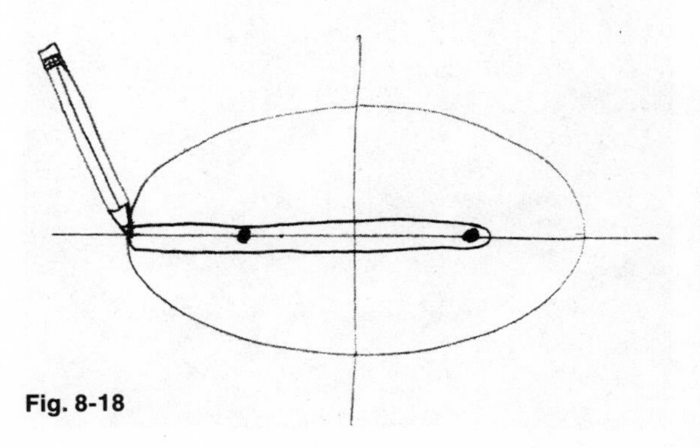
Fig. 8-18

board at these two points. Make a loop of string as shown in Fig. 8-18.

Put your pencil in the string loop and draw the ellipse, keeping the string taut. Does the ellipse match the observed orbit of the planet? What other methods can be used to find if a curve is a good approximation to an ellipse?

You might ask whether checking Kepler's laws for these orbits is just busy-work, since the computer already "knew" Kepler's laws and used them in calculating the orbits. But the computer was *not* given instructions for Kepler's laws. What you are checking is whether Newton's laws lead to motions that fit Kepler's descriptive laws. The computer "knew" (through the program we gave it) only Newton's laws of motion and the inverse-square law of gravitation. This computation is exactly what Newton did, but without the aid of a computer to do the routine work.

FILM LOOP 17 UNUSUAL ORBITS

In this film a modification of the computer program described in "Central forces—iterated blows" is used. There are two sequences: the first shows the effect of a disturbing force on an orbit produced by a central inverse-square force; the second shows an orbit produced by an inverse-cube force.

The word "perturbation" refers to a small variation in the motion of a celestial body caused by the gravitational attraction of another body. For example, the planet Neptune was discovered because of the perturbation it caused in the orbit of Uranus. The main force on Uranus is the gravitational pull of the sun, and the force exerted on it by Neptune causes a perturbation which changes the orbit of Uranus very slightly. By working backward, astronomers were able to predict the position and mass of the unknown planet from its small effect on the orbit of Uranus. This spectacular "astronomy of the invisible" was rightly regarded as a triumph for the Newtonian law of universal gravitation.

Typically a planet's entire orbit rotates slowly, because of the small pulls of other planets and the retarding force of friction due to dust in space. This effect is called "advance of perihelion." (Fig. 8-19.) Mercury's perihelion advances about 500 seconds of arc, ($\frac{1}{7}^\circ$) per century. Most of this was explained by perturbations due to the other planets. However, about 43 seconds per century remained unexplained. When Einstein reexamined the nature of space and time in developing the theory of relativity, he developed a new gravitational theory that modified Newton's theory in cru-

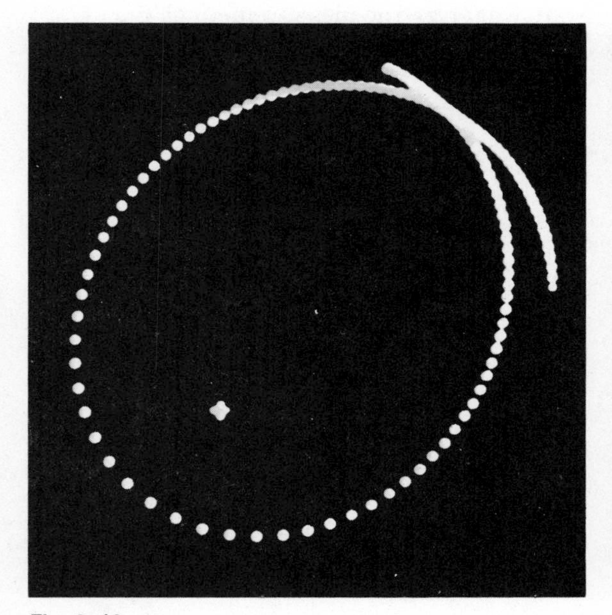

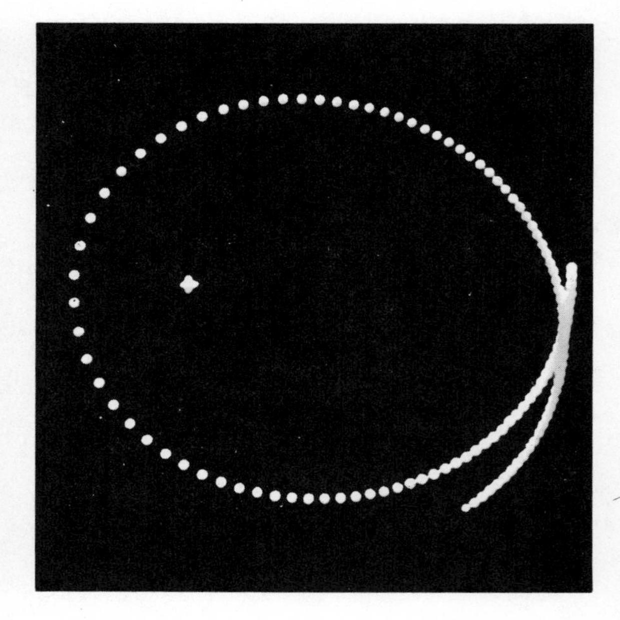

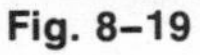
Fig. 8–19

cial ways. Relativity theory is important for bodies moving at high speeds or near massive bodies. Mercury's orbit is closest to the sun and therefore most affected by Einstein's extension of the law of gravitation. Relativity was successful in explaining the extra 43 seconds per century of advance of Mercury's perehelion. But recently this "success" has again been questioned, with the suggestion that the extra 43 seconds may be explained instead by a slight bulge of the sun at its equator.

The first sequence shows the advance of perihelion due to a small force proportional to the distance R, added to the usual inverse-square force. The "dialogue" between operator and computer starts as follows:

PRECESSION PROGRAM WILL USE
  ACCEL = G/(*R***R*) + *P***R*
GIVE ME PERTURBATION *P*.
  P = .66666.
GIVE ME INITIAL POSITION IN AU
  X = 2.
  Y = 0.
GIVE ME INITIAL VELOCITY IN AU/YR
  XVEL = 0.
  YVEL = 3.

The symbol * means multiplication in the Fortran language used in the program. Thus $G/(R*R)$ is the inverse-square force, and $P*R$ is the perturbing force, proportional to R.

In the second part of the film, the force is an inverse-cube force. The orbit resulting from the inverse-cube attractive force, as from most force laws, is not closed. The planet spirals into the sun in a "catastrophic" orbit. As the planet approaches the sun, it speeds up, so points are separated by a large fraction of a revolution. Different initial positions and velocities would lead to quite different orbits.

Man in observation chamber of the 200-inch reflecting telescope on Mt. Palomar.

SATELLITES OF THE PLANETS

		DISCOVERY	AVERAGE RADIUS OF ORBIT	PERIOD OF REVOLUTION			DIAMETER
EARTH:	Moon		238,857 miles	27d	7h	43m	2160 miles
MARS:	Phobos	1877, Hall	5,800	0	7	39	10?
	Deimos	1877, Hall	14,600	1	6	18	5?
JUPITER:	V	1892, Barnard	113,000	0	11	53	150?
	1 (Io)	1610, Galileo	262,000	1	18	28	2000
	II (Europa)	1610, Galileo	417,000	3	13	14	1800
	III (Ganymede)	1610, Galileo	666,000	7	3	43	3100
	IV (Callisto)	1610, Galileo	1,170,000	16	16	32	2800
	VI	1904, Perrine	7,120,000	250	14		100?
	VII	1905, Perrine	7,290,000	259	14		35?
	X	1938, Nicholson	7,300,000	260	12		15?
	XII	1951, Nicholson	13,000,000	625			14?
	XI	1938, Nicholson	14,000,000	700			19?
	VIII	1908, Melotte	14,600,000	739			35?
	IX	1914, Nicholson	14,700,000	758			17?
SATURN:	Mimas	1789, Herschel	115,000	0	22	37	300?
	Enceladus	1789, Herschel	148,000	1	8	53	350
	Tethys	1684, Cassini	183,000	1	21	18	500
	Dione	1684, Cassini	234,000	2	17	41	500
	Rhea	1672, Cassini	327,000	4	12	25	1000
	Titan	1655, Huygens	759,000	15	22	41	2850
	Hyperion	1848, Bond	920,000	21	6	38	300?
	Phoebe	1898, Pickering	8,034,000	550			200?
	Iapetus	1671, Cassini	2,210,000	79	7	56	800
URANUS:	Miranda	1948, Kuiper	81,000	1	9	56	—
	Ariel	1851, Lassell	119,000	2	12	29	600?
	Umbriel	1851, Lassell	166,000	4	3	28	400?
	Titania	1787, Herschel	272,000	8	16	56	1000?
	Oberon	1787, Herschel	364,000	13	11	7	900?
NEPTUNE:	Triton	1846, Lassell	220,000	5	21	3	2350
	Nereid	1949, Kuiper	3,440,000	359	10		200?

THE SOLAR SYSTEM

	RADIUS	MASS	AVERAGE RADIUS OF ORBIT	PERIOD OF REVOLUTION
Sun	6.95×10^8 meters	1.98×10^{30} kilograms	—	—
Moon	1.74×10^6	7.34×10^{22}	3.8×10^8 meters	2.36×10^6 seconds
Mercury	2.57×10^6	3.28×10^{23}	5.79×10^{10}	7.60×10^6
Venus	6.31×10^6	4.83×10^{24}	1.08×10^{11}	1.94×10^7
Earth	6.38×10^6	5.98×10^{24}	1.49×10^{11}	3.16×10^7
Mars	3.43×10^6	6.37×10^{23}	2.28×10^{11}	5.94×10^7
Jupiter	7.18×10^7	1.90×10^{27}	7.78×10^{11}	3.74×10^8
Saturn	6.03×10^7	5.67×10^{26}	1.43×10^{12}	9.30×10^8
Uranus	2.67×10^7	8.80×10^{25}	2.87×10^{12}	2.66×10^9
Neptune	2.48×10^7	1.03×10^{26}	4.50×10^{12}	5.20×10^9
Pluto	?	?	5.9×10^{12}	7.28×10^9

The Project Physics Course

Handbook 3

The Triumph of Mechanics

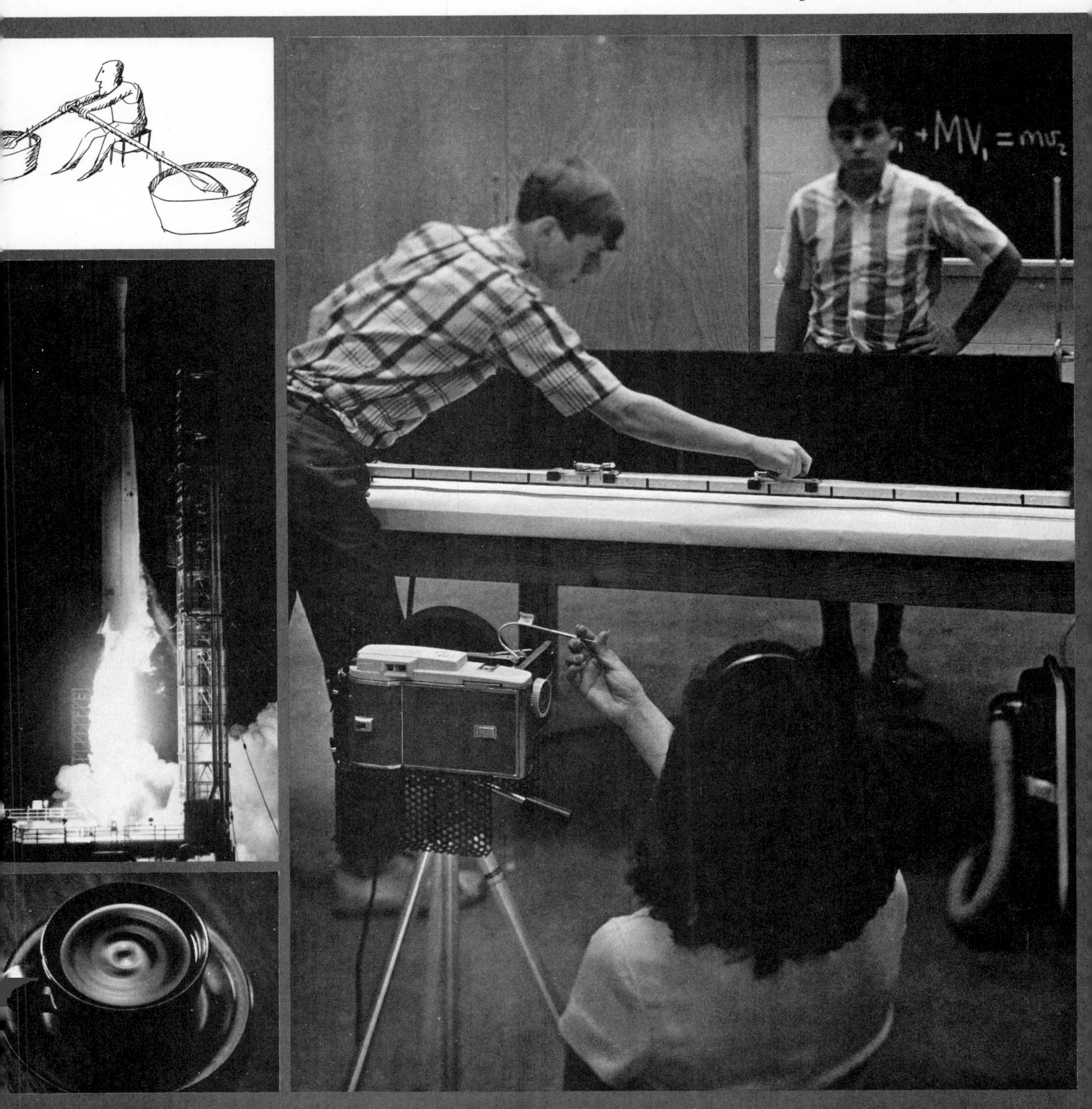

Picture Credits, Unit 3

Cover: (upper left) Cartoon by Charles Gary Solin and reproduced by his permission only. Rocket photograph courtesy of Douglas Missile and Space Systems Division.

P. 141 Fig. 9–2. Photograph courtesy of Damon Corporation, Educational Division.

P. 142 Photo by J. Ph. Charbonnier (Photo Researchers).

P. 145 Wide World Photo.

P. 145 Cartoon Laffs from *True,* Fawcett Publications, Inc., Greenwich, Conn., 1958.

P. 153 Brunswick Corporation.

Pp. 165, 183, 199, 229 (cartoons). By permission of Johnny Hart and Field Enterprises, Inc.

Pp. 171, 206, 207, 211, 229 Cartoons by Charles Gary Solin and reproduced by his permission only.

P. 179 "Black Country," The Mansell Collection, London.

P. 179 Carl Rose, *New Yorker* Magazine, 3/28/70.

P. 180 Photo by the Los Angeles County Air Pollution Control District.

P. 181 Cartoon Laffs from *True,* Fawcett Publications, Inc., Greenwich, Conn., 1958.

P. 194 Linda Cronquist, Wheat Ridge High School, Wheat Ridge, Colorado.

P. 220 Photographs of Chladni plates from "Physics and Music," *Scientific American,* July 1948.

All photographs used with film loops courtesy of National Film Board of Canada.

Photographs of laboratory equipment and of students using laboratory equipment were supplied with the cooperation of the Project Physics staff and Damon Corporation.

Contents HANDBOOK, Unit 3

Chapter 9 Conservation of Mass and Momentum

Chapter 10 Energy

Chapter 11 The Kinetic Theory of Gases

Chapter 12 Waves

Chapter 9 Conservation of Mass and Momentum

EXPERIMENT 22 COLLISIONS IN ONE DIMENSION

In this experiment you will investigate the motion of two objects interacting in one dimension. The interactions (explosions and collisions in the cases treated here) are called one-dimensional because the objects move along a single straight line. Your purpose is to look for quantities or combinations of quantities that remain unchanged before and after the interaction—that is, quantities that are conserved.

Your experimental explosions and collisions may seem not only tame but also artificial and unlike the ones you see around you in everyday life. But this is typical of many scientific experiments, which simplify the situation so as to make it easier to make meaningful measurements and to discover patterns in the observed behavior. The underlying laws are the same for all phenomena, whether or not they are in a laboratory.

Four different ways of observing interactions are described here. You will probably use only one of them. In each method, the friction between the interacting objects and their surroundings is kept as small as possible, so that the objects are a nearly isolated system. Whichever method you do follow, you should handle your results in the way described in the final section: *Analysis of data.*

Method I—Dynamics Carts

"Explosions" are most easily studied with the low-friction dynamics carts. Squeeze the loop of spring steel flat and slip a loop of thread over it, to hold it compressed. Put the compressed loop between two carts on the floor or on a smooth table (Fig. 9-1). When you release the spring by burning the thread, the carts fly apart with velocities that you can measure from a strobe photograph or by any of the techniques you learned in Unit 1.

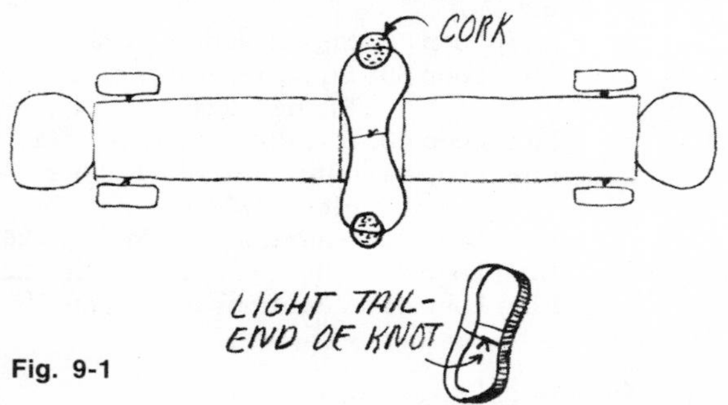

Fig. 9-1

Load the carts with a variety of weights to create simple ratios of masses, say 2 to 1 or 3 to 2. Take data for as great a variety of mass ratios as time permits. Because friction will gradually slow the carts down, you should make measurements on the speeds immediately after the explosion is over (that is, when the spring is through pushing).

Since you are interested only in comparing the speeds of the two carts, you can express those speeds in any units you wish, without worrying about the exact scale of the photograph and the exact strobe rate. For example, you can use distance units measured directly from the photograph (in millimeters) and use time units equal to the time interval between strobe images. If you follow that procedure, the speeds recorded in your notes will be in mm/interval.

Remember that you can get data from the negative of a Polaroid picture as well as from the positive print.

Method II—the Air Track

The air track allows you to observe collisions between objects—"gliders"—that move with almost no friction. You can take stroboscopic photographs of the gliders either with the xenon strobe or, as shown in Fig. 9-2, by using the rotating disk in front of the camera.

The air track has three gliders: two small ones with the same mass, and a larger one which has just twice the mass of a small one. A small and a large glider can be coupled together to make one glider so that you can have collisions between gliders whose masses are in the ratio of 1:1, 2:1, and 3:1. (If you add light sources to the gliders, their masses will no longer be in the same simple ratios. You can find the masses from the measured weights of the glider and light source.)

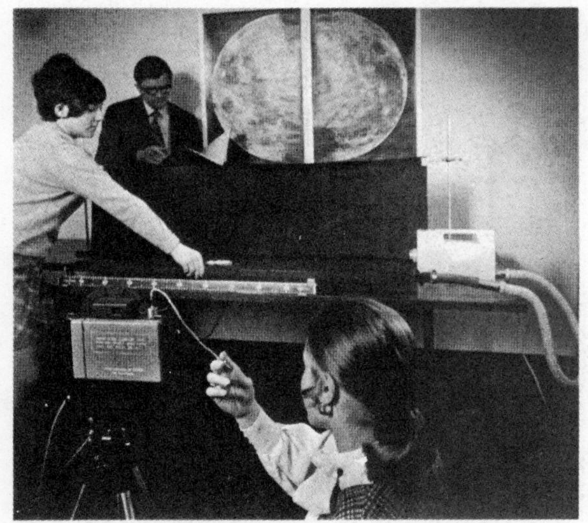

Fig. 9-2

You can arrange to have the gliders bounce apart after they collide (elastic collision) or stick together (inelastic collision). Good technique is important if you are to get good results. Before taking any pictures, try both elastic and inelastic collisions with a variety of mass ratios. Then, when you have chosen one to analyze, rehearse each step of your procedure with your partners before you go ahead.

You can use a good photograph to find the speeds of both carts, before and after they collide. Since you are interested only in comparing the speeds before and after each collision, you can express speeds in any units you wish, without worrying about the exact scale of the photograph or the exact strobe rate. For example, you use distance units measured directly from the photograph (in millimeters) and use time units equal to the time interval between strobe images. If you follow that procedure, the speeds recorded in your notes will be in mm/interval.

Remember that you can get data from the negative of your Polaroid picture as well as from your positive print.

Method III—Film Loops

Film Loops 18, 19, and *20* show one-dimensional collisions that you cannot easily perform in your own laboratory. Since these collisions are shown in slow motion, you can make measurements directly from the pictures projected onto graph paper. Since you are interested only in comparing speeds before and after a collision, you can express speeds in any unit you wish—that is, you can make measurements in any convenient distance and time units.

Notes for these film loops are at the end of this *Handbook* chapter (pages 161 to 165). If you use these loops, read the notes before taking your data.

Method IV—Stroboscopic Photographs

In the activities section of this chapter of this *Handbook* (pp. 149 to 152) you will find stroboscopic photographs of seven one-dimensional collision events. These were photographed during the making of *Film Loops 18, 19,* and *20*. You can express speeds in any units you wish, without worrying about the exact scale of the photograph or the exact strobe rate. For example, you can use distance units measured directly from the photograph (in millimeters) and use time units equal to the time interval between strobe images. Before you take measurements, be sure to read the notes describing what the events were and how the photographs were made.

Analysis of Data

Assemble all your data in a table such as Fig. 9-3. The table should have column headings for the mass of each object, m_A and m_B, the speeds before the interaction, v_A and v_B (for explosions, $v_A = v_B = 0$), and the speeds after the collision, v_A' and v_B'.

Examine your table carefully. Search for quantities or combinations of quantities that remain unchanged before and after the interaction.

Q1 Is *speed* a conserved quantity? That is,

m_A	m_B	v_A	v_B	v_A'	v_B'

Fig. 9-3

does the quantity $(v_A + v_B)$ equal the quantity $(v_A' + v_B')$?

Q2 Consider the direction as well as the speed. Define velocity to the right as positive and velocity to the left as negative. Is *velocity* a conserved quantity?

Q3 If neither speed nor velocity is conserved, try a quantity that combines the mass and velocity of each cart. Compare $(m_A v_A + m_B v_B)$ with $(m_A v_A' + m_B v_B')$ for each interaction. In the same way compare m/v, $m\vec{v}$, $m^2 v$, or any other likely combinations you can think of, before and after interaction.

You should be able to find at least one quantity that is conserved in all your experiments, within the limits of experimental error. Is this quantity conserved in all your classmates' experiments as well?

EXPERIMENT 23 COLLISIONS IN TWO DIMENSIONS

Collisions rarely occur in only one dimension, that is, along a straight line. In billiards, basketball, and tennis, the ball usually rebounds at an angle to its original direction; and ordinary explosions (which can be thought of as collisions in which initial velocities are all zero) send pieces flying off in all directions.

This experiment deals with collisions that occur in two dimensions—that is, in a single plane—instead of along a single straight line. It assumes that you know what momentum is and understand what is meant by "conservation of momentum" in one dimension (as is described in *Text* Sec. 9.3). In this experiment

you will discover a general form of the rule for one dimension that applies also to the conservation of momentum in cases where the parts of the system move in two (or three) dimensions.

Several methods of getting data on two-dimensional collisions are described below, but you will probably want to follow only one method. Whichever method you use, handle your results in the way described in the last section.

Method I—Colliding Pucks

On a carefully leveled glass tray covered with a sprinkling of Dylite spheres, you can make pucks coast with almost uniform speed in any direction. Set one puck motionless in the center of the table and push a second similar one toward it, a little off-center. You can make excellent pictures of the resulting two-dimensional glancing collision with a camera mounted directly above the surface.

To reduce reflection from the glass tray, the photograph should be taken using the xenon stroboscope with the light on one side and almost level with the glass tray. To make each puck's location clearly visible in the photograph, attach a steel ball or a small white Styrofoam hemisphere to its center.

A large puck has twice the mass of a small puck. You can get a greater variety of masses by stacking pucks one on top of the other and fastening them together with tape (but avoid having the collisions cushioned by the tape).

Two people are needed to do the experiment. One experimenter, after some preliminary practice shots, launches the projectile puck while the other experimenter operates the camera. The resulting picture should consist of a series of white dots in a rough "Y" pattern.

Using your picture, measure and record all the speeds before and after collision. Record the masses in each case too. Since you are interested only in comparing speeds, you can use any convenient speed units. You can simplify your work if you record speeds in mm/dot instead of trying to work them out in cm/sec. Because friction does slow the pucks down, find speeds as close to the impact as you can. You can also use the "puck" instead of the kilogram as your unit of mass.

Method II – Colliding Disk Magnets

Disk magnets will also slide freely on Dylite spheres as described in Method I.

The difference here is that the magnets need never touch during the "collision." Since the interaction forces are not really instantaneous as they are for the pucks, the magnets follow *curving* paths during the interaction. Consequently the "before" velocity should be determined as early as possible and the "after" velocities should be measured as late as possible.

Following the procedure described above for pucks, photograph one of these "collisions." Again, small Styrofoam hemispheres or steel balls attached to the magnets should show up in the strobe picture as a series of white dots. Be sure the paths you photograph are long enough so that the dots near the ends form straight lines rather than curves.

Using your photograph, measure and record the speeds and record the masses. Since the interaction forces are not really instantaneous as they are for the pucks, the magnets follow *curving* paths during the interaction. Consequently the "before" velocity should be determined as early as possible and the "after" velocities should be measured as late as possible. You can simplify your work if you record speeds in mm/dot instead of working them out in cm/sec. You can use the disk instead of the kilogram as your unit of mass.

Method III – Film Loops

Several *Film Loops* (*21*, *22*, *23*, *24*, and *25*) show two-dimensional collisions that you cannot conveniently reproduce in the laboratory. Notes on these films appear at the end of this *Handbook* chapter. Project one of the loops on the chalkboard or on a sheet of graph paper. Trace the paths of the moving objects and record their masses and measure their speeds. Then go on to the analysis described in the next section.

Method IV – Stroboscopic Photographs

In the Activities section of this chapter of your *Handbook* (pp. 156 to 158) you will find stroboscopic photographs of seven two-dimensional collision events. These were photographed during the making of *Film Loops 21* through 25. Before you take measurements, be sure to read the notes describing what the events were and how the photographs were made.

Analysis of Data

Whichever procedure you used, you should analyze your results in the following way. Multiply the mass of each object by its before-the-collision speed, and add the products.

Do the same thing for each of the objects in the system after the collision, and add the after-the-collision products together.

Q1 Does the sum before the collision equal the sum after the collision?

If the collision you observed was an explosion of a cluster of objects at rest, the total quantity mass-times-speed before the explosion will be zero. But surely, the mass-times-speed of each of the flying fragments after the explosion is more than zero! "Mass-times-speed" is obviously *not* conserved in an explosion. You probably found it wasn't conserved in the experiments with pucks and magnets, either. You may already have suspected that you ought to be taking into account the *directions* of motion.

To see what *is* conserved, proceed as follows.

Using your measurements construct a drawing like Fig. 9-4, in which you show the directions of motion of all the objects both before and after the collision.

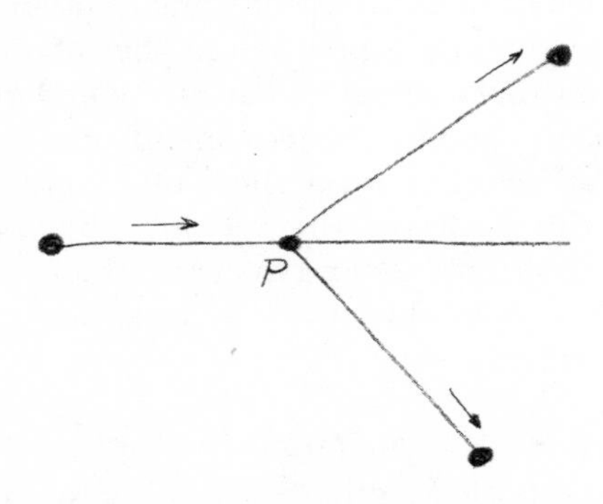

Fig. 9-4

Have all the direction lines *meet at a single point* in your diagram. The actual paths in your photographs will not do so, because the pucks and magnets are large objects instead of points, but you can still draw the *directions* of motion as lines through the single point P.

On this diagram draw a vector arrow whose magnitude (length) is proportional to the mass times the speed of the projectile *before* the collision. (You can use any convenient scale.) In Fig. 9-5, this vector is marked $m_A v_A$. Before an explosion there is no motion at all, and hence, no diagram to draw.

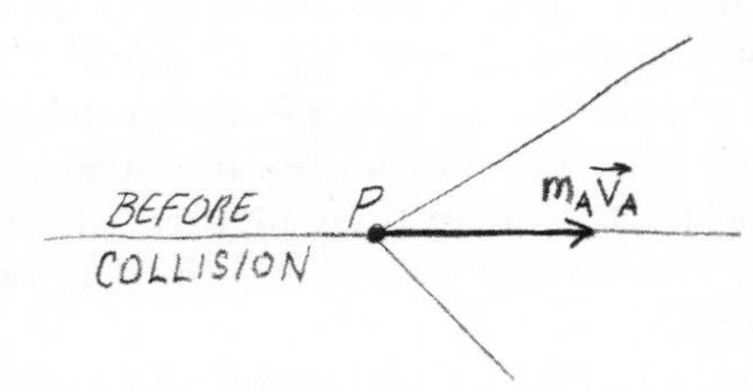

Fig. 9-5

Below your first diagram draw a second one in which you once more draw the directions of motion of all the objects exactly as before. On this second diagram construct the vectors for mass-times-speed for each of the objects leaving P *after* the collision. For the collisions of pucks and magnets your diagram will resemble Fig. 9-6. Now construct the "after the collision" vector sum.

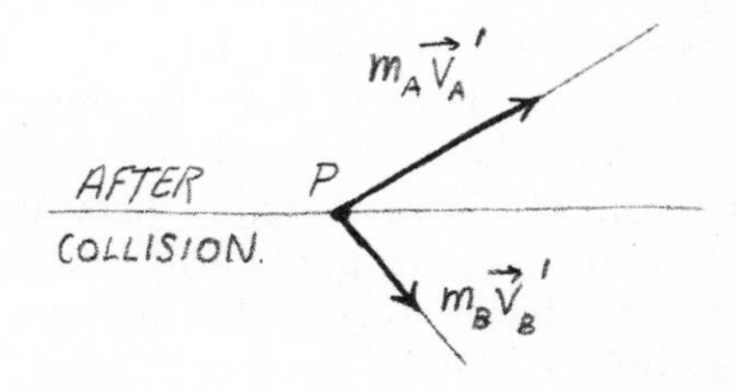

Fig. 9-6

The length of each of your arrows is given by the product of mass and speed. Since each arrow is drawn in the *direction* of the speed, the arrows represent the product of mass and velocity *mv* which is called *momentum*. The vector sums "before" and "after" collision therefore represent the total momentum of the system of objects before and after the collision. If the "before" and "after" arrows are equal, then the total momentum of the system of interacting objects is conserved.

Q2 How does this vector sum compare with the vector sum on your before-the-collision figure? Are they equal within the uncertainty?

Q3 Is the principle of conservation of momentum for one dimension different from that for two, or merely a special case of it? How can the principle of conservation of momentum be extended to three dimensions? Sketch at least one example.

Q4 Write an equation that would express the principle of conservation of momentum for collisions of (a) 3 objects in two dimensions, (b) 2 objects in three dimensions, (c) 3 objects in three dimensions.

A 3,000-pound steel ball swung by a crane against the walls of a condemned building. What happens to the momentum of the ball?

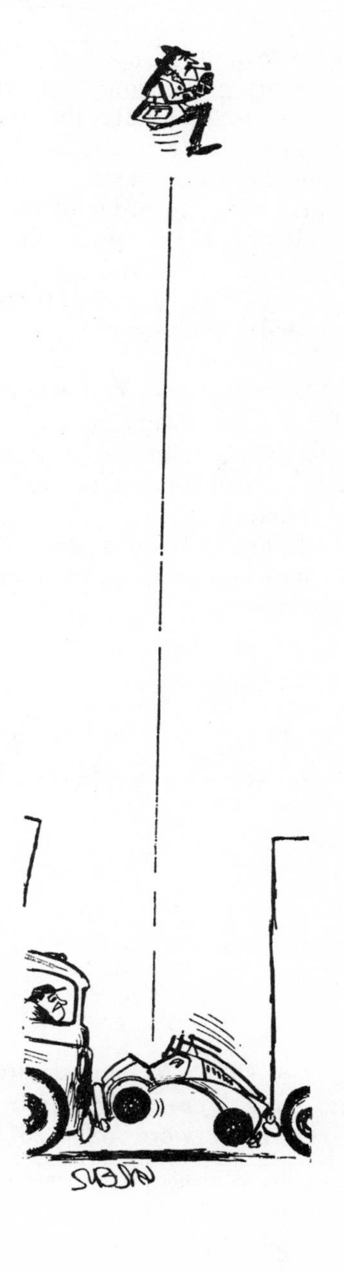

ACTIVITIES

STROBOSCOPIC PHOTOGRAPHS OF ONE-DIMENSIONAL COLLISIONS*

Stroboscopic photographs showing seven different examples of one-dimension collisions appear on the following pages. They are useful both in Chapter 9 of the *Text* for studying momentum and again in Chapter 10 of the *Text* for studying kinetic energy.

For each event you should find the speeds of the balls before and after collision. From the values for mass and speed of each ball, you should calculate the total momentum before and after collision. You will use the same values to calculate the total kinetic energy before and after collision.

You should read section I, before analyzing any of the events, in order to find out what measurements to make and how the collisions were produced. After you have made your measurements, turn to section II for questions to answer about each event.

I. The Measurements You Will Make

To make the necessary measurements you will need a metric ruler marked in millimeters, preferably of transparent plastic with sharp scale markings.

Consult Fig. 9-7 for suggestions on improving your measuring technique before starting your work.

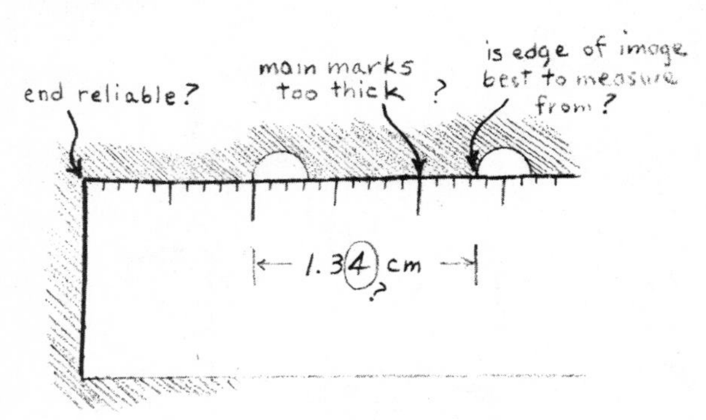

Fig. 9-7

Fig. 9-8 shows schematically that the colliding balls were hung from very long wires. The balls were released from rest,

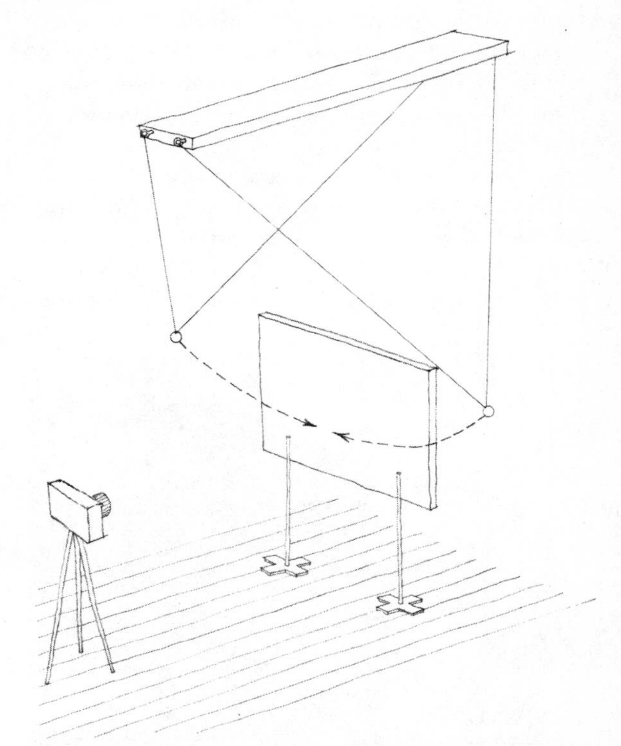

Fig. 9-8 Set-up for photographing one-dimensional collisions.

and their double-wire (bifilar) suspensions guided them to a squarely head-on collision. Stroboscopes illuminated the 3 × 4 ft rectangle that was the field of view of the camera. The stroboscopes are not shown in Fig. 9-8.

Notice the two rods whose tops reach into the field of view. These rods were 1 meter (± 2 millimeters) apart, measured from top center of one rod to top center of the other. The tops of these rods are visible in the photograph on which you will make your measurements. This enables you to convert your measurements to actual distances if you wish. However, it is easier to use the lengths in millimeters measured directly off the photograph if you are merely going to compare momenta.

The balls speed up as they move into the field of view. Likewise, as they leave the field of view, they slow down. Therefore successive displacements on the stroboscopic photograph, each of which took exactly the same time, will

*Reproduced by permission of National Film Board of Canada

not necessarily be equal in length. Check this with your ruler.

As you measure a photograph, number the position of each ball at successive flashes of the stroboscope. Note the interval during which the collision occurred. Identify the clearest time interval for finding the velocity of each ball (a) before the collision and (b) after the collision. Then mark this information close on each side of the interval.

II. Questions to be Answered about Each Event

After you have recorded the masses (or relative masses) given for each ball and have recorded the necessary measurements of velocities, answer the following questions.

1. What is the total momentum of the system of two balls before the collision? Keep in mind here that velocity, and therefore momentum, are vector quantities.
2. What is the total momentum of the system of two balls after the collision?
3. Was momentum conserved within the limits of precision of your measurements?

Event 1

The photographs of this Event 1 and all the following events appear at the end of this activity as Figs. 9-16 to 9-22. This event is also shown as the first example in *Film Loop 18,* "One-Dimensional Collisions I."

Figure 9-9 shows that ball B was initially at rest. After the collision both balls moved off to the left. The balls are made of steel.

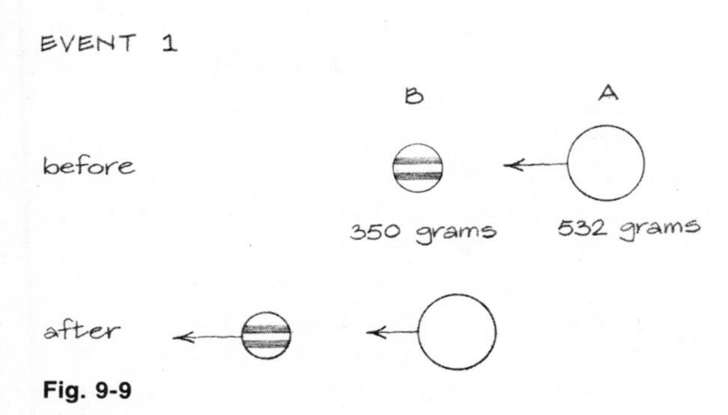

Fig. 9-9

Event 2

This event, the reverse of Event 1, is shown as the second example in *Film Loop 18,* "One-Dimensional Collisions I."

Fig. 9-10 shows that ball B came in from the left and that ball A was initially at rest. The collision reversed the direction of motion of ball B and sent ball A off to the right. (The balls are of hardened steel.)

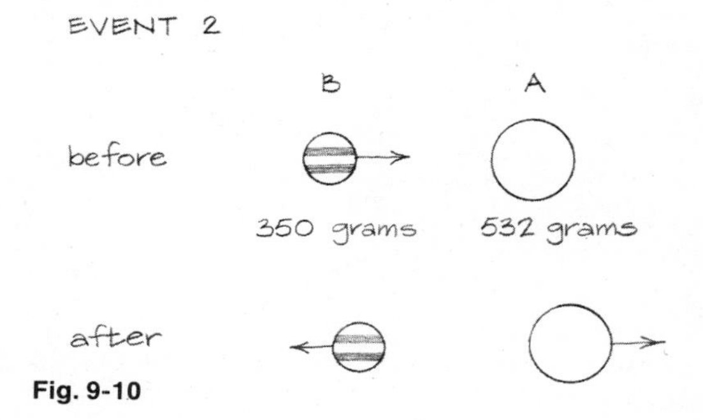

Fig. 9-10

As you can tell by inspection, ball B moved slowly after collision, and thus you may have trouble getting a precise value for its speed. This means that your value for this speed is the least reliable of your four speed measurements. Nevertheless, this fact has only a small influence on the reliability of your value for the total momentum after collision. Can you explain why this should be so?

Why was the direction of motion of ball B reversed by the collision?

If you have already studied Event 1, you will notice that the same balls were used in Events 1 and 2. Check your velocity data, and you will find that the *initial* speeds were roughly equal. Thus, Event 2 was truly the reverse of Event 1. Why, then, was the direction of motion of ball A in Event 1 not reversed although the direction of ball B in Event 2 was reversed?

Event 3

This event is shown as the first example in *Film Loop 19,* "One-Dimensional Collisions II." Event 3 is not recommended unless you also study one of the other events. Event 3 is especially recommended as a companion to Event 4.

Fig. 9-11 shows that a massive ball (A) entered from the left. A less massive ball B came in from the right. The directions of motion of both balls were reversed by the collision. (The balls were made of hardened steel.)

EVENT 3

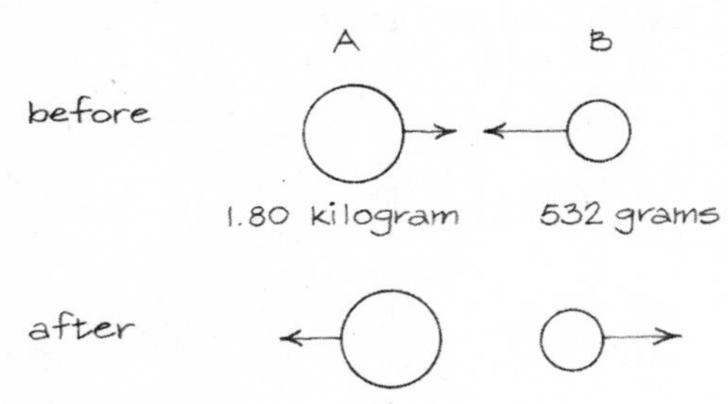

Fig. 9-11

When you compare the momenta before and after the collision you will probably find that they differed by more than any other event so far in this series. Explain why this is so.

Event 4

This event is also shown as the second example in *Film Loop 19,* "One-Dimensional Collisions II."

Fig. 9-12 shows that two balls came in from the left, that ball A was far more massive than ball B, and that ball A was moving faster than ball B before collision. The collision occurred when A caught up with B, increasing B's speed at some expense to its own speed. (The balls were made of hardened steel.)

Each ball moved across the camera's field from left to right on the same line. In order to be able to tell successive positions apart on a stroboscopic photograph, the picture was taken twice. The first photograph shows only the progress of the large ball A because ball B had been given a thin coat of black paint (of negligible mass). Ball A was painted black when the second picture was taken. It will help you to analyze the collision if you actually number white-ball positions at successive stroboscope flashes in each picture.

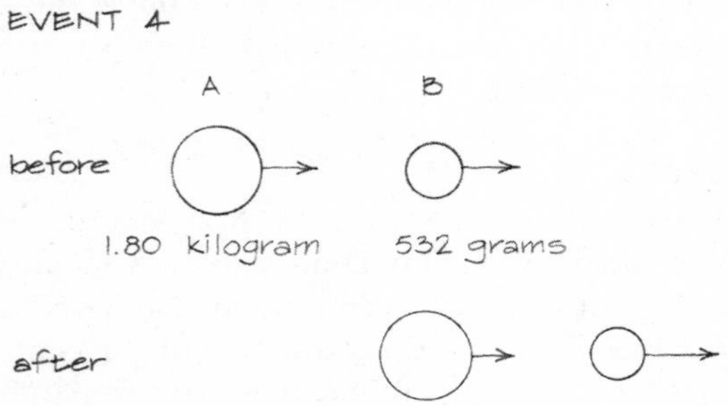

EVENT 4

A B

before

1.80 kilogram 532 grams

after

Fig. 9-12

Event 5

This event is also shown as the first example in *Film Loop 20,* "Inelastic One-Dimensional Collisions." You should find it interesting to analyze this event or Event 6 or Event 7, but it is not necessary to do more than one.

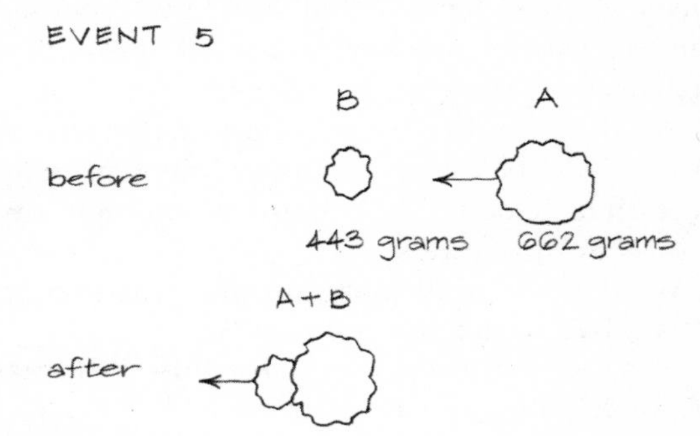

Fig. 9-13

Fig. 9-13 shows that ball A came in from the right, striking ball B which was initially at rest. The balls were made of a soft material (plasticene). They remained lodged together after the collision and moved off to the left as one. A collision of this type is called *"perfectly inelastic."*

Event 6

This event is shown as the second example in *Film Loop 20,* "Inelastic One-Dimensional Collisions."

Fig. 9-14 shows that balls A and B moved in from the right and left, respectively, before collision. The balls were made of a soft material (plasticene). They remained lodged together after the collision and moved off together to the left. This is another "perfectly inelastic" collision, like that in Event 6.

This event was photographed in two parts. The first print shows the conditions before collision, the second print, after collision. Had the picture been taken with the camera shutter

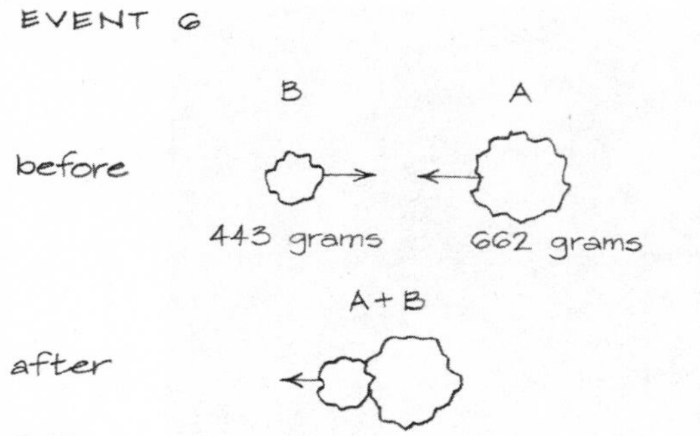

EVENT 6

B A

before

443 grams 662 grams

A+B

after

Fig. 9-14

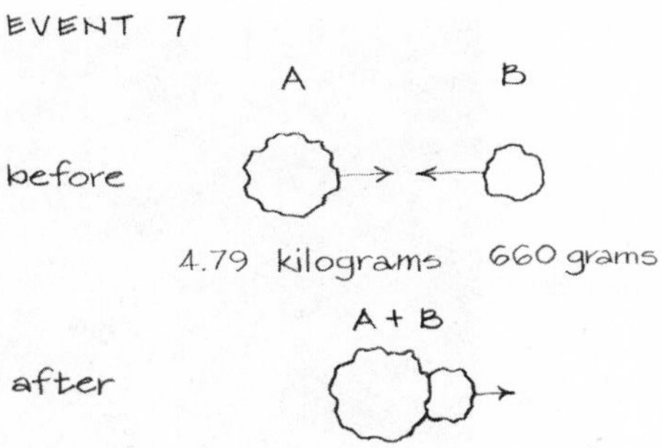

EVENT 7

A B

before

4.79 kilograms 660 grams

A + B

after

Fig. 9-15

open throughout the motion, it would be difficult to take measurements because the combined balls (A + B)—after collision—retraced the path which ball B followed before collision. You can number the positions of each ball before collision at successive flashes of the stroboscope (in the first photo); and you can do likewise for the combined balls (A + B) after the collision in the second photo.

Event 7

Fig. 9-15 shows that balls A and B moved in from opposite directions before collision. The balls are made of a soft material (plasticene). They remain lodged together after collision and move off together to the right. This is another so-called *"perfectly inelastic"* collision.

This event was photographed in two parts. The first print shows the conditions before collision, the second print, after collision. Had the picture been made with the camera shutter open throughout the motion, it would be difficult to take measurements because the combined balls (A + B) trace out the same path as incoming ball B. You can number the positions of each ball before collision at successive flashes of the stroboscope (in the first photograph), and you can do likewise for the combined balls (A + B) after collision in the second photograph.

Photographs of the Events

The photographs of the events are shown in Fig. 9-16 through 9-22.

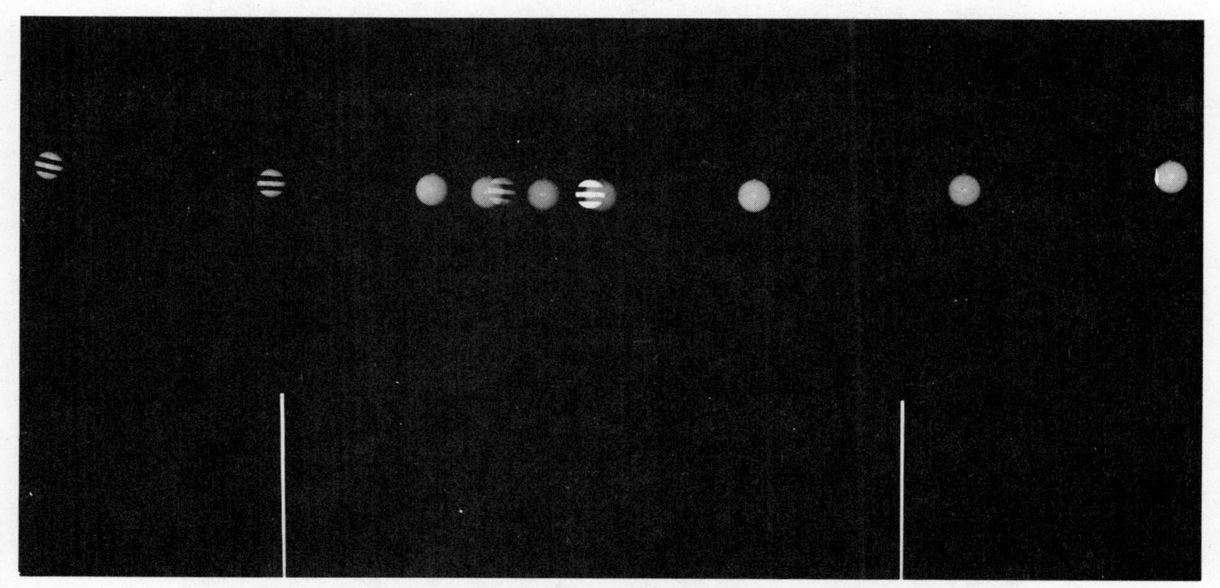

Fig. 9-16 Event 1, 10 flashes/sec

before

after

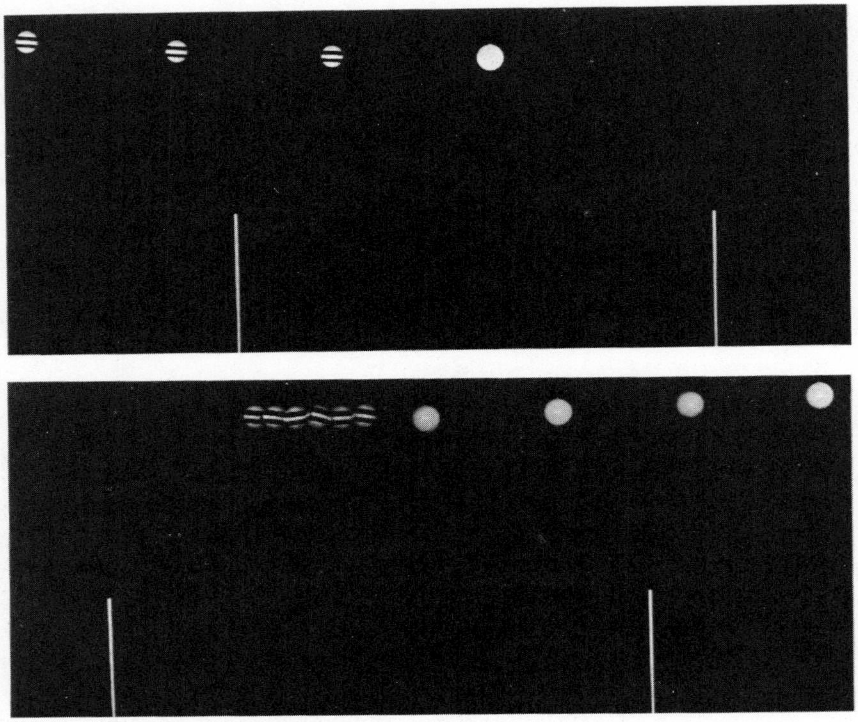

Fig. 9-17 Event 2, 10 flashes/sec

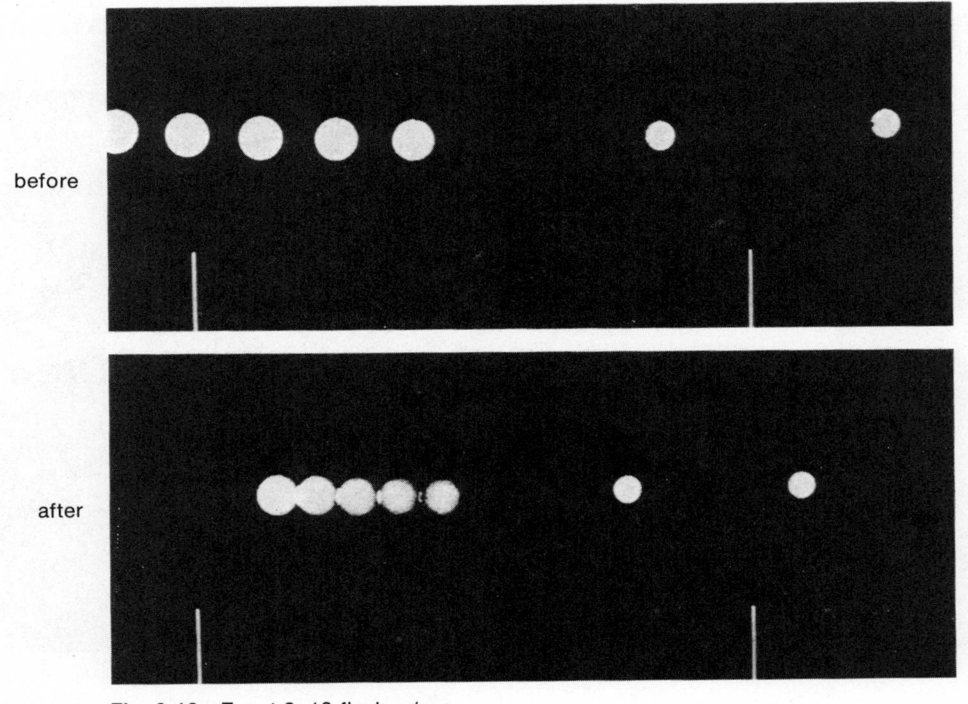

Fig. 9-18 Event 3, 10 flashes/sec

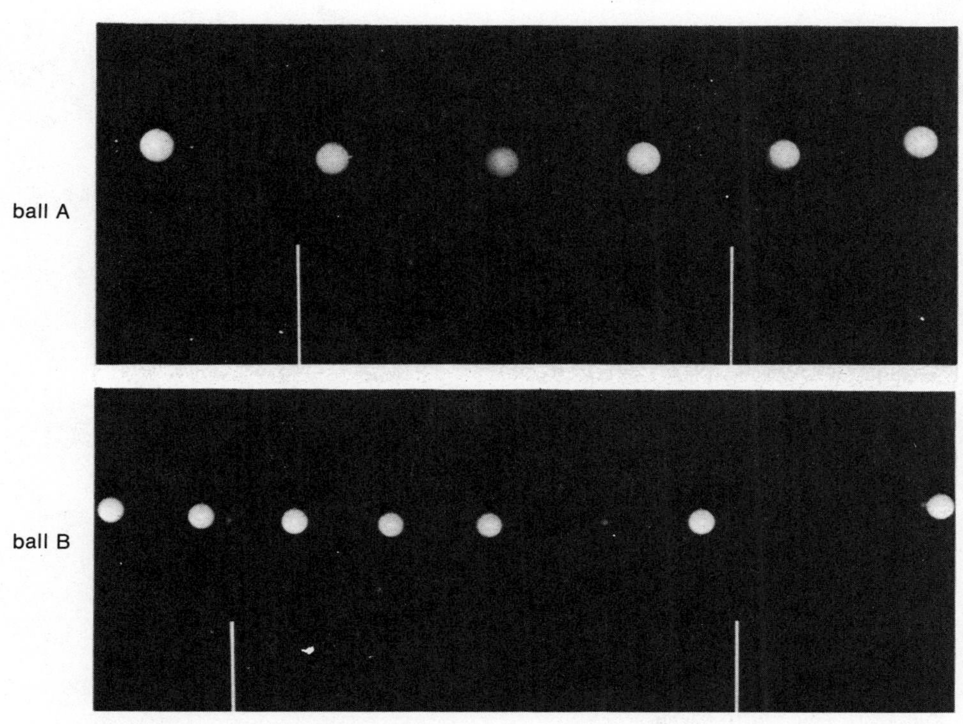

Fig. 9-19 Event 4, 10 flashes/sec

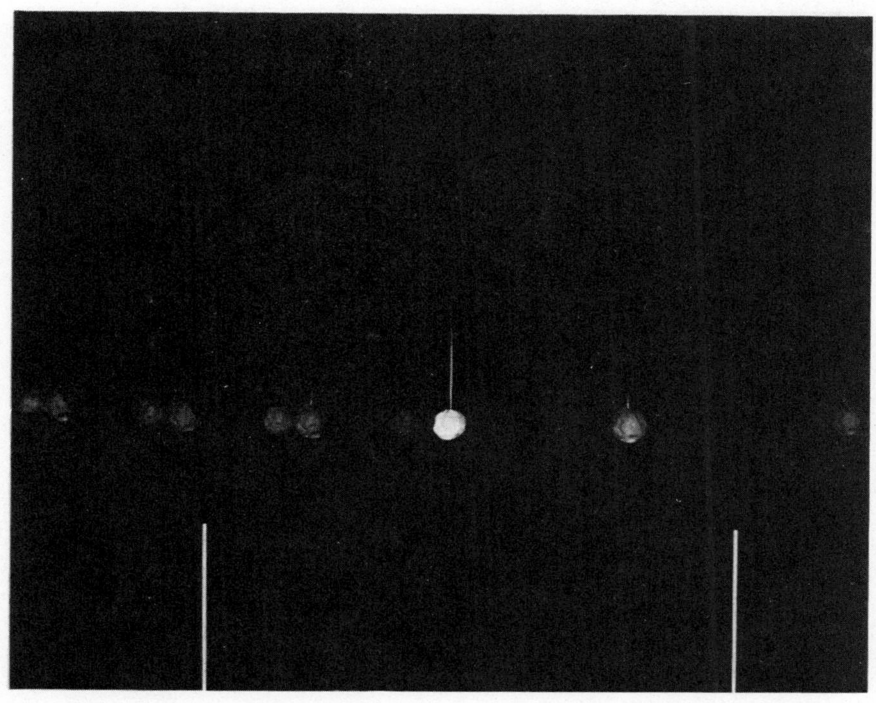

Fig. 9-20 Event 5, 10 flashes/sec

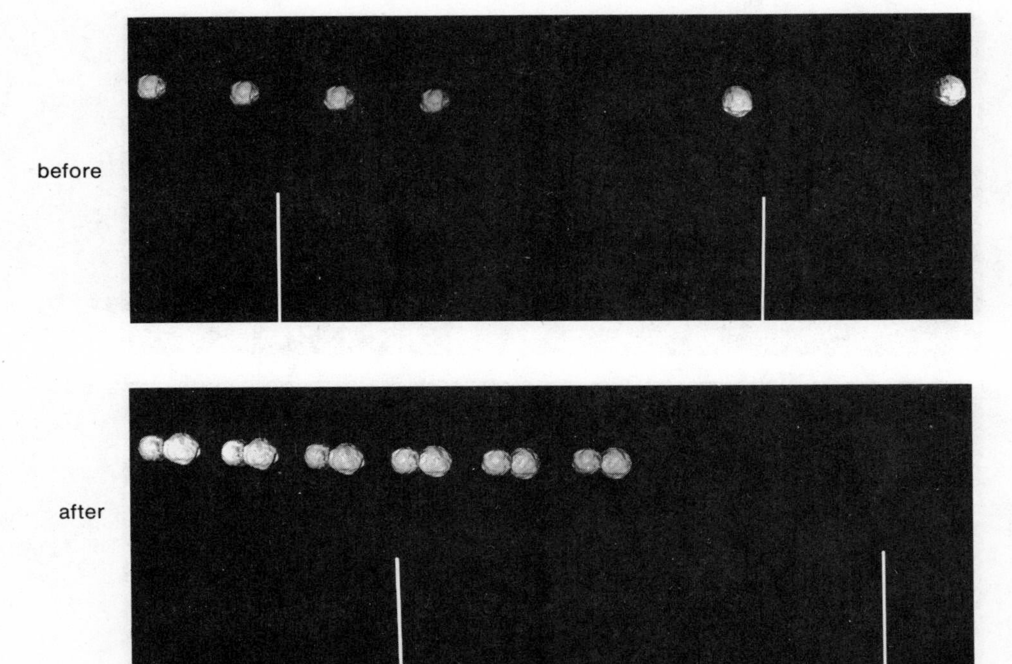

Fig. 9-21 Event 6, 10 flashes/sec

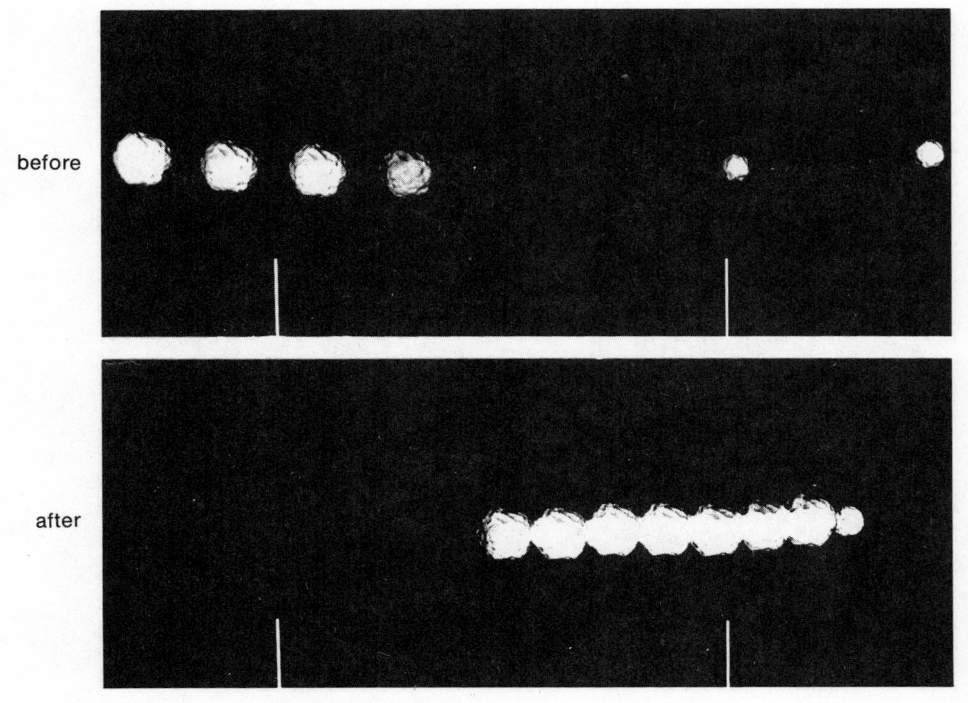

Fig. 9-22 Event 7, 10 flashes/sec

STROBOSCOPIC PHOTOGRAPHS OF TWO-DIMENSIONAL COLLISIONS*

Stroboscopic photographs of seven different two-dimensional collisions in a plane are used in this activity. The photographs (Figs. 9-27 to 9-34) are shown on the pages immediately following the descriptions of these events.

I. Material Needed

1. A transparent plastic ruler, marked in millimeters.
2. A large sheet of paper for making vector diagrams. Graph paper is especially convenient.
3. A protractor and two large drawing triangles are useful for transferring direction vectors from the photographs to the vector diagrams.

II. How the Collisions were Produced

Balls were hung on 10-meter wires, as shown schematically in Fig. 9-23. They were released so as to collide directly above the camera, which was facing upward. Electronic strobe lights (shown in Fig. 9-26) illuminated the rectangle shown in each picture.

Two white bars are visible at the bottom of each photograph. These are rods that had their tips 1 meter (± 2 millimeters) apart in the actual situation. The rods make it possible for you to convert your measurements to the actual distance. It is not necessary to do so, if you choose instead to use actual on-the-photograph distances in millimeters as you may have done in your study of one-dimensional collisions.

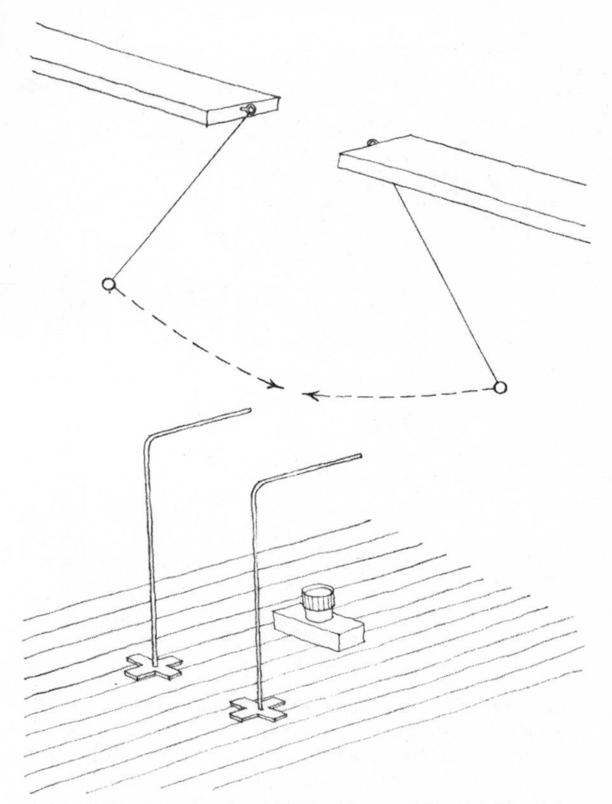

Fig. 9-23 Set-up for photographing two-dimensional collisions.

Since the balls are pendulum bobs, they move faster near the center of the photographs than near the edge. Your measurements, therefore, should be made near the center.

III. A Sample Procedure

The purpose of your study is to see to what extent momentum seems to be conserved in two-dimensional collisions. For this purpose you need to construct vector diagrams.

*Reproduced by permission of National Film Board of Canada

Consider an example: in Fig. 9-24, a 450 g and a 500 g ball are moving toward each other. Ball A has a momentum of 2.4 kg-m/sec, in the direction of the ball's motion. Using the scale shown, you draw a vector 2.4 units long, parallel to the direction of motion of A. Similarly, for ball B you draw a momentum vector of 1.8 units long, parallel to the direction of motion of B

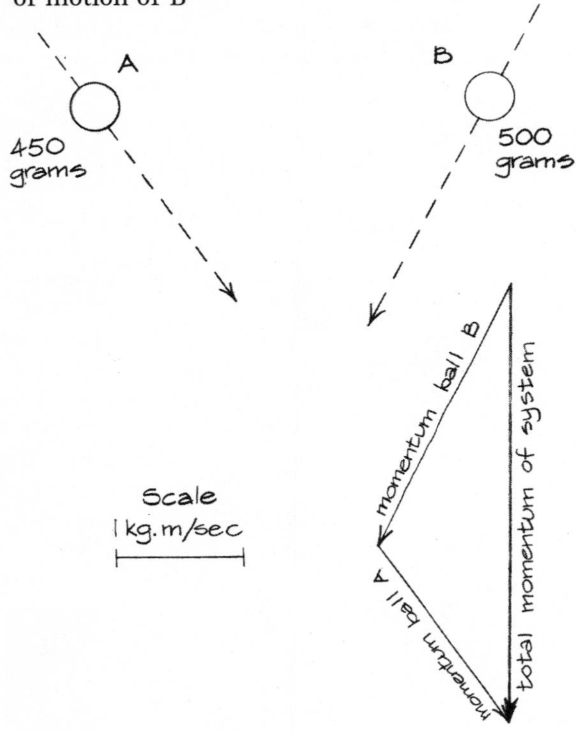

Fig. 9-24 Two balls moving in a plane. Their individual momenta, which are vectors, are added together vectorially in the diagram on the lower right. The vector sum is the total momentum of the system of two balls. (Your own vector drawings should be at least twice this size.)

The system of two balls has a total momentum before the collision equal to the vector sum of the two momentum vectors for A and B.

The total momentum after the collision is also found the same way, by adding the momentum vector for A after the collision to that for B after the collision (see Fig. 9-25).

This same procedure is used for any event you analyze. Determine the momentum (mag-

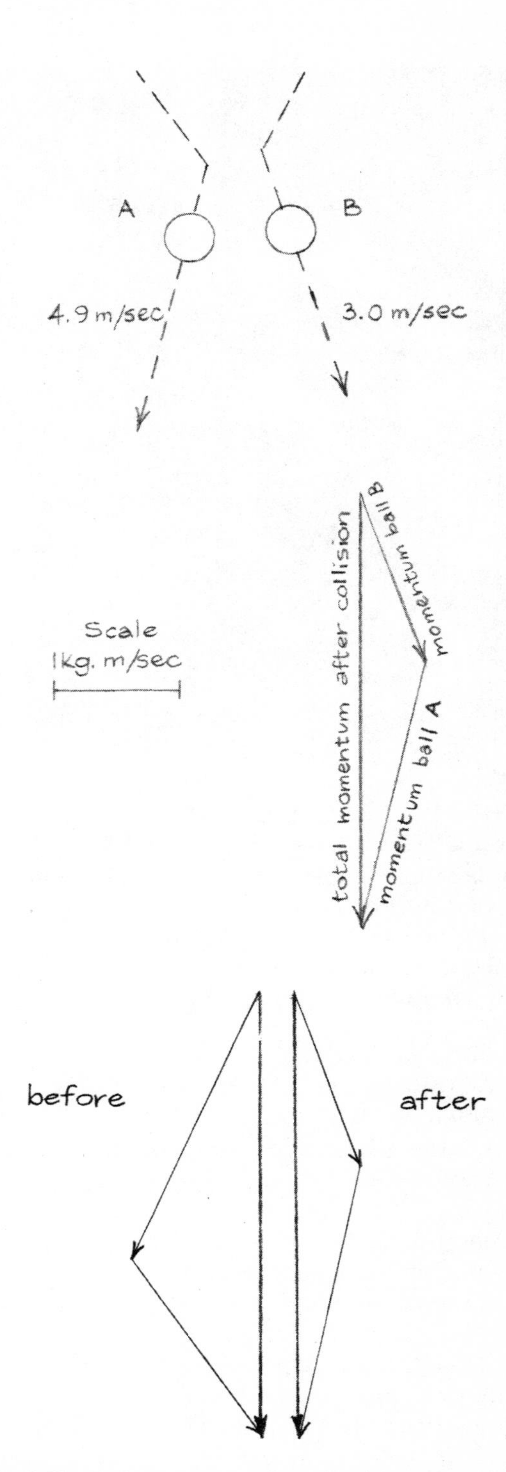

Fig. 9-25 The two balls collide and move away. Their individual momenta after collision are added vectorially. The resultant vector is the total momentum of the system after collision.

nitude *and* direction) for each object in the system before the collision, graphically add them, and then do the same thing for each object after the collision.

For each event that you analyze, consider whether momentum is conserved.

Events 8, 9, 10, and 11

Event 8 is also shown as the first example in *Film Loop 22,* "Two-Dimensional Collisions: Part II," as well as on Project Physics *Transparency T-20.*

Event 10 is also shown as the second example in *Film Loop* 22.

Event 11 is also shown in *Film Loop 21,* "Two-Dimensional Collisions: Part I," and on Project Physics *Transparency T-21.*

These are all elastic collisions. Events 8 and 10, are simplest to analyze because each shows a collision of equal masses. In Events 8 and 9, one ball is initially at rest.

A small sketch next to each photograph indicates the direction of motion of each ball. The mass of each ball and the flash rate are also given.

Events 12 and 13

Event 12 is also shown as the first example in *Film Loop 23,* "Inelastic Two-Dimensional Collisions."

Event 13 is also shown as the second example in *Film Loop* 23. A similar event is shown and analysed in Project Physics *Transparency T-22.*

Since Events 12 and 13 are similar, there is no need to do both.

Events 12 and 13 show inelastic collisions between two plasticene balls that stick together and move off as one compound object after the collision. In 13 the masses are equal; in 12 they are not.

Caution: You may find that the two objects rotate slightly about a common center after the collision. For each image after the collision, you should make marks halfway between the centers of the two objects. Then determine the velocity of this "center of mass," and multiply it by the combined mass to get the total momentum after the collision.

Event 14

Do *not* try to analyze Event 14 unless you have done at least one of the simpler events 8 through 13.

Event 14 is also shown on *Film Loop 24,* "Scattering of a Cluster of Objects."

Figure 9-26 shows the setup used in photographing the scattering of a cluster of balls. The photographer and camera are on the floor, and four electronic stroboscope lights are on tripods in the lower center of the picture.

You are to use the same graphical methods as you used for Events 8 through 13 to see if the conservation of momentum holds for more than two objects. Event 14 is much more complex because you must add seven vectors, rather than two, to get the total momentum after the collision.

In Event 14, one ball comes in and strikes a cluster of six balls of various masses. The balls were initially at rest. Two photographs are included: Print 1 shows only the motion of ball A before the event. Print 2 shows the positions of all seven balls just before the collision and the motion of each of the seven balls after the collision.

Fig. 9-26 Catching the seven scattered balls to avoid tangling in the wires from which they hang. The photographer and the camera are on the floor. The four stroboscopes are seen on tripods in the lower center of the picture.

You can analyze this event in two different ways. One way is to determine the initial momentum of ball A from measurements taken on Print 1, and then compare it to the total final momentum of the system of seven balls from measurements taken on Print 2. The second method is to determine the total final momentum of the system of seven balls on Print 2, predict the momentum of ball A, and then take measurements of Print 1 to see whether ball A had the predicted momentum. Choose one method.

The tops of prints 1 and 2 lie in identical positions. To relate measurements on one print to measurements on the other, measure a ball's distance relative to the top of one picture with a rule; the ball would lie in precisely the same position in the other picture if the two pictures could be superimposed.

There are two other matters you must consider. First, the time scales are different on the two prints. Print 1 was taken at a rate of 5 flashes/second, and Print 2 was taken at a rate of 20 flashes/second. Second, the distance scale may not be exactly the same for both prints. Remember that the distance from the center of the tip of one of the white bars to center of the tip of the other is 1 meter (± 2 mm) in real space. Check this scale carefully on both prints to determine the conversion factor.

The stroboscopic photographs for Events 8 to 14 appear in Figs. 9-27 to 9-34.

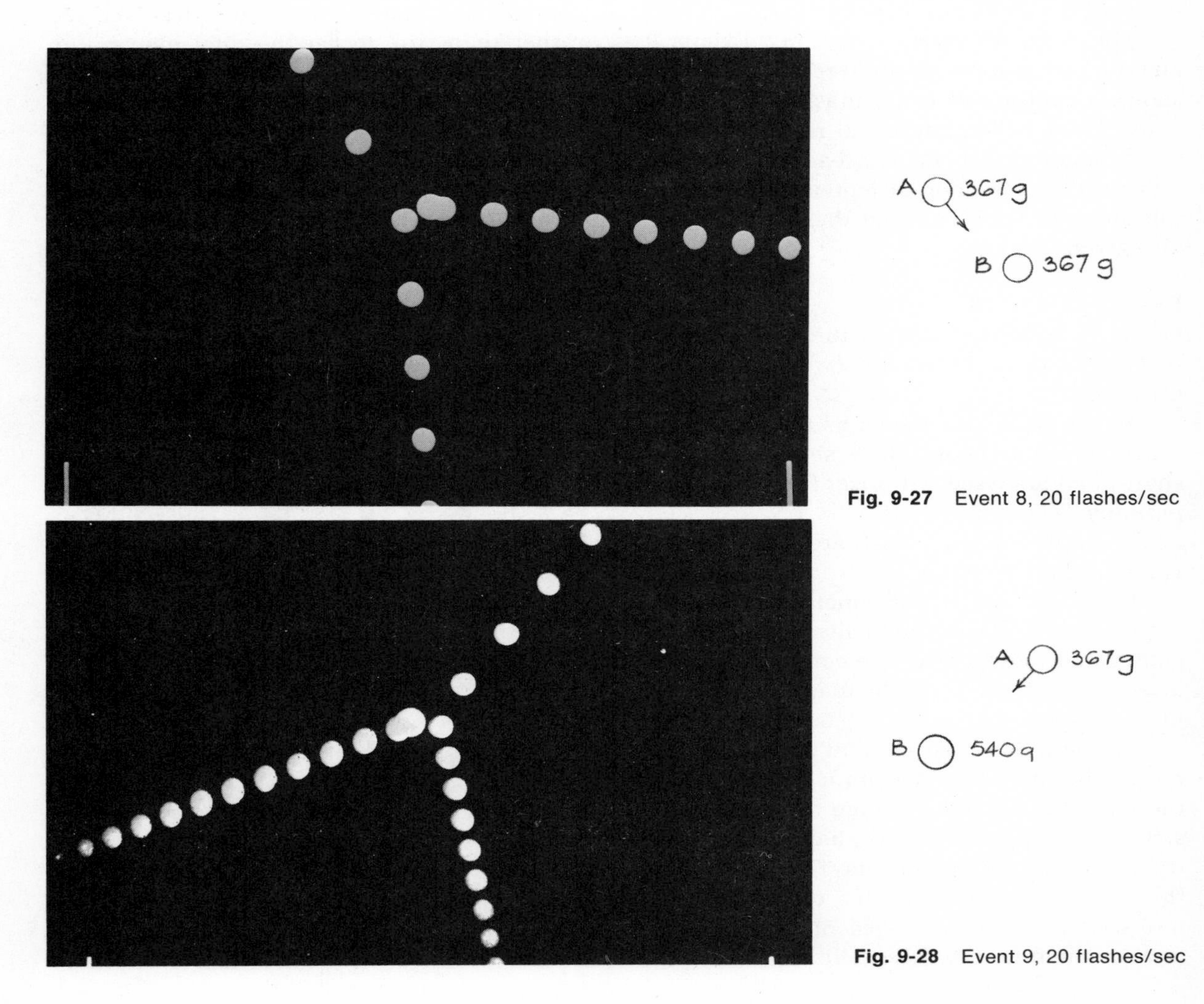

Fig. 9-27 Event 8, 20 flashes/sec

Fig. 9-28 Event 9, 20 flashes/sec

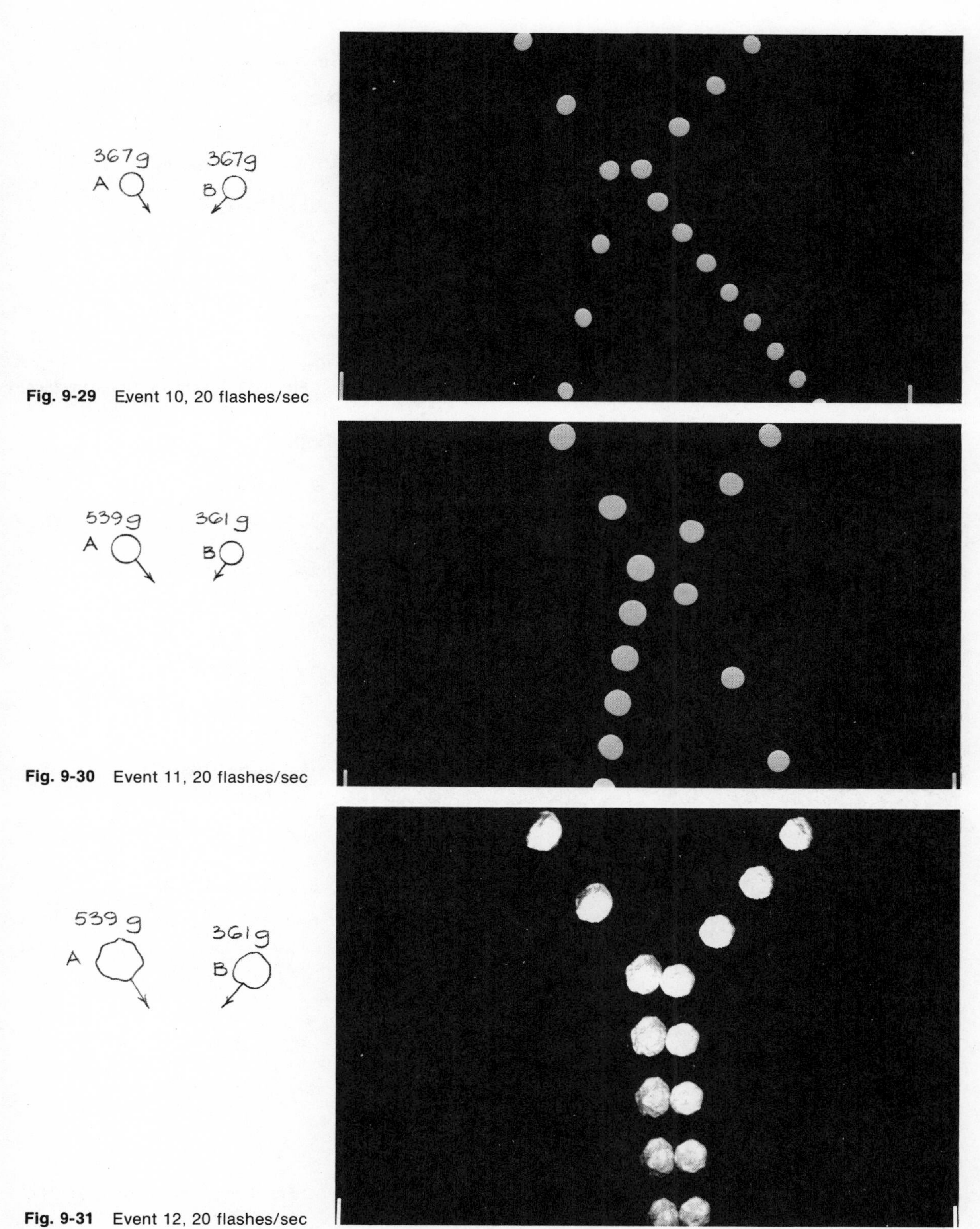

Fig. 9-29 Event 10, 20 flashes/sec

Fig. 9-30 Event 11, 20 flashes/sec

Fig. 9-31 Event 12, 20 flashes/sec

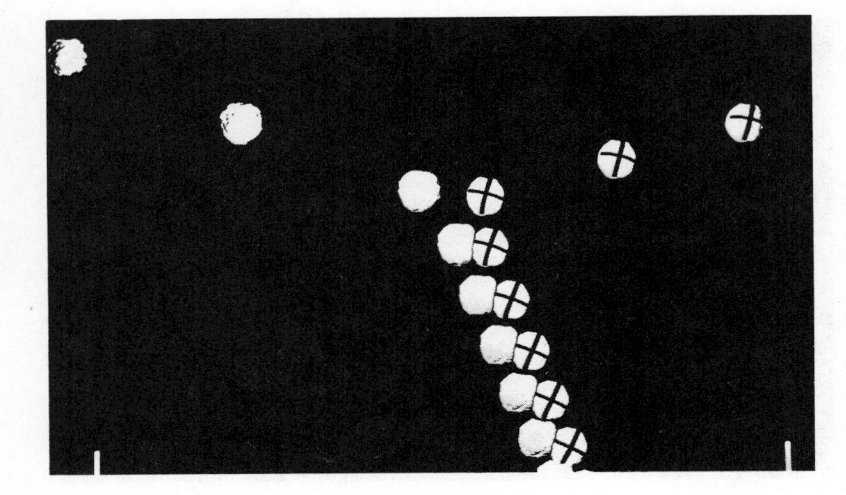

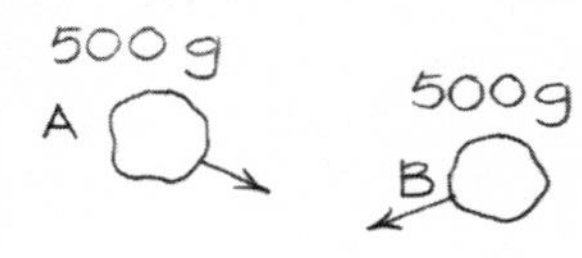

Fig. 9-32 Event 13, 10 flashes/sec

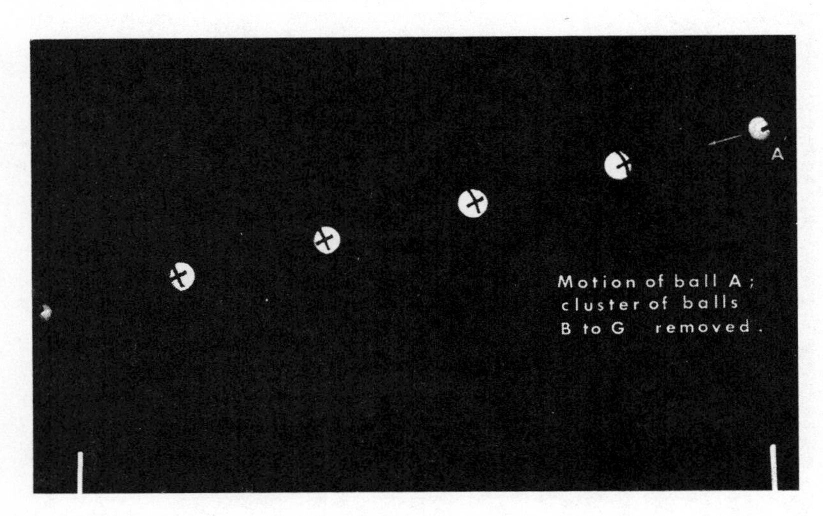

Fig. 9-33 Event 14, print 1, 20 flashes/sec

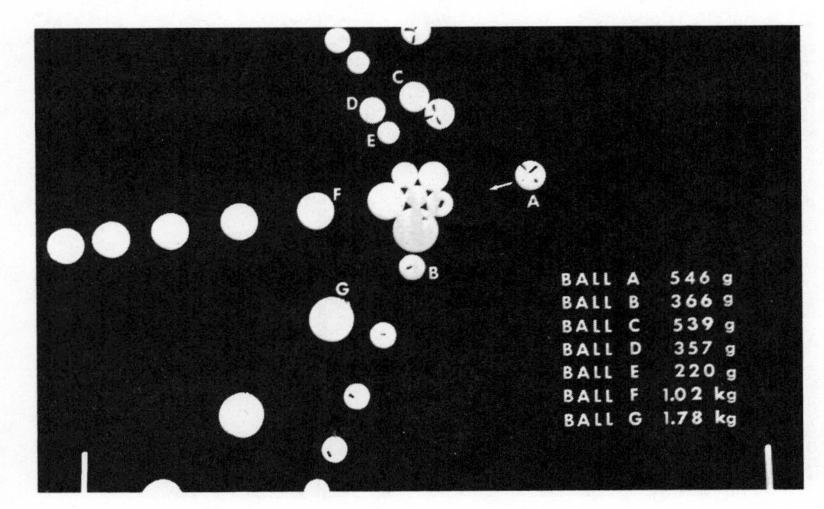

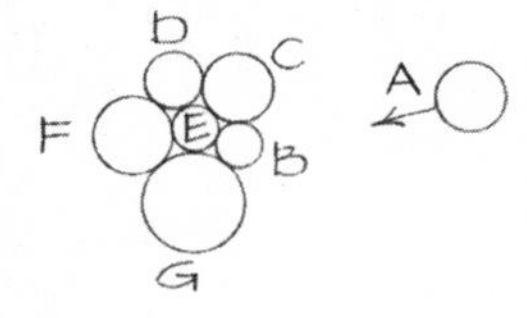

Fig. 9-34 Event 14, print 2, 5 flashes/sec

UNUSUAL CASE OF ELASTIC IMPACT

There is an intriguing case of elastic impact of suspended balls, where one of the balls has three times the mass of the other and both balls are started from rest at equal distances on opposite sides of the center. Can you predict the result? If not, a good activity is to make two pendula with bobs of steel ball bearings, glass marbles, or wooden balls. Each pair must have the mass ratio of three to one. A bifilar (or two-wire) suspension will help to keep the bobs swinging in the same plane.

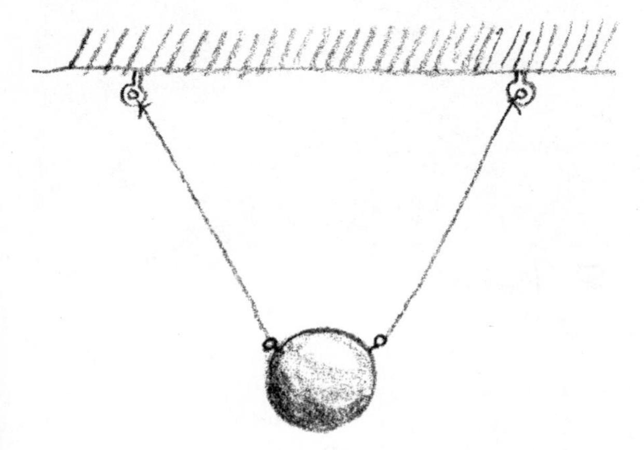

Further suggestions for construction may be found in apparatus catalogs under the headings of momentum, collision, or impact. A complete story will be found in an article, "An Almost Forgotten Case of Elastic Impact," by Harvey B. Lemon, in the *American Journal of Physics,* vol. 3, p. 35 (1935).

IS MASS CONSERVED?

In Sec. 9.1 of the *Text* you read about some of the difficulties in establishing the law of conservation of mass. You can do several different experiments to check this law.

Alka-Seltzer.

You will need the following equipment: Alka-Seltzer tablets; 2-liter flask, or plastic one-gallon jug (such as is used for bleach, distilled water, or duplicating fluid); stopper for flask or jug; warm water; balance (sensitivity better than 0.1 g) spring scale (sensitivity better than 0.5 g).

Balance a tablet and 2-liter flask containing 200-300 cc of water on a sensitive balance. Drop the tablet in the flask. When the tablet disappears and no more bubbles appear, re-adjust the balance. Record any change in mass. If there is a change, what caused it?

Repeat the procedure above, but include the rubber stopper in the initial balancing. Immediately after dropping in the tablet, place the stopper tightly in the mouth of the flask. (The pressure in a 2-liter flask goes up by no more than 20 per cent, so it is not necessary to tape or wire the stopper to the flask. Do not use smaller flasks in which proportionately higher pressure would be built up.) Is there a change in mass? Remove the stopper after all reaction has ceased; what happens? Discuss the difference between the two procedures.

Brightly Colored Precipitate.

You will need: 20 g lead nitrate; 11 g potassium iodide; Erlenmeyer flask, 1000 cc with stopper; test tube, 25 × 150 mm; balance.

Place 400 cc of water in the Erlenmeyer flask, add the lead nitrate, and stir until dissolved. Place the potassium iodide in the test tube, add 30 cc of water, and shake until dissolved. Place the test tube, open and upward, carefully inside the flask and seal the flask with the stopper. Place the flask on the balance and bring the balance to equilibrium. Tip the flask to mix the solutions. Replace the flask on the balance. Does the total mass remain conserved? What *does* change in this experiment?

Magnesium Flash Bulb

On the most sensitive balance you have available, measure the mass of an unflashed magnesium flash bulb. Repeat the measurement several times to make an estimate of the precision of the measurement.

Flash the bulb by connecting it to a battery. Be careful to touch the bulb as little as possible, so as not to wear away any material or leave any fingerprints. Measure the mass of the bulb several times, as before. You can get a feeling for how small a mass change your balance could have detected by seeing how large a piece of tissue paper you have to put on the balance to produce a detectable difference.

EXCHANGE OF MOMENTUM DEVICES

The four situations described below are more complex tests for conservation of momentum, to give you a deeper understanding of the generality of the conservation law and of the importance of your frame of reference. (a) Fasten a section of HO gauge model railroad track to two ring stands as shown in Fig. 9-35. Set one truck of wheels, removed from a car, on the track and from it suspend an object with mass roughly equal to that of the truck. Hold the truck, pull the object to one side, parallel to the track, and release both at the same instant. What happens?

Predict what you think would happen if you released the truck an instant after releasing the object. Try it.

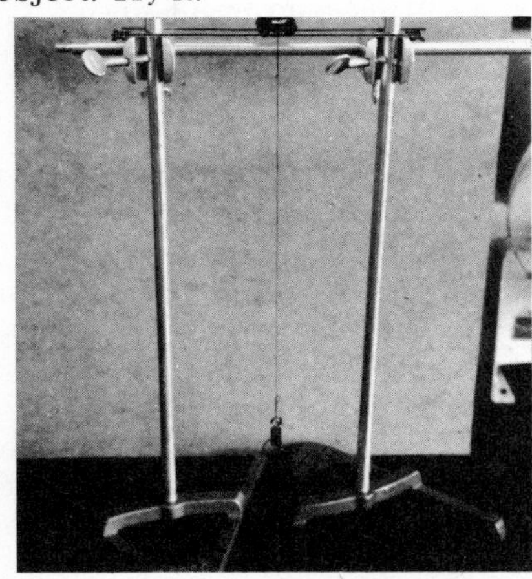

Fig. 9-35

Try increasing the suspended mass. (b) Fig. 9-36 shows a similar situation, using an air track supported on ring stands. An object of 20 g mass was suspended by a 50 cm string from one of the small air-track gliders. (One student trial continued for 166 swings.)

Fig. 9-36

(c) Fasten two dynamics carts together with four hacksaw blades as shown in Fig. 9-37. Push the top one to the right, the bottom to the left, and release them. Try giving the bottom cart a push across the room at the same instant you release them.

What would happen when you released the two if there were 10 or 20 bearing balls or small wooden balls hung as pendula from the top cart? You might find this model useful to

Fig. 9-37

you in Chapter 11. (d) Push two large rubber stoppers onto a short piece of glass tubing or wood (Fig. 9-38). Let the "dumbbell" roll down a wooden wedge so that the stoppers do not touch the table until the dumbbell is almost to the bottom. When the dumbbell touches the table, it suddenly increases its linear momentum as it moves off along the table. Principles of rotational momentum and energy are involved here that are not covered in the *Text*, but even without extending the *Text*, you can deal with the "mysterious" increase in linear momentum when the stoppers touch the table.

Using what you have learned about conservation of momentum, what do you think could account for this increase? (Hint: set the wedge on a piece of cardboard supported on plastic beads and try it.)

Fig. 9-38

FILM LOOPS

FILM LOOP 18
ONE-DIMENSIONAL COLLISIONS I

Two different head-on collisions of a pair of steel balls are shown. The balls hang from long, thin wires that confine each ball's motion to the same circular arc. The radius is large compared with the part of the arc, so the curvature is hardly noticeable. Since the collisions take place along a straight line, they can be called one-dimensional.

In the first example, ball B, weighing 350 grams, is initially at rest. In the second example, ball A, with a mass of 532 grams, is the one at rest.

With this film, you can make detailed measurements on the total momentum and energy of the balls before and after collision. Momentum is a vector, but in this one-dimensional case you need only worry about its sign. Since momentum is the product of mass and velocity, its sign is determined by the sign of the velocity.

You know the masses of the balls. Velocities can be measured by finding the distance traveled in a known time.

After viewing the film, you can decide on what strategy to use for distance and time measurements. One possibility would be to time the motion through a given distance with a stopwatch, perhaps making two lines on the paper. You need the velocity just before and after the collision. Since the balls are hanging from wires, their velocity is not constant. On the other hand, using a small arc increases the chances of distance-time uncertainties. As with most measuring situations, a number of conflicting factors must be considered.

You will find it useful to mark the crosses on the paper on which you are projecting, since this will allow you to correct for projector movement and film jitter. You might want to give some thought to measuring distances. You may use a ruler with marks in millimeters, so you can estimate to a tenth of a millimeter. Is it wise to try to use the zero end of the ruler, or should you use positions in the middle? Should you use the thicker or the thinner marks on the ruler? Should you rely on one measurement, or should you make a number of measurements and average them?

Estimate the uncertainty in distance and time measurements, and the uncertainty in velocity. What can you learn from this about the uncertainty in momentum?

When you compute the total momentum

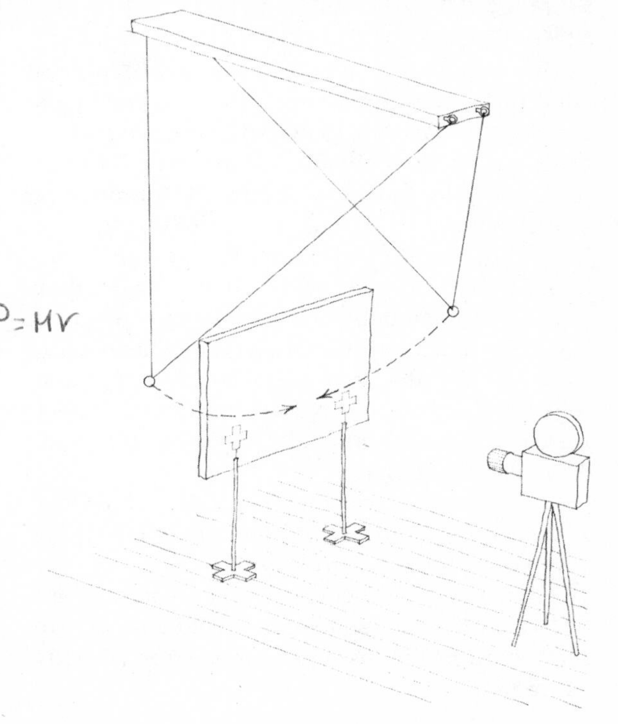

before and after collision (the sum of the momentum of each ball), remember that you must consider the direction of the momentum.

Are the differences between the momentum before and after collision significant, or are they within the experimental error already estimated?

Save the data you collect so that later you can make similar calculations on total kinetic energy for both balls just before and just after collision.

FILM LOOP 19 ONE-DIMENSION COLLISIONS II

Two different head-on collisions of a pair of steel balls are shown, with the same setup as that used in *Film Loop 18,* "One-dimensional collisions I."

In the first example, ball A with a mass of 1.8 kilograms collides head on with ball B, with a mass of 532 grams. In the second example, ball A catches up with ball B. The instructions for *Film Loop 18,* "One-Dimension Collisions I" may be followed for completing this investigation also.

FILM LOOP 20 INELASTIC ONE-DIMENSIONAL COLLISIONS

In this film, two steel balls covered with plasticene hang from long supports. Two collisions are shown. The two balls stick together after colliding, so the collision is "inelastic." In the first example, ball A, weighing 443 grams, is at rest when ball B, with a mass of 662 grams, hits it. In the second example, the same two balls move toward each other. Two other films, "One-Dimensional Collisions I" and "One-Dimensional Collisions II" show collisions where the two balls bounce off each other. What different results might you expect from measurements of an inelastic one-dimensional collision?

The instructions for *Film Loop 18,* "One-Dimensional Collisions I" may be followed for completing this investigation.

Are the differences between momentum before and after collision significant, or are they within the experimental error already estimated?

Save your data so that later you can make similar calculations on total kinetic energy for both balls just before and just after the collision. Is whatever difference you may have obtained explainable by experimental error? Is there a noticeable difference between elastic and inelastic collisions as far as the conservation of kinetic energy is concerned?

FILM LOOP 21 TWO-DIMENSIONAL COLLISIONS I

Two hard steel balls, hanging from long, thin wires, collide. Unlike the collisions in *Film Loops 18* and *20,* the balls do not move along the *same* straight line before or after the collisions. Although strictly the balls do not all move in a plane, as each motion is an arc of a circle, to a good approximation everything occurs in one plane. Hence, the collisions are two-dimensional. Two collisions are filmed in slow motion, with ball A having a mass of 539

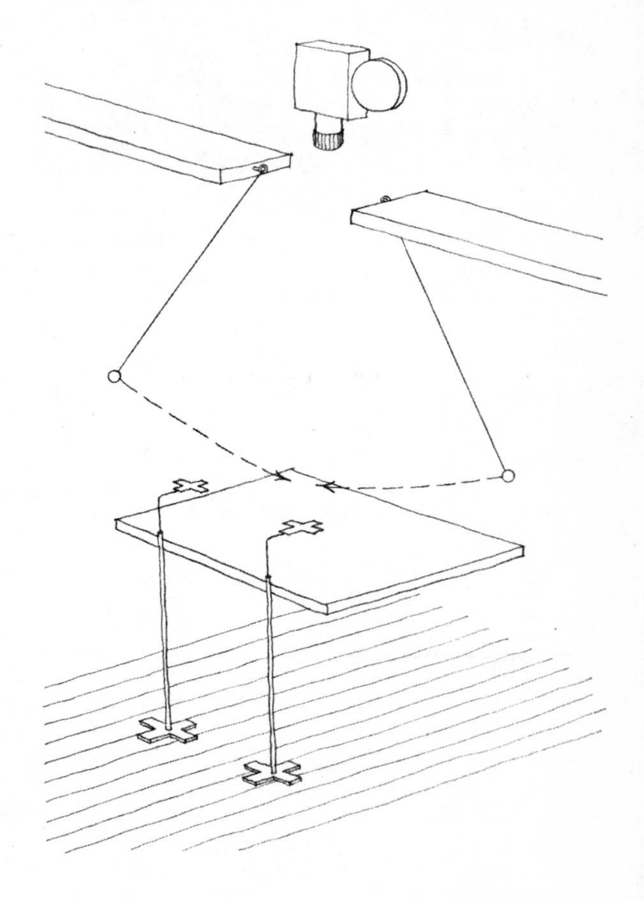

grams, and ball B having a mass of 361 grams. Two more cases are shown in *Film Loop 22*.

Using this film, you can find both the momentum and the kinetic energy of each ball before and after the collision, and thus study total momentum and total kinetic energy conservation in this situation. Thus, you should save your momentum data for later use when studying energy.

Both direction and magnitude of momentum should be taken into account, since the balls do not move on the same line. To find momentum you need velocities. Distance measurements accurate to a fraction of a millimeter and time measurements to about a tenth of a second are suggested, so choose measuring instruments accordingly.

You can project directly onto a large piece of paper. An initial problem is to determine lines on which the balls move. If you make many marks at the centers of the balls, running the film several times, you may find that these do not form a perfect line. This is due both to the inaccuracies in your measurements and to the inherent difficulties of high speed photography. Cameras photographing at a rate of 2,000 to 3,000 frames a second "jitter," because the film moves so rapidly through the camera that accurate frame registration is not possible. Decide which line is the "best" approximation to determine direction for velocities for the balls before and after collision.

You will also need the *magnitude* of the velocity, the speed. One possibility is to measure the time it takes the ball to move across two lines marked on the paper. Accuracy suggests a number of different measurements to determine which values to use for the speeds and how much error is present.

Compare the sum of the momentum before collision for both balls with the total momentum after collision. If you do not know how to add vector diagrams, you should consult your teacher or the Programmed Instruction Booklet *Vectors II*. The momentum of each object is represented by an arrow whose direction is that of the motion and whose length is proportional to the magnitude of the momentum. Then, if the head of one arrow is placed on the tail of the other, moving the line parallel to itself, the vector sum is represented by the arrow which joins the "free" tail to the "free" head.

What can you say about momentum conservation? Remember to consider measurement errors.

FILM LOOP 22 TWO-DIMENSIONAL COLLISIONS II

Two hard steel balls, hanging from long thin wires, collide. Unlike the collisions in *Film Loops 18* and 20, the balls do not move along the *same* straight line before or after the collisions. Although the balls do not strictly all move in a plane, as each motion is an arc of a circle, everything occurs in one plane. Hence, the collisions are two-dimensional. Two collisions are filmed in slow motion, with both balls having a mass of 367 grams. Two other cases are shown in *Film Loop 21*.

Using this film you can find both the kinetic energy and the momentum of each ball before and after the collision, and thus study total momentum and total energy conservation in this situation. Follow the instructions given for *Film Loop 21*, "Two-dimensional collisions I," in completing this investigation.

FILM LOOP 23 INELASTIC TWO-DIMENSIONAL COLLISIONS

Two hard steel balls, hanging from long, thin wires, collide. Unlike the collisions in *Film Loops 18* and *20*, the balls do not move along the *same* straight line before or after the collision. Although the balls do not strictly all move in a plane, as each motion is an arc of a circle, to a good approximation the motion occurs in one plane. Hence, the collisions are two-dimensional. Two collisions are filmed in slow motion. Each ball has a mass of 500 grams. The plasticene balls stick together after collision, moving as a single mass.

Using this film, you can find both the kinetic energy and the momentum of each ball before and after the collision, and thus study total momentum and total energy conservation in this situation. Follow the instructions given for *Film Loop 21,* "Two-dimensional collisions I," in completing this investigation.

FILM LOOP 24 SCATTERING OF A CLUSTER OF OBJECTS

This film and also Film Loop 25 each contain one advanced quantitative problem. We recommend that you do not work on these loops until you have analyzed one of the Events 8 to 13 of the series, *Stroboscopic Still Photographs of Two-Dimensional Collisions,* or one of the examples in the film loops entitled "Two-Dimensional Collisions: Part II," or "Inelastic Two-Dimensional Collisions." All these examples involve two-body collisions, whereas the film here described involves seven objects and *Film Loop 25,* five.

In this film seven balls are suspended from long, thin wires. The camera sees only a small portion of their motion, so the balls all move approximately along straight lines. The slow-motion camera is above the balls. Six balls are initially at rest. A hardened steel ball strikes the cluster of resting objects. The diagram in Fig. 9-39 shows the masses of each of the balls.

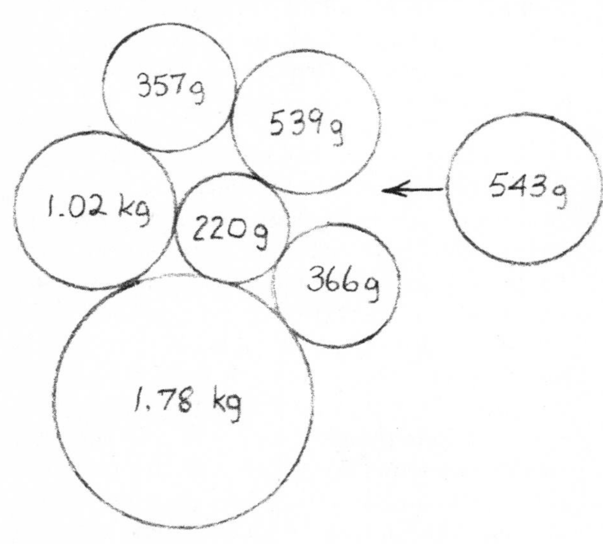

Fig. 9-39

Part of the film is photographed in slow motion at 2,000 frames per second. By projecting this section of the film on paper several times and making measurements of distances and times, you can determine the directions

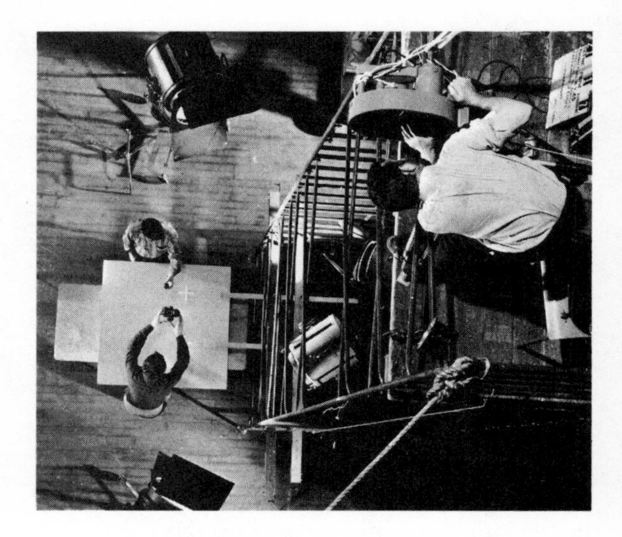

and magnitudes of the velocities of each of the balls. Distance and time measurements are needed. Discussions of how to make such measurements are contained in the Film Notes for one-dimensional and two-dimensional collisions. (See *Film Loops 18* and *21.*)

Compare the total momentum of the system both before and after the collision. Remember that momentum has both direction and magnitude. You can add momenta after collision by representing the momentum of each ball by an arrow, and "adding" arrows geometrically. What can you say about the accuracy of your calculations and measurements? Is momentum conserved? You might also wish to consider energy conservation.

FILM LOOP 25 EXPLOSION OF A CLUSTER OF OBJECTS

Five balls are suspended independently from long thin wires. The balls are initially at rest, with a small cylinder containing gunpowder in the center of the group of balls. The masses and initial positions of the ball are shown in Fig. 9-40. The charge is exploded and each of the balls moves off in an independent direction. In the slow-motion sequence the camera is mounted directly above the resting objects. The camera sees only a small part of the motion, so that the paths of the balls are almost straight lines.

In your first viewing, you may be interested

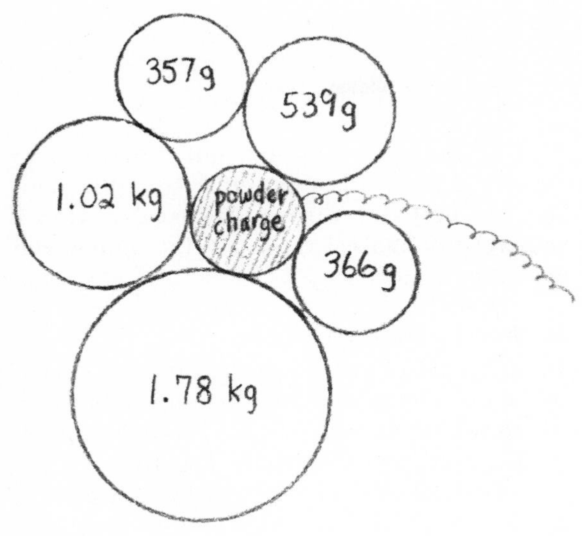

Fig. 9-40

in trying to predict where the "missing" balls will emerge. Several of the balls are hidden at first by the smoke from the charge of powder. All the balls except one are visible for some time. What information could you use that would help you make a quick decision about where this last ball will appear? What physical quantity is important? How can you use this quantity to make a quick estimate? When you see the ball emerge from the cloud, you can determine whether or not your prediction was correct. The animated elliptical ring identifies this final ball toward the end of the film.

You can also make detailed measurements, similar to the momentum conservation measurements you may have made using other Project Physics Film Loops. During the slow-motion sequence find the magnitude and direction of the velocity of each of the balls after the explosion by projecting the film on paper, measuring distances and times. The notes on previous films in this series, *Film Loops 18* and *21,* will provide you with information about how to make such measurements if you need assistance.

Determine the total momentum of all the balls after the explosion. What was the momentum before the explosion? You may find these results slightly puzzling. Can you account for any discrepancy that you find? Watch the film again and pay close attention to what happens during the explosion.

B.C. by John Hart

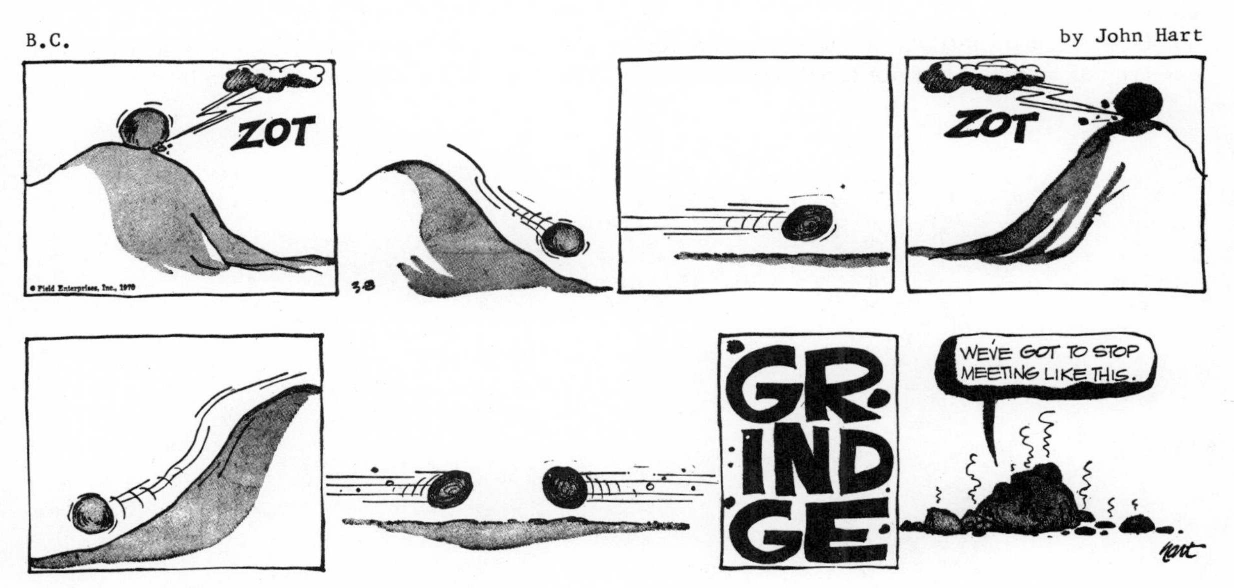

Chapter 10 Energy

EXPERIMENT 24
CONSERVATION OF ENERGY

In the previous experiments on conservation of momentum, you recorded the results of a number of collisions involving carts and gliders having different initial velocities. You found that, within the limits of experimental uncertainty, momentum was conserved in each case. You can now use the results of these collisions to learn about another extremely useful conservation law, the conservation of energy.

Do you have any reason to believe that the product of m and v is the only conserved quantity? In the data obtained from your photographs, look for other combinations of quantities that might be conserved. Find values for m/v, m^2v and mv^2 for each cart before and after collision, to see if the sum of these quantities for both carts is conserved. Compare the results of the elastic collisions with the inelastic ones. Consider the "explosion" too.

Q1 Is there a quantity that is conserved for one type of collision but not for the other?

There are several alternative methods to explore further the answer to this question; you will probably wish to do just one. Check your results against those of classmates who use other methods.

Method I—Dynamics Carts

To take a closer look at the details of an elastic collision, photograph two dynamics carts as you may have done in the previous experiment. Set the carts up as shown in Fig. 10-1.

The mass of each cart is 1 kg. Extra mass is added to make the total masses 2 kg and 4 kg. There is a light source on each cart. So that you can distinguish between the images formed by the two lights, make sure that one of the bulbs is slightly higher than the other.

Place the 2-kg cart at the center of the table and push the other cart toward it from the left. If you use the 12-slot disk on the stroboscope, you should get several images during the time that the spring bumpers are touching. You will need to know which image of the right-hand cart was made at the same instant as a given image of the left-hand cart.

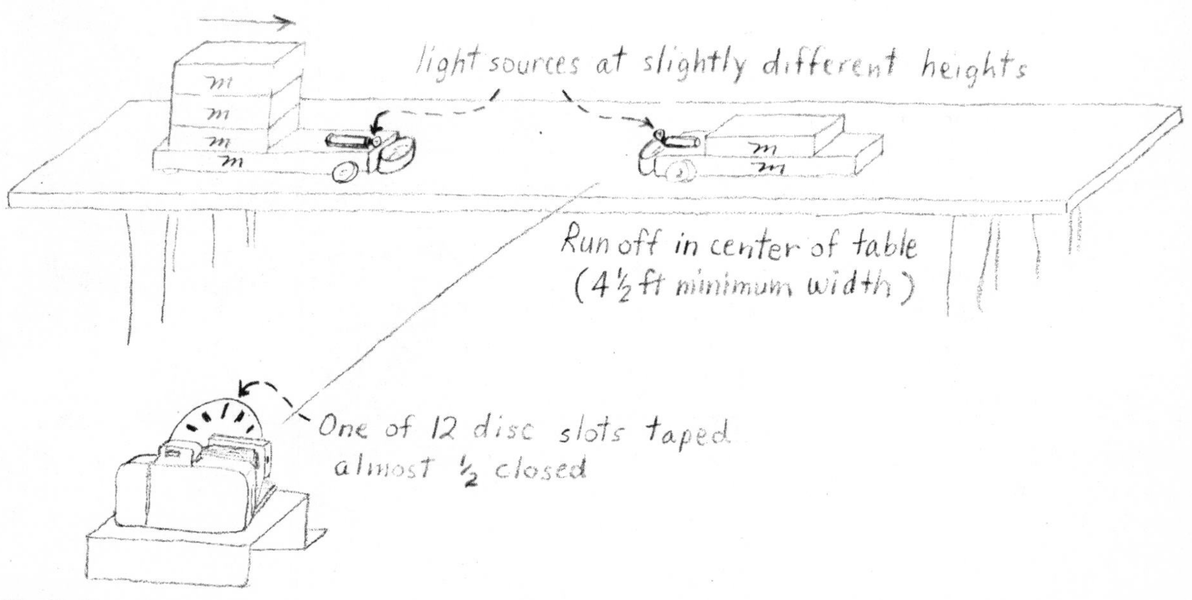

Fig. 10-1

Fig. 10-2

Matching images will be easier if one of the twelve slots on the stroboscope disk should be slightly more than half-covered with tape. (Figure 10-2.) Images formed when that slot is in front of the lens will be fainter than the others.

Compute the values for the momentum, (mv), for each cart for each time interval while the springs were touching, plus at least three intervals before and after the springs touched. List the values in a table, making sure that you pair off the values for the two carts correctly. Remember that the lighter cart was initially at rest while the heavier one moved toward it. This means that the first few values of mv for the lighter cart will be zero.

On a sheet of graph paper, plot the momentum of each cart as a function of time, using the same coordinate axes for both. Connect each set of values with a smooth curve.

Now draw a third curve which shows the sum of the two values of mv, the total momentum of the system, for each time interval.

Q2 Compare the final value of mv for the system with the initial value. Was momentum conserved in the collision?

Q3 What happened to the momentum of the system while the springs were touching—was momentum conserved *during* the collision?

Now compute values for the scalar quantity mv^2 for each cart for each time interval, and add them to your table. On another sheet of graph paper, plot the values of mv^2 for each cart for each time interval. Connect each set of values with a smooth curve.

Now draw a third curve which shows the sum of the two values of mv^2, for each time interval.

Q4 Compare the final value of mv^2 for the system with the initial value. Is mv^2 a conserved quantity?

Q5 How would the appearance of your graph change if you multiplied each quantity by $\frac{1}{2}$? (The quantity $\frac{1}{2}mv^2$ is called the ***kinetic energy*** of the object of mass m and speed v.)

Q6 Compute values for the scalar quantity $\frac{1}{2}mv^2$ for each cart for each time interval. On a sheet of graph paper, plot the kinetic energy of each cart as a function of time, using the same coordinate axes for both.

Now draw a third curve which shows the sum of the two values of $\frac{1}{2}mv^2$, for each time interval. Does the total amount of kinetic energy vary during the collision?

Q7 If you found a change in the total kinetic energy, how do you explain it?

Method II—Magnets

Spread some Dylite spheres (tiny plastic beads) on a glass tray or other hard, flat surface. A disc magnet will slide freely on this low-friction surface. Level the surface carefully.

Strobe photograph of a two-dimensional "collision" between disk magnets. Strobe rate about 30 per second.

Put one magnet puck at the center and push a second one toward it, slightly off center. You want the magnets to repel each other without actually touching. Try varying the speed and direction of the pushed magnet until you find conditions that make both magnets move off after the collision with about equal speeds.

To record the interaction, set up a camera directly above the glass tray (using the motor-strobe mount if your camera does not attach directly to the tripod) and a xenon stroboscope to one side as in Fig. 10-3. Mount a steel ball or Styrofoam hemisphere on the center of each disk with a small piece of clay. The ball will give a sharp reflection of the strobe light.

Take strobe photographs of several interactions. There must be several images before and after the interaction, but you can vary the intial speed and direction of the moving magnet, to get a variety of interactions. Using your photograph, calculate the "before" and "after" speeds of each disk. Since you are interested only in comparing speeds, you can use any convenient units for speed.

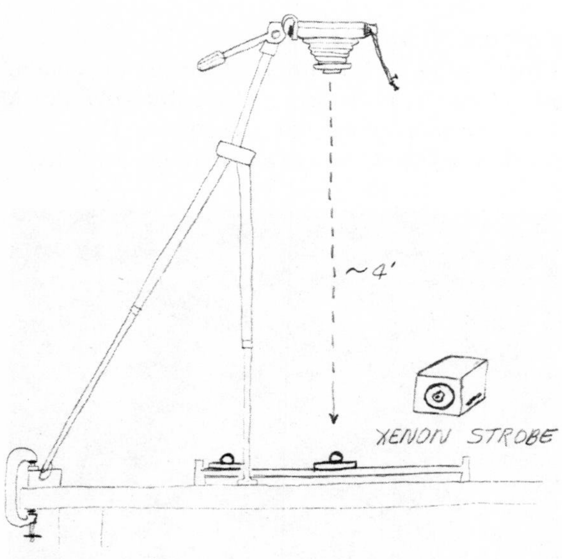

Fig. 10-3

Q8 Is mv^2 a conserved quantity? Is $\frac{1}{2}mv^2$ a conserved quantity?

If you find there has been a decrease in the total kinetic energy of the system of interacting magnets, consider the following: the surface is not perfectly frictionless and a single magnet disc pushed across it will slow down a bit. Make a plot of $\frac{1}{2}mv^2$ against time for a moving puck to estimate the rate at which kinetic energy is lost in this way.

Q9 How much of the loss in $\frac{1}{2}mv^2$ that you observed in the interaction can be due to friction?

Q10 What happens to your results if you consider kinetic energy to be a *vector* quantity?

When the two disks are close together (but not touching) there is quite a strong force between them pushing them apart. If you put the two pucks down on the surface close together and release them, they will fly apart: the kinetic energy of the system has increased.

If you have time to go on, you should try to find out what happens to the total quantity $\frac{1}{2}mv^2$ of the disks while they are close together during the interaction. To do this you will need to work at a fairly high strobe rate, and push the projectile magnet at fairly high speed—without letting the two magnets actually touch, of course. Close the camera shutter before the disks are out of the field of view, so that you can match images by counting backward from the last images.

Now, working backward from the last interval, measure v and calculate $\frac{1}{2}mv^2$ for each puck. Make a graph in which you plot $\frac{1}{2}mv^2$ for each puck against time. Draw smooth curves through the two plots.

Now draw a third curve which shows the sum of the two $\frac{1}{2}mv^2$ values for each time interval.

Q11 Is the quantity $\frac{1}{2}mv^2$ conserved *during* the interaction, that is, while the repelling magnets approach very closely?

Try to explain your observations.

Method III—Inclined Air Tracks

Suppose you give the glider a push at the bottom of an inclined air track. As it moves up the slope it slows down, stops momentarily, and then begins to come back down the track.

Clearly the bigger the push you give the glider (the greater its initial velocity v_i), the higher up the track it will climb before stopping. From experience you know that there is some connection between v_i and d, the distance the glider moves along the track.

According to the *Text*, when a stone is

thrown upward, the kinetic energy that it has initially ($\frac{1}{2}mv_i^2$) is transformed into gravitational potential energy ma_gh) as the stone moves up. In this experiment, you will test to see whether the same relationship applies to the behavior of the glider on the inclined air track. In particular, your task is to find the initial kinetic energy and the increase in potential energy of the air track glider and to compare them.

The purpose of the first set of measurements is to find the initial kinetic energy $\frac{1}{2}mv_i^2$. You cannot measure v_i directly, but you can find it from your calculation of the *average* velocity v_{av} as follows. From *Text* Sec. 2.7 you learned that in the case of uniform acceleration $v_{av} = \frac{1}{2}(v_i + v_f)$, and since $v_f = 0$ at the top of the track, $v_{av} = \frac{1}{2}v_i$ or $v_i = 2v_{av}$. Remember that $v_{av} = \Delta d/\Delta t$, so $v_i = 2\Delta d/\Delta t$; Δd and Δt are easy to measure with your apparatus.

To measure Δd and Δt three people are needed: one gives the glider the initial push, another marks the highest point on the track that the glider reaches, and the third uses a stopwatch to time the motion from push to rest.

Raise one end of the track a few centimeters above the tabletop. The launcher should practice pushing until he can reproduce a push that will send the glider nearly to the raised end of the track.

Record the distance traveled and time taken for several trials, and weigh the glider.

Q12 What is the initial kinetic energy?

To calculate the increase in gravitational potential energy, you must measure the vertical height h through which the glider moves for each push. You will probably find that you need to measure from the tabletop

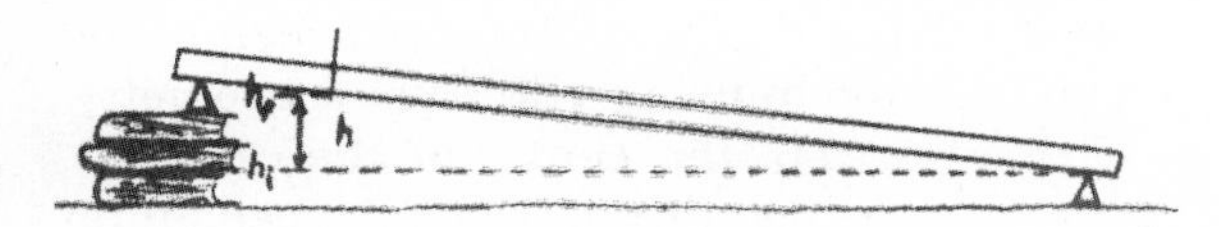

to the track at the intial and final points of the glider's motion (see Fig. 10-4), since $h = h_f - h_i$.

Q13 What is the potential energy increase, the quantity ma_gh for each of your trials?

For each trial, compare the kinetic energy loss with the potential energy increase. Be sure that you use consistent units: m in kilograms, v in meters/second, a_g in meters/second^2, h in meters.

Q14 Are the kinetic energy loss and the potential energy increase equal within your experimental uncertainty?

Q15 Explain the significance of your result.

Here are more things to do if you have time to go on:

(a) See if your answer to Q14 continues to be true as you make the track steeper and steeper.

(b) When the glider rebounds from the rubber band at the bottom of the track it is momentarily stationary—its kinetic energy is zero. The same is true of its gravitational potential energy, if you use the bottom of the truck as the zero level. And yet the glider will rebound from the rubber band (regain its kinetic energy) and go quite a way up the track (gaining gravitational potential energy) before it stops. See if you can explain what happens at the rebound in terms of the conservation of mechanical energy.

(c) The glider does not get quite so far up the track on the second rebound as it did on the first. There is evidently a loss of energy. See if you can measure how much energy is lost each time.

EXPERIMENT 25
MEASURING THE SPEED OF A BULLET

In this experiment you will use the principle of the conservation of momentum to find the speed of a bullet. Sections 9.2 and 9.3 in the *Text* discuss collisions, define momentum, and give you the general equation of the principle of conservation of momentum for two-body collisions: $m_A\vec{v}_A + m_B\vec{v}_B = m_A\vec{v}'_A + m_B\vec{v}'_B$.

The experiment consists of firing a projectile into a can packed with cotton or a heavy block that is free to move horizontally. Since all velocities before and after the collision are in the same direction, you may neglect the vector nature of the equation above and work only with speeds. To avoid subscripts, call the mass of the target M and the much smaller mass of the projectile m. Before impact the target is at rest, so you have only the speed v of the projectile to consider. After impact both target and embedded projectile move with a common speed v'. Thus the general equation becomes

$$mv = (M + m)v'$$

or

$$v = \frac{(M + m)v'}{m}$$

Both masses are easy to measure. Therefore, if the comparatively slow speed v' can be found after impact, you can compute the high speed v of the projectile before impact. There are at least two ways to find v'.

Method I—Air Track

The most direct way to find v' is to mount the target on the air track and to time its motion after the impact. (See Figure 10-5.) Mount a small can, lightly packed with cotton, on an air-track glider. Make sure that the glider will still ride freely with this extra load. Fire a "bullet" (a pellet from a toy gun that has been checked for safety by your teacher) horizontally, parallel to the length of the air track.

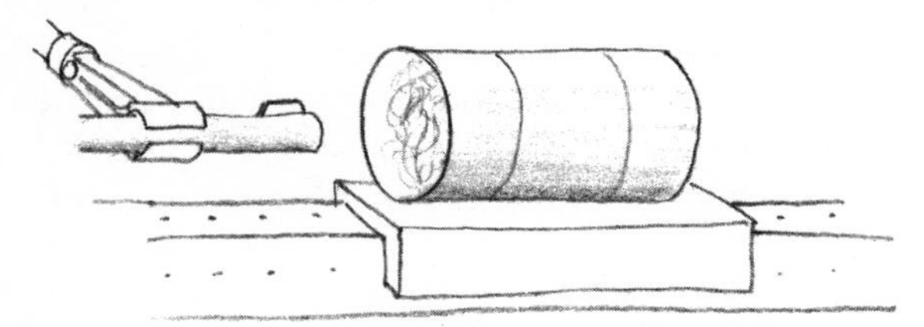

Fig. 10-5

If M is large enough, compared to m, the glider's speed will be low enough so that you can use a stopwatch to time it over a meter distance. Repeat the measurement a few times until you get consistent results.

Q1 What is your value for the bullet's speed?

Q2 Suppose the collision between bullet and can was not completely inelastic, so that the bullet bounced back a little after impact. Would this increase or decrease your value for the speed of the bullet?

Q3 Can you think of an independent way to measure the speed of the bullet? If you can, go on and make the independent measurement. Then see if you can account for any differences between the two results.

Method II—Ballistic Pendulum

This was the original method of determining the speed of bullets, invented in 1742 and is still used in some ordnance laboratories. A movable block is suspended as a freely swinging pendulum whose motion reveals the bullet's speed.

Obtaining the Speed Equation

The collision is inelastic, so kinetic energy is not conserved in the impact. But during the nearly frictionless swing of the pendulum after the impact, mechanical energy is conserved—that is, the increase in gravitational potential energy of the pendulum at the end of its upward swing is equal to its kinetic energy immediately after impact. Written as an equation, this becomes

$$(M + m)a_g h = \frac{(M + m)v'^2}{2}$$

where h is the increase in height of the pendulum bob.

Solving this equation for v' gives:

$$v' = \sqrt{2a_g h}$$

Substituting this expression for v' in the momentum equation above leads to

$$v' = \left(\frac{M + m}{m}\right)\sqrt{2a_g h}$$

Now you have an equation for the speed v of the bullet in terms of quantities that are known or can be measured.

A Useful Approximation.

The change h in vertical height is hard to measure accurately, but the horizontal displacement d may be 10 centimeters or more and can be found easily. Therefore, let's see if h can be replaced by an equivalent expression involving d. The relation between h and d can be found by using a little plane geometry.

In Fig. 10-6, the center of the circle, O, represents the point from which the pendulum is hung. The length of the cords is l.

In the triangle OBC,

$$l^2 = d^2 + (l - h)^2$$

so

$$l^2 = d^2 + l^2 - 2h + h^2$$

and

$$2lh = d^2 + h^2$$

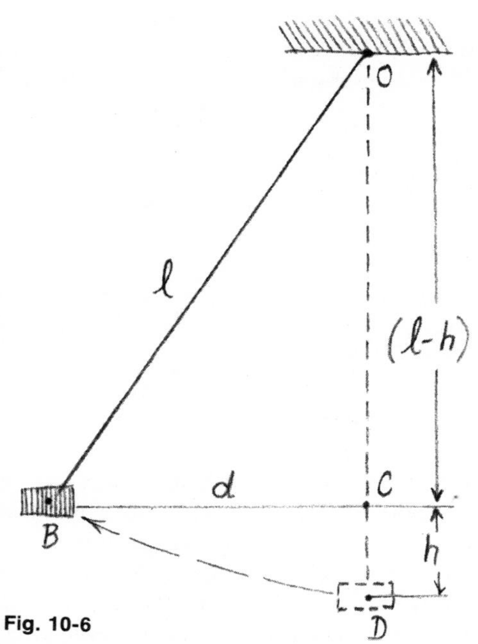

Fig. 10-6

For small swings, h is small compared with l and d, so you may neglect h^2 in comparison with d^2, and write the close approximation

$$2lh = d^2$$

or

$$h = d^2/2l$$

Putting this value of h into your last equation for v above and simplifying gives:

$$v = \frac{(M+m)d}{m}\sqrt{\frac{a_g}{l}}$$

If the mass of the projectile is small compared with that of the pendulum, this equation can be simplified to another good approximation. How?

Finding the Projectile's Speed

Now you are ready to turn to the experiment. The kind of pendulum you use will depend on the nature and speed of the projectile. If you use pellets from a toy gun, a cylindrial cardboard carton stuffed lightly with cotton and suspended by threads from a laboratory stand will do. If you use a good bow and arrow, stuff straw into a fairly stiff corrugated box and hang it from the ceiling. To prevent the target pendulum from twisting, hang it by parallel cords connecting four points on the pendulum to four points directly above them.

To measure d, a light rod (a pencil or a soda straw) is placed in a tube clamped to a stand. The rod extends out of the tube on the side toward the pendulum. As the pendulum swings, it shoves the rod back into the tube so that the rod's final position marks the end of the swing of the pendulum. Of course the pendulum must not hit the tube and there must be sufficient friction between rod and tube so that the rod stops when the pendulum stops. The original rest position of the pendulum is readily found so that the displacement d can be measured.

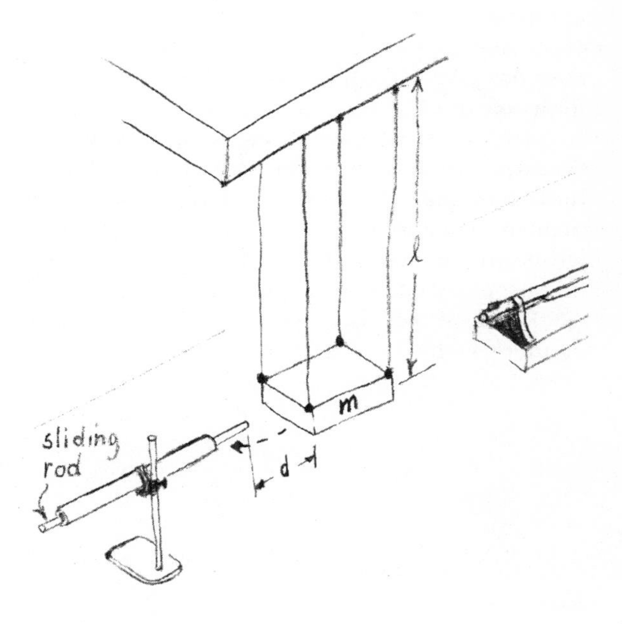

Repeat the experiment a few times to get an idea of how precise your value for d is. Then substitute your data in the equation for v, the bullet's speed.

Q4 What is your value for the bullet's speed?

Q5 From your results, compare the kinetic energy of the bullet before impact with that of the pendulum after impact. Why is there such a large difference in kinetic energy?

Q6 Can you describe an independent method for finding v? If you have time, try it, and explain any difference between the two v values.

EXPERIMENT 26
TEMPERATURE AND THERMOMETERS

You can usually tell just by touch which of two similar bodies is the hotter. But if you want to tell exactly *how* hot something is or to communicate such information to somebody else, you have to find some way of assigning a *number* to "hotness." This number is called temperature, and the instrument used to get this number is a thermometer.

It's not difficult to think of standard units for measuring intervals of time and distance—the day and the foot are both familiar to us. But try to imagine yourself living in an era before the invention of thermometers and temperature scales, that is, before the time of Galileo. How would you describe, and if possible give a number to, the "degree of hotness" of an object?

Any property (such as length, volume, density, pressure, or electrical resistance) that changes with hotness and that can be measured could be used as an indication of temperature; and any device that measures this property could be used as a thermometer.

In this experiment you will be using thermometers based on properties of liquid-expansion, gas-expansion, and electrical resistance. (Other common kinds of thermometers are based on electrical voltages, color, or gas-pressure.) Each of these devices has its own particular merits which make it suitable, from a practical point of view, for some applications, and difficult or impossible to use in others.

Of course it's most important that readings given by two different types of thermometers agree. In this experiment you will make your own thermometers, put temperature scales on them, and then compare them to see how well they agree with each other.

Defining a Temperature Scale

How do you make a thermometer? First, you decide what *property* (length, volume, etc.) of what *substance* (mercury, air, etc.) to use in your thermometer. Then you must decide on two fixed points in order to arrive at the size of a degree. A fixed point is based on a physical phenomenon that always occurs at the same degree of hotness. Two convenient fixed points to use are the melting point of ice and the boiling point of water. On the Celsius (centrigrade) system they are assigned

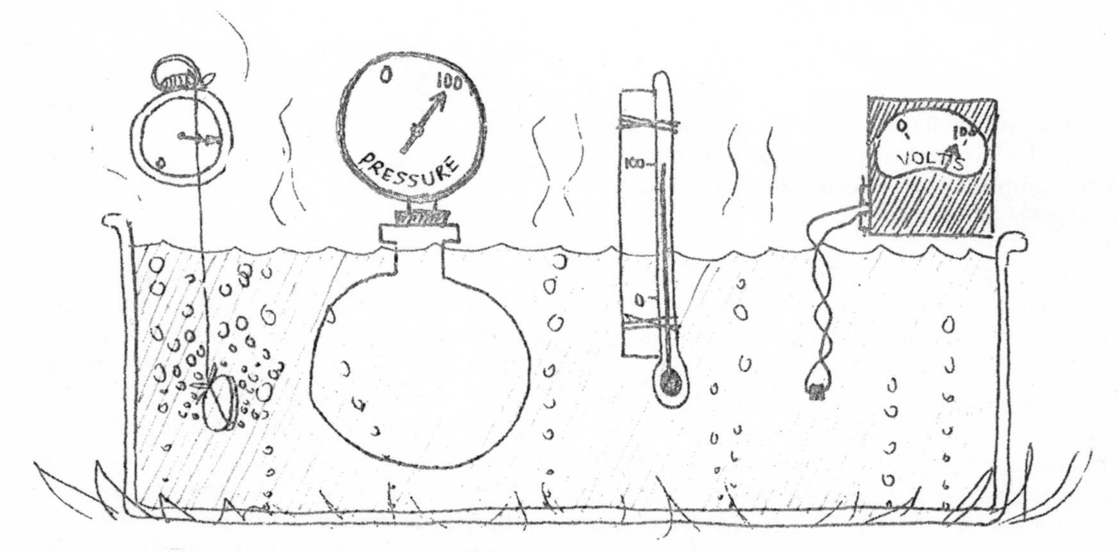

Any quantity that varies with hotness can be used to establish a temperature scale (even the time it takes for an alka seltzer tablet to dissolve in water!). Two "fixed points" (such as the freezing and boiling points of water) are needed to define the size of a degree.

the values 0°C and 100°C at ordinary atmospheric pressure.

When you are making a thermometer of any sort, you have to put a scale on it against which you can read the hotness-sensitive quantity. Often a piece of centimeter-marked tape or a short piece of ruler will do. Submit your thermometer to two fixed points of hotness (for example, a bath of boiling water and a bath of ice water) and mark the positions of the indicator.

The length of the column can now be used to define a temperature scale by saying that equal temperature changes cause equal changes along the scale between the two fixed-point positions. Suppose you marked the length of a column of liquid at the freezing point and again at the boiling point of water. You can now divide the total increase in length into equal parts and call each of these parts "one degree" change in temperature.

On the Celsius scale the degree is 1/100 of the temperature range between the boiling and freezing points of water.

To identify temperatures between the fixed points on a thermometer scale, mark off the actual distance between the two fixed points on the vertical axis of a graph and equal intervals for degrees of temperature on the horizontal axis, as in Fig. 10-7. Then plot the fixed points (x) on the graph and draw a straight line between them.

Now, the temperature on this scale, corresponding to any intermediate position l, can be read off the graph.

Other properties and other substances can

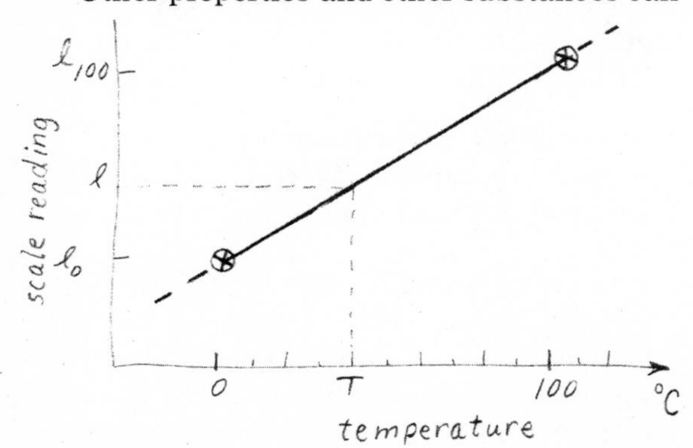

Fig. 10-7

be used (the volume of different gases, the electrical resistance of different metals, and so on), and the temperature scale defined in the same way. All such thermometers will have to agree at the two fixed points—but do they agree at intermediate temperatures?

If different physical properties do not change in the same way with hotness, then the temperature values you read from thermometers using these properties will not agree. Do similar temperature scales defined by different physical properties agree anywhere besides at the fixed points? That is a question that you can answer from this lab experience.

Comparing Thermometers

You will make or be given two "thermometers" to compare. Take readings of the appropriate quantity—length of liquid column, volume of gas, electrical resistance, thermocouple voltage, or whatever—when the devices are placed in an ice bath, and again when they are placed in a boiling water bath. Record these values. Define these two temperatures as 0° and 100° and draw the straight-line graphs that define intermediate temperatures as described above.

Now put your two thermometers in a series of baths of water at intermediate temperatures, and again measure and record the length, volume, resistance, etc. for each bath. Put both devices in the bath at the same time in case the bath is cooling down. Use your graphs to read off the temperatures of the water baths as indicated by the two devices.

Q4 Do the temperatures measured by the two devices agree, or is there consistently a difference between them?

If the two devices do give the same readings at intermediate temperatures, then you could apparently use either as a thermometer. But if they do not agree, you must choose only one of them as a standard thermometer. Give whatever reasons you can for choosing one rather than the other before reading the following discussion. If possible, compare your results with those of classmates using the same or different kinds of thermometers.

There will of course be some uncertainty

in your measurements, and you must decide whether the differences you observe between two thermometers might be due only to this uncertainty.

The relationship between the readings from two different thermometers can be displayed on another graph, where one axis is the reading on one thermometer and the other axis is the reading on the other thermometer. Each bath will give a plot point on this graph. If the points fall along a straight line, then the two thermometer properties must change with both in the same way. If, however, a fairly regularly smooth curve can be drawn through the points, then the two thermometer properties probably depend on hotness in different ways. (Figure 10-8 shows possible results for two thermometers.)

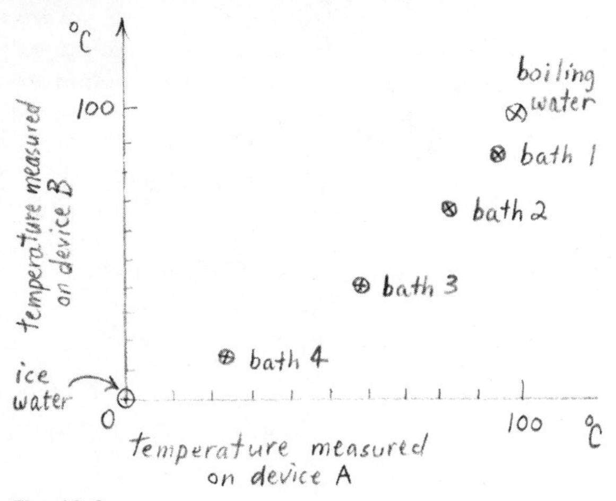

Fig. 10-8

Discussion

If we compare many gas thermometers—at constant volume as well as pressure, and using different gases, and different initial volumes and pressures—we find that they all behave quantitatively in very much the same way with respect to changes in hotness. If a given hotness change causes a 10% increase in the pressure of gas A, then the same change will also cause a 10% increase in gas B's pressure. Or, if the volume of one gas sample decreases by 20% when transferred to a particular cold bath, then a 20% decrease in volume will also be observed in a sample of any other gas. This means that the temperatures read from different gas thermometers all agree.

This sort of close similarity of behavior between different substances is not found as consistently in the expansion of liquids or solids, or in their other properties—electrical resistance, etc.—and so these thermometers do not agree, as you may have just discovered.

This suggests two things. First, that there is quite a strong case for using the change in pressure (or volume) of a gas to *define* the temperature change. Second, the fact that in such experiments all gases do behave quantitatively in the same way suggests that there may be some underlying simplicity in the behavior of gases not found in liquids and solids, and that if one wants to learn more about the way matter changes with temperature, one would do well to start with gases.

EXPERIMENT 27 CALORIMETRY

Speedometers measure speed, voltmeters measure voltage, and accelerometers measure acceleration. In this experiment you will use a device called a calorimeter. As the name suggests, it measures a quantity connected with heat. The calorie, a unit of heat energy, is discussed in the *Text* in Sec. 10.7.

Unfortunately heat energy cannot be measured as directly as some of the other quantities mentioned above. In fact, to measure the heat energy absorbed or given off by a substance you must measure the change in temperature of a second substance chosen as a standard. The heat exchange takes place inside a calorimeter, a container in which measured quantities of materials can be mixed together without an appreciable amount of heat being gained from or lost to the outside.

A Preliminary Experiment

The first experiment will give you an idea of how good a calorimeter's insulating ability really is.

Fill a calorimeter cup about half full of ice water. Put the same amount of ice water with one or two ice cubes floating in it in a second cup. Measure the temperature of the water in each cup, and record the temperatures and the time of observation.

Repeat the observations at about five-minute intervals throughout the period. Between observations, prepare a sheet of graph paper with coordinate axes so that you can plot temperature as a function of time.

Mixing Hot and Cold Liquids

(You can do this experiment while continuing to take readings of the temperature of the water in your three test cups.) You are to make several assumptions about the nature of heat. Then you will use these assumptions to predict what will happen when you mix two samples that are initially at different temperatures. If your prediction is correct, then you can feel some confidence in your assumptions—at least, you can continue to use the assumptions until they lead to a prediction that turns out to be wrong.

First, assume that, in your calorimeter, heat behaves like a fluid that is conserved—that is, it can flow from one substance to another but the total quantity of heat H present in the calorimeter in any given experiment is constant.

This implies that

heat lost by warm object	just equals	heat gained by cold object

Or, in symbols

$$-\Delta H_1 = \Delta H_2$$

Next, assume that, if two objects at different temperatures are brought together, heat will flow from the warmer to the cooler object until they reach the same temperature.

Finally, assume that the amount of heat fluid ΔH which enters or leaves an object is proportional to the change in temperature ΔT and to the mass of the object, m. In symbols,

$$\Delta H = cm\,\Delta T$$

where c is a constant of proportionality that depends on the units—and is different for different substances.

The units in which heat is measured have been defined so that they are convenient for calorimeter experiments. The calorie is defined as the quantity of heat necessary to change the temperature of one gram of water by one Celsius degree. (This definition has to be refined somewhat for very precise work, but it is adequate for your purpose.) In the expression

$$\Delta H = cm\Delta T$$

when m is measured in grams of water and T in Celsius degrees, H will be the number of calories. Because the calorie was defined this way, the proportionality constant *c* has the value 1 cal/g·C° when water is the only substance in the calorimeter. (The calorie is 1/1000 of the kilocalorie – or Calorie – used in the *Text.*)

Checking the Assumptions

Measure and record the mass of two empty plastic cups. Then put about $\frac{1}{3}$ cup of cold water in one and about the same amount of hot water in the other, and record the mass and temperature of each. (Don't forget to subtract the mass of the empty cup.) Now mix the two together in one of the cups, stir *gently* with a thermometer, and record the final temperature of the mixture.

Multiply the change in temperature of the cold water by its mass. Do the same for the hot water.

Q1 What is the product (mass × temperature change) for the cold water?
Q2 What is this product for the hot water?
Q3 Are your assumptions confirmed, or is the difference between the two products greater than can be accounted for by uncertainties in your measurement?

Predicting from the Assumptions

Try another mixture using different quantities of water, for example $\frac{1}{4}$ cup of hot water and $\frac{1}{2}$ cup of cold. Before you mix the two, try to predict the final temperature.

Q4 What do you predict the temperature of the mixture will be?
Q5 What final temperature do you observe?
Q6 Estimate the uncertainty of your thermometer readings and your mass measurements. Is this uncertainty enough to account for the difference between your predicted and observed values?
Q7 Do your results support the assumptions?

Melting

The cups you filled with hot and cold water at the beginning of the period should show a measurable change in temperature by this time. If you are to hold to your assumption of conservation of heat fluid, then it must be that some heat has gone from the hot water into the room and from the room to the cold water.

Q8 How much has the temperature of the cold water changed?
Q9 How much has the temperature of the water that had ice in it changed?

The heat that must have gone from the room to the water-ice mixture evidently did not change the temperature of the water as long as the ice was present. But some of the ice melted, so apparently the heat that leaked in was used to melt the ice. Evidently, heat was needed to cause a "change of state" (in this case, to melt ice to water) even if there was no change in temperature. The additional heat required to melt one gram of ice is called "latent heat of melting." Latent means hidden or dormant. The units are cal/g – there is no temperature unit here because no temperature change is involved in latent heat.

Next, you will do an experiment mixing materials other than liquid water in the calorimeter to see if your assumptions about heat as a fluid can still be used. Two such experiments are described below, "Measuring Heat Capacity" and "Measuring Latent Heat." If you have time for only one of them, choose either one. Finally, do "Rate of Cooling" to complete your preliminary experiment.

Measuring Heat Capacity

(While you are doing this experiment, continue to take readings of the temperature of the water in your three test cups.) Measure the mass of a small metal sample. Put just enough cold water in a calorimeter to cover the sample. Tie a thread to the sample and suspend it in a beaker of boiling water. Measure the temperature of the boiling water.

Record the mass and temperature of the water in the calorimeter.

When the sample has been immersed in the boiling water long enough to be heated uniformly (2 or 3 minutes), lift it out and hold it just above the surface for a few seconds to let the water drip off, then transfer it quickly to the calorimeter cup. Stir gently with a thermometer and record the temperature when it reaches a steady value.

Q10 Is the product of mass and temperature

change the same for the metal sample and for the water?

Q11 If not, must you modify the assumptions about heat that you made earlier in the experiment?

In the expression $\Delta H = cm\Delta T$, the constant of proportionality c (called the "specific heat capacity") may be different for different materials. For water the constant has the value 1 cal/gC°. You can find a value of c for the metal by using the assumption that heat gained by water equals the heat lost by sample. Or, writing subscripts w and s for water and metal sample, $\Delta H_w = -\Delta H_s$.

Then $c_w m_w \Delta t_w = -c_s m_s \Delta t_s$

and $$c_s = \frac{-c_w m_w \Delta t_w}{m_s \Delta t_s}$$

Q12 What is your calculated value for the specific heat capacity c_s for the metal sample you used?

If your assumptions about heat being a fluid are valid, you now ought to be able to predict the final temperature of *any* mixture of water and your material.

Try to verify the usefulness of your value. Predict the final temperature of a mixture of water and a heated piece of your material, using different masses and different initial temperatures.

Q13 Does your result support the fluid model of heat?

Measuring Latent Heat

Use your calorimeter to find the "latent heat of melting" of ice. Start with about $\frac{1}{2}$ cup of water that is a little above room temperature, and record its mass and temperature. Take a small piece of ice from a mixture of ice and water that has been standing for some time; this will assure that the ice is at 0°C and will not have to be warmed up before it can melt. Place the small piece of ice on paper toweling for a moment to dry off water on its surface, and then transfer it quickly to the calorimeter.

Stir gently with a thermometer until the ice is melted and the mixture reaches an equilibrium temperature. Record this temperature and the mass of the water plus melted ice.

Q14 What was the mass of the ice that you added? The heat given up by the warm water is:

$$\Delta H_w = c_w m_w \Delta t_w$$

The heat gained by the water formed by the melted ice is:

$$H_i = c_w m_i \Delta t_i$$

The specific heat capacity c_w is the same in both cases—the specific heat of water.

The heat given up by the warm water first melts the ice, and then heats the water formed by the melted ice. If we use the symbol ΔH_L for the heat energy required to melt the ice, we can write:

$$-\Delta H_w = \Delta H_L + \Delta H_i$$

So the heat energy needed to melt the ice is

$$\Delta H_L = -\Delta H_w - \Delta H_i$$

The latent heat of melting is the heat energy needed *per gram* of ice, so

$$\text{latent heat of melting} = \frac{\Delta H_L}{m}$$

Q15 What is your value for the latent heat of melting of ice?

When this experiment is done with ice made from distilled water with no inclusions of liquid water, the latent heat is found to be 80 calories per gram of ice. How does your result compare with the accepted value?

Rate of Cooling

If you have been measuring the temperature of the water in your three test cups, you should have enough data by now to plot three curves of temperature against time. Mark the temperature of the air in the room on your graph too.

Q16 How does the rate at which the hot water cools depend on its temperature?

Q17 How does the rate at which the cold water heats up depend on its temperature?

Weigh the amount of water in the cups. From the rates of temperature change (degrees/minute) and the masses of water, calculate the rates at which heat leaves or enters the cups at various temperatures. Use this information to estimate the error in your earlier results for latent or specific heat.

ACTIVITIES

STROBOSCOPIC PHOTOGRAPHS OF COLLISIONS

When studying momentum, you may have taken measurements on the one-dimensional and two-dimensional collisions shown in stroboscopic photographs on pages 149-152 and 156-158. If so, you can now easily reexamine your data and compute the kinetic energy $\frac{1}{2}mv^2$ for each ball before and after the interaction. Remember that kinetic energy is a scalar quantity, and so you will use the magnitude of the velocity but not the direction in making your computations. You would do well to study one or more of the simpler events (for example, Events 1, 2, 3, 8, 9, or 10) before attempting the more complex ones involving inelastic collisions or several balls. Also you may wish to review the discussions given earlier for each event.

If you find there is a loss of kinetic energy beyond what you would expect from measurement error, try to explain your results. Some questions you might try to answer are these: How does kinetic energy change as a function of the distance from impact? Is it the same before and after impact? How is energy conservation influenced by the relative speed at the time of collision? How is energy conservation influenced by the angle of impact? Is there a difference between elastic and inelastic interactions in the fraction of energy conserved?

STUDENT HORSEPOWER

When you walk up a flight of stairs, the work you do goes into frictional heating and increasing gravitational potential energy. The $\Delta(PE)_{grav}$, in joules, is the product of your weight in newtons and the height of the stairs in meters. (In foot-pounds, it is your weight in pounds times the height of the stairs in feet.)

Your useful power output is the average rate at which you did the lifting work—that is, the total change in $(PE)_{grav}$, divided by the time it took to do the work.

Walk or run up a flight of stairs and have someone time how long it takes. Determine the total vertical height that you lifted yourself by measuring one step and multiplying by the number of steps.

Calculate your useful work output and your power, in both units of watts and in horsepower. (One horsepower is equal to 550 foot-pounds/sec which is equal to 746 watts.)

STEAM-POWERED BOAT

You can make a steam-propelled boat that will demonstrate the principle of Heron's steam engine (*Text* Sec. 10.5) from a small tooth-powder or talcum-powder can, a piece of candle, a soap dish, and some wire.

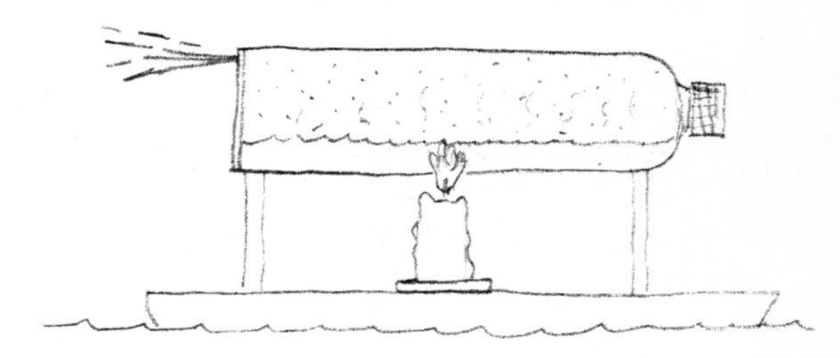

Place the candle in the soap dish. Punch a hole near the edge of the bottom of the can with a needle. Construct wire legs which are long enough to support the can horizontally over the candle and the soap dish. Rotate the can so that the needle-hole is at the top. Half fill the can with water, replace the cover, and place this "boiler" over the candle and light the candle. If this boat is now placed in a large pan of water, it will be propelled across the pan.

Can you explain the operation of this boat in terms of the conservation of momentum? of the conservation of energy?

PROBLEMS OF SCIENTIFIC AND TECHNOLOGICAL GROWTH

The Industrial Revolution of the eighteenth and nineteenth centuries is rich in examples of man's disquiet and ambivalence in the face of technological change. Instead of living among pastoral waterwheel scenes, men began to live in areas with pollution problems as bad or worse than those we face today, as is shown in the scene at Wolverhampton, England in 1866. As quoted in the *Text*, William Blake lamented in "Stanzas from Milton,"

A portrayal of the scene near Wolverhampton, England in 1866, called "Black Country."

CHAPTER 10

And did the Countenance Divine
Shine forth upon our clouded hills?
And was Jerusalem builded here
Among these dark Satanic mills?

Ever since the revolution began, we have profited from advances in technology. But we also still face problems like those of pollution and of displacement of men by machines.

One of the major problems is a growing lack of communication between people working in science and those working in other fields. When C. P. Snow published his book *The Two Cultures and the Scientific Revolution* in 1959, he initiated a wave of debate which is still going on.

In your own community there are probably some pollution problems of which you are aware. Find out how science and technology may have contributed to these problems—and how they can contribute to solutions!

"Honk, honk, honk, honk, honk, honk, honk,

cough, cough, cough, cough, cough, cough, cough,

honk, honk, honk, honk, honk, honk, honk."

Smog trapped by a temperature inversion at approximately 300 feet above the ground over the Los Angeles Basin. Relatively clear air is visible above the base of the temperature inversion.

ENERGY ANALYSIS OF A PENDULUM SWING

According to the law of conservation of energy, the loss in gravitational potential energy of a simple pendulum as it swings from the top of its swing to the bottom is completely transferred into kinetic energy at the bottom of the swing. You can check this with the following photographic method. A one-meter simple pendulum (measured from the support to the *center* of the bob) with a 0.5 kg bob works well. Release the pendulum from a position where it is 10 cm higher than at the bottom of its swing.

To simplify the calculations, set up the camera for 10:1 scale reduction (see page 8 of the Unit I *Handbook*). Two different strobe approaches have proved successful: (1) tape an AC blinky to the bob, or (2) attach an AA cell and bulb to the bob and use a motor-strobe disk in front of the camera lens. In either case you may need to use a two-string suspension to prevent the pendulum bob from spinning while swinging. Make a time exposure for one swing of the pendulum.

You can either measure directly from your print (which should look something like the one in Fig. 10-9), or make pinholes at the center of each image on the photograph and project the hole images onto a larger sheet of paper. Calculate the instantaneous speed v at the bottom of the swing by dividing the distance traveled between the images nearest the bottom of the swing by the time interval between the images. The kinetic energy at the bottom of the swing, $\frac{1}{2}mv^2$, should equal the change in potential energy from the top of the swing to the bottom. If Δh is the difference in vertical height between the bottom of the swing and the top, then

$$v = \sqrt{2a_g \Delta h}$$

If you plot both the kinetic and potential energy on the same graph (using the bottommost point as a zero level for gravitational

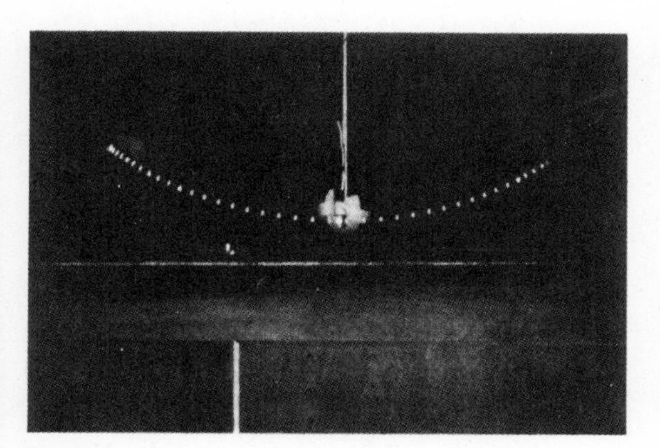

Fig. 10-9

potential energy), and then plot the sum of $KE + PE$, you can check whether total energy is conserved during the entire swing.

"What's this scheme of yours for an economical method of launching a satellite?"

PREDICTING THE RANGE OF AN ARROW

If you are interested in predicting the range of a projectile from the work you do on a slingshot while drawing it back, ask your teacher about it. Perhaps your teacher will do this with you or tell you how to do it yourself.

Another challenging problem is to estimate the range of an arrow by calculating the work done in drawing the bow. To calculate work, you need to know how the force used in drawing the string changed with the string displacement. A bow behaves even less according to Hooke's law than a slingshot; the force vs. displacement graph is definitely not a straight line.

To find how the force depends on string displacement, fasten the bow *securely* in a vise or some solid mounting. Attach a spring balance to the bow and record values of force (in newtons) as the bowstring is drawn back one centimeter at a time from its rest position (*without* having an arrow notched). Or, have someone stand on a bathroom scale, holding the bow, then pull upwards on the string; the force on the string at each position will equal this apparent *loss* of weight.

Now to calculate the amount of work done, plot a force vs. displacement graph. Count squares to find the "area" (force units times displacement units) under the graph; this is the work done on the bow – equal to the elastic potential energy of the drawn bow.

Assume that all the elastic potential energy of the bow is converted into the kinetic energy of the arrow and predict the range of the arrow by the same method used in predicting the range of a slingshot projectile.

A recent magazine article stated that an alert deer can leap out of the path of an approaching arrow when he hears the twang of the bowstring. Under what conditions do you think this is possible?

LEAST ENERGY, AREA, WORK, AND TIME SITUATIONS

Nature is lazy. She often behaves so as to do as little work as possible. Here are four clear examples.

Example 1: A Hanging Chain

Hang a three-foot length of beaded chain, the type used on light sockets, from two points as shown in Fig. 10-10. What shape does the chain assume? At first glance it seems to be a parabola.

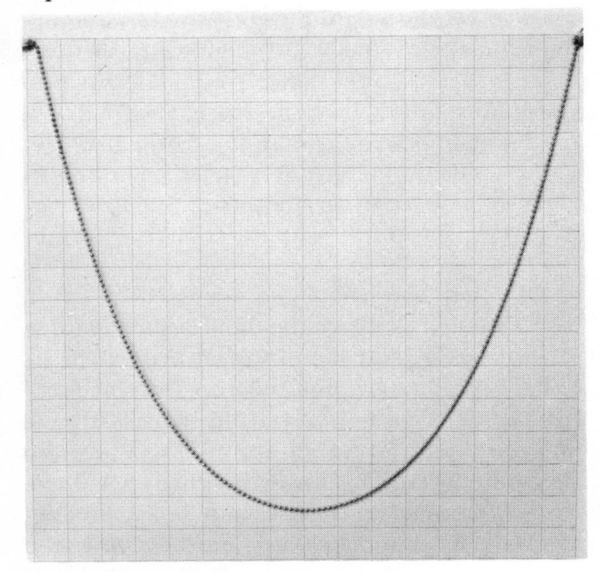

Fig. 10-10

Check whether it is a parabola by finding the equation for the parabola which would go through the vertex and the two fixed points. (See note at the end of this activity.) Determine other points on the parabola by using the equation. Plot them and see whether they match the shape of the chain.

A more interesting question is why the chain assumes this particular shape, which is called a catenary curve. You will recall from *Text* Sec. 10.3 that the gravitational potential energy of a body mass m is defined as ma_gh, where a_g is the acceleration due to gravity, —and h the height of the body above the reference level chosen. Remember that only a *difference* in energy level is meaningful; a different reference level only adds a constant to each value associated with the original reference level. In theory, you could measure the mass of one bead on the chain, measure the height of each bead above the reference level, and total the potential energies for all the beads to get the total potential energy for the whole chain.

In practice that would be quite tedious, so you will use an approximation that will still allow you to get a reasonably good result. (This would be an excellent computer problem.) Draw vertical parallel lines about 1-inch apart on the paper behind the chain (or use graph paper). In each vertical section, make a mark beside the chain in that section (see Fig. 10-11).

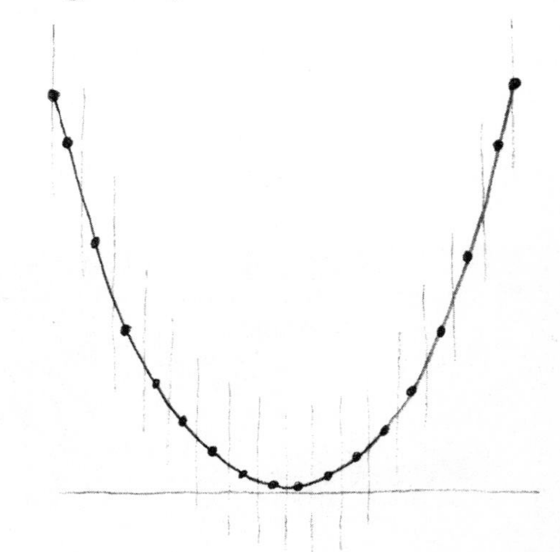

Fig. 10-11

The total potential energy for that section of the chain will be approximately Ma_gh_{av}, where h_{av} is the average height which you marked, and M is the total mass in that section of chain. Notice that near the ends of the chain there are more beads in one horizontal inter-

B.C.
By John Hart
WHY ARE YOU DRAGGING THAT CHAIN AROUND?
DID YOU EVER TRY PUSHING ONE?
10-13
hart

val than there are near the center of the chain. To simplify the solution further, assume that M is always an integral number of beads which you can count.

In summary, for each interval multiply the number of beads by the average height for that interval. Total all these products. This total is a good approximation to the gravitational potential energy of the chain.

After doing this for the freely hanging chain, pull the chain with thumbtacks into such different shapes as those shown in Fig. 10-12. Calculate the total potential energy for each shape. Does the catenary curve (the freely-formed shape) or one of these others have the minimum total potential energy?

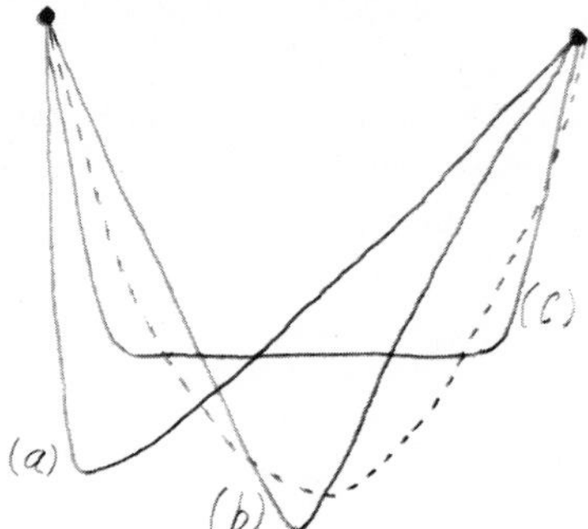
(a)
(b)
(c)

Fig. 10-12

The following example may help you plot the parabola (which you encountered before when you studied projectile motion in Unit I). The vertex in Fig. 10-10 is at (0,0) and the two fixed points are at (−8, 14.5) and (8, 14.5). All parabolas symmetric to the y axis have the formula $y = kx^2$, where k is a constant. For this example you must have $14.5 = k(8)^2$, or $14.5 = 64k$. Therefore, $k = 0.227$, and the equation for the parabola going through the given vertex and two points is $y = 0.227x^2$. By substituting values for x, we calculated a table of x and y values for our parabola and plotted it.

Example 2: Soap Films

When various shapes of wire are dipped into a soap solution, the resulting film always forms so that the total surface area of the film is a minimum. For this minimum surface, the total potential energy due to surface tension is a minimum. In many cases the resulting surface is not at all what you would expect, as shown for the wire cube in Fig. 10-13. An excellent

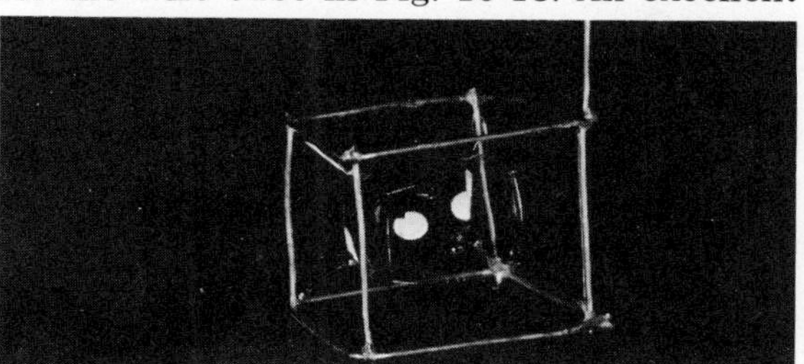

Fig. 10-13 You will find soap films to be a difficult photographic challenge.

source of suggested experiments with soap bubbles, and recipes for good solutions, is the paperback *Soap Bubbles and the Forces that Mould Them,* by C. V. Boys, Doubleday Anchor Books. Also see "The Strange World of Surface Film," *The Physics Teacher,* Sept., 1966. One recipe for a soap solution from *The Physics Teacher,* Sept. 1965 is:

$2\frac{1}{2}$ oz. Joy detergent
8 oz. distilled water
$6\frac{1}{2}$ oz. glycerine

CHAPTER 10

Example 3: A Meandering River

Read the article of the same title in the June, 1966 *Scientific American* which explains that rivers meander in such a way that the work done by the river is a minimum.

Example 4: A Challenging Minimum-time Problem

Suppose that points A and B are placed in a vertical plane as shown in Fig. 10-14, and you want to build a track between the two points so that a ball will roll from A to B in the least

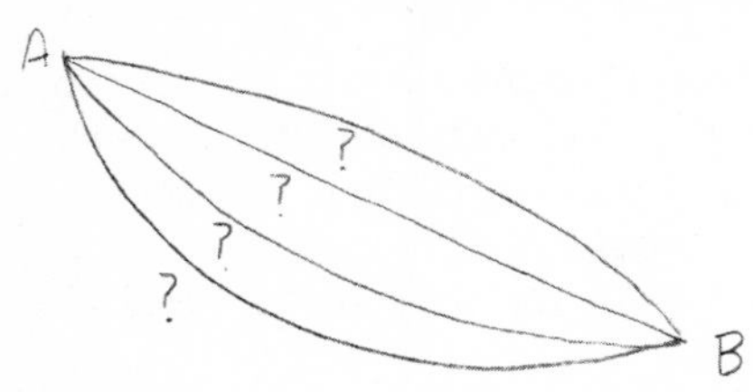

Fig. 10-14

possible time. Should the track be straight or in the shape of a circle, parabola, cycloid, catenary, or some other shape? (See the following section on how to draw a cycloid.) An interesting property of a cycloid, not to be confused with the challenge above, is that no matter where on a cycloidal track you release a ball, it will take the same amount of time to reach the bottom of the track. You may want to build a cycloidal track in order to check this. Don't make the track so steep that the ball slips instead of roll.

A more complete treatment of "The Principle of Least Action" is given in the *Feynman Lectures on Physics, Vol. II, p. 19-1.*

Generating A Cycloid

Cut a hole near the rim of a disc and insert a piece of chalk. A cycloid can be generated on the chalkboard by rolling the disk along the chalk tray.

Another method is to tape an AA cell and bulb to the spoke of a bicycle wheel (Fig. 10-15) and roll the wheel across a darkened room while taking a Polaroid time exposure. Try it with the light near the axle, on the rim and outside the rim. The last can be done by taping a meter stick to the axle and rim, with the

light taped to the outer rim. Roll the wheel along the edge of a table so that the stick end can swing down below the level of the table.

ICE CALORIMETRY

A simple apparatus made up of thermally insulating styrofoam cups can be used for doing some ice calorimetry experiments. Although the apparatus is simple, careful use will give you excellent results. To determine the heat transferred in processes in which heat energy is given off, you will be measuring either the volume of water or the mass of water from a melted sample of ice.

You will need either three cups the same size (8 oz or 6 oz), or two 8 oz and one 6 oz cup. Also have some extra cups ready. One large cup serves as the collector, A, (Fig. 10-16), the second cup as the ice container, I, and the smaller cup (or one of the same size cut back to fit inside the ice container as shown) as the cover, K.

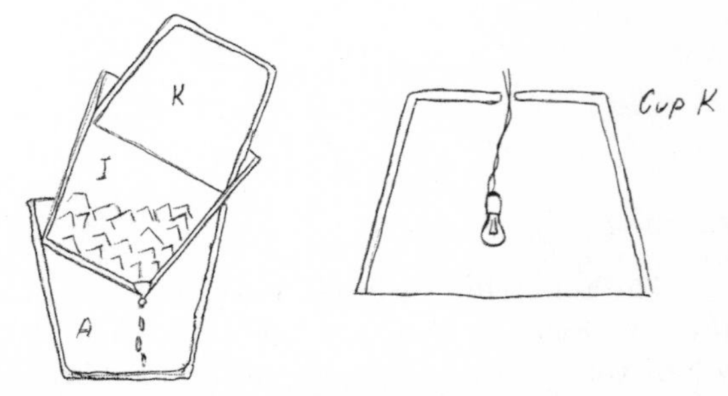

Fig. 10-16 **Fig. 10-17**

Cut a hole about $\frac{1}{4}$-inch in diameter in the bottom of cup I so that melted water can drain out into cup C. To keep the hole from becoming clogged by ice, place a bit of window screening in the bottom of I.

In each experiment, ice is placed in cup I. This ice should be carefully prepared, free of bubbles, and dry, if you want to use the known value of the heat of fusion of ice. However, you can use ordinary crushed ice, and, before doing any of the experiment, determine experimentally the effective heat of melting of this non-ideal ice. (Why should these two values differ?)

In some experiments which require some time to complete (such as Experiment b), you should set up two identical sets of apparatus (same quantity of ice, etc.), except that one does not contain a source of heat. One will serve as a fair measure of the background effect. Measure the amount of water collected in it during the same time, and subtract it from the total amount of water collected in the experimental apparatus, thereby correcting for the amount of ice melted just by the heat leaking in from the room. An efficient method for measuring the amount of water is to place the arrangement on the pan of a balance and lift up cups I and K at regular intervals (about 10 min.) while you weigh A with its contents of melted ice water.

(a) Heat of melting of ice. Fill a cup about $\frac{1}{2}$ to $\frac{1}{3}$ full with crushed ice. (Crushed ice has a larger amount of surface area, and so will melt more quickly, thereby minimizing errors due to heat from the room.) Bring a small measured amount of water (say about 20 cc) to a boil in a beaker or large test tube and pour it over the ice in the cup. Stir briefly with a poor heat conductor, such as a glass rod, until equilibrium has been reached. Pour the ice-water mixture through cup I. Collect and measure the final amount of water (m_f) in C. If m_o is the original mass of hot water at 100°C with which you started, then $m_f - m_o$ is the mass of ice that was melted. The heat energy absorbed by the melting ice is the latent heat of melting for ice, L_i, times the mass of melted ice: $L_i(m_f - m_o)$. This will be equal to the heat energy lost by the boiling water cooling from 100°C to 0°C, so we can write

$$L_i(m_f - m_o) = m_o \Delta T$$

$$\text{and } L_i = \frac{m_o}{m_f - m_o} \, 100\text{C}^\circ$$

Note: This derivation is correct only if there is still some ice in the cup afterwards. If you start with too little ice, the water will come out at a higher temperature.

For crushed ice which has been standing for some time, the value of L_i will vary between 70 and 75 calories per gram.

(b) Heat exchange and transfer by conduction and radiation. For several possible experiments you will need the following additional apparatus. Make a small hole in the bottom of cup K and thread two wires, soldered to a lightbulb, through the hole. A flashlight bulb which operates with an electric current between 300 and 600 milliamperes is preferable; but even a GE #1130 6-volt automobile headlight bulb (which draws 2.4 amps) has been used with success. (See Fig. 10-17) (In Chapter 14 of the *Text*, you will learn how to compute the energy output of the bulb, but for this experiment you need to know only that the energy output is the same for each run.) In each experiment, you are to observe how different apparatus affects heat transfer into or out of the system.

1. Place the bulb in the ice and turn it on for 5 minutes. Measure the ice melted.
2. Repeat 1, but place the bulb above the ice for 5 minutes.

3. and 4: Repeat 1 and 2, but cover the inside of cup K with aluminum foil.

5. and 6: Repeat 3 and 4, but in addition cover the inside of cup I with aluminum foil.

7. Prepare "heat absorbing" ice by freezing water to which you have added a small amount of dye, such as India ink. Repeat any or all of experiments 1 through 6 using this "specially prepared" ice.

Some questions to guide your observations: Does any heat escape when the bulb is immersed in the ice? What arrangement keeps in as much heat as possible?

FILM LOOPS

You may have used one or more of *Film Loops 18* through 25 in your study of momentum. You will find it helpful to view these slow-motion films of one and two-dimensional collisions again, but this time in the context of the study of energy. The data you collected previously will be sufficient for you to calculate the kinetic energy of each ball before and after the collision. Remeber that kinetic energy $\frac{1}{2}mv^2$ is *not* a vector quantity, and hence, you need only use the magnitude of the velocities in your calculations.

On the basis of your analysis you may wish to try to answer such questions as these: Is kinetic energy consumed in such interactions? If not, what happened to it? Is the loss in kinetic energy related to such factors as relative speed, angle of impact, or relative masses of the colliding balls? Is there a difference in the kinetic energy lost in elastic and inelastic collisions?

FILM LOOP 26 FINDING THE SPEED OF A RIFLE BULLET I

In this film a rifle bullet of 13.9 grams is fired into an 8.44 kg log. The log is initially at rest, and the bullet imbeds itself in the log. The two bodies move together after this violent collision. The height of the log is 15.0 centimeters. You can use this information to convert distances to centimeters. The setup is illustrated in Fig. 10-18 and 10-19.

You can make measurements in this film using the extreme slow-motion sequence.

Fig. 10-18

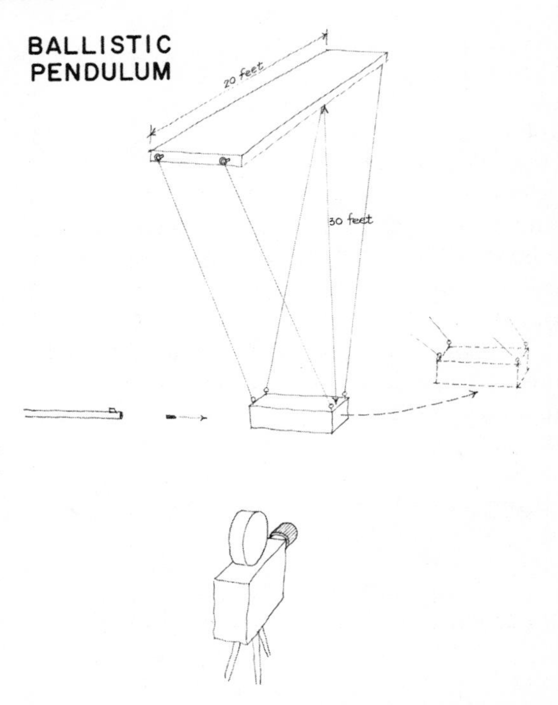

Fig. 10-19 Schematic diagram of ballistic pendulum (not to scale).

The high-speed camera used to film this sequence operated at an average rate of 2850 frames per second; if your projector runs at 18 frames per second, the slow-motion factor is 158. Although there was some variation in the speed of this camera, the average frame rate of 2850 is quite accurate. For velocity measurements in centimeters per second, a convenient unit to use in considering a rifle bullet, convert the apparent time of the film to seconds. Find the exact duration with a timer or a stop-watch by timing the interval from the yellow circle at the beginning to the one at the end of the film. There are 3490 frames in the film, so you can determine the precise speed of the projector.

Project the film onto a piece of white paper or graph paper to make your measurements of distance and time. View the film before making decisions about which measuring instruments to use. As suggested above, you can convert your distance and time measurements to centimeters and seconds.

After measuring the speed of the log after

impact, calculate the bullet speed at the moment when it entered the log. What physical laws do you need for the calculation? Calculate the kinetic energy given to the bullet, and also calculate the kinetic energy of the log after the bullet enters it. Compare these two energies and discuss any differences that you might find. Is kinetic energy conserved?

A final sequence in the film allows you to find a *lower limit* for the bullet's speed. Three successive frames are shown, so the time between each is 1/2850 of a second. The frames are each printed many times, so each is held on the screen. How does this lower limit compare with your measured velocity?

FILM LOOP 27 FINDING THE SPEED OF A RIFLE BULLET II

The problem proposed by this film is that of determining the speed of the bullet just before it hits a log. The wooden log with a mass of 4.05 kilograms is initially at rest. A bullet fired from a rifle enters the log. (Fig. 10-20.) The mass of the bullet is 7.12 grams. The bullet is imbedded in the thick log and the two move together after the impact. The extreme slow-motion sequence is intended for taking measurements.

The log is suspended from thin wires, so that it behaves like a pendulum that is free to swing. As the bullet strikes the log it starts to rise. When the log reaches its highest point, it momentarily stops, and then begins to swing back down. This point of zero velocity is visible in the slow-motion sequence in the film.

Fig. 10-20

The bullet plus the log *after* impact forms a closed system, so you would expect the total amount of mechanical energy of such a system to be conserved. The total mechanical energy is the sum of kinetic energy plus potential energy. If you conveniently take the potential energy as zero at the moment of impact for the lowest position of the log, then the energy at that time is all kinetic energy. As the log begins to move, the potential energy is proportional to the vertical distance above its lowest point, and it increases while the kinetic energy, depending upon the speed, decreases. The kinetic energy becomes zero at the point where the log reverses its direction, because the log's speed is zero at that point. All the mechanical energy at the reversal point is potential energy. Because energy is conserved, the initial kinetic energy at the lowest point should equal the potential energy at the top of the swing. On the basis of this result, write an equation that relates the initial log speed to the final height of rise. You might check this result with your teacher or with other students in the class.

If you measure the vertical height of the rise of the log, you can calculate the log's initial speed, using the equation just derived. What is the initial speed that you find for the log? If you wish to convert distance measurements to centimeters, it is useful to know that the vertical dimension of the log is 9.0 centimeters.

Find the speed of the rifle bullet at the moment it hits the log, using conservation of momentum.

Calculate the kinetic energy of the rifle bullet before it strikes and the kinetic energy of the log plus bullet after impact. Compare the two kinetic energies, and discuss any difference.

FILM LOOP 28 RECOIL

Conservation laws can be used to determine recoil velocity of a gun, given the experimental information that this film provides.

Fig. 10-21

The preliminary scene shows the recoil of a cannon firing at the fort on Ste. Hélène Island, near Montreal, Canada. (Fig. 10-21) The small brass laboratory "cannon" in the rest of the film is suspended by long wires. It has a mass of 350 grams. The projectile has a mass of 3.50 grams. When the firing is photographed in slow motion, you can see a time lapse between the time the fuse is lighted and the time when the bullet emerges from the cannon. Why is this delay observed? The camera used here exposes 8000 frames per second.

Project the film on paper. It is convenient to use a horizontal distance scale in centimeters. Find the bullet's velocity by timing the bullet over a large fraction of its motion. (Only relative values are needed, so it is not necessary to convert this velocity into cm/sec.)

Use momentum conservation to predict the gun's recoil velocity. The system (gun plus bullet) is one dimensional; all motion is along one straight line. The momentum before the gun is fired is zero in the coordinate system in which the gun is at rest. So the momentum of the cannon after collision should be equal and opposite to the momentum of the bullet.

Test your prediction of the recoil velocity by running the film again and timing the gun to find its recoil velocity experimentally. What margin of error might you expect? Do the predicted and observed values agree? Give reasons for any difference you observe. Is kinetic energy conserved? Explain your answer.

FILM LOOP 29 COLLIDING FREIGHT CARS

This film shows a test of freight-car coupling. The collisions, in some cases, were violent enough to break the couplings. The "hammer car" coasting down a ramp, reaches a speed of about 6 miles per hour. The momentary force between the cars is about 1,000,000 pounds. The photograph below (Fig. 10-22) shows

Fig. 10-22 Broken coupling pins from colliding freight cars.

coupling pins that were sheared off by the force of the collision. The slow-motion collision allows you to measure speeds before and after impact, and thus to test conservation of momentum. The collisions are *partially* elastic, as the cars separate to some extent after collision.

The masses of the cars are:

Hammer car: $m_1 = 95{,}000$ kg (210,000 lb)

Target car: $m_2 = 120{,}000$ kg (264,000 lb)

To find velocities, measure the film time for the car to move through a given distance. (You may need to run the film several times.) Use any convenient units for velocities.

Simple timing will give v_1 and v_2. The film was made on a cold winter day and friction was appreciable for the hammer car after collision. One way to allow for friction is to make a velocity time graph, assume a uniform negative acceleration, and extrapolate to the instant after impact.

An example might help. Suppose the hammer car coasts 3 squares on graph paper in 5 seconds after collision, and it coasts 6 squares in 12 seconds after collision. The *average* velocity during the first 5 seconds was v_1 = (3 squares)/(5 sec) = 0.60 squares/sec. The average velocity during any short interval approximately equals the instantaneous velocity at the mid-time of that interval, so the car's velocity was about $v_1 = 0.60$ squares/sec at $t = 2.5$ sec. For the interval 0-12 seconds, the velocity was v_1 = 0.50 squares/sec at t = 6.0 sec. Now plot a graph like that shown in Fig. 10-23. This graph shows by extrapolation that v_1 = 0.67 squares/sec at t = 0, just after the collision.

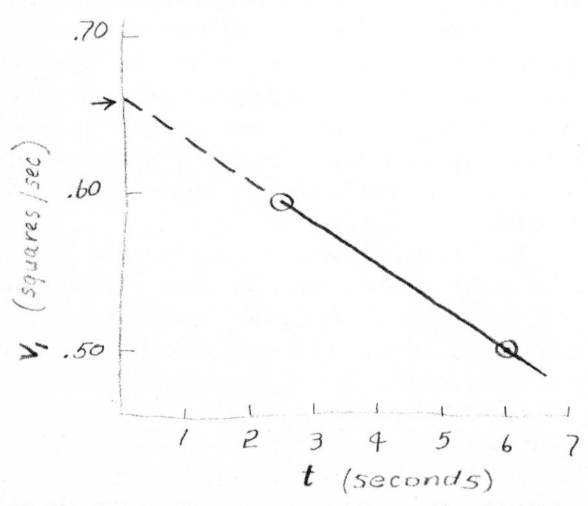

Fig. 10-23 Extrapolation backwards in time to allow for friction in estimating the value of v_1 immediately after the collision.

Compare the total momentum of the system before collision with the total momentum after collision. Calculate the kinetic energy of the freight cars before and after collision. What fraction of the hammer car's original kinetic energy has been "lost"? Can you account for this loss?

FILM LOOP 30 DYNAMICS OF A BILLIARD BALL

The event pictured in this film is one you have probably seen many times—the striking of a ball, in this case a billiard ball, by a second ball. Here, the camera is used to "slow down" time so that you can see details in this event which you probably have never observed. The ability of the camera to alter space and time is important in both science and art. The slow-motion scenes were shot at 3000 frames per second.

The "world" of your physics course often has some simplifications in it. Thus, in your textbook, much of the discussion of mechanics of bodies probably assumes that the objects are point objects, with no size. But clearly these massive billiard balls have size, as do all the things you encounter. For a point particle we can speak in a simple, meaningful way of its position, its velocity, and so on.

But the particles photographed here are billiard balls and not points. What information might be needed to describe their positions and velocities? Looking at the film may suggest possibilities. What motions can you see besides simply the linear forward motion? Watch each ball carefully, just before and just after the collision, watching not only the overall motion of the ball, but also "internal" motions. Can any of these motions be appropriately de-

Billiard balls near impact. The two cameras took side views of the collision, which are not shown in this film loop.

scribed by the word "spin"? Can you distinguish the cases where the ball is rolling along the table, so that there is no slippage between the ball and the table, from the situations where the ball is skidding along the table without rolling? Does the first ball move *immediately* after the collision? You can see that even this simple phenomenon is a good bit more complex than you might have expected.

Can you write a careful verbal description of the event? How might you go about giving a more careful mathematical description?

Using the slow-motion sequence you can make a momentum analysis, at least partially, of this collision. Measure the velocity of the cue ball before impact and the velocity of both balls after impact. Remember that there is friction between the ball and the table, so velocity is *not* constant. The balls have the same mass, so conservation of momentum predicts that

velocity of cue ball just before collision	=	sum of velocities of the balls just after collision

How closely do the results of your measurements agree with this principle? What reasons, considering the complexity of the phenomenon, might you suggest to account for any disagreement? What motions are you neglecting in your analysis?

FILM LOOP 31 A METHOD OF MEASURING ENERGY—NAILS DRIVEN INTO WOOD

Some physical quantities, such as distance, can be measured directly in simple ways. Other concepts can be connected with the world of experience only through a long series of measurements and calculations. One quantity that we often would like to measure is *energy*. In certain situations, simple and reliable methods of determining energy are possible. Here, you are concerned with the energy of a moving object.

This film allows you to check the validity of one way of measuring mechanical energy. If a moving object strikes a nail, the object will lose all of its energy. This energy has some effect, in that the nail is driven into the wood. The energy of the object becomes work done on the nail, driving it into the block of wood.

The first scenes in the film show a construction site. A pile driver strikes a pile over and over again, "planting" it in the ground. The laboratory situation duplicates this situation under more controlled circumstances. Each of the blows is the same as any other because the massive object is always raised to the same height above the nail. The nail is hit ten times. Because the conditions are kept the same, you expect the energy by the impact to be the same for each blow. Hence, the work from each blow is the same. Use the film to find if the distance the nail is driven into the wood is proportional to the energy or work. Or, better, you want to know how you can find the energy if you know the depth of penetration of the nail.

The simplest way to display the measurements made with this film may be to plot the depth of nail penetration versus the number of blows. Do the experimental points that you obtain lie approximately along a straight line? If the line is a good approximation, then the energy is about proportional to the depth of penetration of the nail. Thus, depth of penetration can be used in the analysis of other films to measure the energy of the striking object.

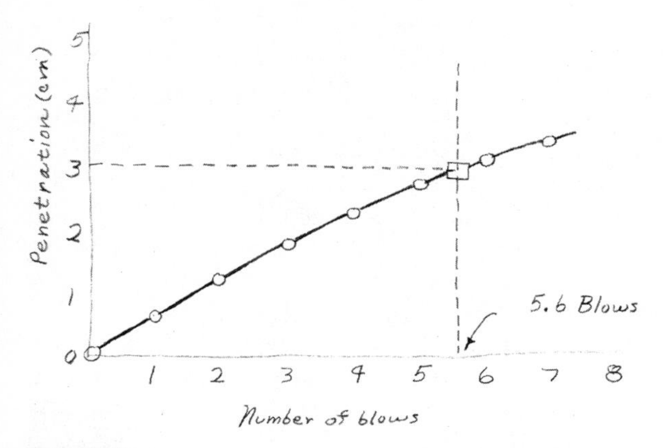

Fig. 10-24

If the graph is not a straight line, you can still use these results to calibrate your energy-measuring device. By use of penetration versus the number of blows, an observed penetration (in centimeters, as measured on the screen), can be converted into a number of blows, and therefore an amount proportional to the work done on the nail, or the energy transferred to the nail. Thus in Fig. 10-24, a penetration of 3 cm signifies 5.6 units of energy.

FILM LOOP 32 GRAVITATIONAL POTENTIAL ENERGY

Introductory physics courses usually do not give a complete definition of potential energy, because of the mathematics involved. Only particular kinds of potential energy, such as gravitational potential energy, are considered.

You may know the expression for the gravitational potential energy of an object near the earth—the product of the weight of the object and its height. The height is measured from a location chosen arbitrarily as the zero level for potential energy. It is almost impossible to "test" a formula without other physics concepts. Here we require a method of measuring energy. The previous *Film Loop 31*, "A Method of Measuring Energy," demonstrated that the depth of penetration of a nail into wood, due to a blow, is a good measure of the energy at the moment of impact of the object.

Although you are concerned with potential energy you will calculate it by first finding kinetic energy. Where there is no loss of energy through heat, the sum of the kinetic energy and potential energy is constant. If you measure potential energy from the point at which the weight strikes the nail, at the moment of striking all the energy will be kinetic energy. On the other hand, at the moment an object is released, the kinetic energy is zero, and all the energy is potential energy. These two must, by conservation of energy, be equal.

Since energy is conserved, you can figure the initial potential energy that the object had from the depth of penetration of the nail by using the results of the measurement connecting energy and nail penetration.

Two types of measurements are possible with this film. The numbered scenes are all photographed from the same position. In the first scenes (Fig. 10-25) you can determine how gravitational potential energy depends upon weight. Objects of different mass fall from the same distance. Project the film on paper and measure the positions of the nailheads before and after the impact of the falling objects.

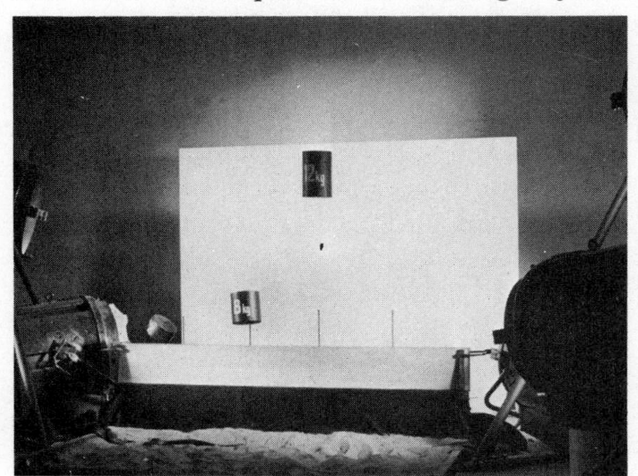

Fig. 10-25

Make a graph relating the penetration depth and the weight ma_g. Use the results of the previous experiment to convert this relation into a relation between gravitational potential energy and weight. What can you learn from this graph? What factors are you holding constant? What conclusions can you reach from your data?

Later scenes (Fig. 10-26), provide information for studying the relationship between gravitational potential energy and position. Bodies of equal mass are raised to different heights and allowed to fall. Study the relationship between the distance of fall and the gravitational potential energy. What graphs might be useful? What conclusion can you reach from your measurements?

Fig. 10-26

Can you relate the results of these measurements with statements in your text concerning gravitational potential energy?

FILM LOOP 33 KINETIC ENERGY

In this film you can test how kinetic energy (KE) depends on speed (v). You measure both KE and v, keeping the mass m constant.

Penetration of a nail driven into wood is a good measure of the work done on the nail, and hence is a measure of the energy lost by whatever object strikes the nail. The speed of the moving object can be measured in several ways.

The preliminary scenes show that the object falls on the nail. Only the speed just before the object strikes the nail is important. The scenes intended for measurement were photographed with the camera on its side, so the body appears to move horizontally toward the nail.

The speeds can be measured by timing the motion of the leading edge of the object as it moves from one reference mark to the other. The clock in the film (Fig. 10-27) is a disk that rotates at 3000 revolutions per minute. Project the film on paper and mark the positions of the clock pointer when the body crosses each reference mark. The time is proportional to the angle through which the pointer turns. The speeds are proportional to the reciprocals of the times, since the distance is the same in each case. Since you are testing only the *form* of the kinetic energy dependence on speed, any convenient unit can be used. Measure the speed for each of the five trials.

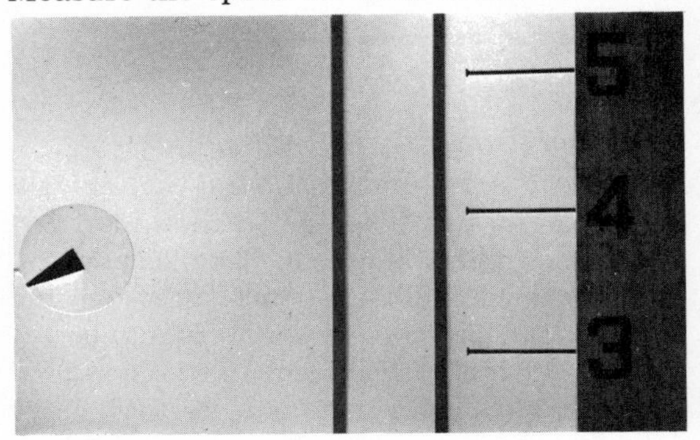

Fig. 10-27

The kinetic energy of the moving object is transformed into the work required to drive the nail into the wood. In *Film Loop 31*, "A Method of Measuring Energy," you relate the work to the distance of penetration. Measure the nail penetration for each trial, and use your results from the previous film.

How does KE depend on v? The conservation law derived from Newton's laws indicates that KE is proportional to v^2, the square of the speed, not to v. Test this by making two graphs. In one graph, plot KE vertically and plot v^2 horizontally. For comparison, plot KE versus v. What can you conclude? Do you have any assurance that a similar relation will hold, if the speeds or masses are very different from those found here? How might you go about determining this?

FILM LOOP 34 CONSERVATION OF ENERGY—POLE VAULT

This quantitative film can help you study conservation of energy. A pole vaulter (mass 68 kg, height 6 ft) is shown, first at normal speed and then in slow motion, clearing a bar at 11.5 ft. You can measure the total energy of the system at two points in time just before the jumper starts to rise and part way up, when the pole has a distorted shape. The total energy of the system is constant, although it is divided differently at different times. Since it takes work to bend the pole, the pole has elastic potential energy when bent. This elastic energy comes from some of the kinetic energy the vaulter has as he runs horizontally before inserting the pole in the socket. Later, the elastic potential energy of the bent pole is transformed into some of the jumper's gravitational potential energy when he is at the top of the jump.

Position 1 The energy is entirely kinetic energy, $\frac{1}{2}mv^2$. To help you measure the runner's speed, successive frames are held as the runner moves past two markers 1 meter apart. Each "freeze frame" represents a time interval of 1/250 sec, the camera speed. Find the runner's average speed over this meter, and then find the kinetic energy. If m is in kg and v is in m/sec, E will be in joules.

Position 2 The jumper's center of gravity is about 1.02 meters above the soles of his feet. Three types of energy are involved at the intermediate positions. Use the stop-frame sequence to obtain the speed of the jumper. The seat of his pants can be used as a reference. Calculate the kinetic energy and gravitational potential energy as already described.

The work done in deforming the pole is stored as elastic potential energy. In the final scene, a chain windlass bends the pole to a shape similar to that which it assumes during the jump in position 2. When the chain is shortened, work is done on the pole: work = (average force) × (displacement). During the cranking sequence, the force varied. The average force can be approximated by adding the initial and final values, found from the scale and dividing by two. Convert this force to newtons. The displacement can be estimated from the number of times the crank handle is pulled. A close-up shows how far the chain moves during a single stroke. Calculate the work done to crank the pole into its distorted shape.

You now can add and find the total energy. How does this compare with the original kinetic energy?

Position 3 Gravitational potential energy is the work done to raise the jumper's center of gravity. From the given data, estimate the vertical rise of the center of gravity as the jumper moves from position (1) to position (3). (His center of gravity clears the bar by about a foot, or 0.3m.) Multiply this height of rise by the jumper's weight to get potential energy. If weight is in newtons and height is in meters, the potential energy will be in joules. A small additional source of energy is in the jumper's muscles: judge for yourself how far he lifts his body by using his arm muscles as he nears the highest point. This is a small correction, so a relatively crude estimate will suffice. Perhaps he pulls with a force equal to his own weight through a vertical distance of $\frac{2}{3}$ of a meter.

How does the initial kinetic energy, plus the muscular energy expended in the pull-up, compare with the final gravitational potential energy? (An agreement to within about 10 per cent is about as good as you can expect from a measurement of this type.)

As a general reference see "Mechanics of the Pole Vault," 16th ed., by Dr. R. V. Ganslen; John Swift & Co., St. Louis, Mo. (1965).

FILM LOOP 35 CONSERVATION OF ENERGY—AIRCRAFT TAKEOFF

The pilot of a Cessna 150 holds the plane at constant speed in level flight, just above the surface of the runway. Then, keeping the throttle fixed, he pulls back on the stick, and the plane begins to rise. With the same throttle setting, he levels off at several hundred feet. At this altitude the aircraft's speed is less than at ground level. You can use this film to make a crude test of energy conservation. The plane's initial speed was constant, indicating that the net force on it was zero. In terms of an approximation, air resistance remained the

same after lift-off. How good is this approximation? What would you expect air resistance to depend on? When the plane rose, its gravitational potential energy increased, at the expense of the initial kinetic energy of the plane. At the upper level, the plane's kinetic energy is less, but its potential energy is greater. According to the principle of conservation of energy, the total energy ($KE + PE$) remained constant, assuming that air resistance and any other similar factors are neglected. But are these negligible? Here is the data concerning the film and the airplane:

Length of plane: 7.5 m (23 ft)
Mass of plane: 550 kg
Weight of plane:
550 kg $\times$ 9.8 m/sec^2 = 5400 newtons (1200 lb)
Camera speed: 45 frames/sec

Project the film on paper. Mark the length of the plane to calibrate distances.

Stop-frame photography helps you measure the speed of 45 frames per second. In printing the measurement section of the film only every third frame was used. Each of these frames was repeated ("stopped") a number of times, enough to allow time to mark a position on the screen. The effect is one of "holding" time, and then jumping a fifteenth of a second.

Measure the speeds in all three situations, and also the heights above the ground. You have the data needed for calculating kinetic energy ($\frac{1}{2}mv^2$) and gravitational potential energy (ma_gh) at each of the three levels. Calculate the total energy at each of the three levels.

Can you make any comments concerning air resistance? Make a table showing (for each level) KE, PE, and E total. Do your results substantiate the law of conservation of energy within experimental error?

Steve Aacker of Wheat Ridge High School, Wheat Ridge, Colorado, seems a bit skeptical about elastic potential energy.

Chapter 11 The Kinetic Theory of Gases

EXPERIMENT 28 MONTE CARLO EXPERIMENT ON MOLECULAR COLLISIONS

A model for a gas consisting of a large number of very small particles in rapid random motion is discussed in Chapter 11.

This experiment will show how from a comparatively small random sample of molecules you can estimate properties of the gas as a whole. The technique is named the Monte Carlo method after that famous (or infamous) gambling casino in Monaco. The experiment consists of two games, both of which involve the concept of randomness. You will probably have time to play only one.

Game I Collision Probability for a Gas of Marbles

In this part of the experiment, you will try to find the diameter of marbles by rolling a "bombarding marble" into an array of "target marbles" placed at random positions on a level sheet of graph paper. The computation of the marble diameter will be based on the proportion of hits and misses. In order to assure randomness in the motion of the bombarding marble, start at the top of an inclined board studded with nails spaced about an inch apart—a sort of pinball machine (Fig. 11-1).

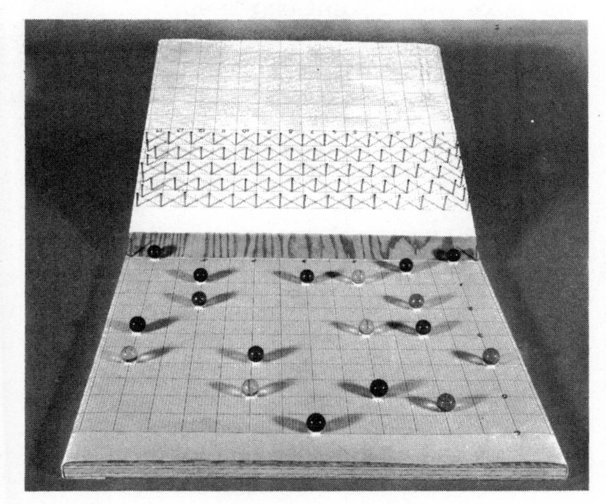

Fig. 11-1

To get a fairly even, yet random, distribution of the bombarding marble's motion, move its release position over one space for each release in the series.

First you need to place the target marbles at random. Then draw a network of crossed grid lines spaced at least two marble diameters apart on your graph paper. (If you are using marbles whose diameters are half an inch, these grid lines should be spaced 1.5 to 2 inches apart.) Number the grid lines as shown in Fig. 11-2.

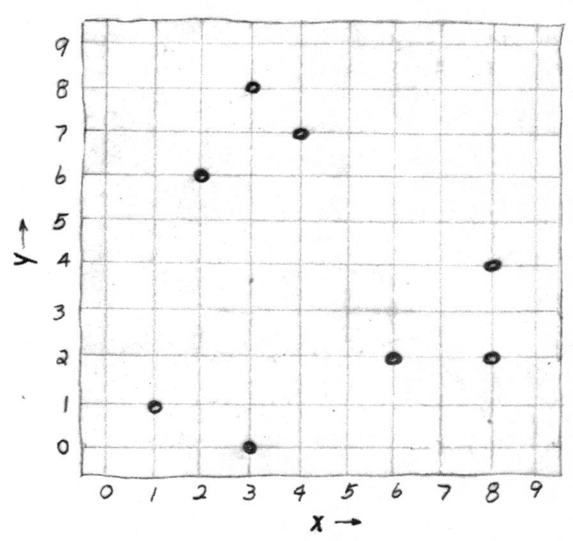

Fig. 11-2 Eight consecutive two-digit numbers in a table of random numbers were used to place the marbles.

One way of placing the marbles at random is to turn to the table of random numbers at the end of this experiment. Each student should start at a different place in the table and then select the next eight numbers. Use the first two digits of these numbers to locate positions on the grid. The first digit of each number gives the x coordinate, the second gives the y coordinate—or vice versa. Place the target marbles in these positions. Books may be placed along the sides of the graph paper and across the bottom to serve as containing walls.

With your array of marbles in place, make about fifty trials with the bombarding marble. From your record of hits and misses compute R, the ratio between the number of runs in which there are one or more hits to the total number of runs. Remember that you are counting "runs with hits," not hits, and hence, several hits in a single run are still counted as "one."

Inferring the size of the marbles. How does the ratio R lead to the diameter of the target object? The theory applies just as well to determining the size of molecules as it does to marbles, although there would be 10^{20} or so molecules instead of 8 "marble molecules."

If there were no target marbles, the bombarding marble would get a clear view of the full width, say D, of the back wall enclosing the array. There could be no hit. If, however, there were target marbles, the 100% clear view would be cut down. If there were N target marbles, each with diameter d, then the clear path over the width D would be reduced by $N \times d$.

It is assumed that no target marble is hiding behind another. (This corresponds to the assumption that the sizes of molecules are extremely small compared with the distances between them.)

The blocking effect on the bombarding marble is greater than just Nd, however. The bombarding marble will miss a target marble only if its center passes more than a distance of one radius on either side of it. (See the drawings below.) This means that a target marble has a blocking effect over twice its diameter (its own diameter plus two radii), so the total blocking effect of N marbles is $2Nd$. Therefore the expected ratio R of hits to total trials is $2\,Nd/D$ (total blocked width to total width). Thus:

$$R = \frac{2Nd}{D}$$

which we can rearrange to give an expression for d:

$$d = \frac{RD}{2N}$$

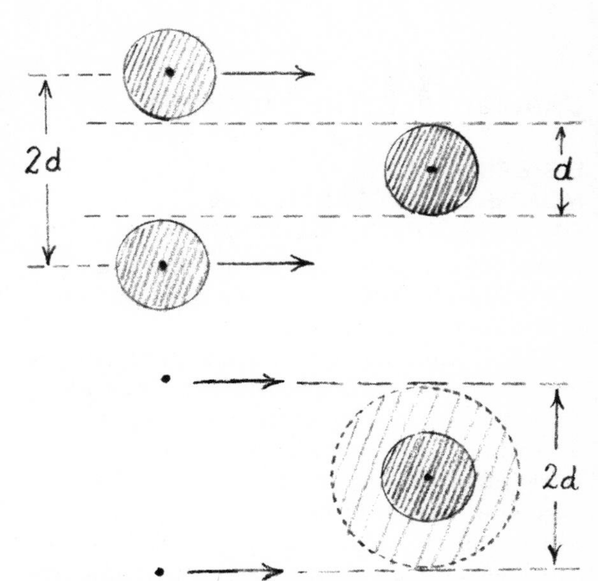

A projectile will clear a target only if it passes outside a center-to-center distance d on either side of it. Therefore, thinking of the projectiles as points, the effective blocking width of the target is $2d$.

Q1 What value do you calculate for the marble diameter?

To check the accuracy of the Monte Carlo method compare the value for d obtained from the formula above with that obtained by direct measurement of the marbles. (For example, line up your eight marbles against a book. Measure the total length of all of them together and divide by eight to find the diameter d of one marble.)

Q2 How well does your experimental prediction agree with this measurement?

Game II Mean Free Path Between Collision Squares

In this part of the experiment you play with blacked-in squares as target molecules in place of marble molecules in a pinball game. On a sheet of graph paper, say 50 units on a side (2,500 squares), you will locate by the Monte Carlo method between 40 and 100 molecules. Each student should choose a different number of molecules.

You will find a table of random numbers (from 0 to 50) at the end of this experiment. Begin anywhere you wish in the table, but then proceed in a regular sequence. Let each pair of numbers be the x and y coordinates of a point on your graph. (If one of the pair is greater than 49, you cannot use it. Ignore it and take the next pair.) Then shade in the squares for which these points are the lower left-hand corners. You now have a random array of square target "molecules."

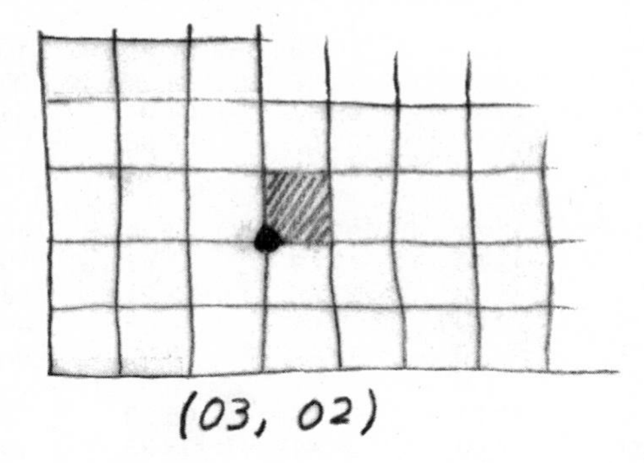

Rules of the game. The way a bombarding particle passes through this array, it is bound to collide with some of the target particles. There are five rules for this game of collision. All of them are illustrated in Fig. 11-3.

(a) The particle can travel only along lines of the graph paper, up or down, left or right. They start at some point (chosen at random) on the left-hand edge of the graph paper. The particle initially moves horizontally from the starting point until it collides with a blackened square or another edge of the graph paper.

(b) If the particle strikes the upper left-hand corner of a target square, it is diverted upward through a right angle. If it should strike a lower left-hand corner it is diverted downward, again through ninety degrees.

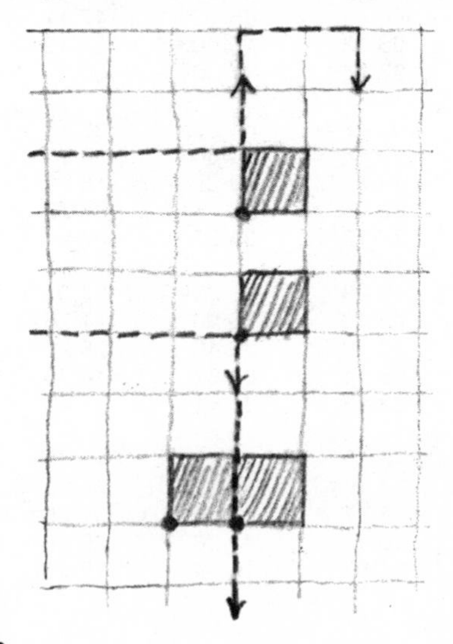

Fig. 11-3

(c) When the path of the particle meets an edge or boundary of the graph paper, the particle is *not* reflected directly back. (Such a reversal of path would make the particle retrace its previous paths.) Rather it moves two spaces to its *right* along the boundary edge before reversing its direction.

(d) There is an exception to rule (c). Whenever the particle strikes the edge so near a corner that there isn't room for it to move two spaces to the right without meeting another edge of the graph paper, it moves two spaces to the *left* along the boundary.

(e) Occasionally two target molecules may occupy adjacent squares and the particle may hit touching corners of the two target molecules at the same time. The rule is that this counts as two hits and the particle goes straight through without changing its direction.

Finding the "mean free path." With these collision rules in mind, trace the path of the particle as it bounces about among the random array of target squares. Count the number of collisions with targets. Follow the path of the particle until you get 51 hits with target squares (collisions with the edge do not count). Next, record the 50 lengths of the paths of the particle between collisions. Distances to and from a boundary should be included, but *not* distances *along* a boundary (the two spaces introduced to avoid backtracking). These 50 lengths are the free paths of the particle. Total them and divide by 50 to obtain the mean free path, L, for your random two-dimensional array of square molecules. In Sec. 11.4 of the *Text*, you saw how Clausius modified this model by giving the particles a definite size. Clausius showed that the average distance L a molecule travels between collisions, the so-called "mean free path," is given by

$$L = \frac{V}{Na}$$

where V is the volume of the gas, N is the number of molecules in that volume, and a is

TABLE OF 1000 RANDOM TWO-DIGIT NUMBERS (FROM 0 to 50)

03 47	44 22	30 30	22 00	00 49	22 17	38 30	23 21	20 11	24 33
16 22	36 10	44 39	46 40	24 02	19 36	38 21	45 33	14 23	01 31
33 21	03 29	08 02	20 31	37 07	03 28	47 24	11 29	49 08	10 39
34 29	34 02	43 28	03 43	43 40	26 08	28 06	50 14	21 44	47 21
32 44	11 05	05 05	05 50	23 29	26 00	09 05	27 31	08 43	04 14
18 18	04 02	48 39	48 22	38 18	15 39	48 34	50 28	37 21	15 09
23 42	31 08	19 30	06 00	20 18	30 24	15 33	10 07	14 29	05 24
35 12	11 12	11 04	01 10	25 39	48 50	24 44	03 47	34 04	44 07
12 13	42 10	40 48	45 44	42 35	41 26	41 10	23 05	06 36	08 43
37 35	12 41	02 02	19 11	06 07	42 31	23 47	47 25	10 43	12 38
16 08	18 39	03 31	49 26	07 12	17 31	17 31	35 07	44 38	40 35
31 16	10 47	38 45	28 40	33 34	24 16	42 38	19 09	41 47	50 41
32 43	45 37	30 38	22 01	30 14	02 17	45 18	29 06	13 27	46 24
27 42	03 09	08 32	24 02	05 49	18 05	22 00	23 02	44 43	43 20
00 39	05 03	49 37	23 22	33 42	26 29	00 20	12 03	10 05	02 39
11 27	39 32	13 30	36 45	09 03	46 40	22 07	03 03	05 39	03 46
35 24	22 49	17 33	35 01	01 32	18 09	47 03	39 41	36 23	19 41
16 20	38 36	29 48	07 27	48 14	34 13	07 48	39 12	20 18	19 42
38 23	33 26	15 29	20 02	21 45	04 31	48 13	23 32	37 30	09 24
45 11	27 07	39 43	13 05	47 45	47 45	00 06	41 18	05 02	03 09
18 00	14 21	49 17	30 37	25 15	04 49	24 19	40 23	24 17	17 16
20 46	06 18	45 07	06 28	49 44	10 08	43 00	38 26	34 41	11 16
05 26	50 25	38 47	39 38	42 45	10 08	16 06	43 18	34 48	27 03
21 19	13 42	16 04	00 18	16 46	13 13	16 29	44 10	29 18	22 45
41 23	03 10	35 30	24 36	38 09	25 21	08 40	20 46	39 14	37 31
34 50	20 14	21 46	38 46	12 27	20 44	46 06	01 41	30 49	18 48
39 43	13 04	24 15	08 22	13 29	04 05	42 29	50 47	01 50	01 48
18 14	04 43	27 46	23 07	19 28	07 10	23 19	41 45	25 27	19 10
09 47	34 45	08 45	25 21	49 21	18 46	16 40	35 14	41 28	41 15
44 17	04 33	15 22	12 45	39 07	34 27	14 47	35 33	42 29	47 47
40 33	42 45	07 08	38 15	08 25	22 06	07 26	32 44	03 42	42 34
33 27	10 45	18 40	11 48	48 03	07 16	32 25	20 25	44 22	39 28
06 09	04 26	14 35	36 03	15 22	02 07	46 48	45 12	47 11	30 19
33 32	34 25	45 17	13 26	03 37	33 35	08 13	15 26	09 18	34 25
42 38	40 01	43 31	30 33	39 11	49 41	27 44	11 39	06 19	47 23
15 06	22 08	50 44	50 11	18 16	00 41	07 47	34 25	28 10	50 03
22 35	49 36	44 21	25 12	19 44	31 51	49 18	40 36	00 27	22 12
31 04	32 17	08 23	38 32	01 47	43 53	44 04	10 27	16 00	16 33
39 00	01 50	07 28	35 02	38 00	46 47	33 29	28 41	09 23	47 48
37 32	07 02	07 48	07 41	22 13	37 27	27 12	34 21	07 04	49 34
05 03	36 07	10 15	21 48	14 44	39 39	15 09	23 23	37 31	00 25
17 37	13 41	13 39	40 14	19 48	34 18	08 18	08 06	44 26	12 45
32 24	24 30	29 13	34 39	27 44	11 20	37 40	36 46	35 22	09 09
07 45	29 12	48 35	05 38	43 11	45 18	28 14	04 37	48 38	43 12
14 08	04 04	18 17	10 33	04 32	27 37	33 42	34 41	07 41	49 14
31 38	08 31	38 30	42 10	08 09	17 32	46 15	15 43	15 31	46 45
42 34	46 31	29 03	08 32	11 06	20 21	24 16	13 17	29 34	42 31
16 00	02 48	10 34	32 14	25 39	29 31	18 37	28 50	07 28	08 24
20 15	60 11	21 31	20 49	07 35	41 16	16 17	43 36	20 26	39 38
00 49	14 10	29 01	49 28	21 30	40 15	01 07	16 04	19 09	36 12

the cross-sectional area of an individual molecule. In this two-dimensional game, the particle was moving over an area A, instead of through a volume V, and was obstructed by targets of width d, instead of cross-sectional area a. A two-dimensional version of Clausius's equation might therefore be:

$$L = \frac{A}{2Nd}$$

where N is the number of blackened square "molecules."

Q3 What value of L do you get from the data for your runs?

Q4 Using the two-dimensional version of Clausius's equation, what value do you estimate for d (the width of a square)?

Q5 How does your calculated value of d compare with the actual value? How do you explain the difference?

By permission of John Hart and Field Enterprises, Inc.

EXPERIMENT 29 BEHAVIOR OF GASES

Air is elastic or springy. You can feel this when you put your finger over the outlet of a bicycle pump and push down on the pump plunger. You can tell that there is some connection between the volume of the air in the pump and the force you exert in pumping, but the exact relationship is not obvious. About 1660, Robert Boyle performed an experiment that disclosed a very simple relationship between gas pressure and volume, but not until two centuries later was the kinetic theory of gases developed, which accounted for Boyle's law satisfactorily.

The purpose of these experiments is not simply to show that Boyle's Law and Gay Lussac's Law (which relates temperature and volume) are "true." The purpose is also to show some techniques for analyzing data that can lead to such laws.

I. Volume and Pressure

Boyle used a long glass tube in the form of a J to investigate the "spring of the air." The short arm of the J was sealed, and air was trapped in it by pouring mercury into the top of the long arm. (Apparatus for using this method may be available in your school.)

A simpler method requires only a small plastic syringe, calibrated in cc, and mounted so that you can push down the piston by piling weights on it. The volume of the air in the syringe can be read directly from the calibrations on the side. The pressure on the air due to the weights on the piston is equal to the force exerted by the weights divided by the area of the face of the piston:

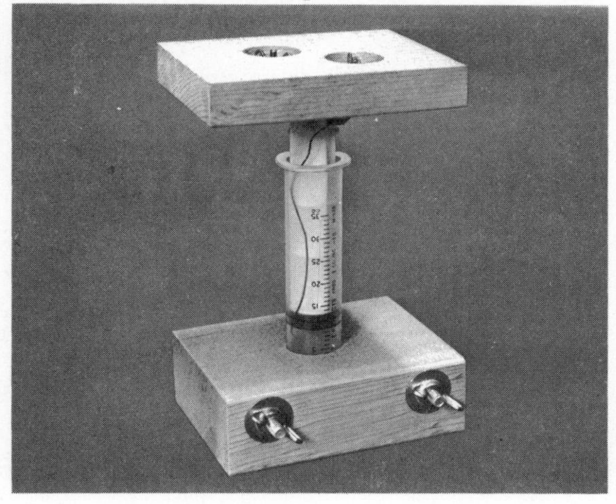

$$P_w = \frac{F_w}{A}$$

Because "weights" are usually marked with the value of their *mass*, you will have to compute the force from the relation $F_{grav} = ma_{grav}$. (It will help you to answer this question before going on: What is the weight, in newtons, of a 0.1 kg mass?)

To find the area of the piston, remove it from the syringe. Measure the diameter ($2R$) of the piston face, and compute its area from the familiar formula $A = \pi R^2$.

You will want to both decrease and increase the volume of the air, so insert the piston about halfway down the barrel of the syringe. The piston may tend to stick slightly. Give it a twist to free it and help it come to its equilibrium position. Then record this position.

Add weights to the top of the piston and each time record the equilibrium position, after you have given the piston a twist to help overcome friction.

Record your data in a table with columns for volume, weight, and pressure. Then remove the weights one by one to see if the volumes are the same with the piston coming up as they were going down.

If your apparatus can be turned over so that the weights pull out on the plunger, obtain more readings this way, adding weights to increase the volume. Record these as negative forces. (Stop adding weights before the piston is pulled all the way out of the barrel!) Again remove the weights and record the values on returning.

Interpreting your results. You now have a set of numbers somewhat like the ones Boyle reported for his experiment. One way to look for a relationship between the pressure P_w and the volume V is to plot the data on graph paper, draw a smooth simple curve through the points, and try to find a mathematical expression that would give the same curve when plotted.

Plot volume V (vertical axis) as a function of pressure P_w (horizontal axis). If you are willing to believe that the relationship between P_w and V is fairly simple, then you should try to draw a simple curve. It need not actually go through all the plot points, but should give an overall "best fit."

Since V decreases as P_w increases, you can tell before you plot it that your curve represents an "inverse" relationship. As a first guess at the mathematical description of this curve, try the simplest possibility, that $1/V$ is proportional to P_w. That is, $1/V \propto P_w$. A graph of proportional quantities is a straight line. If $1/V$ *is* proportional to P_w, then a plot of $1/V$ value against P_w will lie on a straight line.

Add another column to your data table for values of $1/V$ and plot this against P_w.

Q1 Does the curve pass through the origin?

Q2 If not, at what point does your curve cross the horizontal axis? (In other words, what is the value of P_w for which $1/V$ would be zero?) What is the physical significance of the value of P_w?

In Boyle's time, it was not understood that air is really a mixture of several gases. Do you believe you would find the same relationship between volume and pressure if you tried a variety of pure gases instead of air? If there are other gases available in your laboratory, flush out and refill your apparatus with one of them and try the experiment again.

Q3 Does the curve you plot have the same shape as the previous one?

Q4 Is the law that relates volume to pressure the same for all the gases tested in your school?

II. Volume and Temperature

Boyle suspected that the temperature of his air sample had some influence on its volume, but he did not do a quantitative experiment to find the relationship between volume and temperature. It was not until about 1880, when there were better ways of measuring temperature, that this relationship was established.

You could use several kinds of equipment to investigate the way in which volume changes with temperature. Such a piece of equipment is a glass bulb with a J tube of mercury or the syringe described above. Make sure the gas inside is dry and at atmospheric pressure. Immerse the bulb or syringe in a beaker of cold water and record the volume of gas and temperature of the water (as measured on a suitable thermometer) periodically as you slowly heat the water.

A simpler piece of equipment that will give just as good results can be made from a piece of glass capillary tubing.

Equipment note: assembling a constant-pressure gas thermometer

About 6″ of capillary tubing makes a thermometer of convenient size. The dimensions of the tube are not critical, but it is very important that the bore by *dry*. It can be dried by heating, or by rinsing with alcohol and waving it frantically—or better still, by connecting it to a vacuum pump for a few moments.

Filling with air. The dry capillary tube is dipped into a container of mercury, and the end sealed with fingertip as the tube is withdrawn, so that a pellet of mercury remains in the lower end of the tube.

The tube is held at an angle and the end tapped gently on a hard surface until the mercury pellet slides to about the center of the tube.

One end of the tube is sealed with a dab of silicone sealant; some of the sealant will go up the bore, but this is perfectly all right. The sealant is easily set by immersing it in boiling water for a few moments.

Taking measurements. A scale now must be positioned along the completed tube. The scale will be directly over the bore if a stick is placed as a spacer next to the tube and bound together with rubber bands. (A long stick makes a convenient handle.) The zero of the scale should be aligned carefully with the end of the gas column, that is, the end of the silicone seal.

In use, the thermometer should be completely immersed in whatever one wishes to measure the temperature of, and the end tapped against the side of the container gently to allow the mercury to slide to its final resting place.

Filling with some other gas. To use some gas other than air, begin by connecting a short length of rubber tubing to a fairly low-pressure supply of gas. As before, trap a pellet of mercury in the end of a capillary tube, but this time do not tap it to the center. Leave it flat so that it will be pushed to the center by the gas pressure. Open the gas valve slightly for a moment to flush out the rubber tube. With your finger tip closing off the far end of the capillary tube to prevent the mercury being

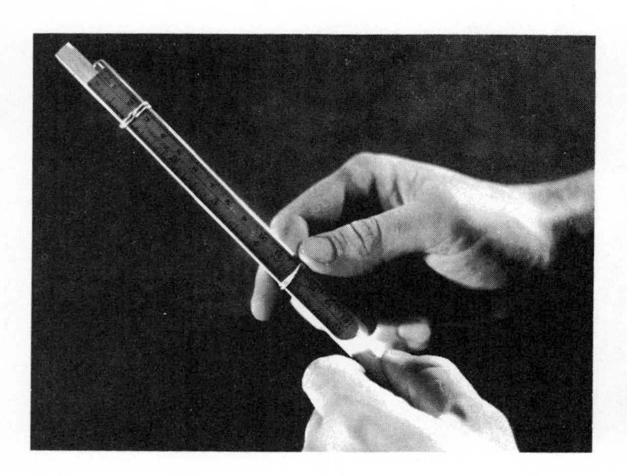

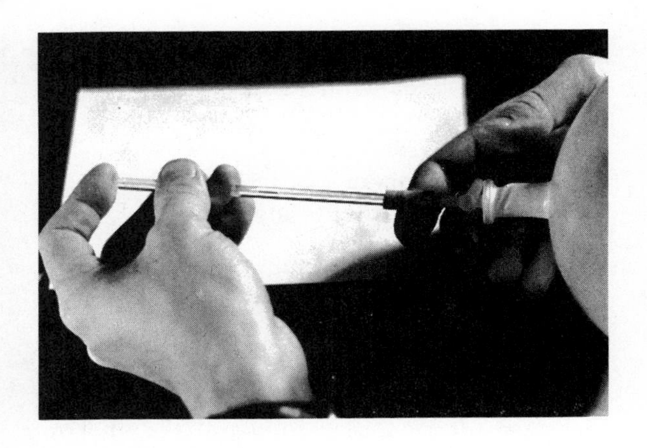

blown out, work the rubber connecting tube over the capillary tube. Open the gas valve slightly, and very cautiously release your finger very slightly for a brief instant until the pellet has been pushed to about the middle of the tube.

Remove from the gas supply, seal off as before (the end that was connected to gas supply), and attach scale. Plot a graph of volume against temperature.

Interpreting your results.

Q5 With any of the methods mentioned here, the pressure of the gas remains constant. If the curve is a straight line, does this "prove" that the volume of a gas at constant pressure is proportional to its temperature?

Q6 Remember that the thermometer you used probably depended on the expansion of a liquid such as mercury or alcohol. Would your graph have been a straight line if a different type of thermometer had been used?

Q7 If you could continue to cool the air, would there be a lower limit to the volume it would occupy?

Draw a straight line as nearly as possible through the points on your *V-T* graph and extend it to the left until it shows the approximate temperature at which the volume would be zero. Of course, you have no reason to assume that gases have this simple linear relationship all the way down to zero volume. (In fact, air would change to a liquid long before it reached the temperature indicated on your graph for zero volume.) However, some gases do show this linear behavior over a wide temperature range, and for these gases the straight line always crosses the *T*-axis at the same point. Since the volume of a sample of gas cannot be less than 0, this point represents the lowest possible temperature of the gases—the "absolute zero" of temperature.

Q8 What value does your graph give for this temperature?

III. Questions for Discussion

Both the pressure and the temperature of a gas sample affect its volume. In these experiments you were asked to consider each of these factors separately.

Q9 Were you justified in assuming that the temperature remained constant in the first experiment as you varied the pressure? How could you check this? How would your results be affected if, in fact, the temperature went up each time you added weight to the plunger?

Q10 In the second experiment the gas was at atmospheric pressure. Would you expect to find the same relationship between volume and temperature if you repeated the experiment with a different pressure acting on the sample?

Gases such as hydrogen, oxygen, nitrogen, and carbon dioxide are very different in their chemical behavior. Yet they all show the same simple relationships between volume, pressure, and temperature that you found in these experiments, over a fairly wide range of pressures and temperatures. This suggests that perhaps there is a simple physical model that will explain the behavior of all gases within these limits of temperature and pressure. Chapter 11 of the *Text* describes just such a simple model and its importance in the development of physics.

ACTIVITIES

DRINKING DUCK

A toy called a Drinking Duck (No. 60,264 in Catalogue 671, about $1.00, Edmund Scientific Co., Barrington, New Jersey 08007) demonstrates very well the conversion of heat energy into energy of gross motion by the processes of evaporation and condensation. The duck will continue to bob up and down as long as there is enough water in the cup to wet his beak.

Rather than dampen your spirit of adventure, we won't tell you how it works. First, see if you can figure out a possible mechanism for yourself. If you can't, George Gamow's book, *The Biography of Physics*, has a very good explanation. Gamow also calculates how far the duck could raise water in order to feed himself. An interesting extension is to replace the water with rubbing alcohol. What do you think will happen?

Lest you think this device useful only as a toy, an article in the June 3, 1967, *Saturday Review* described a practical application being considered by the Rand Corporation. A group of engineers built a 7-foot "bird" using Freon 11 as the working fluid. Their intention was to investigate the possible use of large-size ducks for irrigation purposes in the Nile River Valley.

MECHANICAL EQUIVALENT OF HEAT

By dropping a quantity of lead shot from a measured height and measuring the resulting change in temperature of the lead, you can get a value for the ratio of work units to heat units – the "mechanical equivalent of heat."

You will need the following equipment:

- Cardboard tube
- Stoppers
- Lead shot (1 to 2 kg)
- Thermometer

Close one end of the tube with a stopper, and put in 1 to 2 kg lead shot that has been cooled about 5°C below room temperature. Close the other end of the tube with a stopper in which a hole has been drilled and a thermometer inserted. Carefully roll the shot to this end of the tube and record its temperature. Quickly invert the tube, remove the thermometer, and plug the hole in the stopper. Now invert the tube so the lead falls the full length of the tube and repeat this quickly one hundred times. Reinsert the thermometer and measure the temperature. Measure the average distance the shot falls, which is the length of the tube minus the thickness of the layer of shot in the tube.

If the average distance the shot falls is h and the tube is inverted N times, the work you did raising the shot of mass m is:

$$\Delta W = N \times ma_g \times h$$

The heat ΔH needed to raise the temperature of the shot by an amount ΔT is:

$$\Delta H = cm\Delta T$$

where c is the specific heat capacity of lead, 0.031 cal/gC°.

The mechanical equivalent of heat is $\Delta W/\Delta H$. The accepted experimental value is 4.184 newton-meters per kilocalorie.

A DIVER IN A BOTTLE

Descartes is a name well known in physics. When we graphed motion in *Text* Sec. 1.5, we used Cartesian coordinates, which Descartes introduced. Using Snell's law of refraction, Descartes traced a thousand rays through a sphere and came up with an explanation of the rainbow. He and his astronomer friend Gassendi were a bulwark against Aristotelian physics. Descartes belonged to the generation between Galileo and Newton.

On the lighter side, Descartes is known for a toy called the Cartesian diver which was very popular in the eighteenth century when very elaborate ones were made. To make one, first you will need a column of water. You may find a large cylindrical graduate about the laboratory, the taller the better. If not, you can improvise one out of a gallon jug or any other tall glass container. Fill the container almost to the top with water. Attach a piece of glass tubing that has been fire-polished on each end. Lubricate the glass tubing and the hole in the stopper with water and carefully insert the glass tubing. Fit the rubber stopper into the top of the container as shown in Fig. 11.4.

Next construct the diver. You may limit yourself to pure essentials, namely a small pill bottle or vial, which may be weighted with wire and partially filled with water so it just barely floats *upside down* at the top of the water column. If you are so inclined, you can decorate the bottle so it looks like a real underwater swimmer (or creature, if you prefer). The essential things are that you have a diver that just floats and that the volume of water can be changed.

Now to make the diver perform, just blow momentarily on the rubber tube. According to Boyle's law, the increased pressure (transmitted by the water) decreases the volume of trapped air and water enters the diver. The buoyant force decreases, according to Archimedes' principle, and the diver begins to sink.

If the original pressure is restored, the diver rises again. However, if you are lucky, you will find that as you cautiously make him sink deeper and deeper down into the column of water, he is more and more reluctant to return to the surface as the additional surface pressure is released. Indeed, you may find a depth at which he remains almost stationary. However, this apparent equilibrium, at which his weight just equals the buoyant force, is unstable. A bit above this depth, the diver will freely rise to the surface, and a bit below this depth he will sink to the bottom of the water

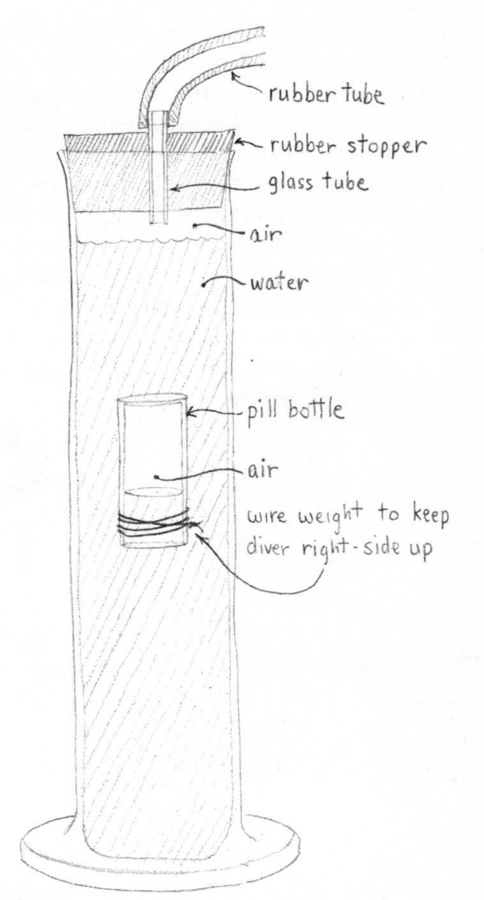

Fig. 11-4

column from which he can be brought to the surface only by vigorous sucking on the tube.

If you are mathematically inclined, you can compute what this depth would be in terms of the atmospheric pressure at the surface, the volume of the trapped air, and the weight of the diver. If not, you can juggle with the volume of the trapped air so that the point of unstable equilibrium comes about halfway down the water column.

The diver raises interesting questions. Suppose you have a well-behaved diver who "floats" at room temperature just halfway down the water column. Where will he "float" if the atmospheric drops? Where will he "float" if the water is cooled or is heated? Perhaps the ideal gas law is not enough to answer this question, and you may have to do a bit of reading about the "vapor pressure" of water.

After demonstrating the performance of your large-scale model by blowing or sucking in the rubber tube, you can mystify your audience by making a small scale model in a bottle. A plastic bottle with flat sides can act like a diaphragm which increases the pressure within as the sides are pushed together. The bottle and diver are tightly sealed. In this case, add a rubber tube leading to a *holeless* stopper. Your classmates blowing as hard as they will cannot make the diver sink; but you, by secretly squeezing the bottle, can make him perform at your command.

ROCKETS

If it is legal to set off rockets in your area, and their use is supervised, they can provide excellent projects for studying conversion from kinetic to potential energy, thrust, etc.

Ask your teacher for instructions on how to build small test stands for taking thrust data to use in predicting the maximum height, range, etc. of the rockets. (Estes Industries, Box 227, Penrose, Colorado 81240, will send a very complete free catalogue and safety rules on request.)

HOW TO WEIGH A CAR WITH A TIRE PRESSURE GAUGE

Reduce the pressure in all four of your auto tires so that the pressure is the same in each and somewhat below recommended tire pressure.

Drive the car onto four sheets of graph paper placed so that you can outline the area of the tire in contact with each piece of paper. The car should be on a reasonably flat surface (garage floor or smooth driveway). The flattened part of the tire is in equilibrium between the vertical force of the ground upward and the downward force of air pressure within.

Measure the air pressure in the tires, and the area of the flattened areas. If you use inch graph paper, you can determine the area in square inches by counting squares.

Pressure P (in pounds per square inch) is defined as F/A, where F is the downward force (in pounds) acting perpendicularly on the flattened area A (in square inches). Since the tire pressure gauge indicates the pressure *above* the normal atmospheric pressure of 15 lb/in^2 you must add this value to the gauge reading. Compute the four forces as pressure times area. Their sum gives the weight of the car.

PERPETUAL-MOTION MACHINES?

You must have heard of "perpetual-motion" machines which, once started, will continue running and doing useful work forever. These proposed devices are inconsistent with laws of thermodynamics. (It is tempting to say that they *violate* laws of thermodynamics—but this implies that laws are rules by which Nature must run, instead of descriptions men have thought up.) We now believe that it is in principle impossible to build such a machine.

But the dream dies hard! Daily there are new proposals. Thus S. Raymond Smedile, in *Perpetual Motion and Modern Research for Cheap Power* (Science Publications of Boston, 1962), maintains that this attitude of "it can't be done" negatively influences our search for new sources of cheap power. His book gives sixteen examples of proposed machines, of which two are shown here.

Number 5 represents a wheel composed of twelve chambers marked A. Each chamber contains a lead ball B, which is free to roll. As the wheel turns, each ball rolls to the lowest level possible in its chamber. As the balls roll out to the right edge of the wheel, they create a preponderance of turning effects on the right side as against those balls that roll toward the hub on the left side. Thus, it is claimed the

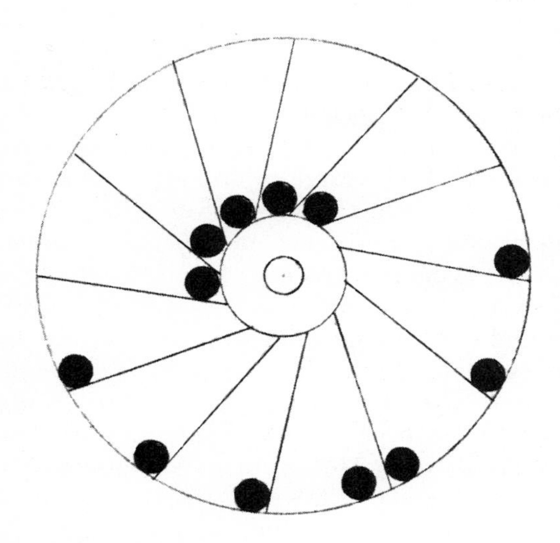

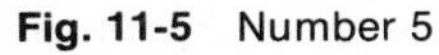

Fig. 11-5 Number 5

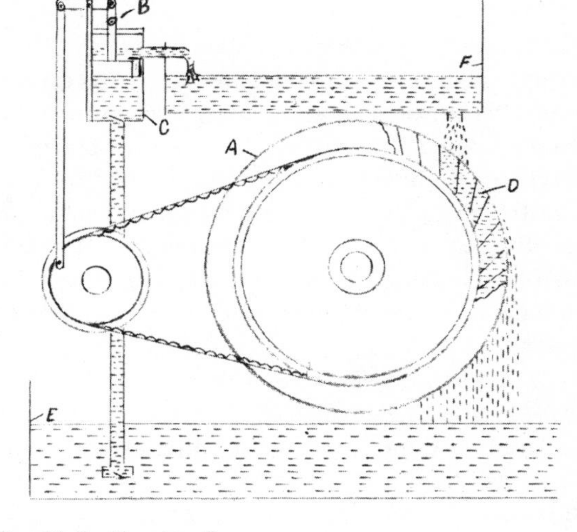

Fig. 11-6 Number 7

wheel is driven clockwise perpetually. If you think this will not work, explain why not.

Number 7 represents a water-driven wheel marked A. D represents the buckets on the perimeter of the waterwheel for receiving water draining from the tank marked F. The waterwheel is connected to pump B by a belt and wheel. As the overshot wheel is operated by water dropping on it, it operates the pump which sucks water into C from which it enters into tank F. This operation is supposed to go on perpetually. If you think otherwise, explain why.

Q1 If such machines would operate, would the conservation laws necessarily be wrong?

Q2 Is the reason that true perpetual motion machines are not found due to "theoretical" or "practical" deficiencies?

The cartoon at the left (and others of the same style which are scattered through the *Handbook*) was drawn in response to some ideas in the Project Physics Course by a cartoonist who was unfamiliar with physics. On being informed that the drawing on the left did not represent conservation because the candle wasn't a closed system, he offered the solution at the right. (Whether a system is "closed" depends, of course, upon what you are trying to conserve.)

FILM LOOP

FILM LOOP 37: REVERSIBILITY OF TIME

It may sound strange to speak of "reversing time." In the world of common experience we have no control over time direction, in contrast to the many aspects of the world that we can modify. Yet physicists have been very much concerned with the reversibility of time; perhaps no other issue so clearly illustrates the imaginative and speculative nature of modern physics.

The camera gives us a way to manipulate time. If you project film backward, the events pictured happen in reverse time order. This film has sequences in both directions, some shown in their "natural" time order and some in reverse order.

The film concentrates on the motion of objects. Consider each scene from the standpoint of time direction: Is the scene being shown as it was taken, or is it being reversed and shown backward? Many sequences are paired, the same film being used in both time senses. Is it always clear which one is forward in time and which is backward? With what types of events is it difficult to tell the "natural" direction?

The Newtonian laws of motion do *not* depend on time direction. Any filmed motion of particles following strict Newtonian laws should look completely "natural" whether seen forward or backward. Since Newtonian laws are "invariant" under time reversal, changing the direction of time, you could not tell by examining a motion obeying these laws whether the sequence is forward or backward. Any motion which could occur forward in time can also occur, under suitable conditions, with the events in the opposite order.

With more complicated physical systems, with extremely large number of particles, the situation changes. If ink were dropped into water, you would have no difficulty in determining which sequence was photographed forward in time and which backward. So certain physical phenomena at least *appear* to be irreversible, taking place in only one time direction. Are these processes *fundamentally* irreversible, or is this only some limitation on human powers? This is not an easy question to answer. It could still be considered, in spite of a fifty-year history, a frontier problem.

Reversibility of time has been used in many ways in twentieth-century physics. For example, an interesting way of viewing the two kinds of charge in the universe, positive and negative, is to think of some particles as "moving" backward in time. Thus, if the electron is viewed as moving forward in time, the positron can be considered as exactly the same particle moving backward in time. This backward motion is equivalent to the forward-moving particle having the *opposite* charge! This was one of the keys to the development of the space-time view of quantum electrodynamics which R. P. Feynman described in his Nobel Prize lecture (*Reader* 6).

For a general introduction to time reversibility, see the Martin Gardner article, "Can Time Go Backward?" originally published in *Scientific American* January, 1967. (*Reader* 6)

Chapter 12 Waves

EXPERIMENT 30
INTRODUCTION TO WAVES

In this laboratory exercise you will become familiar with a variety of wave properties in one- and two-dimensional situations.* Using ropes, springs, Slinkies, or a ripple tank, you can find out what determines the speed of waves, what happens when they collide, and how waves reflect and go around corners.

Waves in a Spring

Many waves move too fast or are too small to watch easily. But in a long "soft" spring you can make big waves that move slowly. With a partner to help you, pull the spring out on a smooth floor to a length of about 20 to 30 feet. Now, with your free hand, grasp the stretched spring two or three feet from the end. Pull the two or three feet of spring together toward the end (Fig. 12-1) and then release it, being careful *not* to let go of the fixed end with your other hand! Notice the single wave, called a pulse, that travels along the spring. In such a *longitudinal* pulse the spring coils move back and forth along the same direction as the wave travels. The wave carries energy, and hence, could be used to carry a message from one end of the spring to the other.

You can see a longitudinal wave more easily if you tie pieces of string to several of the loops of the spring and watch their motion when the spring is pulsed.

A *transverse* wave is easier to see. To make one, practice moving your hand very quickly back and forth at right angles to the stretched spring, until you can produce a pulse that travels down only one side of the spring. This pulse is called "transverse" because the individual coils of wire move at right angles to (transverse to) the length of the spring.

Perform experiments to answer the following questions about transverse pulses.

*Adapted from R. F. Brinckerhoff and D. S. Taft, Modern Laboratory Experiments in Physics, by permission of Science Electronics, Nashua, N.H.

Fig. 12-1

Q1 Does the size of the pulse change as it travels along the spring? If so, in what way?

Q2 Does the pulse reflected from the far end return to you on the same side of the spring as the original pulse, or on the opposite side?

Q3 Does a change in the tension of the spring have any effect on the speed of the pulses? When you stretch the spring farther, in effect you are changing the nature of the *medium* through which the pulses move.

Next observe what happens when waves go from one material into another—an effect called *refraction.* To one end of your spring attach a length of rope or rubber tubing (or a different kind of spring) and have your partner hold the end of this.

Q4 What happens to a pulse (size, shape, speed, direction) when it reaches the boundary between the two media? The far end of your spring is now free to move back and forth at the joint which it was unable to do before because your partner was holding it.

Have your partner detach the extra spring and once more grasp the far end of your original spring. Have him send a pulse on the same side, at the same instant you do, so that the two pulses meet. The interaction of the two pulses is called *interference*.

Q5 What happens (size, shape, speed, direction) when two pulses reach the center of the spring? (It will be easier to see what happens in the interaction if one pulse is larger than the other.)

Q6 What happens when two pulses on opposite sides of the spring meet?

As the two pulses pass on opposite sides of the spring, can you observe a point on the spring that does not move at all?

Q7 From these two observations, what can you say about the displacement caused by the addition of two pulses at the same point?

By vibrating your hand steadily back and forth, you can produce a train of pulses, a *periodic wave*. The distance between any two neighboring crests on such a periodic wave is the *wavelength*. The rate at which you vibrate your hand will determine the *frequency* of the periodic wave. Use a long spring and produce short bursts of periodic waves so you can observe them without interference by reflections from the far end.

Q8 How does the wavelength seem to depend on the frequency?

You have now observed the reflection, refraction, and interference of single waves, or pulses, traveling through different materials. These waves, however, moved only along one dimension. So that you can make a more realistic comparison with other forms of traveling energy, turn to these same wave properties spread out over a two-dimensional surface.

Waves in a Ripple Tank

Set up the ripple tank as shown in Fig. 12-2 and fill it with water to a depth of 6 or 8 mm. The tank must be leveled so that it has equal depths at the four corners.

To see what a single pulse looks like in a ripple tank, gently touch the water with your fingertip—or, better, let a drop of water fall into it from a medicine dropper held only a few millimeters above the surface.

For certain purposes it is easier to study pulses in water if their crests are straight.

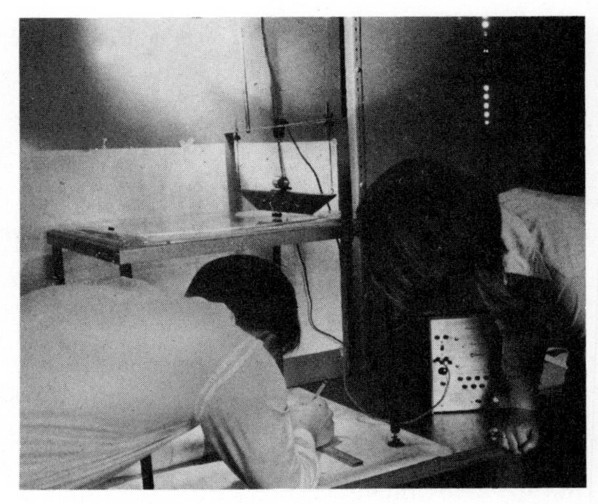

Fig. 12-2 These students are measuring the distance between two nodal lines. Note that the d.c. power cord, attached to the motor, is plugged into the Power Supply holes marked 0-5V / 5 amp. The selector switch above these holes must be in the "normal" position, and the variable resistor knob below the holes will then control the current to the motor.

To generate single straight pulses, place a three-quarter-inch dowel, or a section of a broom handle, along one edge of the tank and roll it backward a fraction of an inch. A periodic wave, a continuous train of pulses, can be formed by rolling the dowel backward and forward with a uniform frequency.

Use straight pulses in the ripple tank to observe reflection, refraction, and diffraction, and circular pulses from paint saucers to observe interference.

To study *reflection,* place a straight barrier across the path of the pulse. Try different angles between the barrier and the incoming pulse.

Q9 What is the relationship between the *direction* of the incoming pulse and the reflected one?

Replace the straight barrier with a curved one.

Q10 What is the shape of the reflected pulse?

Find the point where the reflected pulses run together.

Q11 What happens to the pulse after it converges at this point? At this point—called the *focus*—start a pulse with your finger, or a drop of water.

Q12 What is the shape of the pulse after reflection from the curved barrier?

To study *refraction*, lay a sheet of glass in the center of the tank, supported by coins if necessary, to make a very shallow area.

Q13 What happens to the wave speed at the boundary? Try varying the angle at which the pulse strikes the boundary between deep and shallow water.

Q14 What happens to the wave direction at the boundary?

Q15 How is change in direction related to change in speed?

To study *interference*, arrange two point sources side by side a few centimeters apart. When tapped gently, they should produce two pulses which interact much like the two pulses on the spring. You will see the action of interference better if you vibrate the two point sources continuously with a motor and study the resulting pattern of waves.

Q16 How does changing the wave frequency affect the original waves?

Find regions in the interference pattern where the waves from the two sources cancel and leave the water undisturbed. Find the regions where the two waves add up to create a doubly great disturbance.

Q17 Make a sketch of the interference pattern indicating these regions.

Q18 How does the pattern change as you change the wavelength?

With two-dimensional waves you can observe a new phenomenon—the behavior of a wave when it passes around an obstacle or through an opening. The spreading of the wave into the "shadow" area is called *diffraction*. Place a small barrier in the center of the tank and direct a wave-train at it.

Q19 How does the interaction with the obstacle vary with the wavelength?

Place two long barriers in the tank, leaving a small opening between them.

Q20 How does the angle by which the wave spreads out beyond the opening depend on the size of the opening?

Q21 In what way does the spread of the diffraction pattern depend on the length of the waves?

There are two ways of measuring the wavelength. You should try them both, if possible, and check one measurement against the other.

(a) One way is to place a straight barrier across the center of the tank parallel to the advancing waves. When the distance of the barrier from the generator is properly adjusted, the superposition of the advancing waves and the waves reflected from the barrier will produce *standing waves*. In other words, the reflected waves are at some points reinforcing the original waves, while at other points there is always cancellation. The points of continual cancellation are called *nodes*. As is shown in Sec. 12 of the *Text* and in *Transparency T27*, "Standing Waves," the distance between nodes is one-half wavelength.

(b) The second way to measure the wavelength is to remove the barrier and observe the moving waves with a stroboscope. Adjust the vibrator motor to the lowest frequency that will "freeze" when viewed with the stroboscope. Measure the wavelength by counting the waves along a meter stick laid across the ripple tank.

As explained in Sec. 12 of the *Text*, the wavelength times the frequency of a periodic wave gives the wave *speed*—that is, if v is the wave speed, f is the frequency, and λ is the wavelength, then $v = f\lambda$.

Q22 What value do you find for wave speed? If you have time, compare the speeds for waves of different frequencies.

If you have still more time, find out how wave speed depends on the depth of water.

EXPERIMENT 31 SOUND

You have seen how pulses and waves behave in several different media. In this experiment you compare what you have learned with the behavior of sound.

Part A: Preliminary Experiments

There are three "tracks" through the experiment. You will follow only one of the tracks and then compare your findings with those of other students. Out of the combined experiences of the class, you should be able to form some important conclusions about the nature of sound which you can then test by further experiment.

Track I – Sound

The station for Track I requires an oscillator, a power supply, two small loudspeakers, and a group of materials to be tested. A loudspeaker is the source of audible sound waves, and your ear is the detector. First connect one of the loudspeakers to the output of the oscillator and adjust the oscillator to a frequency of about 4000 cycles per second. Adjust the loudness so that the signal is just audible one meter away from the speaker. The gain-control setting should be low enough to produce a clear, pure tone. Reflections from the floor, tabletop, and hard-surfaced walls may interfere with your observations so set the sources at the edge of a table, and put soft material over any unavoidably close hard surface that could cause reflective interference.

You may find that you can localize sounds better if you make an "ear trumpet" or stethoscope from a small funnel or thistle tube and a short length of rubber tubing (Fig. 12-6). Cover the ear not in use to minimize confusion when you are hunting for nodes and maxima.

Each member of your lab group should have a chance to observe the transmission, absorption, and reflection of sound waves.

Place samples of various materials at your station between the speaker and the receiver to see how they transmit the sound wave. In a table, record your qualitative judgments as best, good, poor, etc.

Test the same materials for their ability

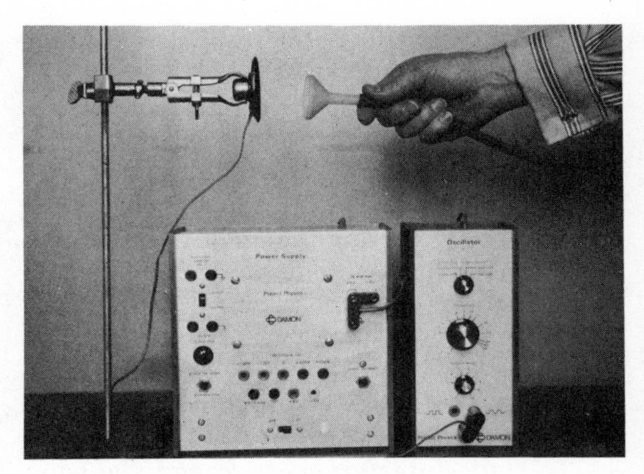

Fig. 12-3 Sound from the speaker can be detected by using a funnel and rubber hose, the end of which is placed to the ear. The Oscillator's banana plug jacks must be inserted into the −8V, +8V and ground holes of the Power Supply. Insert the speaker's plugs into the sine wave–ground recepticles of the Oscillator. Select the audio range by means of the top knob of the Oscillator and then turn on the Power Supply.

to reflect sound and record your results. Be sure that the sound is really being reflected and is not coming to your detector by some other path. You can check how the intensity varies at the detector when you move the reflector and rotate it about a vertical axis (see Fig. 12-4).

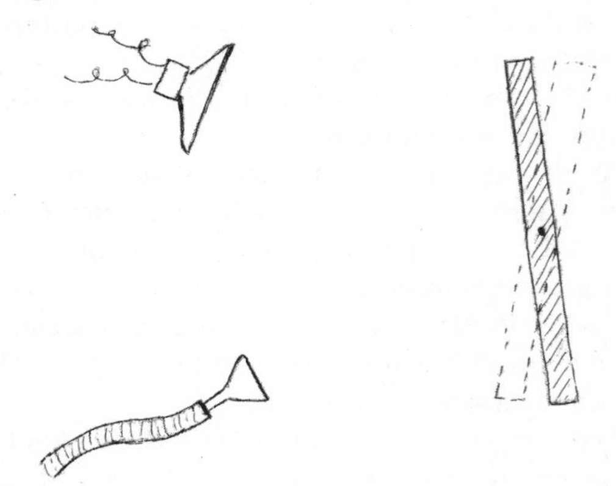

Fig. 12-4

Q1 You may find a material that neither reflects nor transmits well. What happens to the sound in such a material?

If you have available some hard, curved surface (like a hubcap or "snow saucer" or just a sheet of cardboard), try to find a focus. If you have time, you can investigate how the

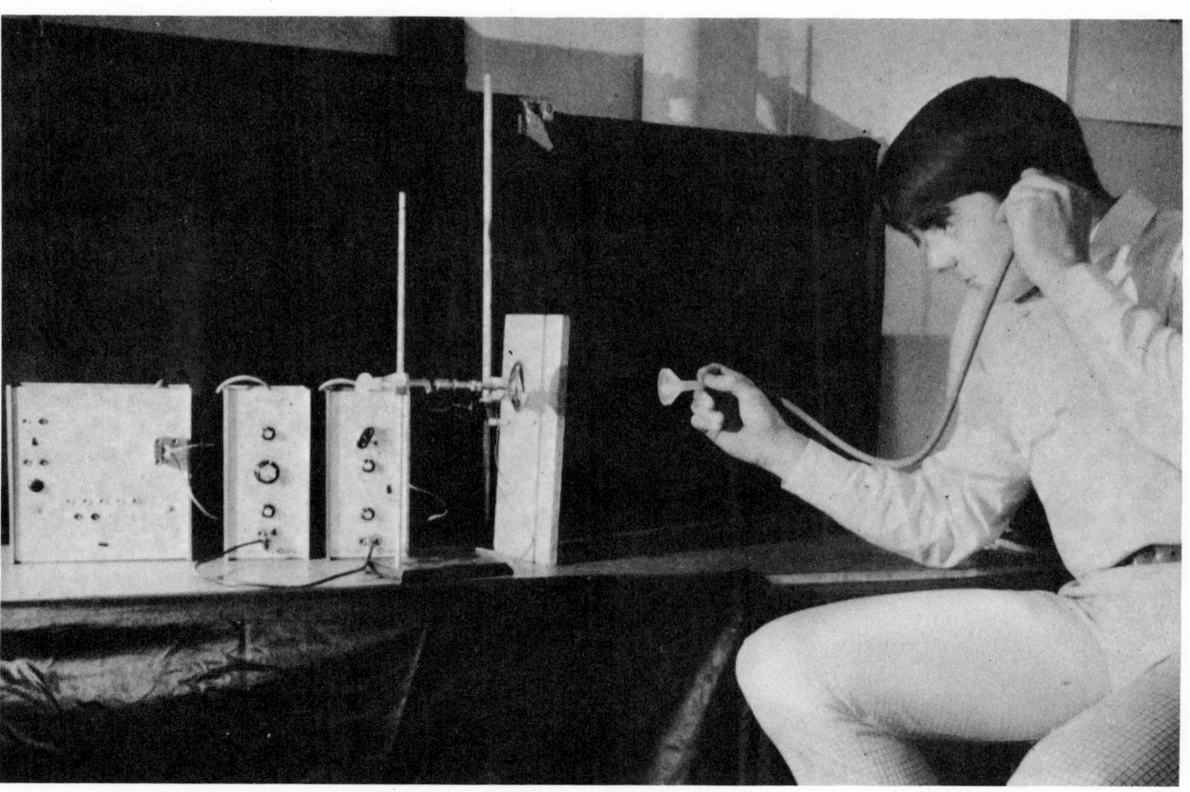

Fig. 12-6 The diffraction of sound is being detected.

position of the focus varies with the position of the source.

Refraction involves the change of velocity of waves as they pass from one material into another. In Experiment 30 you observed the "bending" of a wave front in the ripple tank as it passed from deep water into shallow water. Any kind of wave is refracted when it crosses a boundary between two materials in which the wave's speed is different.

You may observe the refraction of sound waves using a "lens" made of gas. Inflate a spherical balloon with carbon dioxide gas to a diameter of about 4 to 6 inches. Explore the area near the balloon on the side away from the source (see Fig. 12-5). Locate a point where the sound seems loudest, and then remove the balloon.

Q2 Do you notice any difference in loudness when the balloon is in place?

To study *diffraction* around an obstacle, use a thick piece of hard material about 25 cm long, mounted vertically about 25 cm directly in front of the speaker (see Fig. 12-6). Slowly probe the area about 75 cm beyond the obstacle.

Fig. 12-5 Sound is being retracted through the balloon filled with carbon dioxide.

Q3 Do you hear changes in loudness? Is there sound in the "shadow" area? Are there regions of silence where you would expect to hear sound? Does there seem to be any pattern to the areas of minimum sound?

You may also use a large piece of wood placed about 25 cm in front of the speaker with an edge aligned with the center of the source (Fig. 12-7).

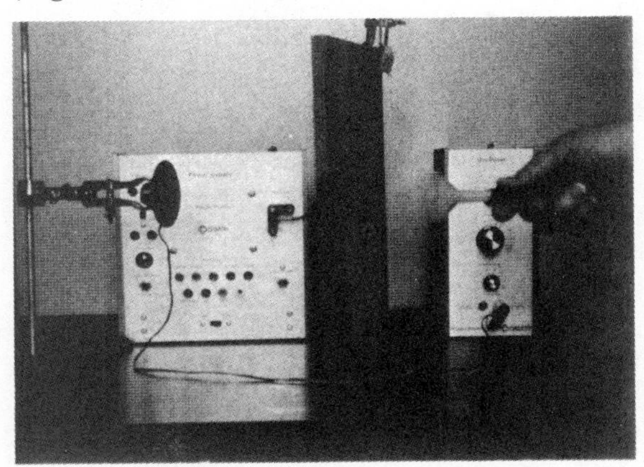

Fig. 12-7 Audible sound being diffracted around a board. A funnel and rubber hose are being used to detect the diffracted waves.

Again, explore the area inside the "shadow" zone and just outside it. Describe the pattern of sound interference that you observe.

Next investigate *standing waves*. In general, standing waves require that a good reflector be set up facing the source several wavelengths in front of it.

Set your loudspeaker about $\frac{1}{2}$ meter above and facing toward a *hard* tabletop or floor or about that distance from a hard, smooth plaster wall or other good sound reflector (see Fig. 12-8). Your ear is *most* sensitive to the changes in intensity of faint sounds, so be sure to keep the volume low.

Explore the space between the source and reflector, listening for changes in loudness. Record the positions of minimum loudness, or at least find the approximate distance between two consecutive minima. These minima are located $\frac{1}{2}$ wavelength apart.

Q4 Does the spacing of the minima depend on the intensity of the wave?

Q5 How does the wavelength change when the frequency is loudest?

Measure the wavelength of sound at several different frequencies.

Q6 What speed do you calculate for the sound waves? Is it different for different frequencies? Does it depend on intensity?

Fig. 12-8 Detecting standing sound waves produced by the interference of waves from the speaker and waves reflected from the table top.

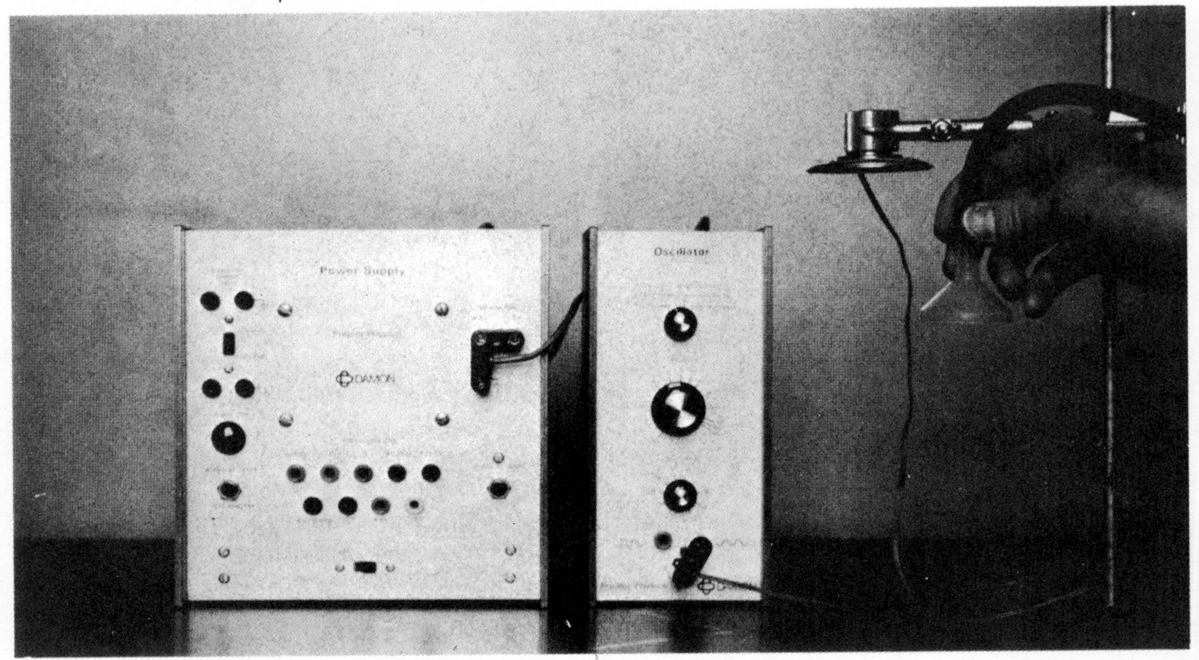

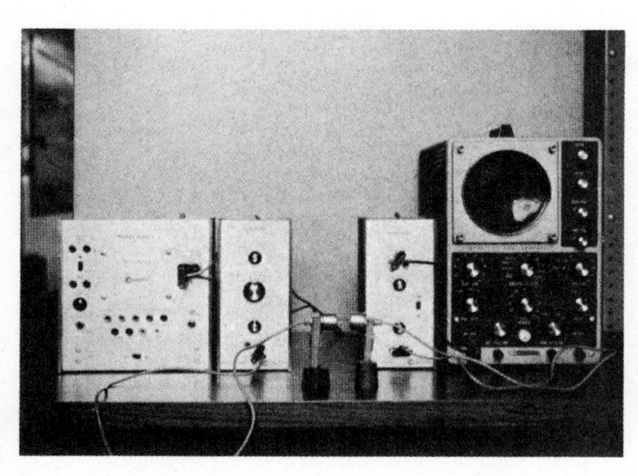

Fig. 12-9 Complete ultrasound equipment. Plug the +8v, −8v, ground jacks from the Amplifier and Oscillator into the Power Supply. Plug the coaxial cable attached to the transducer to the sine wave output of the Oscillator. Plug the coaxial cable attached to a second transducer into the input terminals of the amplifier. Be sure that the shield of the coaxial cable is attached to ground. Turn the oscillator range switch to the 5K-50K position. Turn the horizontal frequency range switch of the oscilloscope to at least 10kHz. Turn on the Oscillator and Power Supply. Tune the Oscillator for maximum reception, about 40 kilocycles.

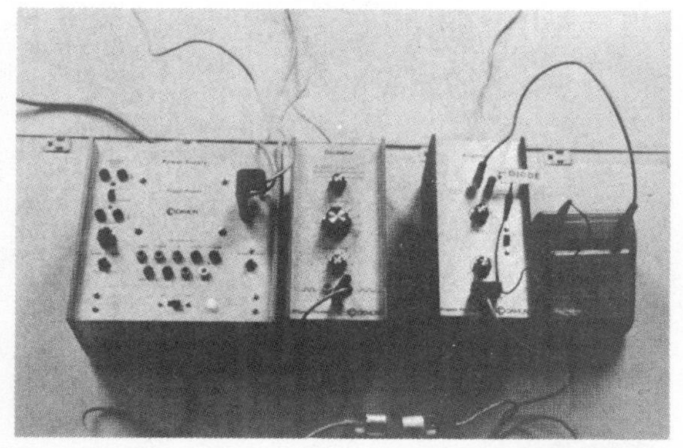

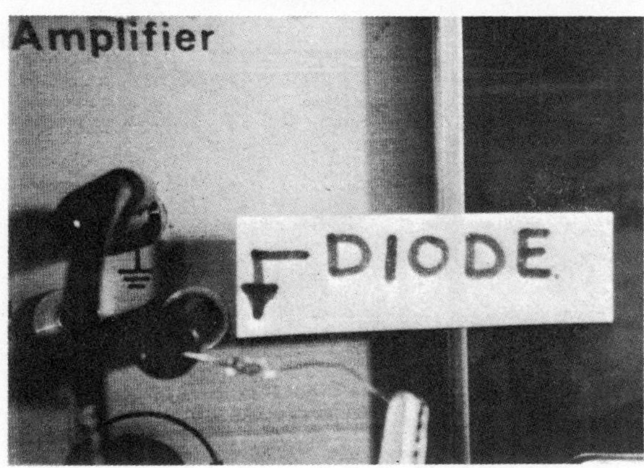

Fig. 12-10 Above, ultra sound transmitter and receiver. The signal strength is displayed on a microammeter connected to the receiver amplifier. Below, a diode connected between the amplifier and the meter, to rectify the output current. The amplifier selector switch should be turned to *ac*. The *gain* control on the amplifier should be adjusted so that the meter will deflect about full-scale for the loudest signal expected during the experiment. The *offset* control should be adjusted until the meter reads zero when there is no signal.

Track II – Ultrasound

The station for Track II requires an oscillator, power supply, and three ultrasonic transducers —crystals that transform electrical impulses into sound waves (or vice versa), and several materials to be tested. The signal from the detecting transducer can be displayed with either an oscilloscope (as in Fig. 12-9) or an amplifier and meter (Fig. 12-10). One or two of the transducers, driven by the oscillator are sources of the ultrasound, while the third transducer is a detector. Before you proceed, have the teacher check your setup and help you get a pattern on the oscilloscope screen or a reading on the meter.

The energy output of the transducer is highest at about 40,000 cycles per second, and the oscillator must be carefully "tuned" to that frequency. Place the detector a few centimeters directly in front of the source and set the oscillator range to the 5-50 kilocycle position. Tune the oscillator carefully around 40,000 cycles/second for maximum deflection of the meter or the scope track. If the signal output is too weak to detect beyond 25 cm, plug the detector transducer into an amplifier and connect the output of the amplifier to the oscilloscope or meter input.

Test the various samples at your station to see how they *transmit* the ultrasound. Record your judgments as best, good, poor, etc. Hold the sample of the material being tested close to the detector.

Test the same materials for their ability to *reflect* ultrasound. Be sure that the ultrasound is really being reflected and is not coming to your detector by some other path. You can check this by seeing how the intensity varies at the detector when you move the reflector.

Make a table of your observations.

Q7 What happens to the energy of ultrasonic waves in a material that neither reflects nor transmits well?

To observe diffraction around an obstacle, put a piece of hard material about 3 cm wide 8 or 10 cm in front of the source (see Fig. 12-11). Explore the region 5-10 cm behind the obstacle.

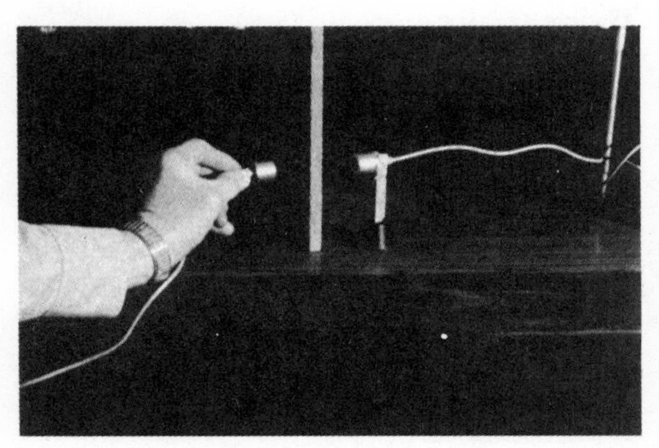

Fig. 12-11 Detecting diffraction of ultrasound around a barrier.

Q8 Do you find any signal in the "shadow" area? Do you find minima in the regions where you would expect a signal to be? Does there seem to be any pattern relating the areas of minimum and maximum signals?

Put a larger sheet of absorbing material 10 cm in front of the source so that the edge obstructs about one-half of the source.

Again probe the "shadow" area and the area near the edge to see if a pattern of maxima and minima seems to appear.

Finally, investigate the standing waves set up between a source and a reflector, such as a hard tabletop or metal plate. Place the source about 10 to 15 cm from the reflector, and probe the space between source and reflector with the detector (see Fig. 12-13).

Q9 Does the spacing of nodes depend on the intensity of the waves?

Find the approximate distance between two consecutive maxima or two consecutive minima. This distance is one half the wavelength.

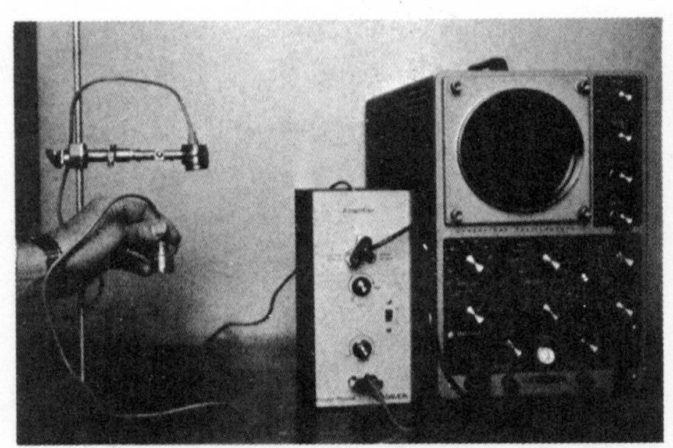

Fig. 12-13 Detecting standing ultrasound waves.

Track III – Measuring Wavelength from Interference Patterns

In Experiment 30 on waves, you set up the ripple tank with two point sources arranged to show interference. Using the same arrangement in this experiment, you will find how to compute the wavelength from measurements of the interference pattern.

You can then explain your method to your classmates working on the other two parts of the experiment. They can adapt the same method to measure the lengths of their sound and ultrasonic waves.

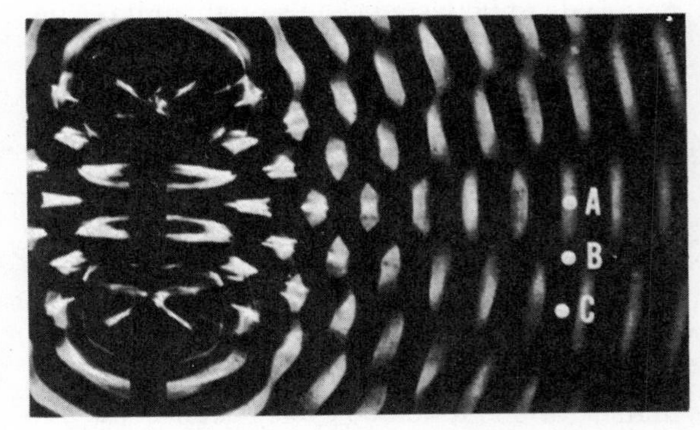

Fig. 12-14 An interference pattern in water. Two point sources vibrating in phase generate waves in a ripple tank. A and C are points of maximum disturbance (in opposite directions) and B is a point of minimum disturbance.

The waves shown in Fig. 12-14 are exactly in phase and of the same frequency as they leave the two point sources. The photograph illustrates a "frozen" or instantaneous view of these waves as they travel out from the two sources.

As you study the pattern of ripples you will notice lines along which the waves cancel almost completely so that the amplitude of the disturbance is almost zero. These lines are called *nodal lines*, or *nodes*. You have already seen nodes in your earlier experiment with standing waves in the ripple tank.

At every point along a node the waves arriving from the two sources are half a wavelength out of step, or "out of phase." This means that for a point (such as B in Fig. 12-15) to be on a line of nodes it must be $\frac{1}{2}$ or $1\frac{1}{2}$ or $2\frac{1}{2}$ . . . wavelengths farther from one source than the other.

Between the lines of nodes are regions of maximum disturbance. Points A and C in Fig. 12-14 are on lines down the center of such regions, called *antinodal lines*. Reinforcement of waves from the two sources is a maximum along these lines.

For reinforcement to occur at a point, the two waves must arrive in step or "in phase." This means that any point on a line of antinodes is a whole number of wavelengths farther from one source than the other. The tan page 120 in Chapt. 12 of the *Text* shows a mathematical argument for the relationship of wavelength to the geometry of the interference pattern. If the distance between the sources is d and the detector is at a comparatively great distance L from the sources, then d, L, and λ are related by the equations

$$\frac{\lambda}{d} = \frac{x}{L}$$

or

$$\lambda = \frac{xd}{L}$$

where x is the distance between neighboring antinodes (or neighboring nodes).

You now have a method for computing the wavelength λ from the distances that you can measure precisely.

Measure x, d, and L in your ripple tank and compute λ. Then measure λ by one of the two independent methods described at the end of Experiment 30 to see if your result is consistent with earlier work.

Q10 Considering the uncertainties of measurement in the three methods, which value do you think is best?

Study carefully just what you have done and prepare to describe your method to your classmates who followed tracks I and II. They may wish to use this method to measure wavelengths of sound.

Part B: Discussion of Results

Groups following Tracks I and II have tried to find evidence for the reflection, refraction, absorption, diffraction, and interference of sound and ultrasound.

Q11 It is possible to suppose that sound and ultrasound consist of streams of particles traveling all the way from source to receiver. Which of the various effects seen by experimenters following Tracks I and II are consistent with this particle idea?

Q12 Which of their observations are consistent with the idea that sound is a wave phenomenon?

Q13 Does the evidence give *conclusive* support to either the wave model or the particle model?

Q14 Besides being inaudible, how does ultrasound differ from sound? What evidence supports this answer?

A further difference will become clear when the method of Track III is adapted to sound and ultrasound. With the help of Track III experimenters, this method, as described in Part C, should now be tried.

Part C: Further Experiments

After a discussion of the wave properties of sound and ultrasound, Track III experimenters should describe their method for finding wavelengths. Track I and Track II experimenters should then return to the lab and try out this new method on their respective equipment. Members of Track III can be observers and consultants at this point.

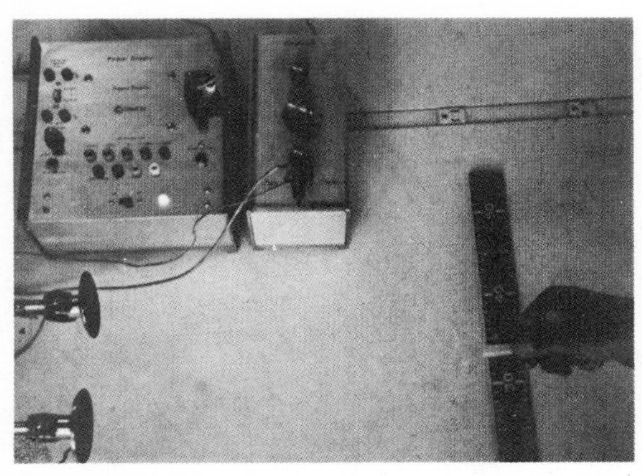

Fig. 12-15

Track I — Sound

Connect the two loudspeakers to the output of the oscillator and mount them at the edge of the table about 25 cm apart. Set the frequency at about 4,000 cycles/sec to produce a high-pitched tone. Keep the gain setting low during the entire experiment to make sure the oscillator is producing a pure tone, and to reduce reflections that would interfere with the experiment. Move your ear or "stethoscope" along a line parallel to, and about 50 cm from, the line joining the sources. Can you detect distinct maxima and minima? Move farther away from the sources; do you find any change in the pattern spacing?

Q15 What effect does a change in the source separation have on the spacing of the nodes?

Q16 What happens to the spacing of the nodes if you change the frequency of the sound? To make this experiment quantitative, work out for yourself a procedure similar to that used with the ripple tank. (Fig. 12-15.)

Q17 Measure the separation d of the source centers and the distance x between nodes and use this data to calculate the wavelength λ.

Q18 The relationship between speed v, wavelength λ, and frequency f is $v = \lambda f$. The oscillator dial gives a rough indication of the frequency (and your teacher can advise you on how to use an oscilloscope to make precise frequency settings). Using your best estimate of λ, calculate the speed of sound. If you have time, extend your data to answering the following questions:

Q19 Does the speed of the sound waves depend on the intensity of the wave?

Q20 Does the speed depend on the frequency?

Track II — Ultrasound

For sources, connect two transducers to the output of the oscillator and set them about 5 cm apart. Set the oscillator switch to the 5-50 kilocycle position. For a detector, connect a third transducer to an oscilloscope or amplifier and meter as described in Part A of the experiment. Then tune the oscillator for maximum signal from the detector when it is held near one of the sources (about 40,000 cycles/sec). Move the detector along a line parallel to and about 25 cm in front of a line connecting the sources. Do you find distinct maxima and minima? Move closer to the sources. Do you find any change in the pattern spacing?

Q21 What effect does a change in the separation of the sources have on the spacing of the nulls?

To make this experiment quantitative, work out a procedure for yourself similar to that used with the ripple tank (Fig. 12-16).

Q22 Measure the separation d of the source centers and use this data to calculate the wavelength λ.

Q23 The relationship between speed v, wavelength λ, and frequency f is $v = \lambda f$. Using your best estimate of λ, calculate the speed of sound.

Q24 Does the speed of the sound waves depend on the intensity of the wave?

Fig. 12-16

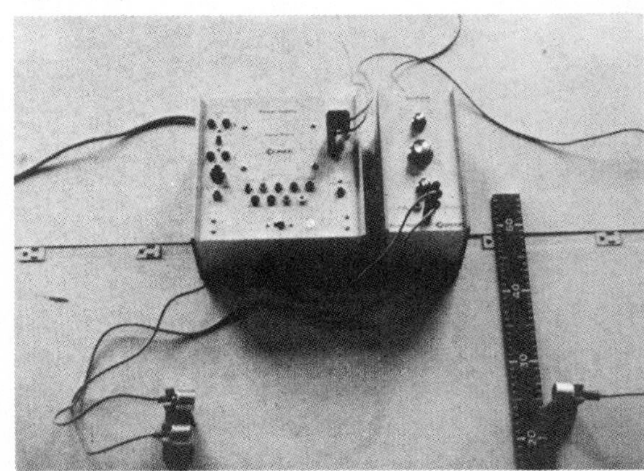

ACTIVITIES

STANDING WAVES ON A DRUM AND A VIOLIN

You can demonstrate many different patterns of standing waves on a rubber membrane using a method very similar to that used in *Film Loop* 44, "Vibrations of A Drum." If you have not yet seen this loop, view it if possible before setting up the demonstration in your lab.

Fig. 12-17 shows the apparatus in action, producing one pattern of standing waves.

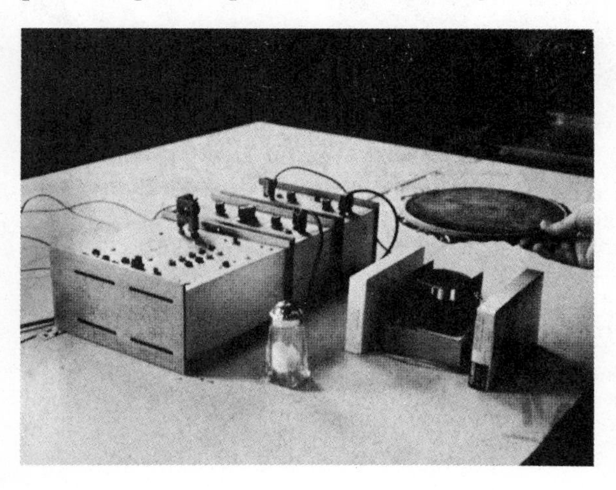

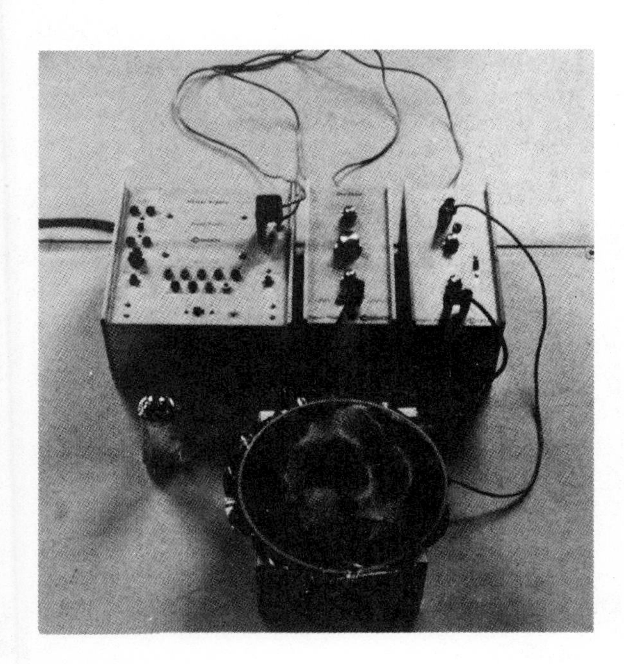

Fig. 12-17

The drumhead in the figure is an ordinary 7-inch embroidery hoop with the end of a large balloon stretched over it. If you make your drumhead in this way, use as large and as strong a balloon as possible, and cut its neck off with scissors. A flat piece of sheet rubber (dental dam) gives better results, since even tension over the entire drumhead is much easier to maintain if the rubber is not curved to begin with. Try other sizes and shapes of hoops, as well as other drumhead materials.

A 4-inch, 45-ohm speaker, lying under the drum and facing upward toward it, drives the vibrations. Connect the speaker to the output of an oscillator. If necessary, amplify the oscillator output.

Turn on the oscillator and sprinkle salt or sand on the drumhead. If the frequency is near one of the resonant frequencies of the surface, standing waves will be produced. The salt will collect along the nodes and be thrown off from the antinodes, thus outlining the pattern of the vibration. Vary the frequency until you get a clear pattern, then photograph or sketch the pattern and move on to the next frequency where you get a pattern.

When the speaker is centered, the vibration pattern is symmetrical around the center of the surface. In order to get antisymmetric modes of vibration, move the speaker toward the edge of the drumhead. Experiment with the spacing between the speaker and the drumhead until you find the position that gives the clearest pattern; this position may be different for different frequencies.

If your patterns are distorted, the tension of the drumhead is probably not uniform. If you have used a balloon, you may not be able to remedy the distortion, since the curvature of the balloon makes the edges tighter than the center. By pulling gently on the rubber, however, you may at least be able to make the tension even all around the edge.

A similar procedure, used 150 years ago and still used in analyzing the performance of violins, is shown in these photos reprinted from *Scientific American*, "Physics and Music."

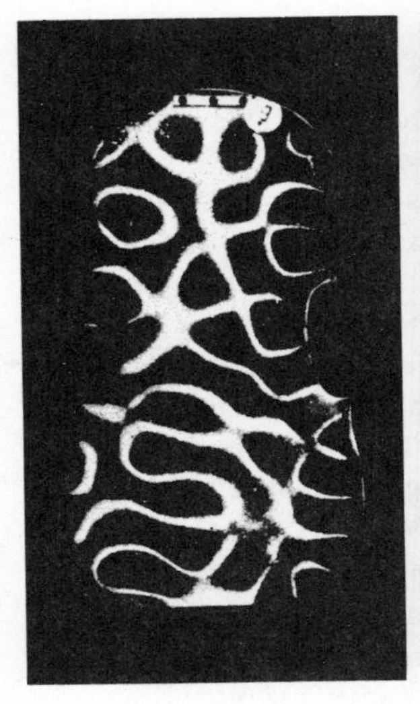

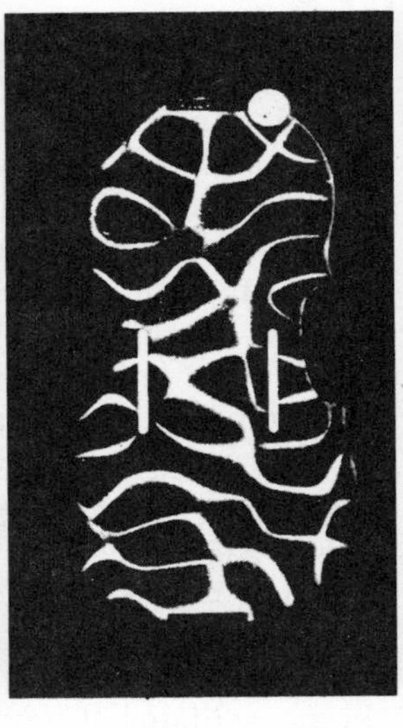

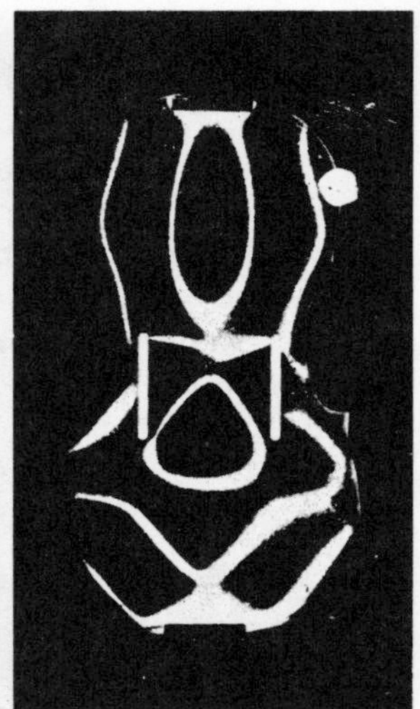

Chladni Plates indicate the vibration of the body of a violin. These patterns were produced by covering a violin-shaped brass plate with sand and drawing a violin bow across its edge. When the bow caused the plate to vibrate, the sand concentrated along quiet nodes between the vibrating areas. Bowing the plate at various points, indicated by the round white marker, produces different frequencies of vibration and different patterns. Low tones produce a pattern of a few large areas; high tones a pattern of many small areas. Violin bodies have a few such natural modes of vibration which tend to strengthen certain tones sounded by the strings. Poor violin bodies accentuate squeaky top notes. This sand-and-plate method of analysis was devided 150 years ago by the German acoustical physicist Earnst Chladni.

MOIRE PATTERNS

You are probably noticing a disturbing visual effect from the patterns in Figs. 12-18 and 12-19 below. Much of "op art" depends on similar effects, many of which are caused by moiré patterns.

If you make a photographic negative of the pattern in Fig. 12-18 or Fig. 12-19 and place it on top of the figure, you can use it to study the interference pattern produced by two point sources. The same thing is done on Transparency 28, Two Slit Interference.

Fig. 12-18

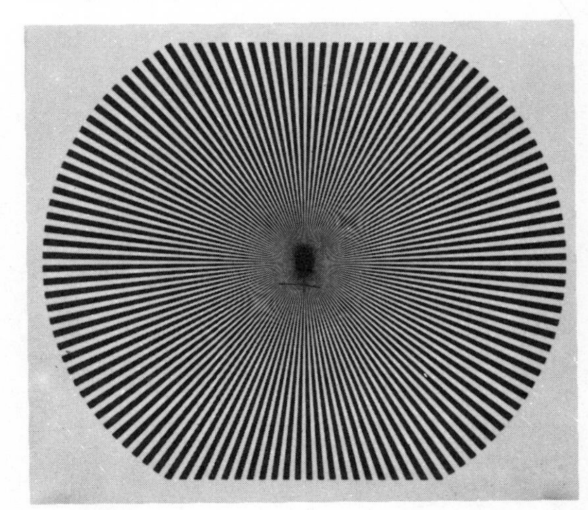

Fig. 12-19

Long before op art, there was an increasing number of scientific applications of moiré patterns. Because of the great visual changes caused by very small shifts in two regular overlapping patterns, they can be used to make measurements to an accuracy of +0.0000001%. Some specific examples of the use of moiré patterns are visualization of two- or multiple-source interference patterns, measurement of small angular shifts, measurements of diffusion rates of solids into liquids, and representations of electric, magnetic, and gravitation fields. Some of the patterns created still cannot be expressed mathematically.

Scientific American, May 1963, has an excellent article; "Moiré Patterns" by Gerald Oster and Yasunori Nishijima. *The Science of Moiré Patterns,* a book by G. Oster, is available from Edmund Scientific Co., Barrington, N.J. Edmund also has various inexpensive sets of different patterns, which save much drawing time, and that are much more precise than hand-drawn patterns.

MUSIC AND SPEECH ACTIVITIES

(a) Frequency ranges: Set up a microphone and oscilloscope so you can display the pressure variations in sound waves. Play different instruments and see how "high C" differs on them.

(b) Some beautiful oscilloscope patterns result when you display the sound of the new computermusic records which use sound-synthesizers instead of conventional instruments.

(c) For interesting background, see the following articles in *Scientific American*: "Physics and Music," July 1948; "The Physics of Violins," November 1962; "The Physics of Wood Winds," October 1960; and "Computer Music," December 1959.

(d) The Bell Telephone Company has an interesting educational item, which may be available through your local Bell Telephone office. A $33\frac{1}{3}$ LP record, "The Science of Sounds," has ten bands demonstrating different ideas about sound. For instance, racing cars demonstrate the Doppler shift, and a soprano, a piano, and a factory whistle all sound alike when overtones are filtered out electronically. The record is also available on the Folkways label FX 6136.

MEASUREMENT OF THE SPEED OF SOUND

For this experiment you need to work outside in the vicinity of a large flat wall that produces a good echo. You also need some source of loud pulses of sound at regular intervals, about one a second or less. A friend beating on a drum or something with a higher pitch will do. The important thing is that the time between one pulse and the next doesn't vary, so a metronome would help. The sound source should be fairly far away from the wall, say a couple of hundred yards in front of it.

Stand somewhere between the reflecting wall and the source of pulses. You will hear both the direct sound and the sound reflected from the wall. The direct sound will reach you first because the reflected sound must travel the additional distance from you to the wall and back again. As you approach the wall, this additional distance decreases, as does the time interval between the direct sound and the echo. Movement away from the wall increases the interval.

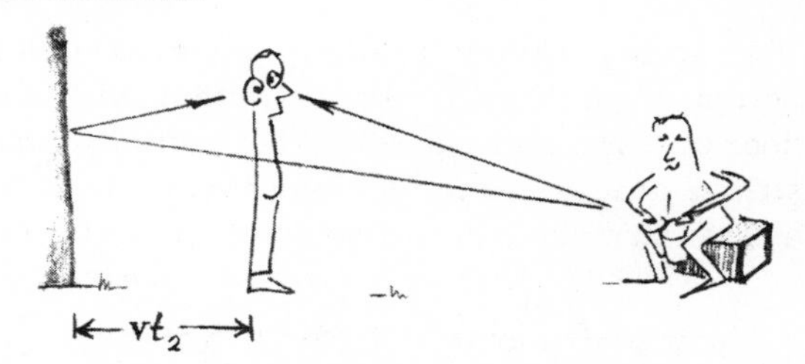

If the distance from the source to the wall is great enough, the added time taken by the echo to reach you can amount to more than the time between drum beats. You will be able to find a position at which you hear the *echo* of one pulse at the same time you hear the *direct* sound of the next pulse. Then you know that the sound took a time equal to the interval between pulses to travel from you to the wall and back to you.

Measure your distance from the source. Find the time interval between pulses by measuring the time for a large number of pulses. Use these two values to calculate the speed of sound.

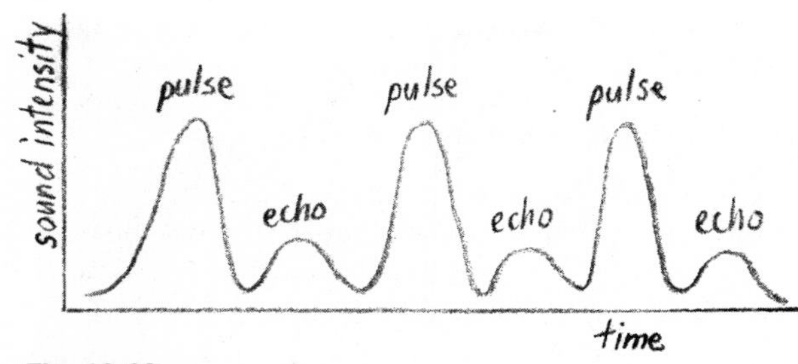

Fig. 12-20

(If you cannot get far enough away from the wall to get this synchronization, increase the speed of the sound source. If this is impossible, you may be able to find a place where you hear the echoes exactly halfway between the pulses. (Fig. 12-20.) You will hear a pulse, when an echo, then the next pulse. Adjust your position so that these three sounds seem equally spaced in time. At this point you know that the time taken for the return trip from you to the wall and back is equal to *half* the time interval between pulses.)

MECHANICAL WAVE MACHINES

Several types of mechanical wave machines are described below. They help a great deal in understanding the various properties of waves.

(a) Slinky

The spring called a Slinky behaves much better when it is freed of friction with the floor or table. Hang a Slinky horizontally from strings at least three feet long tied to rings on a wire stretched from two solid supports. Tie strings to the Slinky at every fifth spiral for proper support. (See Fig. 12-21.)

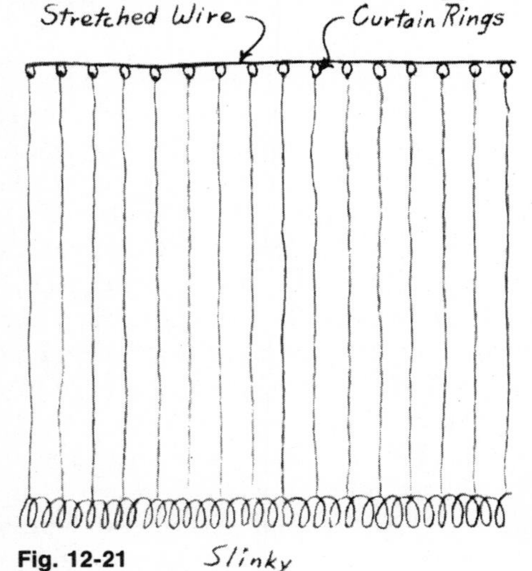

Fig. 12-21

Fasten one end of the Slinky *securely* and then stretch it out to about 20 or 30 feet. By holding onto a ten-foot piece of string tied to the end of the Slinky, you can illustrate "open-ended" reflection of waves.

See Experiment 30 for more details on demonstrating the various properties of waves.

(b) Rubber Tubing and Welding Rod

Clamp both ends of a four-foot piece of rubber tubing to a table so it is under slight tension. Punch holes through the tubing every inch with a hammer and nail. (Put a block of wood under the tubing to protect the table.)

Put enough one-foot lengths of welding rod for all the holes you punched in the tubing. Unclamp the tubing, and insert one rod in each of the holes. Hang the rubber tubing vertically, as shown below, and give its lower end a twist to demonstrate transverse waves. Performance and visibility are improved by adding weights to the ends of the rods or to the lower end of the tubing.

(c) A Better Wave Machine

An inexpensive paperback, *Similiarities in Wave Behavior,* by John N. Shive of Bell Telephone Laboratories, has instructions for building a better torsional wave machine than that described in (b) above. The book is available from Garden State-Novo, Inc., 630 9th Avenue, New York, N.Y. 10036.

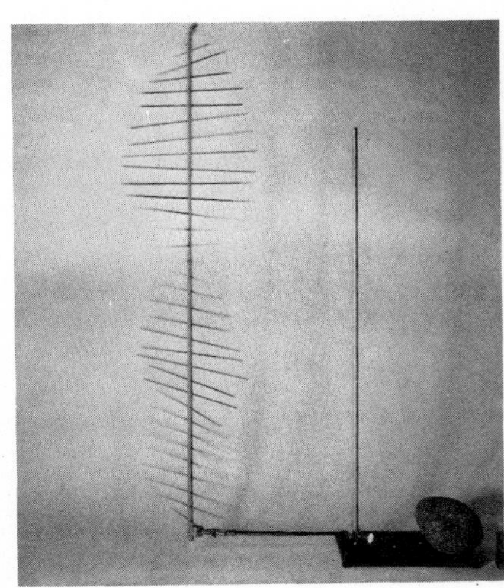

FILM LOOPS

FILM LOOP 38 SUPERPOSITION

Using this film, you study an important physical idea—superposition. The film was made by photographing patterns displayed on the face of the cathode ray tube (CRT) of an oscilloscope, similar to a television set. You may have such an oscilloscope in your laboratory.

Still photographs of some of these patterns appearing on the CRT screen are shown in Figs. 12-22 to 12-26. The two patterns at the top of the screen are called *sinusoidal.* They are not just any wavy lines, but lines generated in a precise fashion. If you are familiar with the sine and cosine functions, you will recognize them here. The sine function is the special case where the origin of the coordinate system is located where the function is zero and starting to rise. No origin is shown, so it is arbitrary as to whether one calls these sine curves, cosine curves, or some other sinusoidal type. What physical situations might lead to curves of this type? (You might want to consult books of someone else about simple harmonic oscillators.) Here the curves are produced by electronic circuits which generate an electrical voltage changing in time so as to cause the curve to be displayed on the cathode ray tube. The oscilloscope operator can adjust the magnitudes and phases of the two top functions.

The bottom curve is obtained by a point-by-point adding of the top curves. Imagine a horizontal axis going through each of the two top curves, and positive and negative distances measured vertically from this axis. The bottom curve is at each point the algebraic sum of the two points above it on the top curves, as measured from their respective axes. This point-by-point algebraic addition, when applied to actual waves, is called superposition.

Two cautions are necessary. First, you are not seeing waves, but *models* of waves. A wave is a disturbance that propagates in time, but, at least in some of the cases shown, there is no propagation. A model always has some limitations. Second, you should not think that all waves are sinusoidal. The form of whatever is propagating can be any shape. Sinusoidal waves constitute only one important class of

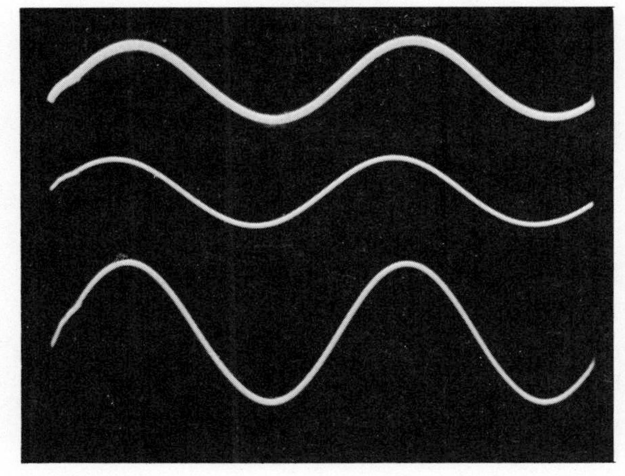

Fig. 12-22

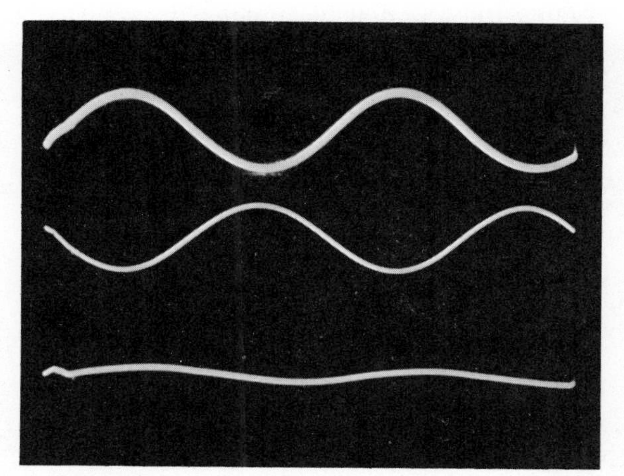

Fig. 12-23

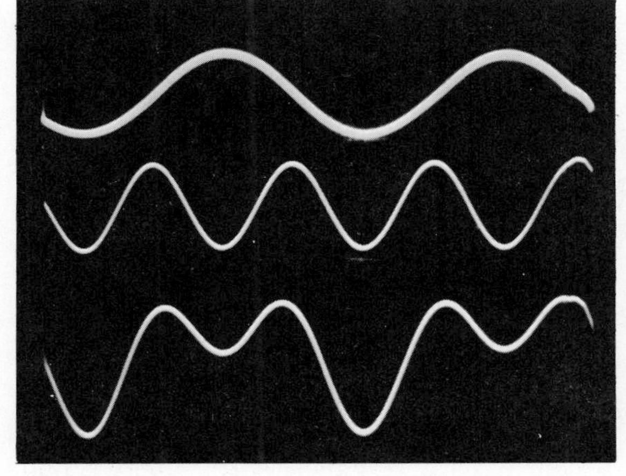

Fig. 12-24

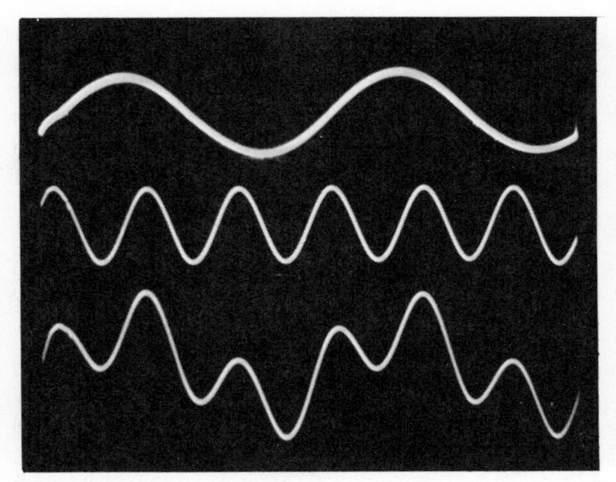

Fig. 12-25

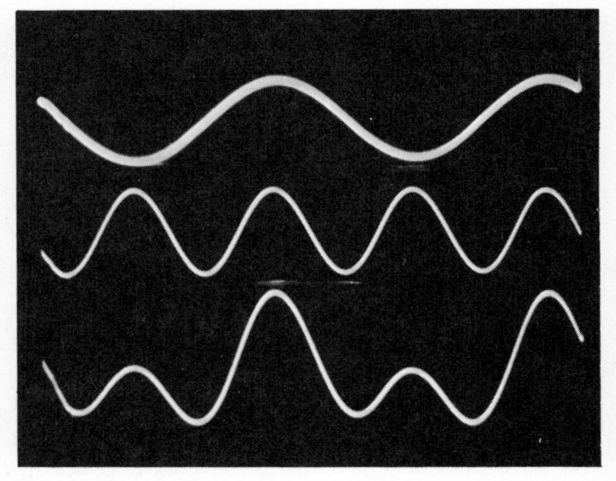

Fig. 12-26

waves. Another common wave is the pulse, such as a sound wave produced by a sharp blow on a table. The pulse is *not* a sinusoidal wave.

Several examples of superposition are shown in the film. If, as approximated in Fig. 12-22, two sinusoidal curves of equal period and amplitude are in phase, both having zeroes at the same places, the result is a double-sized function of the same shape. On the other hand, if the curves are combined out of phase, where one has a positive displacement while the other one has a negative displacement, the result is *zero* at each point (Fig. 12-23). If functions of different periods are combined (Figs. 12-24, 12-25, and 12-26), the result of the superposition is not sinusoidal, but more complex in shape. You are asked to interpret both verbally and quantitatively, the superpositions shown in the film.

FILM LOOP 39 STANDING WAVES ON A STRING

Tension determines the speed of a wave traveling down a string. When a wave reaches a fixed end of a string, it is reflected back again. The reflected wave and the original wave are superimposed or added together. If the tension (and therefore the speed) is just right, the resulting wave will be a "standing wave." Certain nodes will always stand still on the string. Other points on the string will continue to move in accordance with superposition. When the tension in a vibrating string is adjusted, standing waves can be set up when the tension has one of a set of "right" values.

In the film, one end of a string is attached to a tuning fork with a frequency of 72 vibrations per second. The other end is attached to a cylinder. The tension of the string is adjusted by sliding the cylinder back and forth.

Several standing wave patterns are shown. For example, in the third mode the string vibrates in 3 segments with 2 nodes (points of no motion) between the nodes at each end. The nodes are half a wavelength apart. Between the nodes are points of maximum possible vibration called antinodes.

You tune the strings of a violin or guitar by changing the tension on a string of fixed length, higher tension corresponding to higher pitch. Different notes are produced by placing a finger on the string to shorten the vibrating part. In this film the frequency of vibration of a string is fixed, because the string is always driven at 72 vib/sec. When the frequency remains constant, the wavelength changes as the tension is adjusted because velocity depends on tension.

A high-speed snapshot of the string at any time would show its instantaneous shape. Sections of the string move, except at the nodes. The eye sees a blurred or "time exposure" superposition of string shapes because of the frequency of the string. In the film, this blurred effect is reproduced by photographing at a slow rate: Each frame is exposed for about 1/15 sec.

Some of the vibration modes are photographed by a stroboscopic method. If the string vibrates at 72 vib/sec and frames are exposed in the camera at the rate of 70 times per sec, the string seems to go through its complete cycle of vibration at a slower frequency when projected at a normal speed. In this way, a slow-motion effect is obtained.

FILM LOOP 40 STANDING WAVES IN A GAS

Standing waves are set up in air in a large glass tube. (Fig. 12-28.) The tube is closed at one end by an adjustable piston. A loudspeaker at the other end supplies the sound wave. The speaker is driven by a variable-frequency oscillator and amplifier. About 20 watts of audio-power are used, giving notice to everyone in a large building that filming is in progress! The waves are reflected from the piston.

Fig. 12-28

A standing wave is formed when the frequency of the oscillator is adjusted to one of several discrete values. Most frequencies do *not* give standing waves. Resonance is indicated in each mode of vibration by nodes and antinodes. There is always a node at the fixed end (where air molecules cannot move) and an antinode at the speaker (where air is set into motion). Between the fixed end and the speaker there may be additional nodes and antinodes.

The patterns can be observed in several ways, two of which are used in the film. One method of making visible the presence of a stationary acoustic wave in the gas in the tube is to place cork dust along the tube. At resonance the dust is set into violent agitation by the movement of air at the antinodes; the dust remains stationary at the nodes where the air is not moving. In the first part of the film, the dust shows standing wave patterns for these frequencies:

Frequency (vib/sec)	Number of half wavelengths
230	1.5
370	2.5
530	3.5
670	4.5
1900	12.5

The pattern for $f = 530$ is shown in Fig. 12-29. From node to node is $\frac{1}{2}\lambda$, and the length of the pipe is $3\lambda + \frac{1}{2}\lambda$ (the extra $\frac{1}{2}\lambda$ is from the speaker antinode to the first node). There are, generally, $(n + \frac{1}{2})$ half-wavelengths in the fixed length, so $\lambda \propto 1/(n + \frac{1}{2})$. Since $f \propto 1/\lambda$, $f \propto (n + \frac{1}{2})$. Divide each frequency in the table by $(n + \frac{1}{2})$ to find whether the result is reasonably constant.

In all modes the dust remains motionless near the stationary piston which is a node.

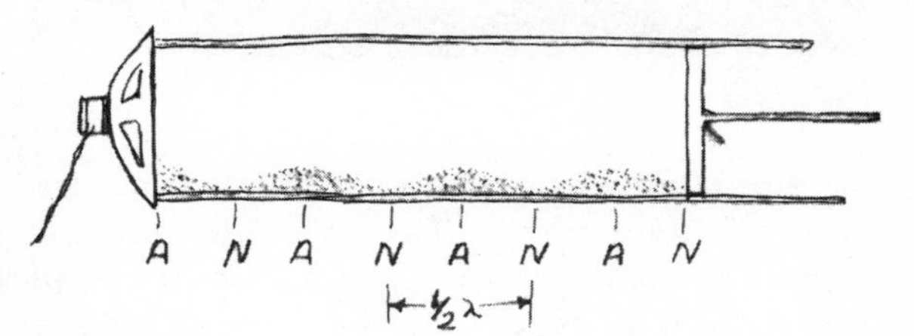

Fig. 12-29

In the second part of the film nodes and antinodes are made visible by a different method. A wire is placed in the tube near the top. This wire is heated electrically to a dull red. When a standing wave is set up, the wire is cooled at the antinodes, because the air carries heat away from the wire when it is in vigorous motion. So the wire is cooled at antinodes and glows less. The bright regions correspond to nodes where there are no air currents. The oscillator frequency is adjusted to give several standing wave patterns with successively smaller wavelengths. How many nodes and antinodes are there in each case? Can you find the frequency used in each case?

FILM LOOP 41 VIBRATIONS OF A RUBBER HOSE

You can generate standing waves in many physical systems. When a wave is set up in a medium, it is usually reflected at the boundaries. Characteristic patterns will be formed, depending on the shape of the medium, the frequency of the wave, and the material. At certain points or lines in these patterns there is no vibration, because all the partial waves passing through these points just manage to cancel each other through superposition.

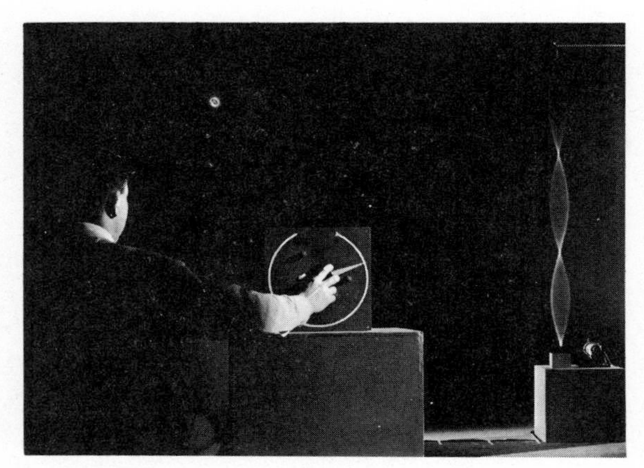

Fig. 12-29

Standing-wave patterns only occur for certain frequencies. The physical process selects a *spectrum* of frequencies from all the possible ones. Often there are an infinite number of such discrete frequencies. Sometimes there are simple mathematical relations between the selected frequencies, but for other bodies the relations are more complex. Several films in this series show vibrating systems with such patterns.

This film uses a rubber hose, clamped at the top. Such a stationary point is called a node. The bottom of the stretched hose is attached to a motor whose speed is increased during the film. An eccentric arm attached to the motor shakes the bottom end of the hose. Thus this end moves slightly, but this motion is so small that the bottom end also is a node.

The motor begins at a frequency below that for the first standing-wave pattern. As the motor is gradually speeded up, the amplitude of the vibrations increase until a well-defined steady loop is formed between the nodes. This loop has its maximum motion at the center. The pattern is half a wavelength long. Increasing the speed of the motor leads to other harmonics, each one being a standing-wave pattern with both nodes and antinodes, points of maximum vibration. These resonances can be seen in the film to occur only at certain sharp frequencies. For other motor frequencies, no such simple pattern is seen. You can count as many as eleven loops with the highest frequency case shown.

It would be interesting to have a sound track for this film. The sound of the motor is by no means constant during the process of increasing the frequency. The stationary resonance pattern corresponds to points where the motor is running much more quietly, because the motor does not need to "fight" against the hose. This sound distinction is particularly noticeable in the higher harmonics.

If you play a violin cello, or other stringed instrument, you might ask how the harmonies observed in this film are related to musical properties of vibrating strings. What can be done with a violin string to change the frequency of vibration? What musical relation exists between two notes if one of them is twice the frequency of the other?

What would happen if you kept increasing the frequency of the motor? Would you expect to get arbitrarily high resonances, or would something "give?"

FILM LOOP 42 VIBRATIONS OF A WIRE

This film shows standing-wave patterns in thin but stiff wires. The wave speed is determined by the wire's cross section and by the elastic constants of the metal. There is no external tension. Two shapes of wire, straight and circular, are used.

The wire passes between the poles of a strong magnet. When a switch is closed, a steady electric current from a battery is set up in one direction through the wire. The interaction of this current and the magnetic field leads to a downward force on the wire. When the direction of the current is reversed, the force on the wire is upward. Repeated rapid reversal of the current direction can make the wire vibrate up and down.

The battery is replaced by a source of variable frequency alternating current whose frequency can be changed. When the frequency is adjusted to match one of the natural frequencies of the wire, a standing wave builds up. Several modes are shown, each excited by a different frequency.

The first scenes show a staright brass wire, 2.4 mm in diameter (Fig. 12-30). The "boundary conditions" for motion require that, in any mode, the fixed end of the wire is a node and the free end is an antinode. (A horizontal plastic rod is used to support the wire at another node.) The wire is photographed in two ways: in a blurred "time exposure," as the eye sees it, and in "slow motion," simulated through stroboscopic photography.

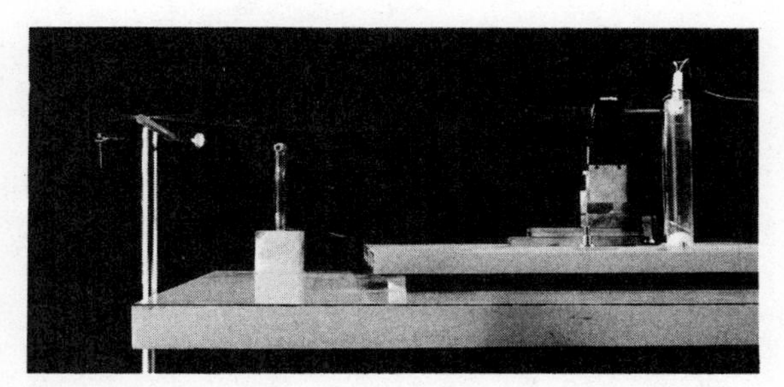

Fig. 12-30

Study the location of the nodes and antinodes in one of the higher modes of vibration. They are not equally spaced along the wire, as for vibrating string (see *Film Loop 39*). This is because the wire is stiff whereas the string is perfectly flexible.

In the second sequence, the wire is bent into a horizontal loop, supported at one point (Fig. 12-31). The boundary conditions require a node at this point; there can be additional nodes, equally spaced around the ring. Several modes are shown, both in "time exposure" and in "slow motion." To some extent the vibrating circular wire is a helpful model for the wave behavior of an electron orbit in an atom such as hydrogen; the discrete modes correspond to discrete energy states for the atom.

Fig. 12-31

FILM LOOP 43 VIBRATIONS OF A DRUM

The standing-wave patterns in this film are formed in a stretched circular rubber membrane driven by a loudspeaker. The loudspeaker is fed large amounts of power, about 30 watts, more power than you would want to use with your living room television set or phonograph. The frequency of the sound can be changed electronically. The lines drawn on the membrane make it easier for you to see the patterns.

The rim of the drum cannot move, so in all cases it must be a nodal circle, a circle that does not move as the waves bounce back and forth on the drum. By operating the camera at a frequency only slightly different from the resonant frequency, a stroboscopic effect enables you to see the rapid vibrations as if in slow motion.

In the first part of the film, the loudspeaker is directly under the membrane, and the vibratory patterns are symmetrical. In the fundamental harmonic, the membrane rises and falls as a whole. At a higher frequency, a second circular node shows up between the center and the rim.

In the second part of the film, the speaker is placed to one side, so that a different set of modes, asymmetrical modes, are generated in the membrane. You can see an antisymmetrical mode where there is a node along the diameter, with a hill on one side and a valley on the other.

Various symmetric and antisymmetric

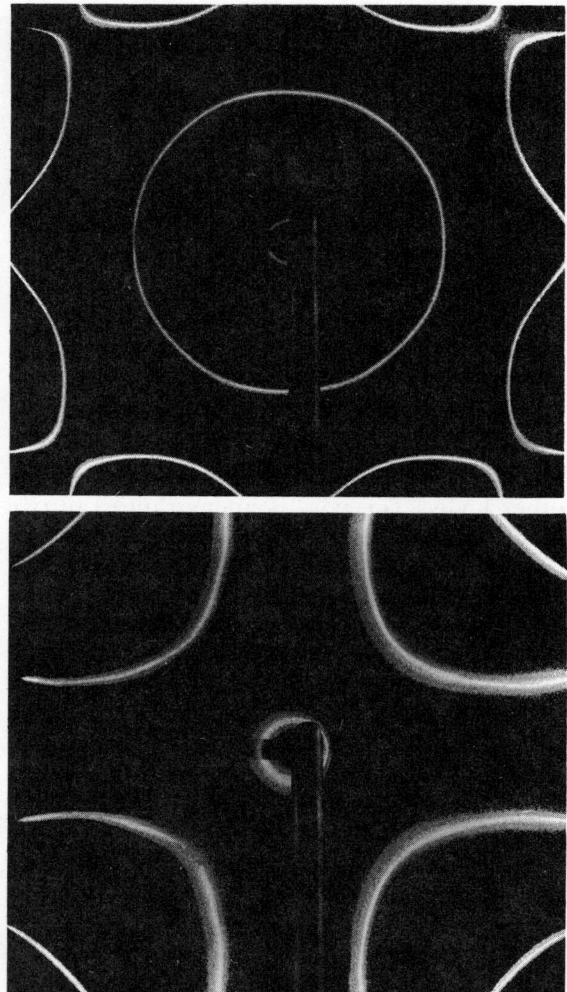

vibration modes are shown. Describe each mode, identifying the nodal lines and circles.

In contrast to the one-dimensional hose in *Film Loop 41,* there is no *simple* relation of the resonant frequencies for this two-dimensional system. The frequencies are *not* integral multiples of any basic frequency. There is a relation between values in the frequency spectrum, but it is more complex than that for the hose.

FILM LOOP 44 VIBRATIONS OF A METAL PLATE

The physical system in this film is a square metal plate. The various vibrational modes are produced by a loudspeaker, as with the vibrating membrane in *Film Loop 44.* The metal plate is clamped at the center, so that point is always a node for each of the standing-wave patterns. Because this is a stiff metal plate, the vibrations are too slight in amplitude to be seen directly. The trick used to make the patterns visible is to sprinkle sand along the plates. This sand is jiggled away from the parts of the plates which are in rapid motion, and tends to fall along the nodal lines, which are not moving. The beautiful patterns of sand are known as Chladni figures. These patterns have often been much admired by artists. These and similar patterns are also formed when a metal plate is caused to vibrate by means of a violin bow, as seen at the end of this film, and in the Activity, "Standing Waves on a Drum and a Violin."

Not all frequencies will lead to stable patterns. As in the case of the drum, these harmonic frequencies for the metal plate obey complex mathematical relations, rather than the simple arithmetic progression seen in a one-dimensional string. But again they are discrete events. As the frequency spectrum is scanned, only at certain sharp well-defined frequencies are these elegant patterns produced.

CHAPTER 12

The Project Physics Course

Handbook **4**

Light and Electromagnetism

Picture Credits, Unit 4

Cover: (upper left) Cartoon by Charles Gary Solin and reproduced by his permission only; (diffraction pattern) Cagner, Francon and Thrierr, *Atlas of Optical Phenomena,* 1962, Springer-verlag OHG, Berlin.

P. 241 (cartoon) By permission of Johnny Hart and Field Enterprises, Inc.

Pp. 243, 246, 254, 268, 277 (cartoons) By Charles Gary Solin and reproduced by his permission only.

P. 268 Collage, "Physics" by Bob Lillich.

P. 269 Burndy Library.

All photographs and notes with film loops courtesy of the National Film Board of Canada.

Photographs of laboratory equipment and of students using laboratory equipment were supplied with the cooperation of the Project Physics staff and Damon Corporation.

Contents HANDBOOK, Unit 4

Chapter 13 Light

EXPERIMENT 32
REFRACTION OF A LIGHT BEAM

You can easily demonstrate the behavior of a light beam as it passes from one transparent material to another. All you need is a semi-circular plastic dish, a lens, a small light source, and a cardboard tube. The light source from the Millikan apparatus (Unit 5) and the telescope tube with objective lens (Units 1 and 2) will serve nicely.

Making a Beam Projector

To begin with, slide the Millikan apparatus light source over the end of the telescope tube (Fig. 13-1). When you have adjusted the bulb-lens distance to produce a parallel beam of light, the beam will form a spot of constant size on a sheet of paper moved toward and away from it by as much as two feet.

Make a thin flat light beam by sticking two pieces of black tape about 1 mm apart over the lens end of the tube, creating a slit. Rotate the bulb filament until it is parallel to the slit.

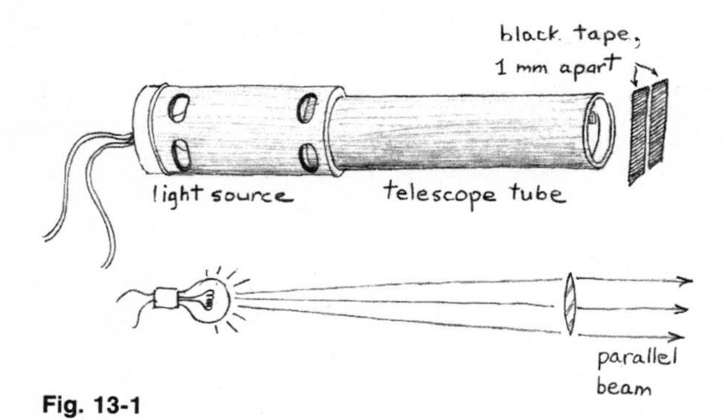

Fig. 13-1

When this beam projector is pointed slightly downward at a flat surface, a thin path of light falls across the surface. By directing the beam into a plastic dish filled with water, you can observe the path of the beam emerging into the air. The beam direction can be measured precisely by placing protractors inside and outside the dish, or by placing the dish on a sheet of polar graph paper. (Fig. 13-2)

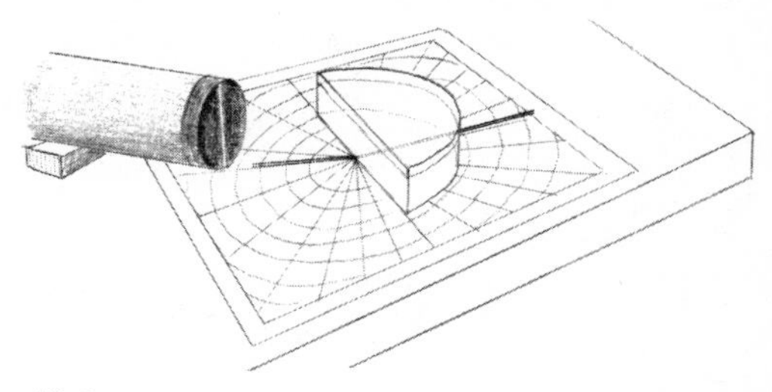

Fig. 13-2

Behavior of a Light Beam at the Boundary Between Two Media

Direct the beam at the center of the flat side of the dish, keeping the slit vertical. Tilt the projector until you can see the path of light both before it reaches the dish and after it leaves the other side.

To describe the behavior of the beam, you need a convenient way of referring to the angle the beam makes with the boundary. In physics, the system of measuring angles relative to a surface assigns a value of 0° to the perpendicular or straight-in direction. The angle at which a beam strikes a surface is called *angle of incidence*; it is the number of degrees away from the straight-in direction. Similarly, the angle at which a refracted beam leaves the boundary is called the *angle of refraction*. It is measured as the deviation from the straight-out direction. (Fig. 13-3)

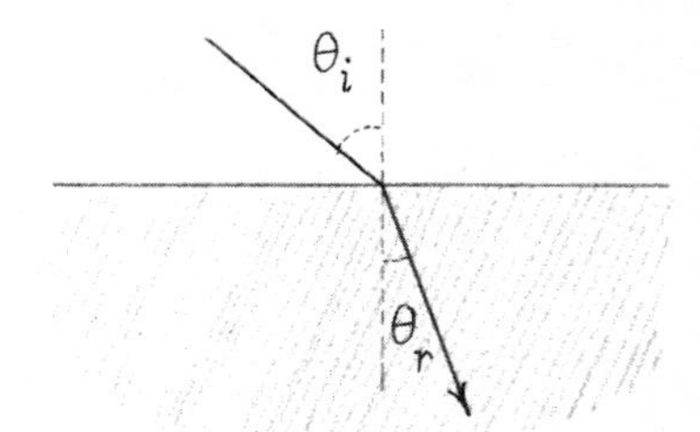

Fig. 13-3

Note the direction of the refracted beam for a particular angle of incidence. Then direct the beam perpendicularly into the rounded side of the dish where the refracted beam came out. (Fig. 13-4) At what angle does the beam now come out on the flat side? Does reversing the path like this have the same kind of effect for all angles?

Q1 Can you state a general rule about the

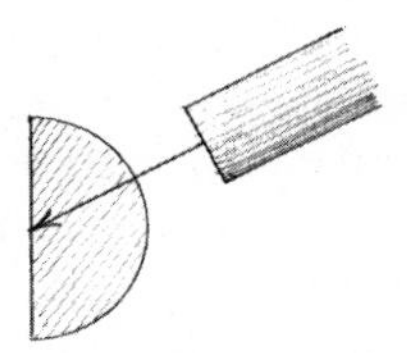
Fig. 13-4

passage of light beams through the medium?
Q2 What happens to the light beam when it reaches the edge of the container along a radius?

Change the angle of incidence and observe how the angles of the reflected and refracted beams change. (It may be easiest to leave the projector supported in one place and to rotate the sheet of paper on which the dish rests.) You will see that the angle of the *reflected* beam is always equal to the angle of the incident beam, but the angle of the *refracted* beam does not change in so simple a fashion.

Refraction Angle and Change in Speed

Change the angle of incidence in 5° steps from 0° to 85°, recording the angle of the refracted beam for each step. As the angles in air get larger, the beam in the water begins to spread, so it becomes more difficult to measure its direction precisely. You can avoid this difficulty by directing the beam into the round side of the dish instead of into the flat side. This will give the same result since, as you have seen, the light path is reversible.

Q3 On the basis of your table of values, does the angle in air seem to increase in proportion to the angle in water?

Q4 Make a plot of the angle in air against the angle in water. How would you describe the relation between the angles?

According to both the simple wave and simple particle models of light, it is not the ratio of angles in two media that will be constant, but the ratio of the *sines* of the angles. Add two columns to your data table and, referring to a table of the sine function, record the sines of the angles you observed. Then plot the sine of the angle in water against the sine of the angle in air.

Q5 Do your results support the prediction made from the models?

Q6 Write an equation that describes the relationship between the angles.

According to the wave model, the ratio of the sines of the angles in two media is the same as the ratio of the light speeds in the two media.

Q7 According to the wave model, what do your results indicate is the speed of light in water?

Color Differences

You have probably observed in this experiment that different colors of light are not refracted by the same amount. (This effect is called *dispersion.*) This is most noticeable when you direct the beam, into the round side of the dish, at an angle such that the refracted beam leaving the flat side lies very close to the flat side. The different colors of light making up the white beam separate quite distinctly.

Q8 What color of light is refracted most?

Q9 Using the relation between sines and speeds, estimate the difference in the speeds of different colors of light in water.

Other Phenomena

In the course of your observations you probably have observed that for some angles of incidence no refracted beam appears on the other side of the boundary. This phenomenon is called *total internal reflection.*

Q10 When does total internal reflection occur?

By immersing blocks of glass or plastic in the water, you can observe what happens to the beam in passing between these media and water. (Liquids other than water can be used, but be sure you don't use one that will dissolve the plastic dish!) If you lower a smaller transparent container upside-down into the water so as to trap air in it, you can observe what happens at another water-air boundary. (Fig. 13-5) A round container so placed will show what effect an air-bubble in water has on light.

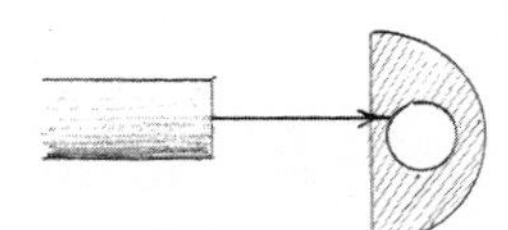
Fig. 13-5

Q11 Before trying this last suggestion, make a sketch of what you think will happen. If your prediction is wrong, explain what happened.

EXPERIMENT 33 YOUNG'S EXPERIMENT —THE WAVELENGTH OF LIGHT

You have seen how ripples on a water surface are diffracted, spreading out after having passed through an opening. You have also seen wave interference when ripples, spreading out from two sources, reinforce each other at some places and cancel out at others.

Sound and ultrasound waves behave like water waves. These diffraction and interference effects are characteristic of all wave motions. If light has a wave nature, must it not also show diffraction and interference effects?

You may shake your head when you think about this. If light is diffracted, this must mean that light spreads around corners. But you learned in Unit 2 that "light travels in straight lines." How can light both spread around corners and move in straight lines?

Simple Tests of Light Waves

Have you ever noticed light spreading out after passing through an opening or around an obstacle? Try this simple test: Look at a narrow light source several meters away from you. (A straight-filament lamp is best, but a single fluorescent tube far away will do.) Hold two fingers in front of one eye and parallel to the light source. Look at the light through the gap between them. (Fig. 13-6) Slowly squeeze your fingers together to decrease the width of the gap. What do you see? What happens to the light as you reduce the gap between your fingers to a very narrow slit?

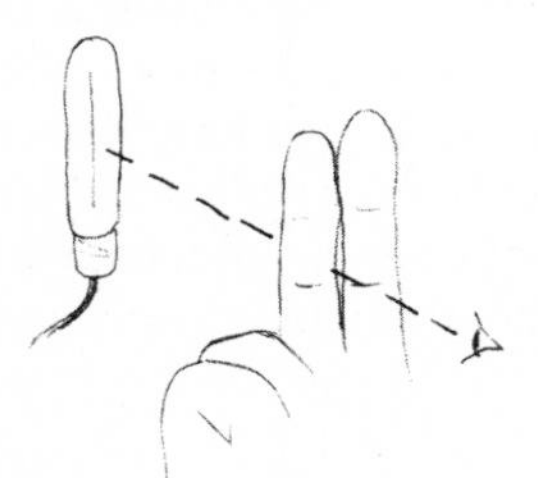

Fig. 13-6

Evidently light *can* spread out in passing through a very narrow opening between your fingers. For the effect to be noticeable, the opening must be small in comparison to the wavelength. In the case of light, the opening must be much smaller than those used in the ripple tank, or with sound waves. This suggests that light is a wave, but that it has a much shorter wavelength than the ripples on water, or sound or ultrasound in the air.

Do light waves show interference? Your immediate answer might be "no." Have you ever seen dark areas formed by the cancellation of light waves from two sources?

As with diffraction, to see interference you must arrange for the light sources to be small and close to each other. A dark photographic negative with two clear lines or slits running across it works very well. Hold up this film in front of one eye with the slits parallel to a narrow light source. You should see evidence of interference in the light coming from the two slits.

Two-slit Interference Pattern

To examine this interference pattern of light in more detail, fasten the film with the double slit on the end of a cardboard tube, such as the telescope tube without the lens. Make sure that the end of the tube is light-tight, except for the two slits. (It helps to cover most of the film with black tape.) Stick a piece of translucent "frosted" tape over the end of a narrower tube that fits snugly inside the first one. Insert this end into the wider tube, as shown in Fig. 13-7.

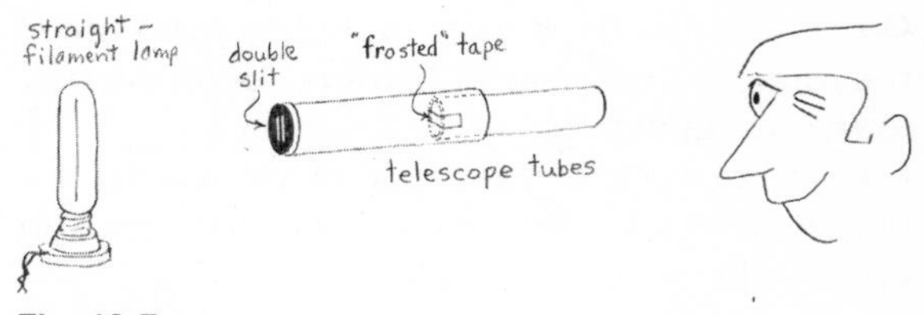

Fig. 13-7

Set up your double tube at least 5 feet away from the narrow light source with the slits parallel to the light source. With your eye about a foot away from the open end of the tube, focus your eye on the *screen*. There on the screen is the interference pattern formed by light from the two slits.

Q1 Describe how the pattern changes as you move the screen farther away from the slits.

Q2 Try putting different colored filters in

front of the double slits. What are the differences between the pattern formed in blue light and the pattern formed in red or yellow light?

Measurement of Wavelength

Remove the translucent tape screen from the inside end of the narrow tube. Insert a magnifying eyepiece and scale unit in the end toward your eye and look through it at the light. (See Fig. 13-8) What you see is a magnified view of the interference pattern in the plane of the scale. Try changing the distance between the eyepiece and the double slits.

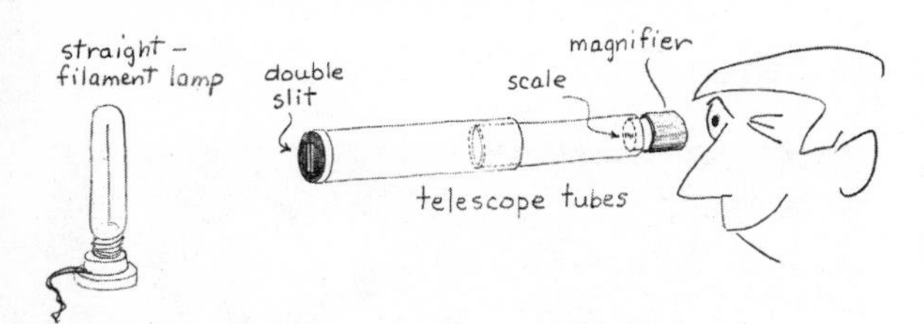

Fig. 13-8

In Experiment 31, you calculated the wavelength of sound from the relationship

$$\lambda = \frac{x}{\ell} d$$

The relationship was derived on page 120 of *Text* Chapter 12. There it was derived for water waves from two in-phase sources, but the mathematics is the same for any kind of wave. (Use of two closely-spaced slits gives a reasonably good approximation to in-phase sources.)

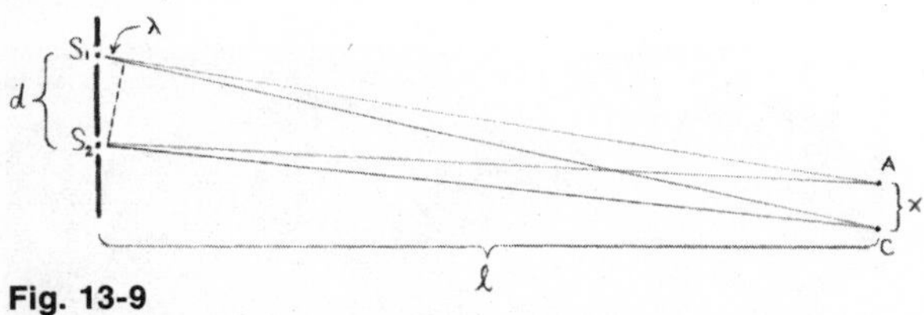

Fig. 13-9

Use the formula to find the wavelength of the light transmitted by the different colored filters. To do so, measure x, the distance between neighboring dark fringes, with the measuring magnifier (Fig. 13-10). (Remember that the smallest divisions on the scale are 0.1 mm.) You can also use the magnifier to measure d, the distance between the two slits. Place the

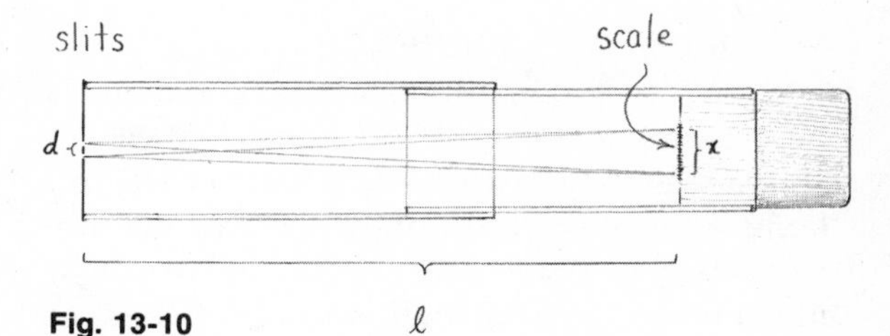

Fig. 13-10

film against the scale and then hold the film up to the light.) In the drawing, ℓ is the distance from the slits to the plane of the pattern you measure.

The speed of light in air is approximately 3×10^8 meters/second. Use your measured values of wavelength to calculate the approximate light *frequencies* for each of the colors you used.

Discussion

Q3 Why couldn't you use the method of "standing waves" (Experiment 31, "Sound") to measure the wavelength of light?

Q4 Is there a contradiction between the statement, "Light consists of waves" and the statement, "Light travels in straight lines"?

Q5 Can you think of a common experience in which the wave nature of light is noticeable?

Suggestions for Some More Experiments

1. Examine light diffracted by a circular hole instead of by a narrow slit. The light source should now be a small point, such as a distant flashlight bulb. Look also for the interference effect with light that passes through two small circular sources—pinholes in a card—instead of the two narrow slits. (Thomas Young used circular openings rather than slits in his original experiment in 1802.)

2. Look for the diffraction of light by an obstacle. For example, use straight wires of various diameters, parallel to a narrow light source. Or use circular objects such as tiny spheres, the head of a pin, etc., and a point source of light. You can use either method of observation—the translucent tape screen, or the magnifier. You may have to hold the magnifier fairly close to the diffracting obstacle.

Instructions on how to photograph some of these effects are in the activities that follow.

ACTIVITIES

THIN FILM INTERFERENCE

Take two *clean* microscope slides and press them together. Look at the light they reflect from a source (like a mercury lamp or sodium flame) that emits light at only a few definite wavelengths. What you see is the result of interference between light waves reflected at the two inside surfaces which are almost, but not quite, touching. (The thin film is the layer of air between the slides.)

This phenomenon can also be used to determine the flatness of surfaces. If the two inside surfaces are planes, the interference fringes are parallel bands. Bumps or depressions as small as a fraction of a wavelength of light can be detected as wiggles in the fringes. This method is used to measure very small distances in terms of the known wavelength of light of a particular color. If two very flat sides are placed at a slight angle to each other, an interference band appears for every wavelength of separation. (Fig. 13-11)

How could this phenomenon be used to measure the thickness of a very fine hair or very thin plastic?

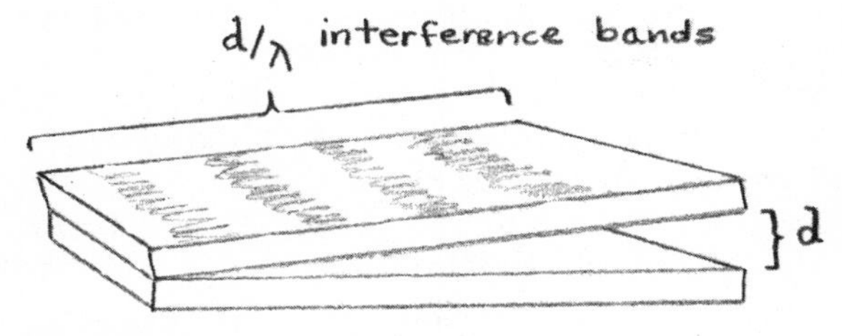

Fig. 13-11

An alternative is to focus an ordinary camera on "infinity" and place it directly behind the magnifier, using the same setup as described in Suggestions for Some More Experiments on page 237.

HANDKERCHIEF DIFFRACTION GRATING

Stretch a linen or cotton handkerchief of good quality and look through it at a distant light source, such as a street light about one block away. You will see an interesting diffraction pattern. (A window screen or cloth umbrella will also work.)

PHOTOGRAPHING DIFFRACTION PATTERNS

Diffraction patterns like those pictured here can be produced in your lab or at home. The photos in Figs. 13-12 and 13-13 were produced with the setup diagrammed in Fig. 13-14.

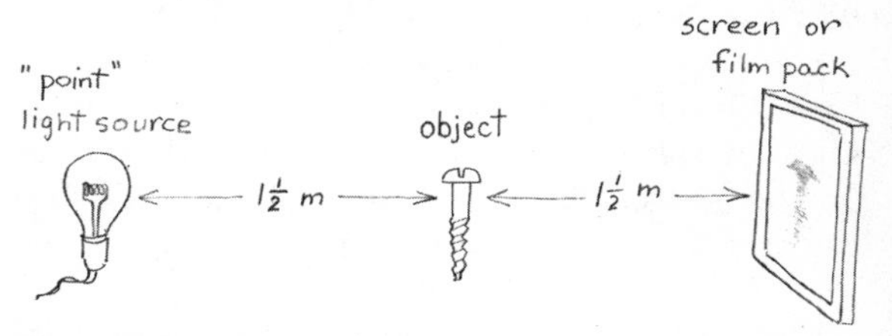

Fig. 13-12

Fig. 13-13

Fig. 13-14

To photograph the patterns, you must have a darkroom or a large, light-tight box. Figure 13-13 was taken using a Polaroid 4 × 5 back on a Graphic press camera. The lens was removed, and a single sheet of 3000-ASA-speed Polaroid film was exposed for 10 seconds; a piece of cardboard in front of the camera was used as a shutter.

As a light source, use a $1\frac{1}{2}$-volt flashlight bulb and AA cell. Turn the bulb so the end of the filament acts as a point source. A red (or blue) filter makes the fringes sharper. You can see the fringes by examining the shadow on the screen with the 10x magnifier. Razor blades, needles, or wire screens make good objects.

POISSON'S SPOT

A bright spot can be observed in a photograph of the center of some shadows, like that shown in the photograph, Fig. 13-15. To see this,

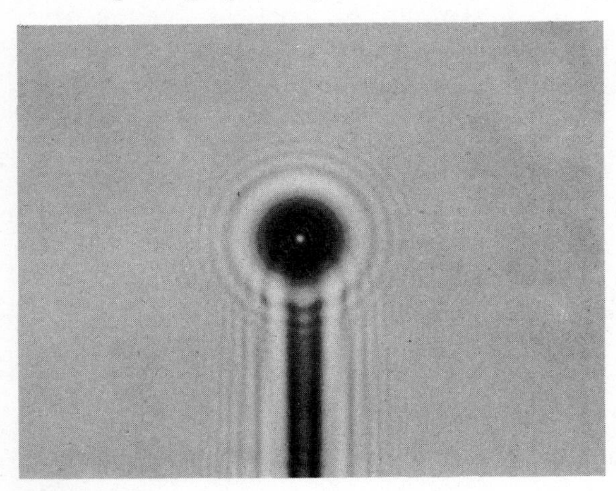

Fig. 13-15

set up a light source, obstacle, and screen as shown in Fig. 13-16. Satisfactory results require complete darkness. Try a two-second exposure with Polaroid 3000-ASA-film.

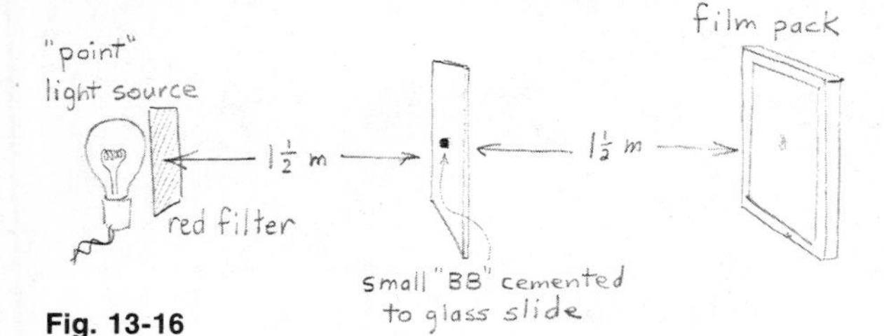

Fig. 13-16

PHOTOGRAPHIC ACTIVITIES

The number of photography activities is limitless, so we shall not try to describe many in detail. Rather, this is a collection of suggestions to give you a "jumping-off" point for classroom displays, demonstrations, and creative work.

(a) History of photography

Life magazine, December 23, 1966, had an excellent special issue on photography. How the world's first trichromatic color photograph was made by James Clerk Maxwell in 1861 is described in the Science Study Series paperback, *Latent Image,* by Beaumont Newhall. Much of the early history of photography in the United States is discussed in *Mathew Brady,* by James D. Horan, Crown Publishers.

(b) Schlieren photography

For a description and instructions for equipment, see *Scientific American,* February 1964, p. 132-3.

(c) Infrared photography

Try to make some photos like that shown on page 14 of your Unit 4 *Text.* Kodak infrared film is no more expensive than normal black and white film, and can be developed with normal developers. If you have a 4 × 5 camera with a Polaroid back, you can use 4 × 5 Polaroid infrared film sheets. You may find the Kodak Data Book M-3, "Infrared and Ultraviolet Photography," very helpful.

COLOR

One can easily carry out many intriguing experiments and activities related to the physical, physiological, and psychological aspects of color. Some of these are suggested here.

(a) Scattered light

Add about a quarter-teaspoon of milk to a drinking glass full of water. Set a flashlight about two feet away so it shines into the glass. When you look through the milky water toward the light, it has a pale orange color. As you move around the glass, the milky water appears to change color. Describe the change and explain what causes it.

(b) The rainbow effect

The way in which rainbows are produced can be demonstrated by using a glass of water as a large cylindrical raindrop. Place the glass on a piece of white paper in the early morning or late afternoon sunlight. To make the rainbow more visible, place two books upright, leaving a space a little wider than the glass between them, so that the sun shines on the glass but the white paper is shaded (Fig. 13-17). The rainbow will be seen on the backs of the books. What is the relationship between the arrangement of colors of the rainbow and the side of the glass that the light entered? This and other interesting optical effects are described in *Science for the Airplane Passenger,* by Elizabeth A. Wood, Houghton-Mifflin Co., 1968.

Fig. 13-17

(c) Color vision by contrast (Land effect)
Hook up two small lamps as shown in Fig. 13-18. Place an obstacle in front of the screen

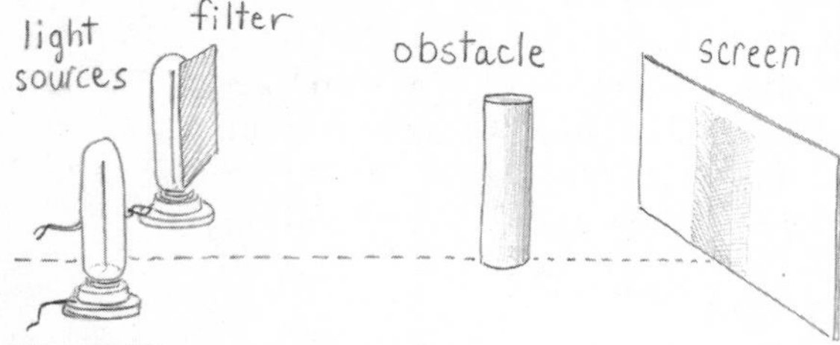

Fig. 13-18

so that adjacent shadows are formed on the screen. Do the shadows have any tinge of color? Now cover one bulb with a red filter and notice that the other shadow appears green by contrast. Try this with different colored filters and vary the light intensity by moving the lamps to various distances.

(d) Land two-color demonstrations
A different and interesting activity is to demonstrate that a full-color picture can be created by simultaneously projecting two black-and-white transparencies taken through a red and a green filter. For more information see *Scientific American,* May 1959; September 1959; and January 1960.

POLARIZED LIGHT

The use of polarized light in basic research is spreading rapidly in many fields of science. The laser, our most intense laboratory source of polarized light, was invented by researchers in electronics and microwaves. Botanists have discovered that the direction of growth of certain plants can be determined by controlling the polarization form of illumination, and zoologists have found that bees, ants, and various other creatures routinely use the polarization of sky light as a navigational "compass." High-energy physicists have found that the most modern particle accelerator, the synchrotron, is a superb source of polarized x-rays. Astronomers find that the polarization of radio waves from planets and from stars offers important clues as to the dynamics of those bodies. Chemists and mechanical engineers are finding new uses for polarized light as an analytical tool. Theoreticians have discovered shortcut methods of dealing with polarized light algebraically. From all sides, the onrush of new ideas is imparting new vigor to this classical subject.

A discussion of many of these aspects of the nature and application of polarized light, including activities such as those discussed below, can be found in *Polarized Light,* by W. A. Shurcliff and S. S. Ballard, Van Nostrand Momentum Book #7, 1964.

(a) Detection
Polarized light can be detected directly by the unaided human eye, provided one knows what to look for. To develop this ability, begin by staring through a sheet of Polaroid at the sky for about ten seconds. Then quickly turn the polarizer 90° and look for a pale yellow brush-shaped pattern similar to the sketch in Fig. 13-19.

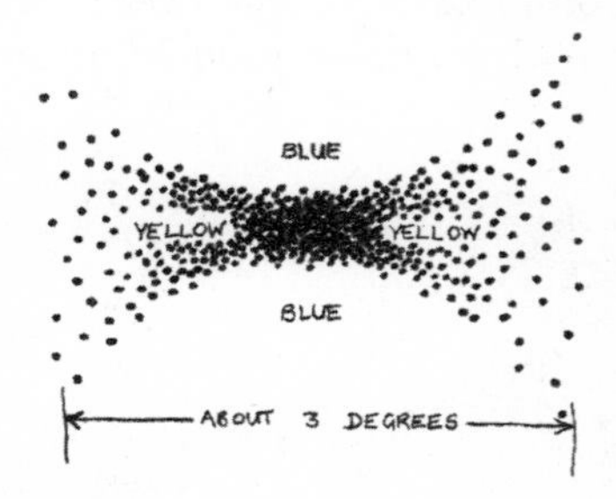

Fig. 13-19

The color will fade in a few seconds, but another pattern will appear when the Polaroid is again rotated 90°. A light blue filter behind the Polaroid may help.

How is the axis of the brush related to the direction of polarization of light transmitted by the Polaroid? (To determine the polarization direction of the filter, look at light reflected from a horizontal non-metallic surface, such as a table top. Turn the Polaroid until the reflected light is brightest. Put tape on one edge of the Polaroid parallel to the floor to show the direction of polarization.) Does the axis of the yellow pattern always make the same angle with the axis of polarization?

Some people see the brush most clearly when viewed with circularly polarized light. To make a circular polarizer, place a piece of Polaroid in contact with a piece of cellophane with its axis of polarization at a 45° angle to the fine stretch lines of the cellophane.

(b) Picket fence analogy

At some time you may have had polarization of light explained to you in terms of a rope tied to a fixed object at one end, and being shaken at the other end. In between, the rope passes through two picket fences (as in Fig. 13-20), or through two slotted pieces of cardboard. This analogy suggests that when the slots are parallel the wave passes through, but when the slots are perpendicular the waves are stopped. (You may want to use a rope and slotted boards to see if this really happens.)

Place two Polaroid filters parallel to each other and turn one so that it blacks out the light completely. Then place a third filter between the first two, and rotate it about the axis of all three. What happens? Does the picket fence analogy still hold?

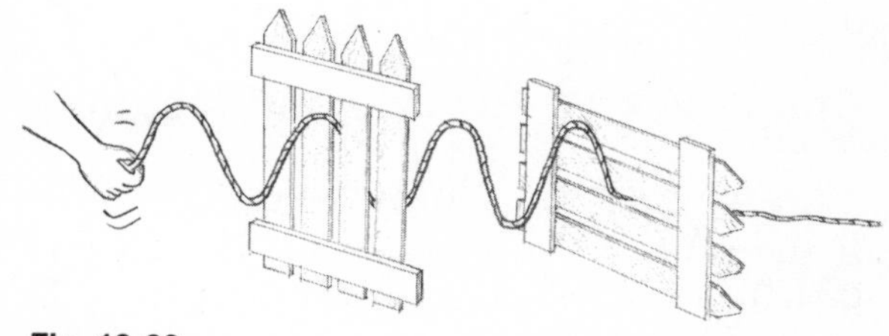

Fig. 13-20

A similar experiment can be done with microwaves using parallel strips of tinfoil on cardboard instead of Polaroid filters.

MAKE AN ICE LENS

Dr. Clawbonny, in Jules Verne's *The Adventures of Captain Hatteras,* was able to light a fire in −48° weather (thereby saving stranded travelers) by shaping a piece of ice into a lens and focusing it on some tinder. If ice is clear, the sun's rays pass through with little scattering. You can make an ice lens by freezing water in a round-bottomed bowl. Use boiled, distilled water, if possible, to minimize problems due to gas bubbles in the ice. Measure the focal length of the lens and relate this length to the radius of the bowl. (Adapted from *Physics for Entertainment,* Y. Perelman, Foreign Languages Publishing House, Moscow, 1936.)

B.C. **By John Hart**

Chapter 14 Electric and Magnetic Fields

EXPERIMENT 34 ELECTRIC FORCES I

If you walk across a carpet on a dry day and then touch a metal doorknob, a spark may jump across between your fingers and the knob. Your hair may crackle as you comb it. You have probably noticed other examples of the electrical effect of rubbing two objects together. Does your hair ever stand on end after you pull off your clothes over your head? (This effect is particuarly strong if the clothes are made of nylon, or another synthetic fiber.)

Small pieces of paper are attracted to a plastic comb or ruler that has been rubbed on a piece of cloth. Try it. The attractive force is often large enough to lift scraps of paper off the table, showing that it is stronger than the gravitational force between the paper and the entire earth!

The force between the rubbed plastic and the paper is an electrical force, one of the four basic forces of nature.

In this experiment you will make some observations of the nature of the electrical force. If you do the next experiment, Electric Forces II, you will be able to make quantitative measurements of the force.

Forces between Electrified Objects

Stick an 8-inch length of transparent tape to the tabletop. Press the tape down well with your finger leaving an inch or so loose as a handle. Carefully remove the tape from the table by pulling on this loose end, preventing the tape from curling up around your fingers.

To test whether or not the tape became electrically charged when you stripped it from the table, see if the non-sticky side will pick up a scrap of paper. Even better, will the paper jump up from the table to the tape?

Q1 Is the tape charged? Is the paper charged?

So far you have considered only the effect of a charged object (the tape) on an uncharged object (the scrap of paper). What effect does a charged object have on another charged object? Here is one way to test it.

Charge a piece of tape by sticking it to the table and peeling it off as you did before. Suspend the tape from a horizontal wood rod, or over the edge of the table. (Don't let the lower end curl around the table legs.)

Now charge a second strip of tape in the same way and bring it close to the first one. It's a good idea to have the two non-sticky sides facing.

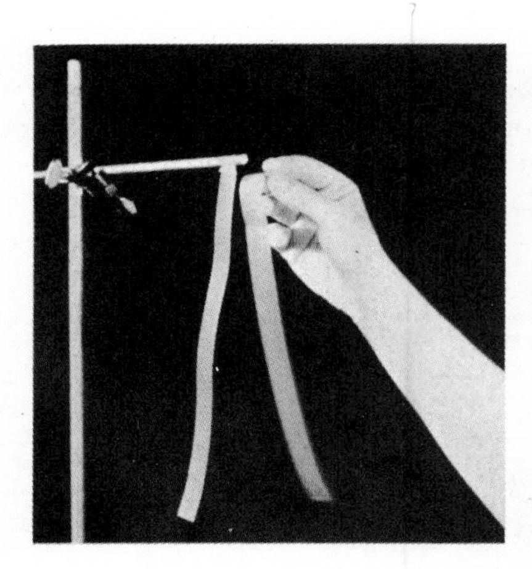

Q2 Do the two tapes affect each other? What kind of force is it – attractive or repulsive?

Hang the second tape a few inches away from the first one. Proceed as before and electrify a third piece of tape. Observe the reaction between this and your first two tapes. Record all your observations. Leave only the first tape hanging from its support—you will need it again shortly. Discard the other two tapes.

Stick down a new piece of tape (A) on the table and stick another tape (B) over it. Press them down well. Peel the stuck-together tapes from the table. To remove the net charge the pair will have picked up, run the nonsticky side of the pair over a water pipe or your lips. Check the pair with the original test strip to be sure the pair is electrically neutral. Now carefully pull the two tapes apart.

Q3 As you separated the tapes did you notice any interaction between them (other than that due to the adhesive)?

Q4 Hold one of these tapes in each hand and bring them slowly towards each other (nonsticky sides facing). What do you observe?

Q5 Bring first one, then the other of the tapes near the original test strip. What happens?

Mount A and B on the rod or table edge to serve as test strips. If you have rods of plastic, glass, or rubber available, or a plastic comb, ruler, etc., charge each one in turn by rubbing on cloth or fur and bring it close to A and then B.

Although you can't prove it from the results of a limited number of experiments, there seem to be only two classes of electrified objects. No one has ever produced an electrified object that either attracts or repels *both A and B* (where A and B are themselves electrified objects. The two classes are called positive (+) and negative (−). Write out a general statement summarizing how all members of the same class behave with each other (attract, repel, or remain unaffected by) and with all members of the other class.

A Puzzle

Your system of two classes of electrified objects was based on observations of the way charged objects interact. But how can you account for the fact that a charged object (like a rubbed comb) will attract an *un*charged object (like a scrap of paper)? Is the force between a charged body (either + or −) and an uncharged body always attractive, always repulsive, or is it sometimes one, sometimes the other?

Q6 Can you explain how a force arises between charged and uncharged bodies and why it is always the way it is? The clue here is the fact that the negative charges can move about slightly—even in materials called non-conductors, like plastic and paper (see Sec. 14.5 *Text*).

EXPERIMENT 35
ELECTRIC FORCES II—COULOMB'S LAW

You have seen that electrically charged objects exert forces on each other, but so far your observations have been qualitative; you have looked but not measured. In this experiment you will find out how the amount of electrical force between two charged bodies depends on the amounts of charge and on the separation of the bodies. In addition, you will experience some of the difficulties in using sensitive equipment.

The electric forces between charges that you can conveniently produce in a laboratory are small. To measure them at all requires a sensitive balance.

Constructing the Balance

(If your balance is already assembled, you need not read this section—go on to "Using the balance.") A satisfactory balance is shown in Fig. 14-1.

Coat a small foam-plastic ball with a conducting paint and fix it to the end of a plastic sliver or toothpick by stabbing the pointed end into the ball. Since it is very important that the plastic be clean and dry (to reduce leakage of charge along the surface); *handle the plastic slivers as little as possible, and then only with clean, dry fingers.* Push the sliver into one end of a soda straw leaving at least an inch of plastic exposed, as shown at the top of Fig. 14-2.

Next, fill the plastic support for the balance with glycerin, or oil, or some other liquid. Cut a *shallow* notch in the top of the straw about 2 cm from the axle on the side away from the sphere—see Fig. 14-2.

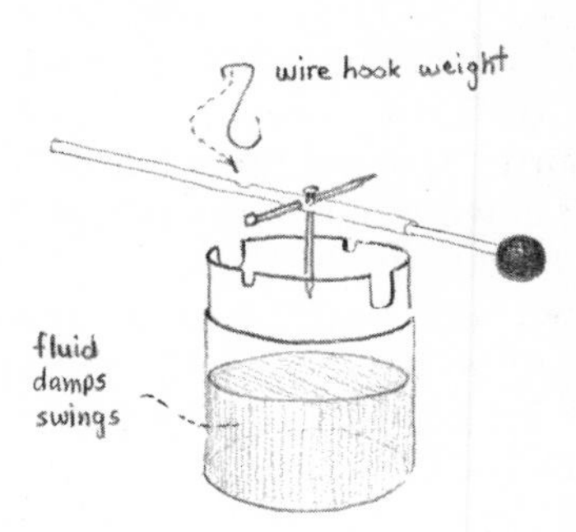

Fig. 14-2

Locate the balance point of the straw, ball, and sliver unit. Push a pin through the straw at this point to form an axle. Push a second pin through the straw directly in front of the axle and perpendicular to it. (As the straw rocks back and forth, this pin moves through the fluid in the support tube. The fluid reduces the swings of the balance.) Place the straw on the support, the pin hanging down inside the vial. Now adjust the balance, by sliding the plastic sliver slightly in or out of the straw, until the straw rests horizontally. If necessary, stick small bits of tape to the straw to make it balance. Make sure the balance can swing freely while making this adjustment.

Finally, cut five or six small, equal lengths of thin, bare wire (such as #30 copper). Each should be about 2 cm long, and they *must all be as close to the same length as you can make them.* Bend them into small hooks (Fig. 14-2) which can be hung over the notch in the straw or hung from each other. These are your "weights."

Mount another coated ball on a pointed plastic sliver and fix it in a clamp on a ring stand, as shown in Fig. 14-1.

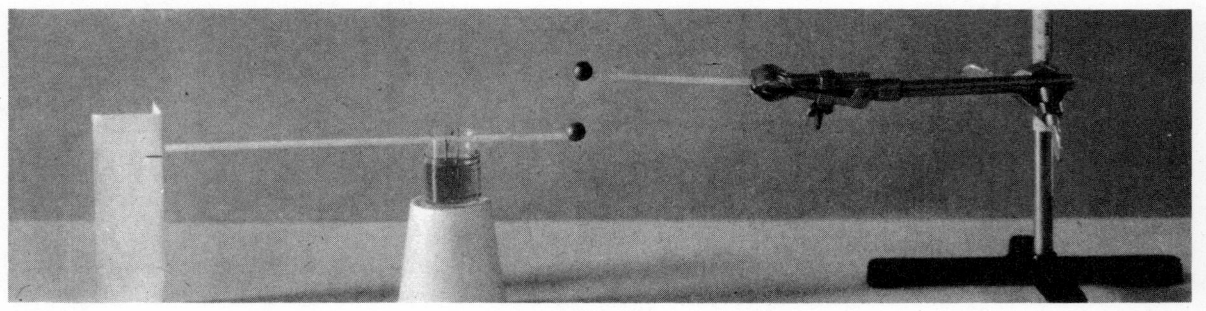

Fig. 14-1

Using the Balance

Charge both balls by wiping them with a rubbed plastic strip. Then bring the ring-stand ball down from above toward the balance ball.

Q1 What evidence have you that there is a force between the two balls?

Q2 Can you tell that it is a force due to the charges?

Q3 Can you compare the size of electrical force between the two balls with the size of gravitational force between them?

Your balance is now ready, but in order to do the experiment, you need to solve two technical problems. During the experiment you will adjust the position of the ring-stand sphere so that the force between the charged spheres is balanced by the wire weights. The straw will then be horizontal. First, therefore, you must check quickly to be sure that the straw *is* balanced horizontally each time. Second, *measure* the distance between the centers of the two balls, yet you cannot put a ruler near the charged balls, or its presence will affect your results. And if the ruler isn't close to the spheres, it is very difficult to make the measurement accurately.

Here is a way to make the measurement. With the balance in its horizontal position, you can record its balanced position with a mark on a folded card placed near the end of the straw (at least 5 cm away from the charges). (See Fig. 14-1.)

How can you avoid the parallax problem? Try to devise a method for measuring the distance between the centers of the spheres. Ask your teacher if you cannot think of one.

You are now ready to make measurements to see how the force between the two balls depends on their separation and on their charge.

Doing the Experiment

From now on, work as quickly as possible but move carefully to avoid disturbing the balance or creating air currents. It is not necessary to wait for the straw to stop moving before you record its position. When it is swinging slightly, but *equally,* to either side of the balanced position, you can consider it balanced.

Charge both balls, touch them together briefly, and move the ring-stand ball until the straw is returned to the balanced position. The weight of one hook now balances the electric force between the charged spheres at this separation. Record the distance between the balls.

Without recharging the balls, add a second hook and readjust the system until balance is again restored. Record this new position. Repeat until you have used all the hooks—but don't reduce the air space between the balls to less than $\frac{1}{2}$ cm. Then quickly retrace your steps by removing one (or more) hooks at a time and raising the ring-stand ball each time to restore balance.

Q4 The separations recorded on the "return trip" may not agree with your previous measurements with this same number of hooks. If they do not, can you suggest a reason why?

Q5 Why must you not recharge the balls between one reading and the next?

Interpreting Your Results

Make a graph of your measurements of force F against separation d between centers. Clearly F and d are inversely related; that is, F increases as d decreases. You can go further to find the relationship between F and d. For example, it might be $F \propto 1/d$, $F \propto 1/d^2$, or $F \propto 1/d^3$, etc.

Q6 How would you test which of these best represents your results?

Q7 What is the actual relationship between F and d?

Further Investigation

In another experiment you can find how the force F varies with the charges on the spheres, when d is kept constant.

Charge both balls and then touch them together briefly. Since they are nearly identical, it is assumed that when touched, they will share the total charge almost equally.

Hang four hooks on the balance and move the ring-stand ball until the straw is in the balanced position. Note this position.

Touch the upper ball with your finger to discharge the ball. If the two balls are again brought into contact, the charge left on the balance ball will be shared equally between the two balls.

Q8 What is the charge on each ball now (as a fraction of the original charge)?

Return the ring-stand ball to its previous position and find how many hooks you must remove to restore the balance.

Q9 Can you state this result as a mathematical relationship between quantity of charge and magnitude of force?

Q10 Consider why you had to follow two precautions in doing the experiment:

(a) Why can a ruler placed too close to the charge affect results?

(b) Why was it suggested that you get the spheres no closer than about $\frac{1}{2}$ cm?

Q11 How might you modify this experiment to see if Newton's third law applies to these electric forces?

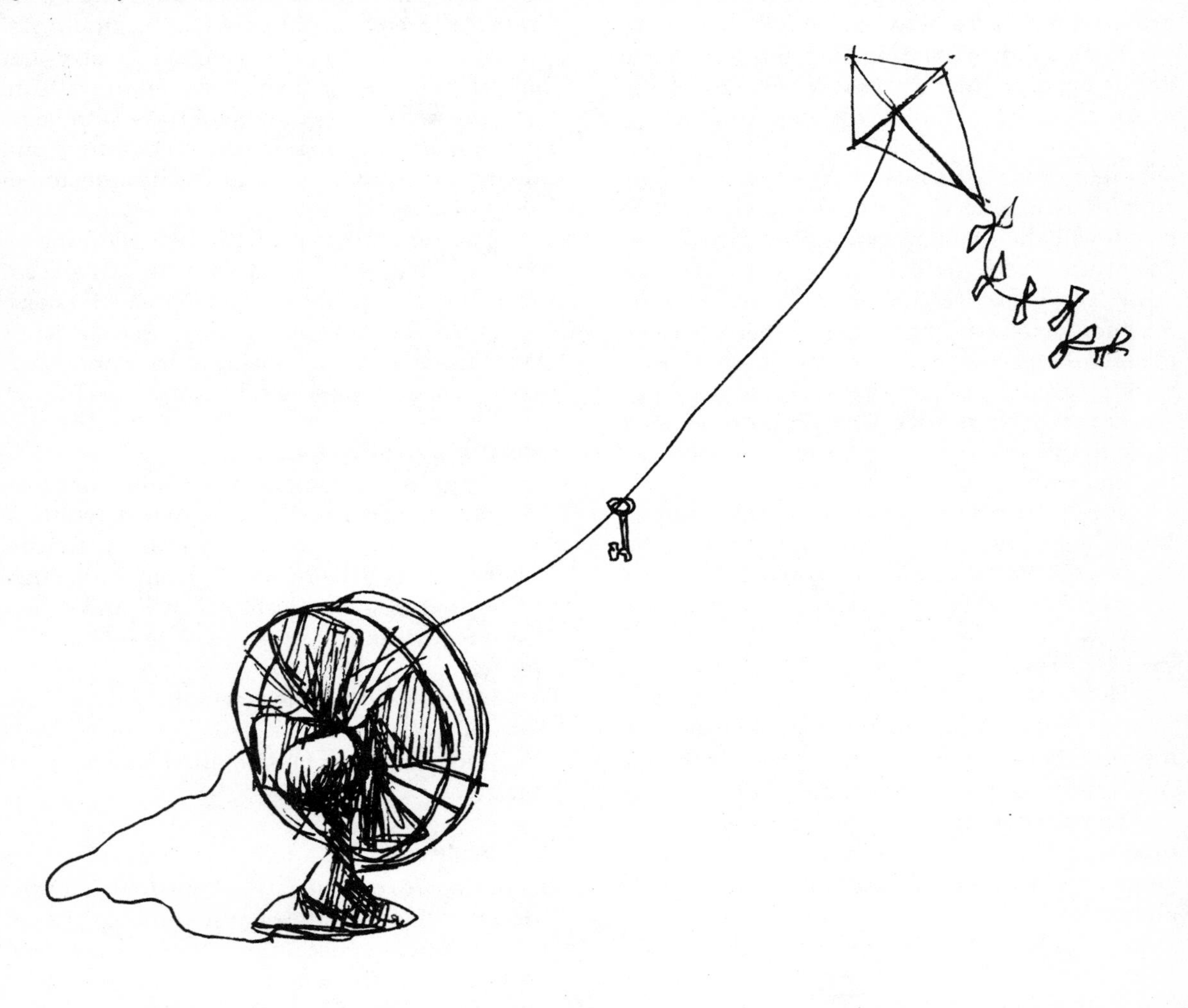

EXPERIMENT 36
FORCES ON CURRENTS

If you did Experiment 35, you used a simple but sensitive balance to investigate how the electric force between two charged bodies depends on the distance between them and on the amount of charge. In this and the next experiment you will examine a related effect: the force between *moving* charges—that is, between electric currents. You will investigate the effect of the magnitudes and the directions of the currents. Before starting the experiment you should have read the description of Oersted and Ampère's work (*Text* Sec. 14.11 and Sec. 14.12).

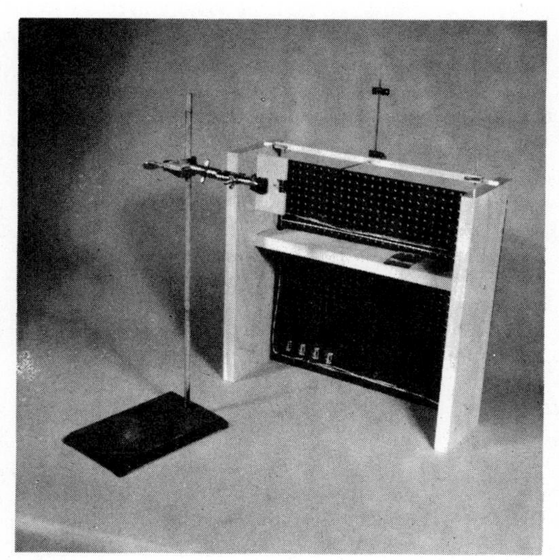

Fig. 14-3

The apparatus for these experiments (like that in Fig. 14-3) is similar in principle to the balance apparatus you used to measure electric forces. The current balance measures the force on a horizontal rod suspended so that it is free to move in a horizontal direction at right angles to its length. You can study the forces exerted by a magnetic field on a current by bringing a magnet up to this rod while there is a current in it. A force on the current-carrying rod causes it to swing away from its original position.

You can also pass a current through a fixed wire parallel to the pivoted rod. Any force exerted on the rod by the current in the fixed wire will again cause the pivoted rod to move. You can measure these forces simply by measuring the counter force needed to return the rod to its original position.

Adjusting the Current Balance

This instrument is more complicated than those most of you have worked with so far. Therefore it is worthwhile spending a little time getting to know how the instrument operates before you start taking readings.

1. You have three or four light metal rods bent into ⊔ or ⌐⊔⌐ shapes. These are the movable "loops." Set up the balance with the longest loop clipped to the pivoted horizontal bar. Adjust the loop so that the horizontal part of the loop hangs level with the bundle of wires (the fixed coil) on the pegboard frame. Adjust the balance on the frame so that the loop and coil are parallel as you look down at them. They should be at least five centimeters apart. Make sure the loop swings freely.

2. Adjust the "counterweight" cylinder to balance the system so that the long pointer arm is approximately horizontal. Mount the ⊏-shaped plate (zero-mark indicator) in a clamp and position the plate so that the zero line is opposite the horizontal pointer (Fig. 14-4). (If you are using the equipment for the first time, draw the zero-index line yourself.)

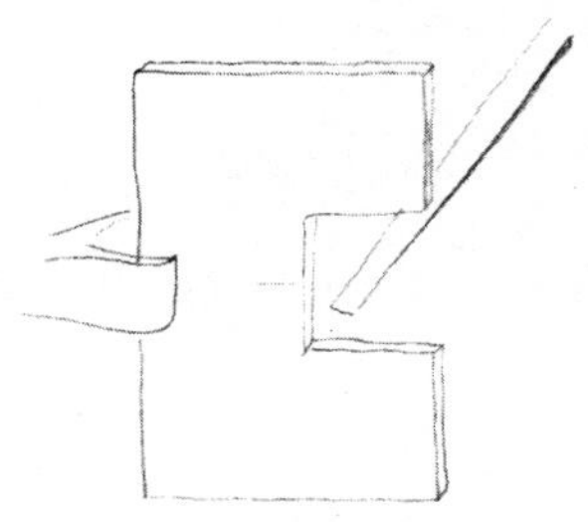

Fig. 14-4 Set the zero mark level with the pointer when there is current in the balance loop and no current in the fixed coil. (See large photo on *Handbook 4* cover.)

3. Now set the balance for maximum sensitivity. To do this, move the sensitivity clip up the vertical rod (Fig. 14-5) until the loop slowly swings back and forth. These oscillations may take as much as four or five seconds per swing. If the clip is raised too far, the balance may

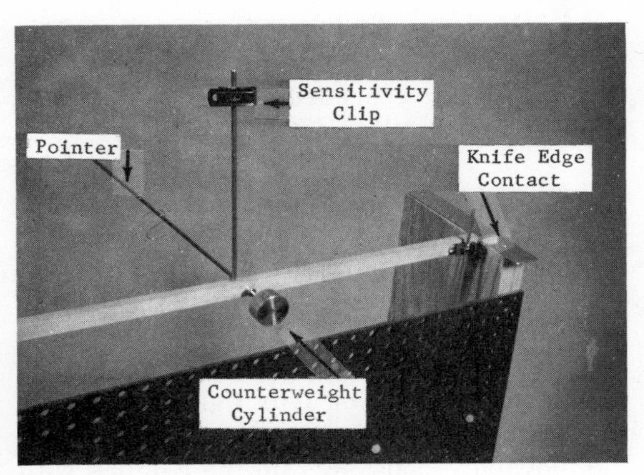

Fig. 14-5

become unstable and flop to either side without "righting" itself.

4. Connect a 6V/5 amp max power supply that can supply up to 5 amps through an ammeter to one of the flat horizontal plates on which the pivots rest. Connect the other plate to the other terminal of the power supply. (Fig. 14.6.)

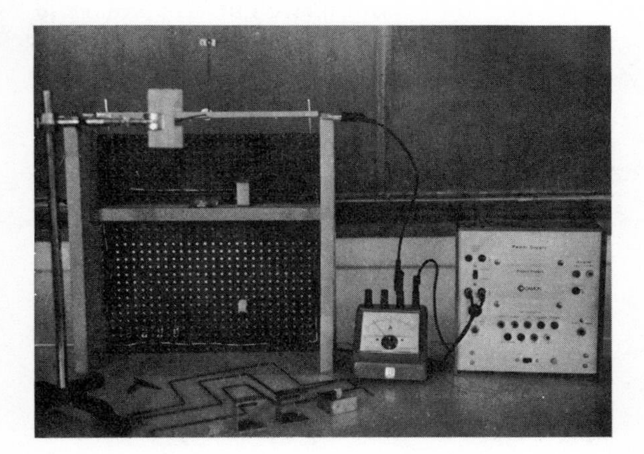

Fig. 14-6

To limit the current and keep it from tripping the circuit-breaker, it may be necessary to put one or two 1-ohm resisters in the circuit. (If your power supply does not have variable control, it should be connected to the plate through a rheostat.)

5. Set the variable control for minimum current, and turn on the power supply. If the ammeter deflects the wrong way, interchange the leads to it. Slowly increase the current up to about 4.5 amps.

6. Now bring a small magnet close to the pivoted conductor.

Q1 How must the magnet be placed to have the biggest effect on the rod? What determines the direction in which the rod swings?

You will make quantitative measurements of the forces between magnet and current in the next experiment, "Currents, Magnets, and Forces." The rest of this experiment is concerned with the interaction between two currents.

7. Connect a similar circuit—power supply, ammeter, and rheostat (if no variable control on the power supply)—to the fixed coil on the vertical pegboard—the bundle of ten wires, not the single wire. The two circuits (fixed coil and movable hook) must be independent. Your setup should now look like the one shown in Fig. 14.7. Only one meter is actually *required*, as you can move it from one circuit to the other as needed. It is, however, more convenient to work with two meters.

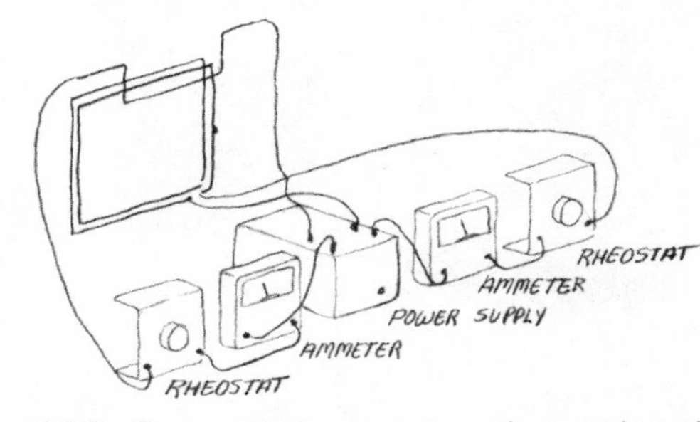

Fig. 14-7 Current balance connections using rheostats when variable power supply is not available.

8. Turn on the currents in both circuits and check to see which way the pointer rod on the balance swings. It should move *up*. If it does not, see if you can make the pointer swing *up* by changing something in your setup.

Q2 Do currents flowing in the same direction attract or repel each other? What about currents flowing in opposite directions?

9. Prepare some "weights" from the thin wire given to you. You will need a set that contains wire lengths of 1 cm, 2 cm, 5 cm, and 10 cm. You may want more than one of each but you can make more as needed during the experiment. Bend them into small S-shaped hooks so that they can hang from the notch on the

pointer or from each other. This notch is the same distance from the axis of the balance as the bottom of the loop so that when there is a force on the horizontal section of the loop, the total weight F hung at the notch will equal the magnetic force acting horizontally on the loop. (See Fig. 14-8.)

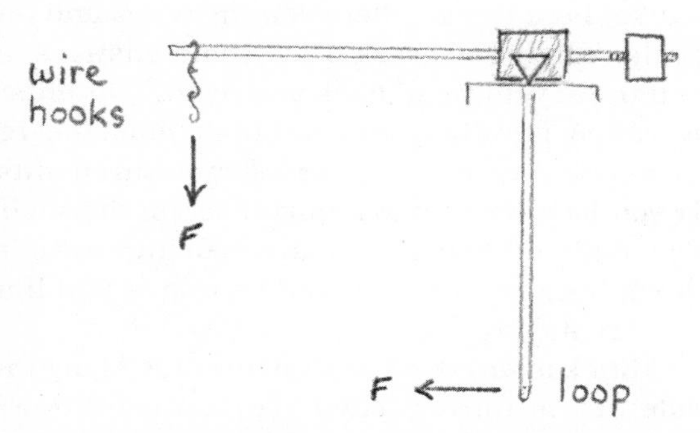

Fig. 14-8 Side view of a balanced loop. The distance from the pivot to the wire hook is the same as the distance to the horizontal section of the loop, so the weight of the additional wire hooks is equal in magnitude to the horizontal magnetic force on the loop.

These preliminary adjustments are common to all the investigations. But from here on there are separate instructions on three different experiments. Different members of the class will investigate how the force depends upon:

(a) the *current* in the wires,
(b) the *distance* between the wires, or
(c) the *length* of one of the wires.

When you have finished your experiment—(a), (b), or (c)—read the section "For class discussion."

(a) How Force Depends on Current in the Wires

By keeping a constant separation between the loop and the coil, you can investigate the effects of varying the currents. Set the balance on the frame so that, as you look down at them, the loop and the coil are parallel and about 1.0 cm apart.

Set the current in the balance loop to about 3 amps. Do not change this current throughout the experiment. With this current in the balance loop and no current in the fixed coil, set the zero-mark in line with the pointer rod.

Starting with a relatively small current in the fixed coil (about 1 amp), find how many centimeters of wire you must hang on the pointer notch until the pointer rod returns to the zero mark.

Record the current I_f in the fixed coil and the length of wire added to the pointer arm. The weight of wire is the balancing force F.

Increase I_f step by step, checking the current in the balance loop as you do so until you reach currents of about 5 amps in the fixed coil.

Q3 What is the relationship between the current in the fixed coil and the force on the balance loop? One way to discover this is to plot force F against current I_f. Another way is to find what happens to the balancing force when you double, then triple, the current I_f.

Q4 Suppose you had held I_f constant and measured F as you varied the current in the balance loop I_b. What relationship do you think you would have found between F and I_b? Check your answer experimentally (say, by doubling I_b) if you have time.

Q5 Can you write a symbolic expression for how F depends on *both* I_f and I_b? Check your answer experimentally (say by doubling both I_f and I_b), if you have time.

Q6 How do you convert the force, as measured in centimeters of wire hung on the pointer arm, into the conventional units for force in newtons?

(b) How Force Varies With the Distance Between Wires

To measure the distance between the two wires, you have to look down. Put a scale on the wooden shelf below the loop. Because there is a gap between the wires and the scale, the number you read on the scale changes as you move your head back and forth. This effect is called parallax, and it must be reduced if you are to get good measurements. If you look down into a mirror set on the shelf, you can tell when you are looking straight down because the wire and its image will line up. Try it. (Fig. 14-9.)

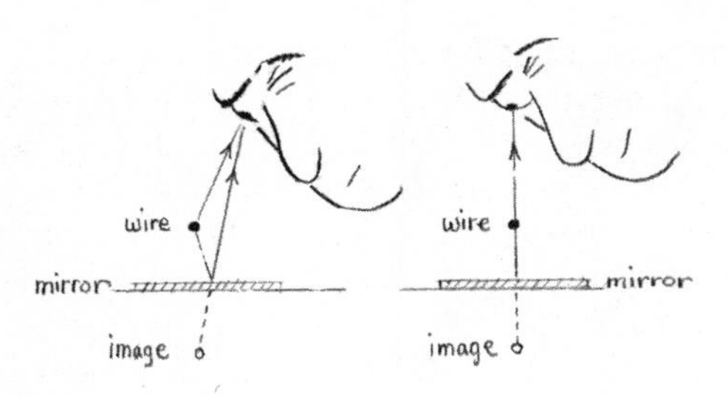

Fig. 14-9 Only when your eye is perpendicularly above the moving wire will it line up with its reflection in the mirror.

Stick a length of centimeter tape along the side of the mirror so that you can sight down and read off the distance between one edge of the fixed wire and the corresponding edge of the balance loop. Set the zero mark with a current I_b of about 4.5 amps in the balance loop and no current I_f in the fixed coil. Then adjust the distance to about 0.5 cm.

Begin the experiment by adjusting the current passing through the fixed coil to 4.5 amps. Hang weights on the notch in the pointer arm until the pointer is again at the zero position. Record the weight and distance carefully.

Repeat your measurements for four or five greater separations. Between each set of measurements make sure the loop and coil are still parallel; check the zero position, and see that the currents are still 4.5 amps.

Q7 What is the relationship between the force E on the balance loop and the distance d between the loop and the fixed coil? One way to discover this is to find some function of d (such as $1/d^2$, $1/d$, d^3, etc.) which gives a straight line when plotted against F. Another way is to find what happens to the balancing force F when you double, then triple, the distance d.

Q8 If the force on the balance loop is F, what is the force on the fixed coil?

Q9 Can you convert the force, as measured in centimeters of wire hung on the pointer arm, into force in newtons?

(c) How Force Varies With the Length of One of the Wires

By keeping constant currents I_f and I_b and a constant separation d, you can investigate the effects of the *length* of the wires. In the current balance setup it is the bottom, horizontal section of the loop which interacts most strongly with the coil and loops with several different lengths of horizontal segment are provided.

To measure the distances between the two wires, you have to look down on them. Put a scale on the wooden shelf below the loop. Because there is a gap between the wires and the scale, what you read on the scale changes as you move your head back and forth. This effect is called parallax, and parallax must be reduced if you are to get good measurements. If you look down into a mirror set on the shelf, you can tell when you are looking straight down because the wire and its image will line up. Try it. (Fig. 14-9.)

Stick a length of centimeter tape along the side of the mirror. Then you can sight down and read off the distance between one edge of the fixed wire and the corresponding edge of the balance loop. Adjust the distance to about 0.5 cm. With a current I_b of about 4.5 amps in the balance loop and no current I_f in the fixed coil, set the pointer at the zero mark.

Begin the experiment by passing 4.5 amps through both the balance loop and the fixed coil. Hang weights on the notch in the pointer in the pointer arm until the pointer is again at the zero position.

Record the value of the currents, the distance between the two wires, and the weights added.

Turn off the currents, and carefully remove the balance loop by sliding it out of the holding clips. (Fig. 14-10.) Measure the length l of the horizontal segment of the loop.

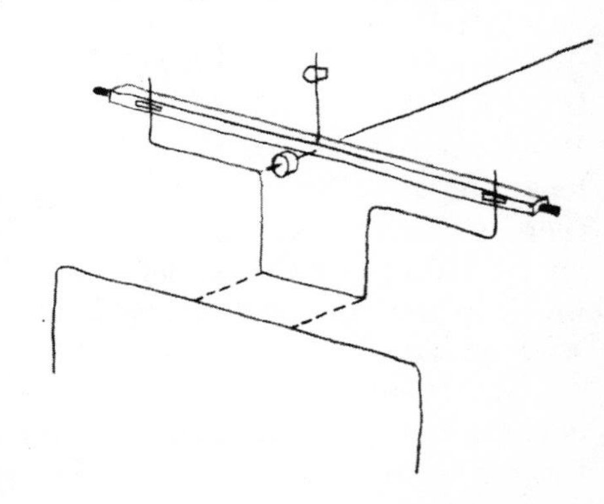

Fig. 14-10

Insert another loop. Adjust it so that it is level with the fixed coil and so that the distance between loop and coil is just the same as you had before. This is important. The loop must also be parallel to the fixed coil, both as you look down at the wires from above and as you look at them from the side. Also reset the clip on the balance for maximum sensitivity. Check the zero position, and see that the currents are still 4.5 amps.

Repeat your measurements for each balance loop.

Q10 What is the relationship between the length l of the loop and the force F on it? One way to discover this is to find some function of l (such as l, l^2, $1/l$, etc.) that gives a straight line when plotted against F. Another way is to find what happens to F when you double l.

Q11 Can you convert the force, as measured in centimeters of wire hung on the pointer arm, into force in newtons?

Q12 If the force on the balance loop is F, what is the force on the fixed coil?

For Class Discussion

Be prepared to report the results of your particular investigation to the rest of the class. As a class you will be able to combine the individual experiments into a single statement about how the force varies with current, with distance, and with length. In each part of this experiment, one factor was varied while the other two were kept constant. In combining the three separate findings into a single expression for force, you are assuming that the effects of the three factors are *independent*. For example, you are assuming doubling one current will *always* double the force—*regardless* of what constant values d and l have.

Q13 What reasons can you give for assuming such a simple independence of effects? What could you do experimentally to support the assumption?

Q14 To make this statement into an equation, what other facts do you need—that is, to be able to predict the force (in newtons) existing between the currents in two wires of given length and separation?

EXPERIMENT 37
CURRENTS, MAGNETS, AND FORCES

If you did the last experiment, "Forces on Currents," you found how the force between two wires depends on the current in them, their length, and the distance between them. You also know that a nearby magnet exerts a force on a current-carrying wire. In this experiment you will use the current balance to study further the interaction between a magnet and a current-carrying wire. You may need to refer back to the notes on Experiment 36 for details on the equipment.

In this experiment you will *not* use the fixed coil. The frame on which the coil is wound will serve merely as a convenient support for the balance and the magnets.

Attach the longest of the balance loops to the pivotal horizontal bar, and connect it through an ammeter to a variable source of current. Hang weights on the pointer notch until the pointer rod returns to the zero mark. (See Fig. 14-11.)

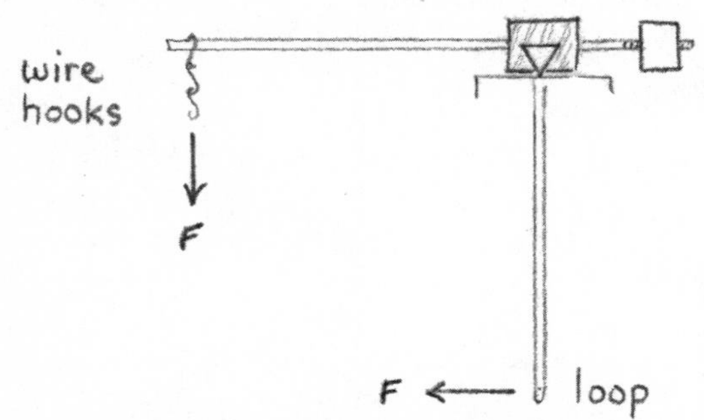

Fig. 14-11 Side view of a balanced loop. Since the distance from the pivot to the wire hooks is the same as the distance to the horizontal section of the loop, the weight of the additional wire hooks is equal to the horizontal magnetic force on the loop.

(a) How the Force Between Current and Magnet Depends On the Current

1. Place two small ceramic magnets on the inside of the iron yoke. Their orientation is important; they must be turned so that the two near faces attract each other when they are moved close together. (Careful: Ceramic magnets are brittle. They break if you drop them.) Place the yoke and magnet unit on the platform so that the balance loop passes through the center of the region between the ceramic magnets. (Fig. 14-12.)

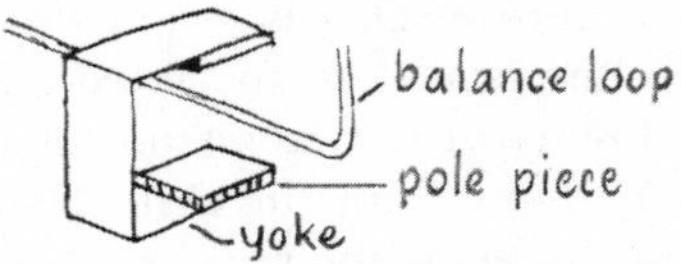

Fig. 14-12 Each magnet consists of a yoke and a pair of removable ceramic-magnet pole pieces.

2. Check whether the horizontal pointer moves *up* when you turn on the current. If it moves down, change something (the current? the magnets?) so that the pointer does swing up.

3. With the current off, mark the zero position of the pointer arm with the indicator. Adjust the current in the coil to about 1 amp. Hang wire weights in the notch of the balance arm until the pointer returns to the zero position.

Record the current and the total balancing weight. Repeat the measurements for at least four greater currents. Between each pair of readings check the zero position of the pointer arm.

Q1 What is the relationship between the current I_b and the resulting force F that the magnet exerts on the wire? (Try plotting a graph.)

Q2 If the magnet exerts a force on the current, do you think the current exerts a force on the magnet? How would you test this?

Q3 How would a stronger or a weaker magnet affect the force on the current? If you have time, try the experiment with different magnets or by doubling the number of pole pieces. Then plot F against I_b on the same graph as in *Q1* above. How do the plots compare?

(b) How the Force Between a Magnet and a Current Depends On the Length of the Region of Interaction

1. Place two small ceramic magnets on the inside of the iron yoke to act as pole pieces (Fig. 14-12). (Careful. Ceramic magnets are brittle. They break if you drop them.) Their orientation is important; they must be turned so that the two near faces attract each other when they are moved close together. Place

the yoke and magnet unit on the platform so that the balance loop passes through the center of the region between the ceramic magnets (Fig. 14-12).

Place the yoke so that the balance loop passes through the center of the magnet and the pointer moves *up* when you turn on the current. If the pointer moves down, change something (the current? the magnets?) so that the pointer does swing up.

With the current off, mark the zero position of the pointer with the indicator.

2. Hang ten or fifteen centimeters of wire on the notch in the balance rod, and adjust the current to return the pointer rod to its zero position. Record the current and the total length of wire, and set aside the magnet for later use.

3. Put a second yoke and pair of pole pieces in position, and see if the balance is restored. You have changed neither the current nor the length of wire hanging on the pointer. Therefore, if balance is restored, this magnet must be of the same strength as the first one. If it is not, try other combinations of pole pieces until you have two magnets of the same strength. If possible, try to get three matched magnets.

4. Now you are ready for the important test. Place two of the magnets on the platform at the same time (Fig. 14-13). To keep the magnets from affecting each other's field appreciably, they should be at least 10 cm apart. Of course each magnet must be positioned so the pointer is deflected upward. With the current just what it was before, hang wire weights in the notch until the balance is restored.

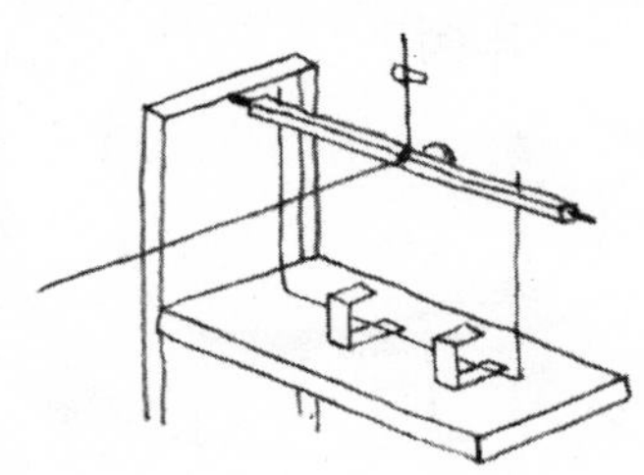

Fig. 14-13

If you have three magnet units, repeat the process using three units at a time. Again, keep the units well apart.

Interpreting Your Data

Your problem is to find a relationship between the length l of the region of interaction and the force F on the wire.

You may not know the exact length of the region of interaction between magnet and wire for a single unit. It certainly extends beyond the region between the two pole pieces. But the force decreases rapidly with distance from the magnets and so as long as the separate units are far from each other, neither will be influenced by the presence of the other. You can then assume that the total length of interaction with two units is double that for one unit.

Q4 How does F depend on l?

(c) A Study of the Interaction Between the Earth and an Electric Current

The magnetic field of the earth is much weaker than the field near one of the ceramic magnets, and the balance must be adjusted to its maximum sensitivity. The following sequence of detailed steps will make it easier for you to detect and measure the small forces on the loop.

1. Set the balance, with the longest loop, to maximum sensitivity by sliding the sensitivity clip to the top of the vertical rod. The sensitivity can be increased further by adding a second clip—but be careful not to make the balance top heavy so that it flops over and won't swing.

2. With no current in the balance loop, align the zero mark with the end of the pointer arm.

3. Turn on the current and adjust it to about 5 amps. Turn off the current and let the balance come to rest.

4. Turn on the current, and observe carefully: Does the balance move when you turn the current on? Since there is no current in the fixed oil, and there are no magnets nearby, any force acting on the current in the loop must be due to an interaction between it and the earth's magnetic field.

5. To make measurements of the force on the loop, you must set up the experiment so that the pointer swings up when you turn on the current. If the pointer moves down, try to find a way to make it go up. (If you have trouble, consult your teacher.) Turn off the

current, and bring the balance to rest. Mark the zero position with the indicator.

6. Turn on the current. Hang weights on the notch, and adjust the current to restore balance. Record the current and the length of wire on the notch. Repeat the measurement of the force needed to restore balance for several different values of current.

If you have time, repeat your measurements of force and current for a shorter loop.

Interpreting Your Data

Try to find the relationship between the current I_b in the balance loop and the force F on it. Make a plot of F against I_b.

Q5 How can you convert your weight unit (say, cm of wire) into newtons of force?

Q6 What force (in newtons) does the earth's magnetic field in your laboratory exert on a current I_b in the loop?

For Class Discussion

Different members of the class have investigated how force F between a current and a magnet varies with current I and with the length of the region of interaction with the current l. It should also be clear that in any statement that describes the force on a current due to a magnet, you must include another term that takes into account the "strength" of the magnet.

Be prepared to report to the class the results of your own investigations and to help formulate an expression that includes all the relevant factors investigated by different members of the class.

Q7 The strength of a magnetic field can be expressed in terms of the force exerted on a wire carrying 1 amp when the length of the wire interacting with the field is 1 meter. Try to express the strength of the magnetic field of your magnet yoke or of the earth's magnetic field in these units, newtons per ampere-meter. (That is, what force would the fields exert on a horizontal wire 1 meter long carrying a current of 1 amp?)

In using the current balance in this experiment, all measurements were made in the zero position—when the loop was at the very bottom of the swing. In this position a vertical force will not affect the balance. So you have measured only *horizontal* forces on the bottom of the loop.

But, since the force exerted on a current by a magnetic field is always at right angles to the field, you have therefore measured only the *vertical* component of the magnetic fields. From the symmetry of the magnet yoke, you might guess that the field is entirely vertical in the region directly between the pole pieces. But the earth's magnetic field is exactly vertical only at the magnetic poles. (See the drawing on page 68 of the *Text*.)

Q8 How would you have to change the experiment to measure the horizontal component of the earth's magnetic field?

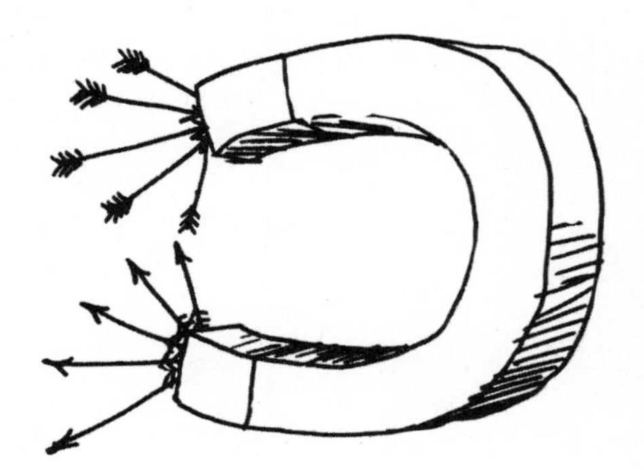

EXPERIMENT 38
ELECTRON BEAM TUBE

If you did the experiment "Electric Forces II – Coulomb's Law," you found that the force on a test charge, in the vicinity of a second charged body, decreases rapidly as the distance between the two charged bodies is increased. In other words, the *electric field* strength due to a single small charged body decreases with distance from the body. In many experiments it is useful to have a region where the field is uniform, that is, a region where the force on a test charge is the same at all points. The field between two closely spaced parallel, flat, oppositely charged plates is very nearly uniform (as is suggested by the behavior of fibers aligned in the electric field between two plates shown in Fig. 14-14).

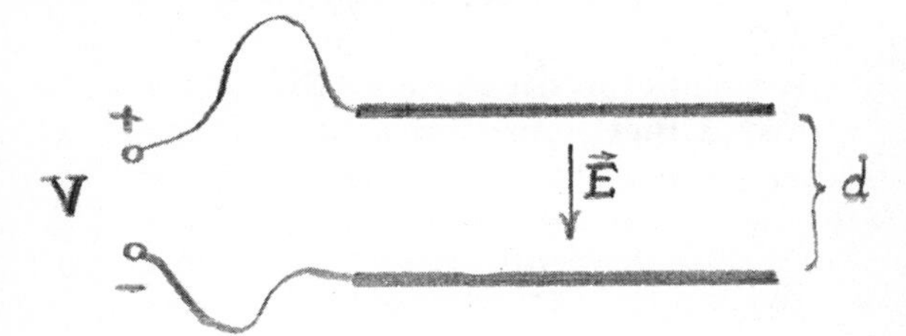

Fig. 14-14 The field between two parallel flat plates is uniform. $E = V/d$ where V is the potential difference (volts) between the two plates.

The nearly uniform magnitude E depends upon the potential difference between the plates and upon their separation d:

$$E = \frac{V}{d}$$

Besides electric forces on charged bodies, you found if you did either of the previous two experiments with the current balance, you found that there is a force on a current-carrying wire in a magnetic field.

Free Charges

In this experiment the charges will not be confined to a foam-plastic ball or to a metallic conductor. Instead they will be free charges – free to move through the field on their own in air at low pressure.

You will build a special tube for this experiment. The tube will contain a filament wire and a metal can with a small hole in one end. Electrons emitted from the heated filament are accelerated toward the positively charged can and some of them pass through the hole into the space beyond, forming a beam of electrons. It is quite easy to observe how the beam is affected by electric and magnetic fields.

When one of the air molecules remaining in the partially evacuated tube is struck by an electron, the molecule emits some light. Molecules of different gases emit light of different colors. (Neon gas, for example, glows red.) The bluish glow of the air left in the tube shows the path of the electron beam.

Building Your Electron Beam Tube

Full instructions on how to build the tube are included with the parts. Note that one of the plates is connected to the can. The other plate must not touch the can.

After you have assembled the filament and plates on the pins of the glass tube base, you can see how good the alignment is if you look in through the narrow glass tube. You should be able to see the filament across the center of the hole in the can. Don't seal the header in the tube until you have checked this alignment. Then leave the tube undisturbed overnight while the sealant hardens.

Fig. 14-15

Operating the Tube

With the power supply turned *OFF*, connect the tube as shown in Figs. 14-15 and 14-16. The low-voltage connection provides current to heat the filament and make it emit electrons. The ammeter in this circuit allows you to keep a close check on the current and avoid burning out the filament. Be sure the 0-6V control is turned down to 0.

The high-voltage connection provides the field that accelerates these electrons toward the can. Let the teacher check the circuits before you proceed further.

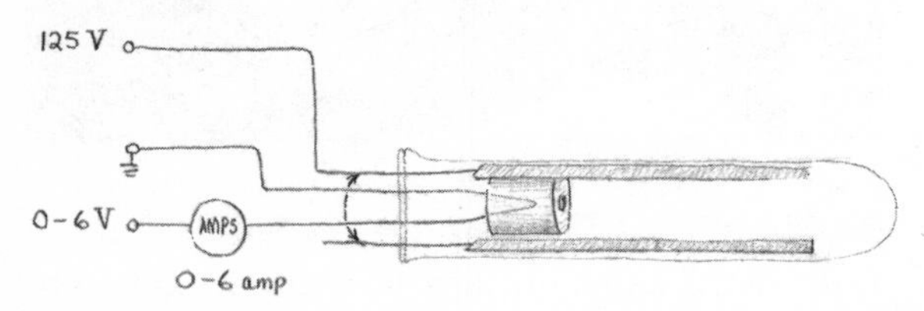

Fig. 14-16 The pins to the two plates are connected together, so that they will be at the same potential and there will be no electric field between them.

Turn on the vacuum pump and let it run for several minutes. If you have done a good job putting the tube together, and if the vacuum pump is in good condition, you should not have much difficulty getting a glow in the region where the electron beam comes through the hole in the can.

You should work with the faintest glow that you can see clearly. Even then, it is important to keep a close watch on the brightness of the glow. There is an appreciable current from the filament wire to the can. As the residual gas gets hotter, it becomes a better conductor increasing the current. The increased current will cause further heating, and the process can build up – the back end of the tube will glow intensely blue-white and the can will become red hot. You must immediately reduce the current to prevent the gun from being destroyed. *If the glow in the back end of the tube begins to increase noticeably, turn down the filament current very quickly, or turn off the power supply altogether.*

Deflection by an Electric Field

When you get an electron beam, try to deflect it in an electric field by connecting the deflecting plate to the ground terminal (See Fig. 14-17), you will put a potential difference between the plates equal to the accelerating voltage. Other connections can be made to get other voltages, but check your ideas with your teacher before trying them.

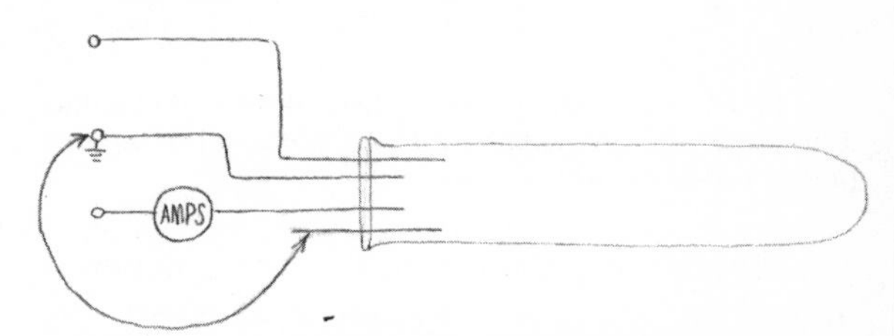

Fig. 14-17 Connecting one deflecting plate to ground will put a potential difference of 125 V between the plates.

Q1 Make a sketch showing the direction of the electric field and of the force on the charged beam. Does the deflection in the electric field confirm that the beam consists of negatively charged particles?

Deflection by a Magnetic Field

Now try to deflect the beam in a magnetic field, using the yoke and magnets from the current balance experiments.

Q2 Make a vector sketch showing the direction of the magnetic field, the velocity of the electrons, and the force on them.

Balancing the Electric and Magnetic Effects

Try to orient the magnets so as to cancel the effect of an electric field between the two plates, permitting the charges to travel straight through the tube.

Q3 Make a sketch showing the orientation of the magnetic yoke relative to the plates.

The Speed of the Charges

As explained in Chapter 14 of the *Text*, the magnitude of the magnetic force is qvB, where q is the electron charge, v is its speed, and B is the magnetic field strength. The magnitude of the electric force is qE, where E is the strength of the electric field. If you adjust the voltage on the plates until the electric force just balances the magnetic force, then $qvB = qE$ and therefore $v = B/E$.

Q4 Show that B/E will be in speed units if B is expressed in newtons/ampere-meter and E is expressed in newtons/coulomb. Hint: Remember that 1 ampere = 1 coulomb/sec.

If you knew the value of B and E, you could calculate the speed of the electron. The value of E is easy to find, since in a uniform field between parallel plates, $E = V/d$, where V is the potential difference between the plate (in volts) and d is the separation of the plates (in meters). (The unit volts/meter is equivalent to newtons/coulomb.)

A rough value for the strength of the magnetic field between the poles of the magnet-yoke can be obtained as described in the experiment, Currents, Magnets, and Forces.

Q5 What value do you get for E (in volts per meter)?

Q6 What value did you get for B (in newtons per ampere-meter)?

Q7 What value do you calculate for the speed of the electrons in the beam?

An Important Question

One of the questions facing physicists at the end of the nineteenth century was to decide on the nature of these "cathode rays" (so-called because they are emitted from the negative electrode or cathode). One group of scientists (mostly German) thought that cathode rays were a form of radiation, like light, while others (mostly English) thought they were streams of particles. J. J. Thomson at the Cavendish Laboratory in Cambridge, England did experiments much like the one described here that showed that the cathode rays behaved like particles: the particles now called electrons.

These experiments were of great importance in the early development of atomic physics. In Unit 5 you will do an experiment to determine the ratio of the charge of an electron to its mass.

In the following activities for Chapter 14, you will find some suggestions for building other kinds of electron tubes, similar to the ones used in radios before the invention of the transistor.

ACTIVITIES

DETECTING ELECTRIC FIELDS

Many methods can be used to explore the shape of electric fields. Two very simple ones are described here.

Gilbert's Versorium

A sensitive electric "compass" is easily constructed from a toothpick, a needle, and a cork. An external electric field induces surface charges on the toothpick. The forces on these induced charges cause the toothpick to line up along the direction of the field.

To construct the versorium, first bend a flat toothpick into a slight arc. When it is mounted horizontally, the downward curve at the ends will give the toothpick stability by lowering its center of gravity below the pivot point of the toothpick. With a small nail, drill a hole at the balance point almost all the way through the pick. Balance the pick horizontally on the needle, being sure it is free to swing like a compass needle. Try bringing charged objects near it.

For details of Gilbert's and other experiments, see Holton and Roller, *Foundations of Modern Physical Science,* Chapter 26.

Charged Ball

A charged pithball (or conductor-coated plastic foam ball) suspended from a stick on a thin insulating thread, can be used as a rough indicator of fields around charged spheres, plates, and wires.

Use a point source of light to project a shadow of the thread and ball. The angle between the thread and the vertical gives a rough measure of the forces. Use the charged pithball to explore the nearly uniform field near a large charged plate suspended by tape strips, and the $1/r$ drop-off of the field near a long charged wire.

Plastic strips rubbed with cloth are adequate for charging well insulated spheres, plates, or wires. (To prevent leakage of charge from the pointed ends of a charged wire, fit the ends with small metal spheres. Even a smooth small blob of solder at the ends should help.)

VOLTAIC PILE

Cut twenty or more disks of each of two different metals. Copper and zinc make a good combination. (The round metal "slugs" from electrical outlet-box installations can be used for zinc disks because of their heavy zinc coating.) Pennies and nickels or dimes will work, but not as well. Cut pieces of filter paper or paper towel to fit in between each pair of two metals in contact. Make a pile of the metal disks and the salt-water soaked paper, as Volta did. Keep the pile in order; for example, copper-paper-zinc, copper-paper-zinc, etc. Connect copper wires to the top and bottom ends of the pile. Touch the free ends of wires with two fingers on one hand. What is the effect? Can you increase the effect by moistening your fingers? In what other ways can you increase the effect? How many disks do you need in order to light a flashlight bulb?

If you have metal fillings in your teeth, try biting a piece of aluminum foil. Can you explain the sensation? (Adapted from *History of Science Cases for High Schools,* Leo E. Klopfer, Science Research Associates, 1966.)

AN 11¢ BATTERY

Using a penny (95 percent copper) and a silver dime (90 percent silver) you can make an 11¢ battery. Cut a one-inch square of filter paper or paper towel, dip it in salt solution, and place it between the penny and the dime. Connect the penny and the dime to the terminals of a galvanometer with two lengths of copper wire. Does your meter indicate a current? Will the battery also produce a current with the penny and dime in direct dry contact?

MEASURING MAGNETIC FIELD INTENSITY

Many important devices used in physics experiments make use of a uniform magnetic field of known intensity. Cyclotrons, bubble chambers, and mass spectrometers are examples. Use the current balance described in Experiments 35 and 36. Measure the magnetic field intensity in the space between the pole faces of two ceramic disk magnets placed close together. Then when you are learning about radioactivity in Unit 6 you can observe the deflection of beta particles as they pass through this space, and determine the average energy of the particles.

Bend two strips of thin sheet aluminum or copper (not iron), and tape them to two disk magnets as shown in the drawing below.

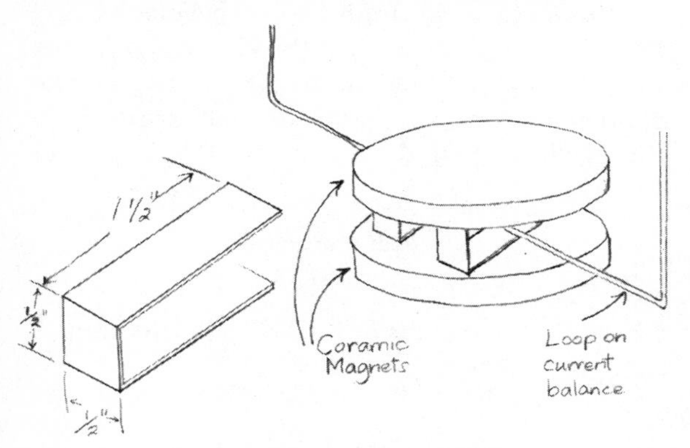

Be sure that the pole faces of the magnets are parallel and are attracting each other (unlike poles facing each other). Suspend the movable loop of the current balance midway between the pole faces. Determine the force needed to restore the balance to its initial position when a measured current is passed through the loop. You learned in Experiment 36 that there is a simple relationship between the magnetic field intensity, the length of the part of the loop which is in the field, and the current in the loop. It is $F = BIl$, where F is the force on the loop (in newtons), B is the magnetic field intensity (in newtons per ampere-meter), I is the current (in amperes), and l is the length (in meters) of that part of the current-carrying loop which is actually in the field. With your current balance, you can measure F, I, and l, and thus compute B.

For this activity, you make two simplifying assumptions which are not strictly true but which enable you to obtain reasonably good values for B: (a) the field is fairly uniform throughout the space between the poles, and (b) the field drops to zero outside this space. With these approximations you can use the diameter of the magnets as the quantity in the above expression.

Try the same experiment with two disk magnets above and two below the loop. How does B change? Bend metal strips of different shapes so you can vary the distance between pole faces. How does this affect B?

An older unit of magnetic field intensity still often used is the *gauss*. To convert from one unit to the other, use the conversion factor, 1 newton /ampere-meter = 10^4 gauss.

Save your records from this activity so you can use the same magnets for measuring beta deflection in Unit 6.

MORE PERPETUAL MOTION MACHINES

The diagrams in Figs. 14-18 and 14-19 show two more of the perpetual motion machines discussed by R. Raymond Smedile in his book, *Perpetual Motion and Modern Research for Cheap Power*. (See also p. 206 of Unit 3 *Handbook*.) What is the weakness of the argument for each of them? (Also see "Perpetual Motion Machines," Stanley W. Angrist, *Scientific American*, January, 1968.)

In Fig. 14-18, *A* represents a stationary wheel around which is a larger, movable wheel, *E*. On stationary wheel *A* are placed three magnets marked *B* in the position shown in the drawing. On rotary wheel *E* are placed eight magnets marked *D*. They are attached to eight levers and are securely hinged to wheel

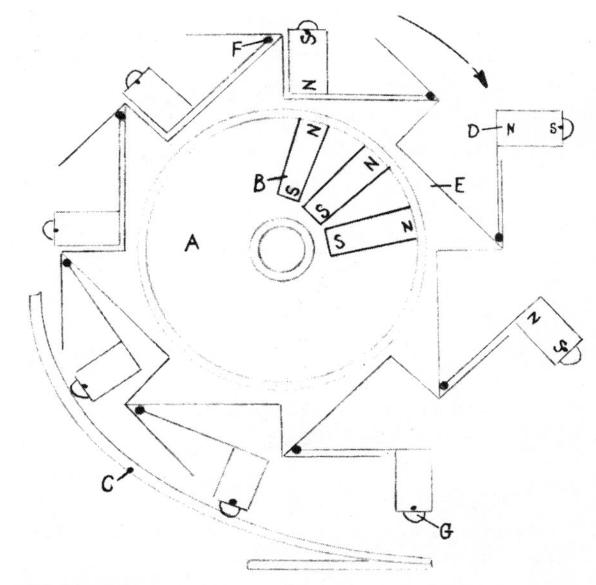

Fig. 14-18

E at the point marked F. Each magnet is also provided with a roller wheel, G, to prevent friction as it rolls on the guide marked C.

Guide C is supposed to push each magnet toward the hub of this mechanism as it is being carried upward on the left-hand side of the mechanism. As each magnet rolls over the top, the fixed magnets facing it cause the magnet on the wheel to fall over. This creates an overbalance of weight on the right of wheel E and thus perpetually rotates the wheel in a clockwise direction.

In Figure 14-19, A represents a wheel in which are placed eight hollow tubes marked E. In each of the tubes is inserted a magnet, B, so that it will slide back and forth. D represents a stationary rack in which are anchored five magnets as shown in the drawing. Each magnet is placed so that it will repel the magnets in wheel A as it rotates in a clockwise direction. Since the magnets in stationary rack D will repel those in rotary wheel A, this will cause a perpetual overbalance of magnet weight on the right side of wheel A.

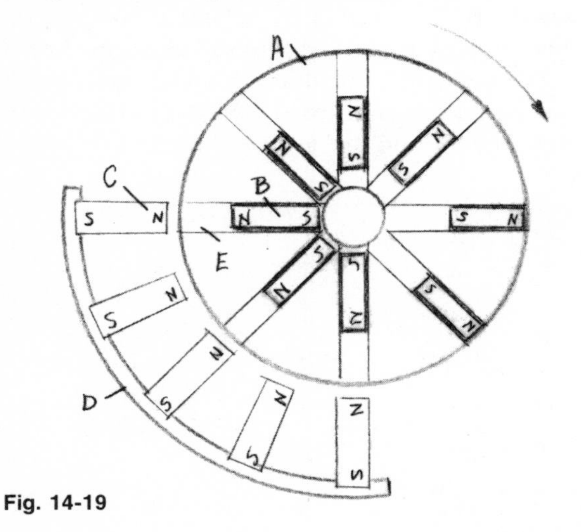

Fig. 14-19

ADDITIONAL ACTIVITIES USING THE ELECTRON BEAM TUBE

1. Focusing the Electron Beam

A current in a wire coiled around the electron tube will produce inside the coil a magnetic field parallel to the axis of the tube. (Ring-shaped magnets slipped over the tube will produce the same kind of field.) An electron moving directly along the axis will experience no force – its velocity is parallel to the magnetic field. But for an electron moving perpendicular to the axis, the field is perpendicular to its velocity – it will therefore experience a force ($F = qvB$) at right angles to both velocity and field. If the curved path of the electron remains in the uniform field, it will be a circle. The centripetal force $F = mv^2/R$ that keeps it in the circle is just the magnetic force qvB, so

$$qvB = \frac{mv^2}{R}$$

where R is the radius of the orbit. In this simple case, therefore,

$$R = \frac{mv}{qB}$$

Suppose the electron is moving down the tube only slightly off axis, in the presence of a field parallel to the axis (Fig. 14-20a). The electron's velocity can be thought of as made up of two components: an axial portion of v_a and a transverse portion (perpendicular to the axis) v_t (Fig. 14-20b). Consider these two com-

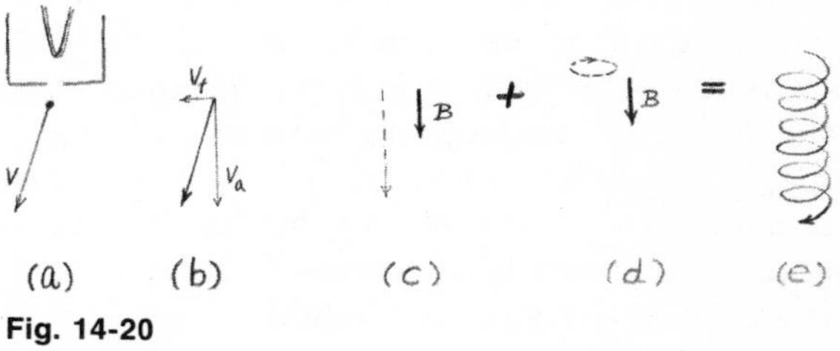

(a) (b) (c) (d) (e)

Fig. 14-20

ponents of the electron's velocity independently. You know that the axial component will be unaffected – the electron will continue to move down the tube with speed v_a (Fig. 14-20c). The transverse component, however, is perpendicular to the field, so the electron will also move in a circle (Fig. 14-20d). In this case,

$$R = \frac{mv_t}{qB}$$

The resultant motion—uniform speed down the axis plus circular motion perpendicular to the axis—is a helix, like the thread on a bolt (Fig. 14-20e).

In the absence of any field, electrons traveling off-axis would continue toward the edge of the tube. In the presence of an axial magnetic field, however, the electrons move down the tube in helixes—they have been focused into a beam. The radius of this beam depends on the field strength B and the transverse velocity v_t.

Wrap heavy-gauge copper wire, such as #18, around the electron beam tube (about two turns per centimeter) and connect the tube to a low-voltage (3-6 volts), high-current source to give a noticeable focusing effect. Observe the shape of the glow, using different coils and currents. (Alternatively, you can vary the number and spacing of ring magnets slipped over the tube to produce the axial field.)

2. Reflecting the Electron Beam

If the pole of a very strong magnet is brought near the tube (with great care being taken that it doesn't pull the iron mountings of the tube toward it), the beam glow will be seen to spiral more and more tightly as it enters stronger field regions. If the field lines diverge enough, the path of beam may start to spiral back. The reason for this is suggested in SG 14.32 in the *Text*.

This kind of reflection operates on particles in the radiation belt around the earth as the approach of the earth's magnetic poles. (See drawing at end of *Text* Chapter 14.) Such reflection is what makes it possible to hold tremendously energetic charged particles in magnetic "bottles." One kind of coil used to produce a "bottle" field appears in the Unit 5 *Text*.

3. Diode and Triode Characteristics

The construction and function of some electronic vacuum tubes is described in the next activity, "Inside a Radio Tube." In this section are suggestions for how you can explore some characteristics of such tubes with your electron beam tube materials.

These experiments are performed at accelerating voltages below those that cause ionization (a visible glow) in the electron beam tube.

(a) *Rectification*

Connect an ammeter between the can and high-voltage supply to show the direction of the current, and to show that there is a current only when the can is at a higher potential than the filament (Fig. 14.21).

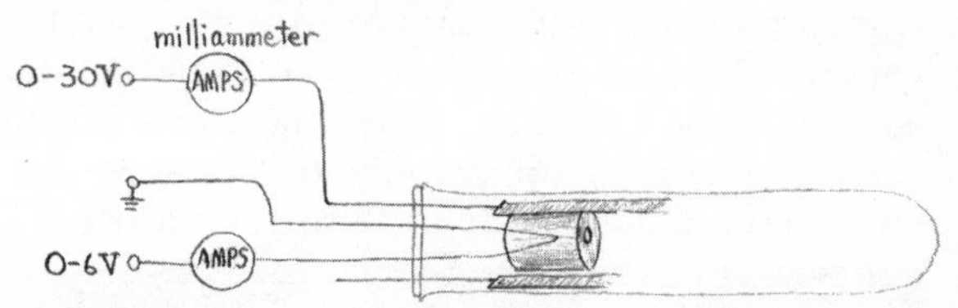

Fig. 14-21

Note that there is a measurable current at voltages far below those needed to give a visible glow in the tube. Then apply an alternating potential difference between the can and filament (for example, from a Variac). Use an oscilloscope to show that the can is alternately above and below filament potential. Then connect the oscilloscope across a resistor in the plate circuit to show that the current is only in one direction. (See Fig. 14-22.)

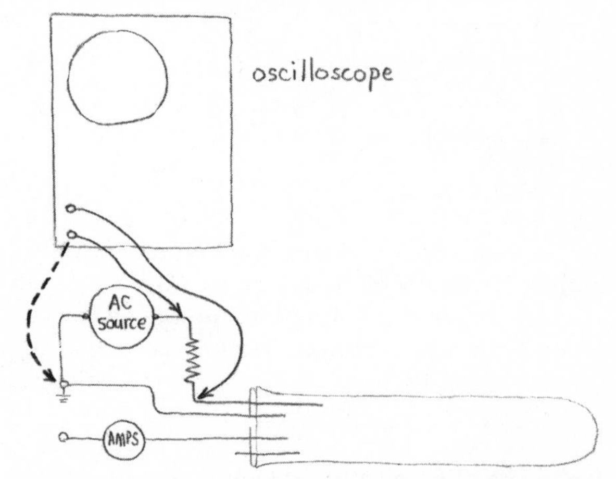

Fig. 14-22 The one-way-valve (rectification) action of a diode can be shown by substituting an AC voltage source for the DC accelerating voltage, and connecting a resistor (about 1000 ohms) in series with it. When an oscilloscope is connected as shown by the solid lines above, it will indicate the current in the can circuit. When the one wire is changed to the connection shown by the dashed arrow, the oscilloscope will indicate the voltage on the can.

(b) *Triode*
The "triode" in the photograph below was made with a thin aluminum sheet for the plate and nichrome wire for the grid. The filament is the original one from the electron beam tube kit, and thin aluminum tubing from a hobby shop was used for the connections to plate and grid. (For reasons lost in the history of vacuum tubes, the can is usually called the "plate." It is interesting to plot graphs of plate current versus filament heating current, and plate current versus voltage. Note that these characteristic curves apply only to voltages too low to produce ionization.) With such a triode,

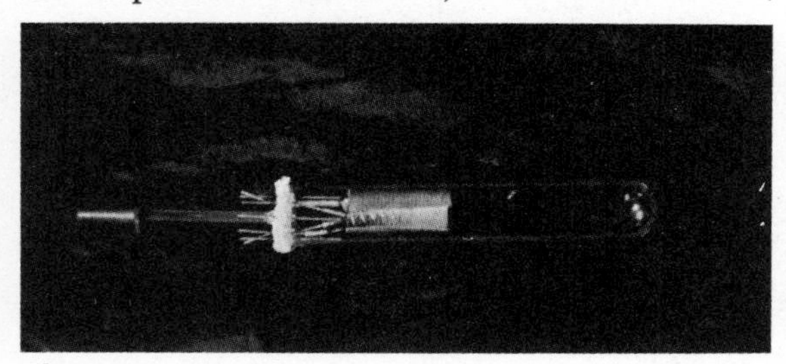

Fig. 14-23

you can plot curves showing triode characteristics: plate current against grid voltage, plate current against plate voltage.

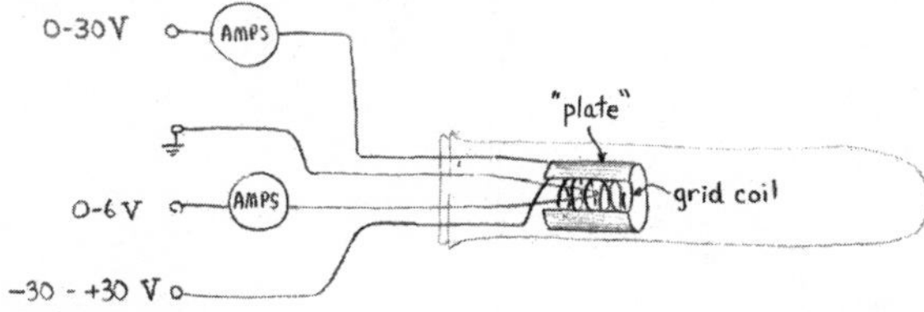

Fig. 14-24

You can also measure the voltage amplification factor, which describes how large a change in plate voltage is produced by a change in grid voltage. More precisely, the amplification factor,

$$\mu = \frac{-\Delta V_{\text{plate}}}{\Delta V_{\text{grid}}}$$

when the plate current is kept constant.

Change the grid voltage by a small amount, then adjust the plate voltage until you have regained the original plate current. The magnitude of the ratio of these two voltage changes is the amplification factor. (Commercial vacuum tubes commonly have amplification factors as high as 500.) The tube gave

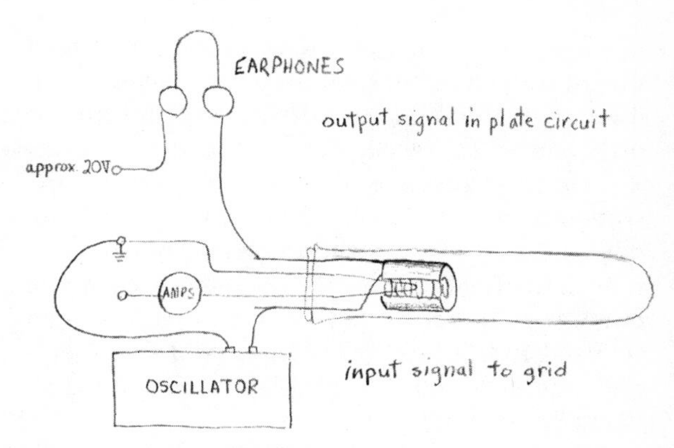

Fig. 14-25 An amplifying circuit

noticeable amplification in the circuit shown in Fig. 14-25 and Fig. 14-26.

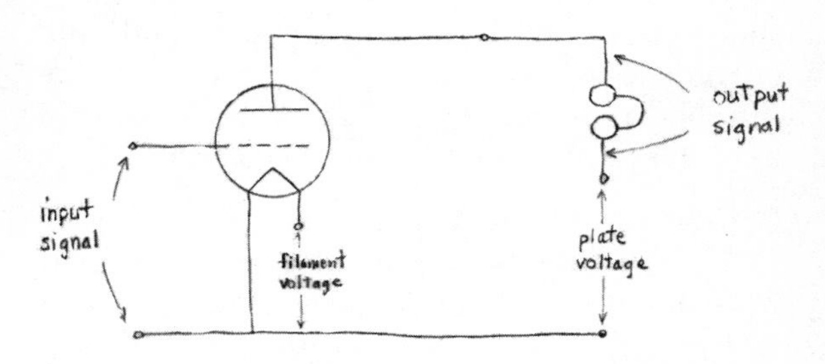

Fig. 14-26 Schematic diagram of amplifying circuit

TRANSISTOR AMPLIFIER

The function of a PNP or NPN transistor is very similar to that of a triode vacuum tube (although its operation is not so easily described). Fig. 14-27 shows a schematic transistor circuit that is analogous to the vacuum tube circuit shown in Fig. 14-26. In both cases, a small input signal controls a large output current.

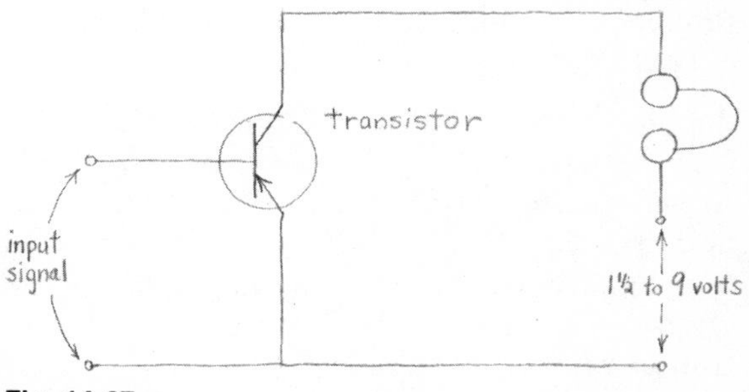

Fig. 14-27

Some inexpensive transistors can be bought at almost any radio supply store, and almost any PNP or NPN will do. Such stores also usually carry a variety of paper-back books that give simplified explanations of how transistors work and how you can use cheap components to build some simple electronic equipment.

INSIDE A RADIO TUBE

Receiving tubes, such as those found in radio and TV sets, contain many interesting parts that illustrates important physical and chemical principles.

Choose some discarded glass tubes at least two inches high. Your radio-TV serviceman will probably have some he can give you. Look up the tube numbers in a receiving tube manual and, if possible, find a triode. (The *RCA Vacuum Tube Manual* is available at most radio-TV supply stores for a couple of dollars.) WARNING: Do *not* attempt to open a TV picture tube!!! Even small TV picture tubes are very dangerous if they burst.

Examine the tube and notice how the internal parts are connected to the pins by wires located in the bottom of the tube. The glass-to-metal seal around the pins must maintain the high vacuum inside the tube. As the tube heats and cools, the pins and glass must expand together. The pins are made of an iron-nickel alloy whose coefficient of expansion is close to that of the glass. The pins are coated with red copper oxide to bond the metal to the glass.

The glass envelope was sealed to the glass base after the interior parts were assembled. Look at the base of the envelope and you can see where this seal was made. After assembly, air was drawn from the tube by a vacuum pump, and the tube was sealed. The sealing nib is at the top of miniature tubes and very old tubes, and is covered by the aligning pin at the bottom of octal-base tubes.

The silvery material spread over part of the inside of the tube is called the *getter*. This coating (usually barium or aluminum) was vaporized inside the tube after sealing to absorb some more of the gas remaining in the envelope after it was pumped out.

To open the tube, spread several layers of newspaper on a flat surface. Use goggles and gloves to prevent injury. With a triangular file, make a small scratch near the bottom. Wrap the tube with an old cloth, hold the tube with the scratch up on the table, and tap the tube with pliers or the file until it breaks. Unwrap the broken tube carefully, and examine the pieces. The getter film will begin to change as soon as it is exposed to air. In a few minutes it will be a white, powdery coating of barium or aluminum oxide.

Protruding from the bottom of the cage assembly are the wires to the pins. Identify the filament leads (there are 2 or 3)—very fine wires with a white ceramic coating. Cut the other wires with diagonal cutters, but leave the filament leads intact. Separate the cage from the base, and slide out the filament.

The tube components in the cage are held in alignment with each other by the mica washers at the top and bottom of the cage. The mica also holds the cage in place inside the envelope.

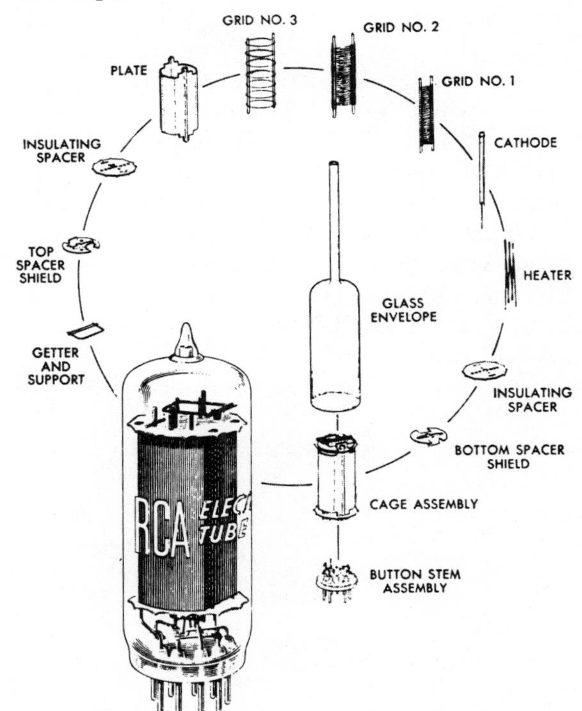

"Exploded" diagram of a 3-grid ("pentode") vacuum tube.

CHAPTER 14

To separate the components, examine the ends that protrude through the mica washer, and decide what to do before the mica can be pulled off the ends. It may be necessary to twist, cut, or bend these parts in order to disassemble the cage. When you have completed this operation, place the components on a clean piece of paper, and examine them one at a time.

Mica was chosen as spacer material for its high strength, high electrical resistance, and the fact that it can withstand high temperatures. White mica consists of a complex compound of potassium, aluminum, silicon and oxygen in crystal form. Mica crystals have very weak bonds between the planes and so they can be split into thin sheets. Try it!

The small metal cylinder with a white coating is the *cathode*. It is heated from inside by the filament. At operating temperature, electrons are "boiled off."

The coating greatly increases the number of electrons emitted from the surface. When you wait for your radio or TV to warm up, you are waiting for the tube cathodes to warm up to an efficient emitting temperature.

The ladder-like arrangement of very fine wire is called the *grid*. The electrons that were boiled off the cathode must pass through this grid. Therefore the current in the tube is very sensitive to the electric field around the grid.

Small changes in the voltage of the grid can have a large effect on the flow of electrons through the tube. This controlling action is the basis for amplification and many other tube operations.

The dark cylinder that formed the outside of the cage is the *plate*. In an actual circuit, the plate is given a positive voltage relative to the cathode in order to attract the electrons emitted by the cathode. The electrons strike the plate and most give up their kinetic energy to the plate, which gets very hot. The plate is a dark color in order to help dissipate this heat energy. Often, the coating is a layer of carbon, which can be rubbed off with your finger.

It is interesting to open different kinds of tubes and see how they differ. Some have more than one cage in the envelope, multiple grids, or beam-confining plates. The *RCA Tube Manual* is a good source for explanations of different tube types and their operation.

(Adapted from "Looking Inside a Vacuum Tube," *Chemistry*, Sept. 1964.)

AN ISOLATED NORTH MAGNETIC POLE?

Magnets made of a rather soft rubber-like substance are available in some hardware stores. Typical magnets are flat pieces 20 mm $\times$ 25 mm and about 5 mm thick, with a magnetic north pole on one 20 $\times$ 25 mm surface and south pole on the other. They may be cut with a sharp knife.

Suppose you cut six of these so that you have six square pieces 20 mm on the edge. Then level the edges on the S side of each piece so that the pieces can be fitted together to form a hollow cube with all the N sides facing outward. The pieces repel each other strongly and may be either glued (with rubber cement) or tied together with thread.

Do you now have an isolated north pole—that is, a north pole all over the outside (and south pole on the inside)?

Is there a magnetic field directed outward from all surfaces of the cube?

(Adapted from *The Physics Teacher*, March 1966.)

CHAPTER 14

Chapter 15 Faraday and the Electrical Age

ACTIVITIES

FARADAY DISK DYNAMO

You can easily build a disk dynamo similar to the one shown at the bottom of page 80 in Unit 4 *Text*. Cut an 8-inch-diameter disk of sheet copper. Drill a hole in the center of the disk, and put a bolt through the hole. Run a nut up tight against the disk so the disk will not slip on the bolt. Insert the bolt in a hand drill and clamp the drill in a ringstand so the disk passes through the region between the poles of a large magnet. Connect one wire of a 100-microamp dc meter to the metal part of the drill that doesn't turn. Tape the other wire to the magnet so it brushes lightly against the copper disk as the disk is spun between the magnet poles.

Frantic cranking can create a 10-microamp current with the magnetron magnet shown above. If you use one of the metal yokes from the current balance, with three ceramic magnets on each side of the yoke, you may be able to get the needle to move from the zero position just noticeably.

The braking effect of currents induced in the disk can also be noticed. Remove the meter, wires, and magnet. Have one person crank while another brings the magnet up to a position such that the disk is spinning between the magnet poles. Compare the effort needed to turn the disk with and without the magnet over the disk.

If the disk will coast, compare the coasting times with and without the magnet in place. (If there is too much friction in the hand drill for the disk to coast, loosen the nut and spin the disk by hand on the bolt.)

GENERATOR JUMP ROPE

With a piece of wire about twice the length of a room, and a sensitive galvanometer, you can generate an electric current using only the earth's magnetic field. Connect the ends of the wire to the meter. Pull the wire out into a long loop and twirl half the loop like a jump rope. As the wire cuts across the earth's magnetic field, a voltage is generated. If you do not have a sensitive meter on hand, connect the input of one of the amplifiers, and connect the amplifier to a less sensitive meter.

How does the current generated when the axis of rotation is along a north-south line compare with that current generated with the same motion along an east-west line? What does this tell you about the earth's magnetic field? Is there any effect if the people stand on two landings and hang the wire (while swinging it) down a stairwell?

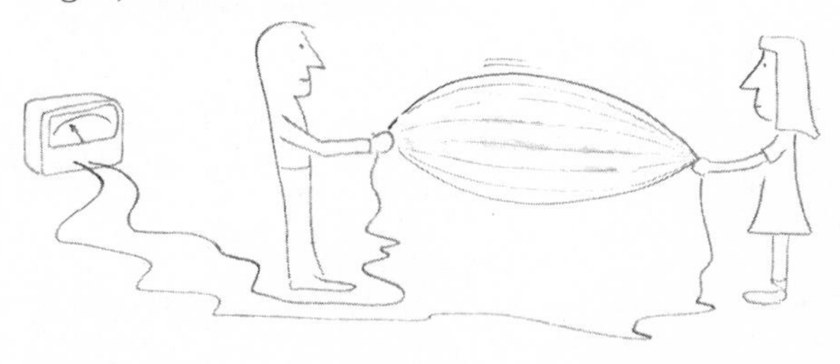

SIMPLE METERS AND MOTORS

You can make workable current meters and motors from very simple parts:

2 ceramic magnets } (from current
1 steel yoke } balance kit)
1 no. 7 cork }
1 metal rod, about 2 mm in diameter and 12 cm long (a piece of bicycle spoke, coat hanger wire, or a large finishing nail will do)
1 block of wood, about 10 cm × 5 cm × 1 cm
About 3 yards of insulated no. 30 copper magnet wire
2 thumb tacks
2 safety pins
2 carpet tacks or small nails } (for meter only)
1 white card 4″ × 5″ }
Stiff black paper, for pointer }
Electrical insulating tape (for motor only)

Meter

To build a meter, follow the steps below paying close attention to the diagrams. Push the rod through the cork. Make the rotating coil or *armature* by winding about 20 turns of wire around the cork, keeping the turns parallel to the rod. Leave about a foot of wire at both ends (Fig. 15-1).

Use nails or carpet tacks to fix two safety

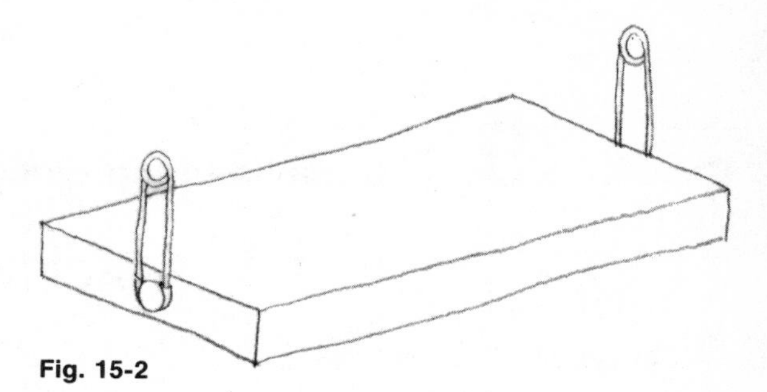

Fig. 15-2

pins firmly to the ends of the wooden-base block (Fig. 15-2).

Make a pointer out of the black paper, and push it onto the metal rod. Pin a piece of white card to one end of the base. Suspend the armature between the two safety pins from the free ends of wire into two loose coils, and attach them to the base with thumb tacks. Put the two ceramic magnets on the yoke (unlike poles facing), and place the yoke around the armature (Fig. 15-3). Clean the insulation off the ends of the leads, and you are ready to connect your meter to a low-voltage dc source.

Calibrate a scale in volts on the white card using a variety of known voltages from dry cells or from a low-voltage power supply, and your meter is complete. Minimize the parallax problem by having your pointer as close to the scale as possible.

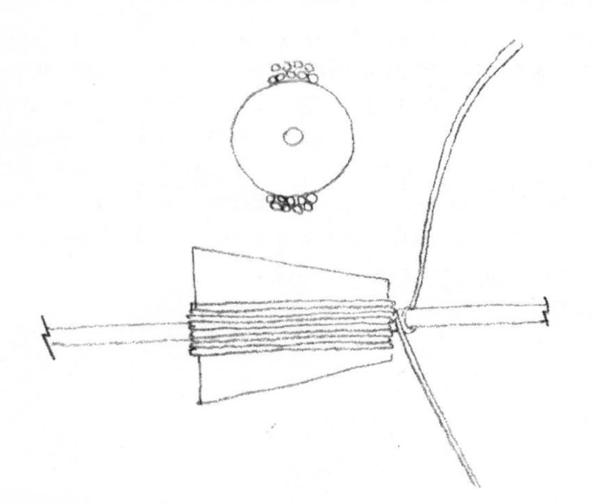

Fig. 15-1

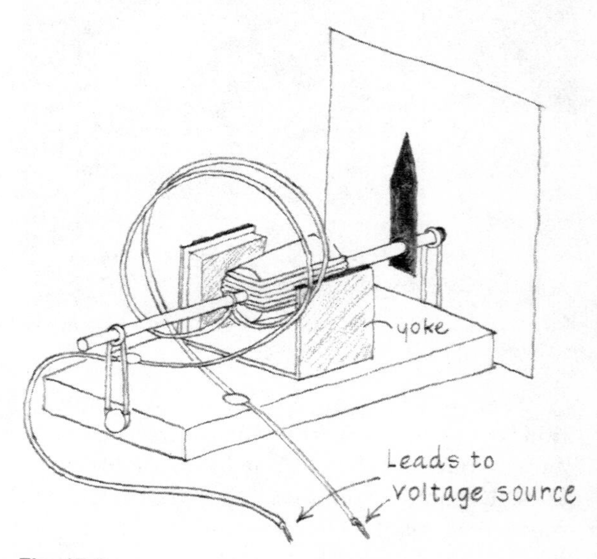

Fig. 15-3

CHAPTER 15

Motor

To make a motor, wind an armature as you did for the meter. Leave about 6 cm of wire at each end, from which you carefully scrape the insulation. Bend each into a loop and then twist into a tight pigtail. Tape the two pigtails along opposite sides of the metal rod (Fig. 15-4).

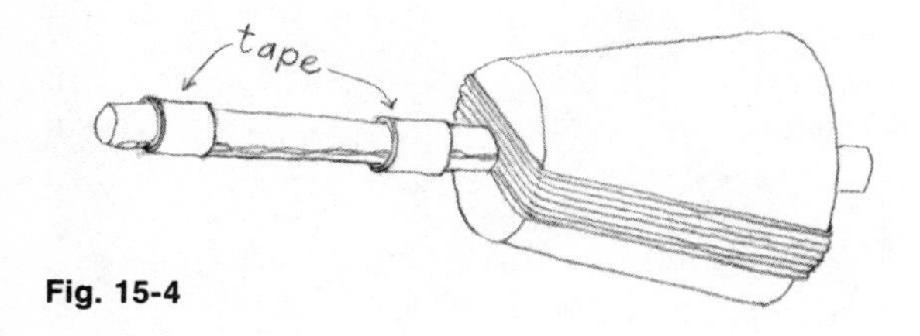

Fig. 15-4

Fix the two safety pins to the base as for the meter, and mount the coil between the safety pins.

The leads into the motor are made from two pieces of wire attached to the baseboard with thumb tacks at points X (Fig. 15-5).

Place the magnet yoke around the coil. The coil should spin freely (Fig. 15-6).

Connect a 1.5-volt battery to the leads. Start the motor by spinning it with your finger. If it does not start, check the contacts between leads and the contact wires on the rod. You may not have removed all the enamel from the wires. Try pressing lightly at points A (Fig. 15-5) to improve the contact. Also check to see that the two contacts touch the armature wires at the same time.

SIMPLE MOTOR-GENERATOR DEMONSTRATION

With two fairly strong U-magnets and two coils, which you wind yourself, you can prepare a simple demonstration showing the principles of a motor and generator. Wind two flat coils of magnet wire 100 turns each. The cardboard tube from a roll of paper towels makes a good form. Leave about $\frac{1}{2}$ meter of wire free at each end of the coil. Tape the coil so it doesn't unwind when you remove it from the cardboard tube.

Adapted from *A Sourcebook for the Physical Sciences*, Joseph and others; Harcourt, Brace and World, 1961, p. 529.

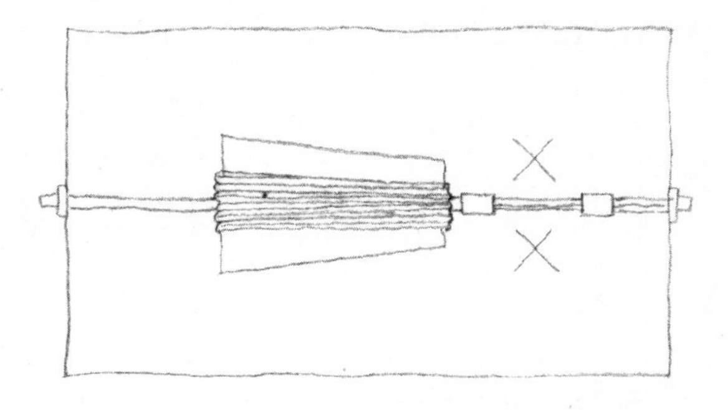

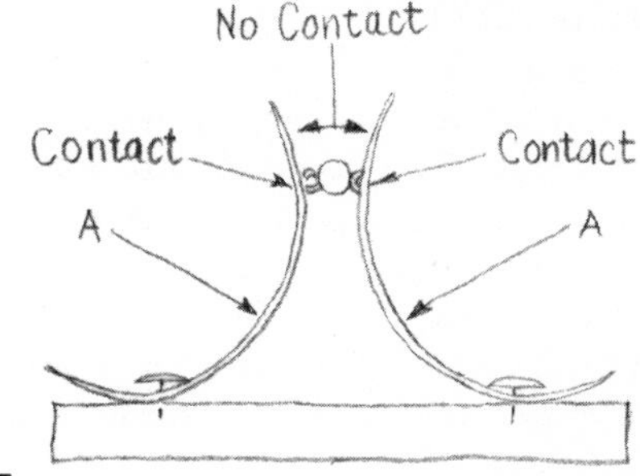

Fig. 15-5

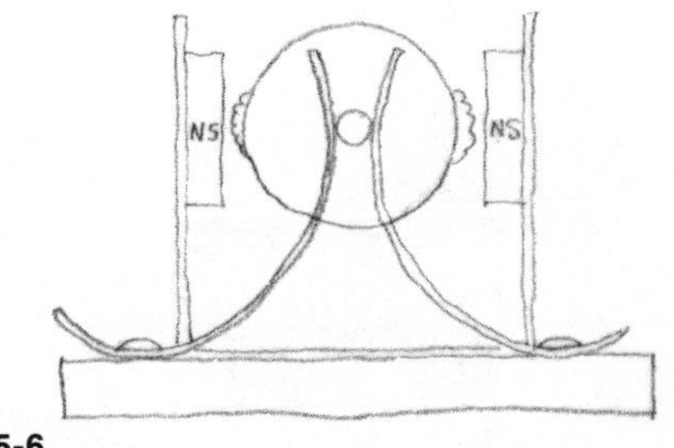

Fig. 15-6

Hang the coils from two supports as shown below so the coils pass over the poles of two U-magnets set on the table about one meter apart. Connect the coils together. Pull one coil to one side and release it. What happens to the other coil? Why? Does the same thing happen if the coils are not connected to each other? What if the magnets are reversed?

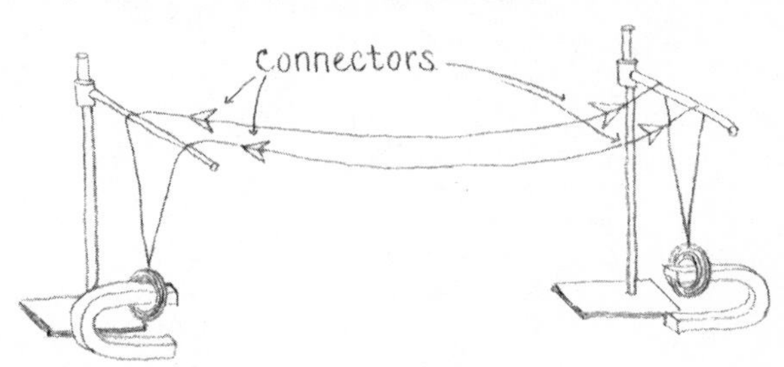

Try various other changes, such as turning one of the magnets over while both coils are swinging, or starting with both coils at rest and then sliding one of the magnets back and forth.

If you have a sensitive galvanometer, it is interesting to connect it between the two coils.

PHYSICS COLLAGE

Many of the words used in physics class enjoy wide usage in everyday language. Cut "physics words" out of magazines, newspapers, etc., and make your own collage. You may wish to take on a more challenging art problem by trying to give a visual representation of a physical concept, such as speed, light, or waves. The *Reader 1* article, "Representation of Movement," may give you some ideas.

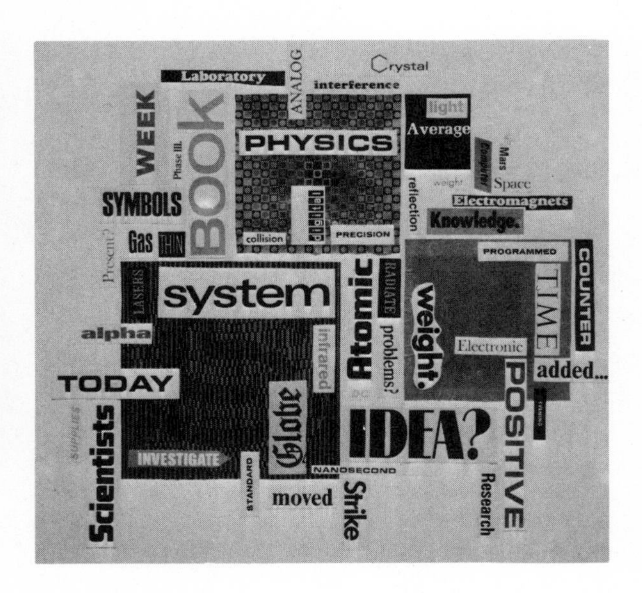

BICYCLE GENERATOR

The generator on a bicycle operates on the same basic principle as that described in the *Text*, but with a different, and extremely simple, design. Take apart such a generator and see if you can explain how it works. Note: You may not be able to reassemble it.

LAPIS POLARIS, MAGNES

The etching (right) shows a philosopher in his study surrounded by the scientific equipment of his time. In the left foreground in a basin of water, a natural magnet or lodestone floating on a stick of wood orients itself north and south. Traders from the great Mediterranean port of Amalfi probably introduced the floating compass, having learned of it from Arab mariners. An Amalfi historian, Flavius Blondus, writing about 1450 A.D., indicates the uncertain origin of the compass, but later historians in repeating this early reference warped it and gave credit for the discovery of the compass to Flavius.

Can you identify the various devices lying around the study? When do you think the etching was made? (If you have some background in art, you might consider whether your estimate on the basis of scientific clues is consistent with the style of the etching.)

2.
LAPIS POLARIS, MAGNES.
Lapis reclusit iste Flauio abditum Poli suum hunc amorem, at ipse nauitæ.

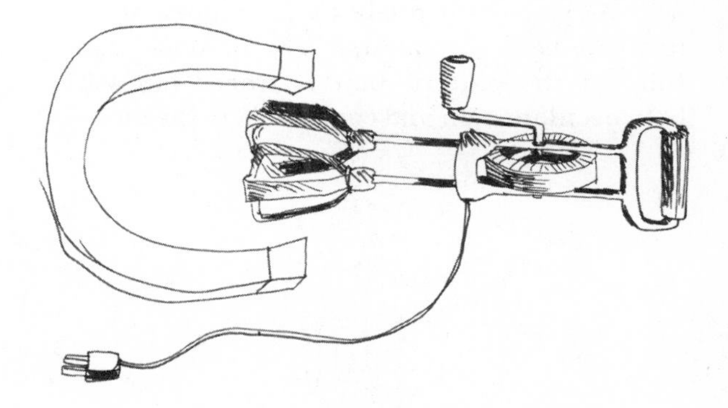

Chapter 16 Electromagnetic Radiation

EXPERIMENT 39 WAVES AND COMMUNICATION

Having studied many kinds and characteristics of waves in Units 3 and 4 of the *Text*, you are now in a position to see how they are used in communications. Here are some suggestions for investigations with equipment that you have probably already seen demonstrated. The following notes assume that you understand how to use the equipment. If you do not, then do not go on until you consult your teacher for instructions. Although different groups of students may use different equipment, all the investigations are related to the same phenomena—how we can communicate with waves.

A. *Turntable Oscillators*

Turn on the oscillator with the pen attached to it. (See p. 310.) Turn on the chart recorder, but do not turn on the oscillator on which the recorder is mounted. The pen will trace out a sine curve as it goes back and forth over the moving paper. When you have recorded a few inches, turn off the oscillator and bring the pen to rest in the middle of the paper. Now turn on the second oscillator at the same rate that the first one was going. The pen will trace out a similar sine curve as the moving paper goes back and forth under it. The wavelengths of the two curves are probably very nearly, but not exactly, equal.

Q1 What do you predict will happen if you turn on *both* oscillators? Try it. Look carefully at the pattern that is traced out with both oscillators on and compare it to the curves previously drawn by the two oscillators running alone.

Change the wavelength of one of the components slightly by putting weights on one of the platforms to slow it down a bit. Then make more traces from other pairs of sine curves. Each trace should consist, as in Fig. 16-1, of three parts: the sine curve from one oscillator; the sine curve from the other oscillator; and the composite curve from both oscillators.

Q2 According to a mathematical analysis of the addition of sine waves, the wavelength of the envelope (λ_e in Fig. 16-1) will *increase* as the wavelengths of the two components (λ_1, λ_2) become more nearly equal. Do your results confirm this?

Q3 If the two wavelengths λ_1 and λ_2 were exactly equal, what pattern would you get when both turntables were turned on; that is, when the two sine curves were superposed? What else would the pattern depend on, as well as λ_1 and λ_2?

As the difference between λ_1 and λ_2 gets smaller, λ_e gets bigger. You can thus detect a very small difference in the two wavelengths by examining the resultant wave for changes in amplitude that take place over a relatively long distance. This method, called the method of beats, provides a sensitive way of comparing two oscillators, and of adjusting one until it has the same frequency as the other.

This method of beats is also used for tuning musical instruments. If you play the same note on two instruments that are not quite in tune, you can hear the beats. And the more nearly in tune the two are, the lower

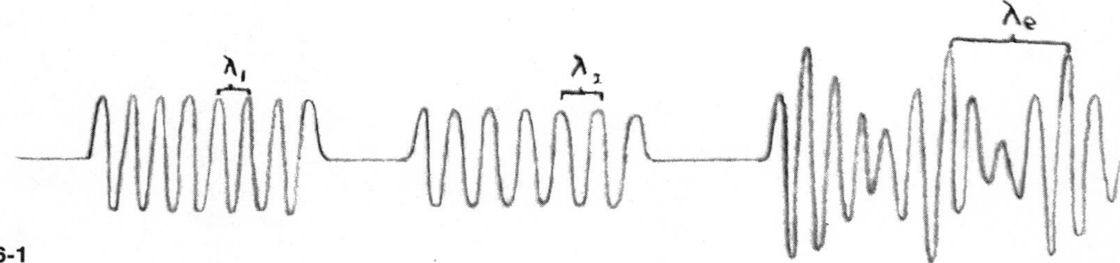

Fig. 16-1

the frequency of the beats. You might like to try this with two guitars or other musical instruments (or two strings on the same instrument).

In radio communication, a signal can be transmitted by using it to modulate a "carrier" wave of much higher frequency. (See part *E* for further explanation of modulation.) A snapshot of the modulated wave looks similar to the beats you have been producing, but it results from one wave being used to control the amplitude of the other, not from simply adding the waves together.

B. Resonant Circuits

You have probably seen a demonstration of how a signal can be transmitted from one tuned circuit to another. (If you have not seen the demonstration, you should set it up for yourself using the apparatus shown in Fig. 16.2.)

Fig. 16-2 Two resonant circuit units. Each includes a wire coil and a variable capacitor. The unit on the right has an electric cell and ratchet to produce pulses of oscillation in its circuit.

This setup is represented by the schematic drawing in Fig. 16-3.

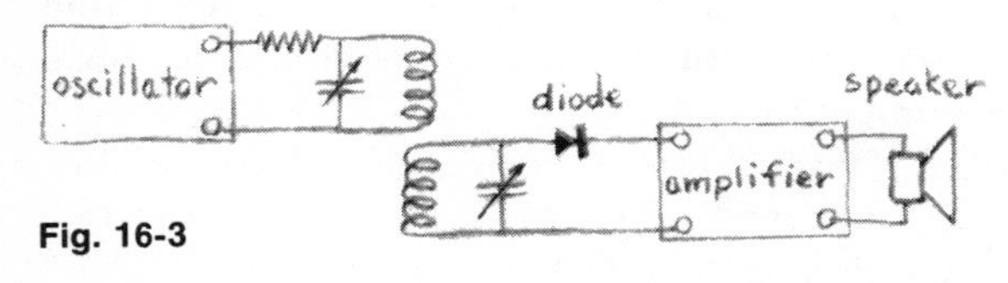

Fig. 16-3

The two coils have to be quite close to each other for the receiver circuit to pick up the signal from the transmitter.

Investigate the effect of changing the position of one of the coils. Try turning one of them around, moving it farther away, etc.

Q4 What happens when you put a sheet of metal, plastic, wood, cardboard, wet paper, or glass between the two coils?

Q5 Why does an automobile always have an outside antenna, but a home radio does not?

Q6 Why is it impossible to communicate with a submerged submarine by radio?

You have probably learned that to transmit a signal from one circuit to another the two circuits must be tuned to the same frequency. To investigate the range of frequencies obtainable with your resonant circuit, connect an antenna (length of wire) to the resonant receiving circuit, in order to increase its sensitivity, and replace the speaker by an oscilloscope (Fig. 16-4). Set the oscilloscope to "Internal Sync" and the sweep rate to about 100 kilocycles/sec.

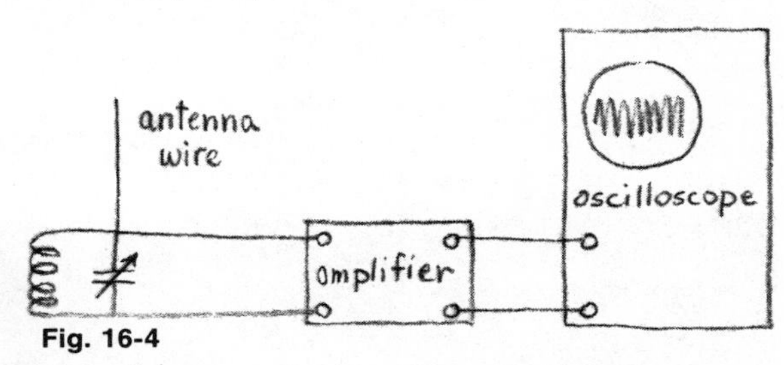

Fig. 16-4

Q7 Change the setting of the variable capacitor ( ) and see how the trace on the oscilloscope changes. Which setting of the capacitor gives the highest frequency? which setting the lowest? By how much can you change the frequency by adjusting the capacitor setting?

When you tune a radio you are usually, in the same way, changing the setting of a variable capacitor to tune the circuit to a different frequency.

The coil also plays a part in determining the resonant frequency of the circuit. If the coil has a different number of turns, a different setting of the capacitor would be needed to get the same frequency.

C. Elementary Properties of Microwaves

With a microwave generator, you can investigate some of the characteristics of short waves in the radio part of the electromagnetic spectrum. In Experiment 30, "Introduction to Waves" and Experiment 31, "Sound," you explored the behavior of several different kinds of waves. These earlier experiments

contain a number of ideas that will help you show that the energy emitted by your microwave generator is in the form of waves.

Refer to your notes on these experiments. Then, using the arrangements suggested there or ideas of your own, explore the transmission of microwaves through various materials as well as microwave reflection and refraction. Try to detect their diffraction around obstacles and through narrow openings in some material that is opaque to them. Finally, if you have two transmitters available or a metal horn attachment with two openings, see if you can measure the wavelength using the interference method of Experiment 31. Compare your results with students doing the following experiment (*D*) on the interference of reflected microwaves.

D. Interference of Reflected Microwaves

With microwaves it is easy to demonstrate interference between direct radiation from a source and radiation reflected from a flat surface, such as a metal sheet. At points where the direct and reflected waves arrive in phase, there will be maxima and at points where they arrive $\frac{1}{2}$-cycle out of phase, there will be minima. The maxima and minima are readily found by moving the detector along a line perpendicular to the reflector. (Fig. 16-5.)

Q8 Can you state a rule with which you could predict the positions of maxima and minima?

By moving the detector back a ways and scanning again, etc., you can sketch out lines of maxima and minima.

Q9 How is the interference pattern similar to what you have observed for two-source radiation?

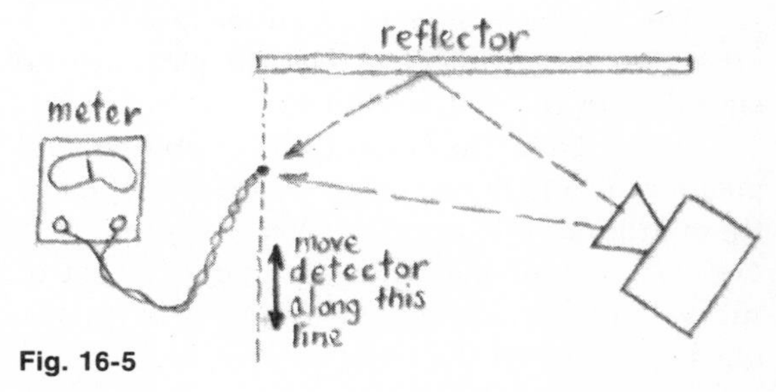

Fig. 16-5

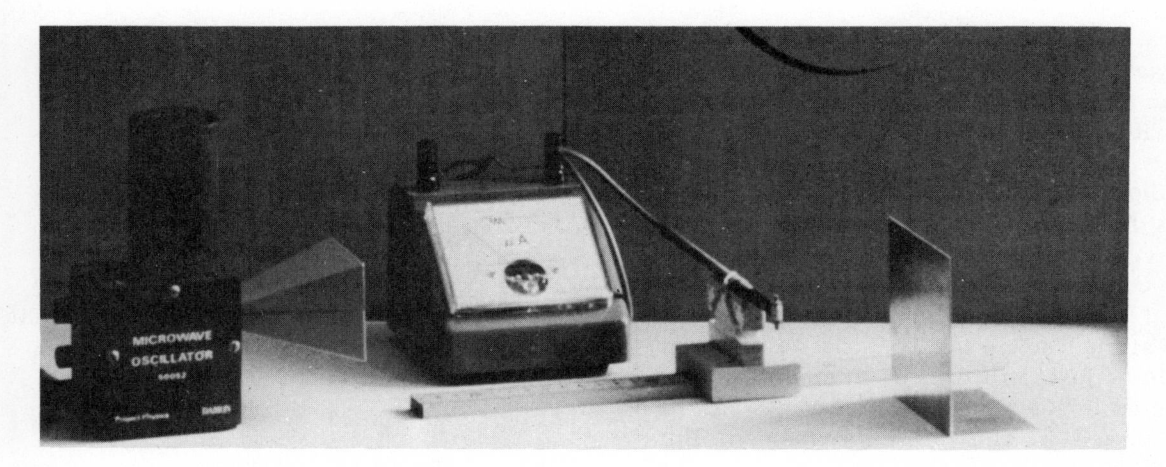

Standing microwaves will be set up if the reflector is placed exactly perpendicular to the source. (As with other standing waves, the nodes are $\frac{1}{2}$ wavelength apart.) Locate several nodes by moving the detector along a line between the source and reflector, and from the node separation calculate the wavelength of the microwaves.

Q10 What is the wavelength of your microwaves?

Q11 Microwaves, like light, propagate at 3 × 10 m/sec. What is the frequency of your microwaves? Check your answer against the chart of the electromagnetic spectrum given on page 113 of *Text* Chapter 16.

The interference between direct and reflected radio waves has important practical consequences. There are layers of partly ionized (and therefore electrically conducting) air, collectively called the ionosphere, that surrounds the earth roughly 30 to 300 kilometers above its surface. One of the layers at about 300 km is a good reflector for radio waves, so it is used to bounce radio messages to points that, because of the curvature of the earth, are too far away to be reached in a straight line.

If the transmitting tower is 100 meters high, then, as shown roughly in Fig. 16-6, point A – the farthest point that the signal can reach directly in flat country – is 35 kilometers (about 20 miles) away. But by reflection from the ionosphere, a signal can reach around the corner to B and beyond.

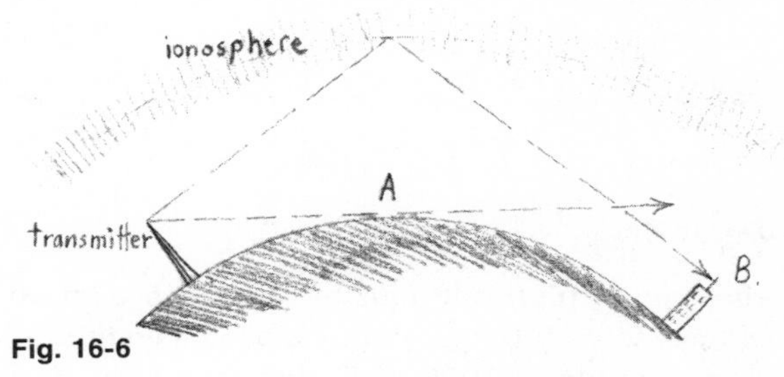

Fig. 16-6

Sometimes both a direct and a reflected signal will arrive at the same place and interference occurs; if the two are out of phase and have identical amplitudes, the receiver will pick up nothing. Such destructive interference is responsible for radio fading. It is complicated by the fact that the height of the ionosphere and the intensity of reflection from it vary during the day with the amount of sunlight.

The setup in Fig. 16-7 is a model of this situation. Move the reflector (the "ionosphere") back and forth. What happens to the signal strength?

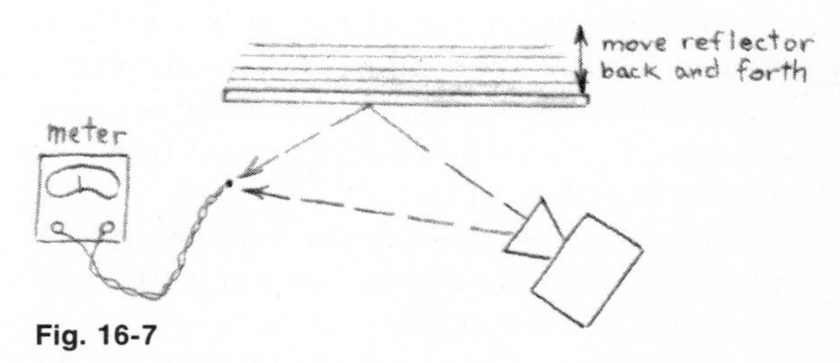

Fig. 16-7

There can also be multiple reflections – the radiation can bounce back and forth between earth and ionosphere several times on its way from, say, New York to Calcutta, India. Perhaps you can simulate this situation too with your microwave equipment.

E. Signals and Microwaves

Thus far you have been learning about the behavior of microwaves of a single frequency and constant amplitude. A *signal* can be added to these waves by changing their amplitude at the transmitter. The most obvious way would be just to turn them on and off as represented in Fig. 16-8. Code messages can be transmitted in this primitive fashion. But

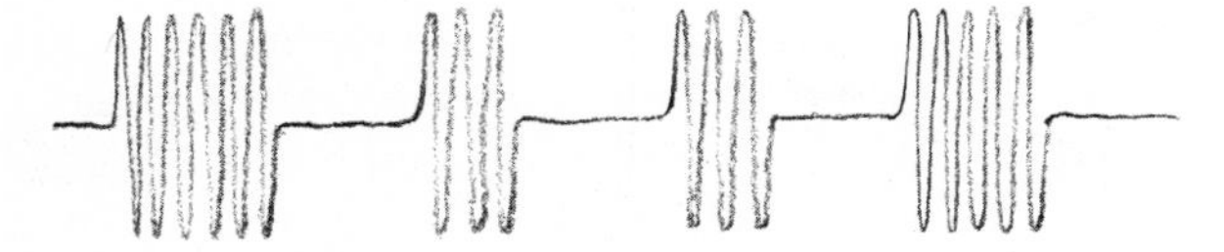

Fig. 16-8

the wave amplitude can be varied in a more elaborate way to carry music or voice signals. For example, a 1000 cycle/sec sine wave fed into part of the microwave transmitter will cause the amplitude of the microwave to vary smoothly at 1000 cyc/sec.

Controlling the amplitude of the transmitted wave like this is called *amplitude modulation*; Fig. 16-9A represents the unmodulated microwave, Fig. 16-9B represents a modulating signal, and Fig. 16-9C the modulated microwave. The Damon microwave oscillator has an input for a modulating signal. You can modulate the microwave output with a variety of signals, for example, with an audio-frequency oscillator or with a microphone and amplifier.

The microwave detector probe is a one-way device—it passes current in only one direction. If the microwave reaching the probe is represented in C, then the electric signal from the probe will be like in D.

You can see this on the oscilloscope by connecting it to the microwave probe (through an amplifier if necessary).

The detected modulated signal from the probe can be turned into sound by connecting an amplifier and loudspeaker to the probe. The

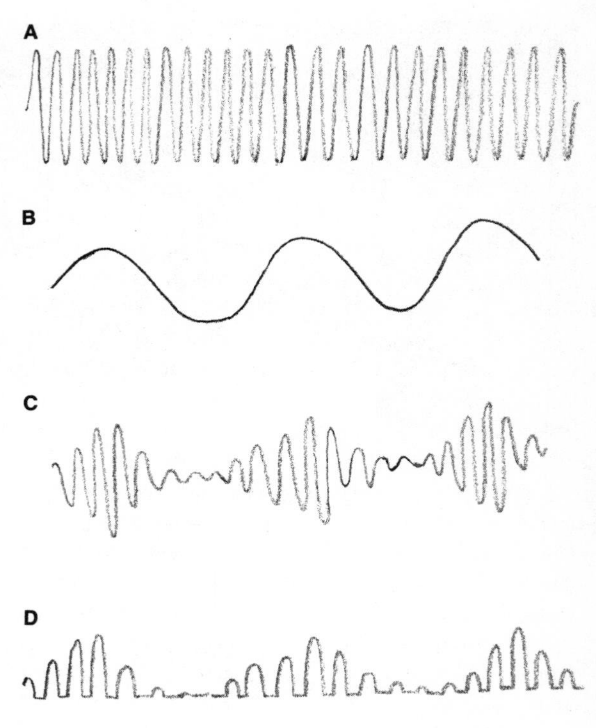

Fig. 16-9

speaker will not be able to respond to the 10^9 individual pulses per second of the "carrier" wave, but only to their averaged effect, represented by the dotted line in E. Consequently, the sound output of the speaker will correspond very nearly to the modulating signal.

Q12 Why must the carrier frequency be much greater than the signal frequency?

Q13 Why is a higher frequency needed to transmit television signals than radio signals? (The highest frequency necessary to convey radio sound information is about 12,000 cycles per second. The electron beam in a television tube completes one picture of 525 lines in 1/30 of a second, and the intensity of the beam should be able to vary several hundred times during a single line scan.)

ACTIVITIES

MICROWAVE TRANSMISSION SYSTEMS

Microwaves of about 6-cm wavelength are used to transmit telephone conversations over long distances. Because microwave radiation has a limited range (they are not reflected well by the ionosphere), a series of relay stations have been erected about 30 miles apart across the country. At each station the signal is detected and amplified before being retransmitted to the next one. If you have several microwave generators that can be amplitude modulated, see if you can put together a demonstration of how this system works. You will need an audio-frequency oscillator (or microphone), amplifier, microwave generator and power supply, detector, another amplifier, and a loudspeaker, another microwave generator, another detector, a third amplifier, and a loudspeaker.

SCIENCE AND THE ARTIST—THE STORY BEHIND A NEW SCIENCE STAMP

The sciences and the arts are sometimes thought of as two distinct cultures with a yawning gulf between them. Perhaps to help bridge that gulf, the Post Office Department held an unprecedented competition among five artists for the design of a postage stamp honoring the sciences.

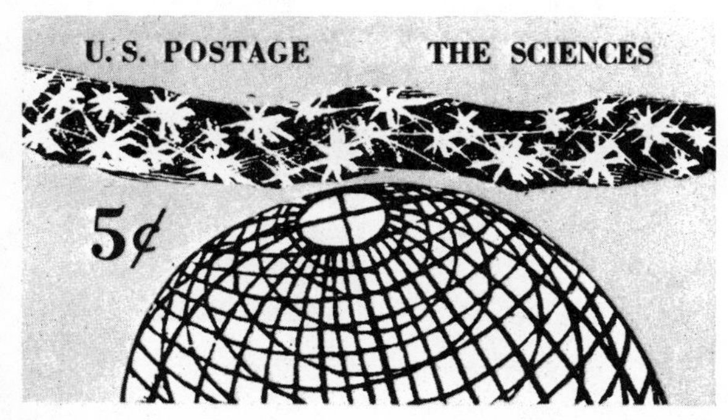

The winning design by Antonio Frasconi. The stamp was printed in light blue and black.

The winning stamp, issued on October 14, 1963, commemorates the 100th anniversary of the National Academy of Sciences (NAS). This agency was established during the Civil War with the objective that it "shall, whenever called upon by any department of government, investigate, experiment, and report upon any subject of science or of art."

To celebrate the NAS anniversary, the late President John F. Kennedy addressed the members of the academy and their distinguished guests from foreign scientific societies. After emphasizing present public recognition of the importance of pure science, the President pointed out how the discoveries of science are forcing the nations to cooperate:

> "Every time you scientists make a major invention, we politicians have to invent a new institution to cope with it—and almost invariably these days it must be an international institution."

As examples of these international institutions, he cited the International Atomic Energy Agency, the treaty opening Antarctica to world scientific research, and the Intergovernmental Oceanographic Commission.

In the scientific sessions marking the NAS anniversary, the latest views of man and matter and their evolution were discussed, as well as the problem of financing future researches—the proper allocation of limited funds to space research and medicine, the biological and the physical sciences.

The stamp competition was initiated by the National Gallery of Art, Washington, D.C. A jury of three distinguished art specialists invited five American artists to submit designs. The artists were chosen for their "demonstrated appropriateness to work on the theme of science." They were Josef Albers, Herbert Bayer, Antonio Frasconi, Buckminster Fuller, and Bradbury Thompson.

Albers and Bayer both taught at the Bauhaus (Germany), a pioneering design center that welded modern industrial know-how to the insights of modern art. (The Bauhaus is probably best known for its development of tubular steel furniture.) Albers has recently published what promises to be the definitive work on color—*Interaction of Color* (Yale University Press). Bayer is an architect as well as an artist; he has designed several of the buildings for the Institute of Humanistic Studies in Aspen, Colorado. Buckminster Fuller, engineer, designer, writer, and in-

Albers

Bayer

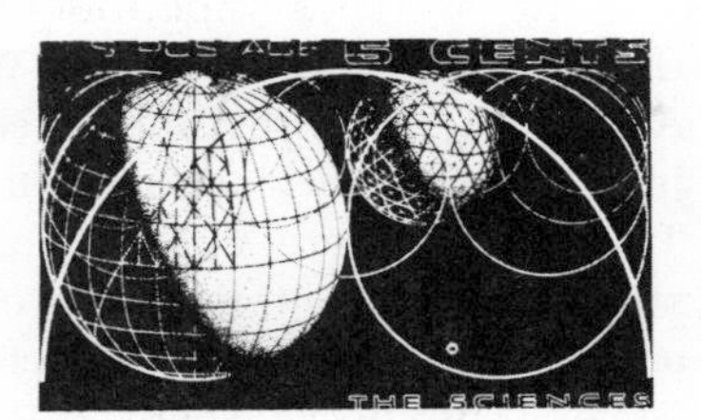

Fuller

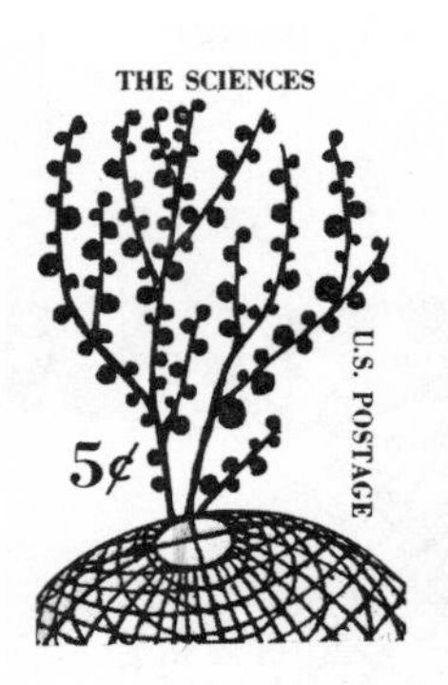

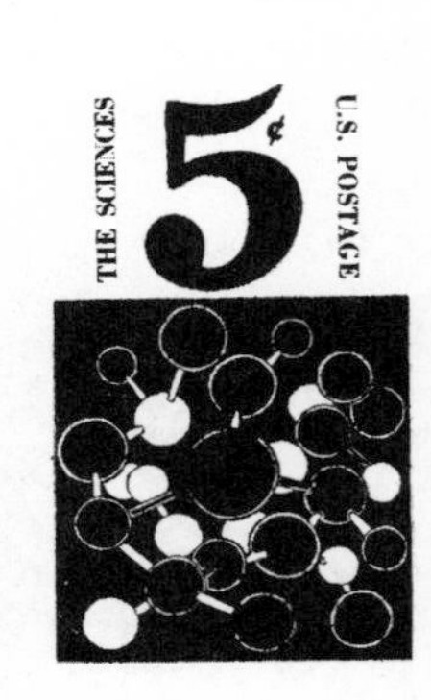

Frasconi

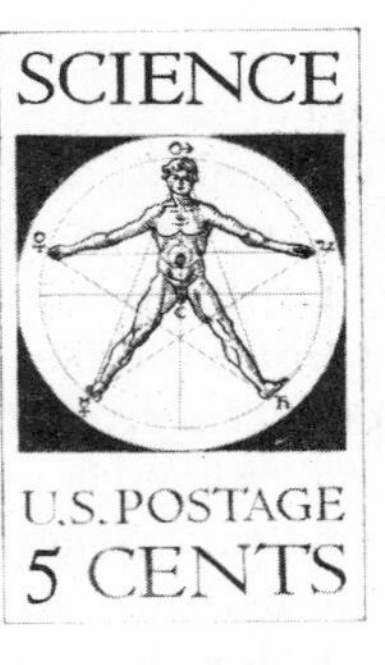

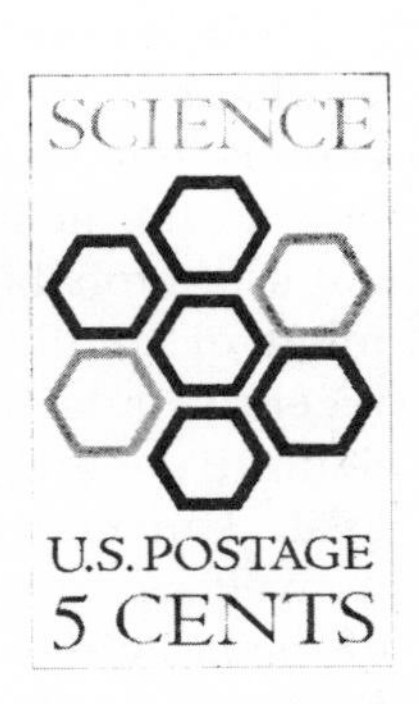

Thompson

ventor, is the creator of the geodesic dome, the Dymaxion three-wheeled car, and Dymaxion map projection. Frasconi, born in Uruguay, is particularly known for his woodcuts; he won the 1960 Grand Prix Award at the Venice Film Festival for his film, *The Neighboring Shore*. Bradbury Thompson is the designer for a number of publications including *Art News*.

Ten of the designs submitted by the artists are shown on the opposite page. The Citizen's Stamp Advisory Committee chose four of the designs from which former Postmaster General J. Edward Day chose the winner. The winning design by Frasconi, depicts a stylized representation of the world, above which is spread the sky luminescent with stars.

Which design do you feel most effectively represents the spirit and character of science, and why? If you are not enthusiastic about any of the stamps, design your own.

BELL TELEPHONE SCIENCE KITS

Bell Telephone Laboratories have produced several kits related to topics in Unit 4. Your local Bell Telephone office may be able to provide a limited number of them for you free. A brief description follows.

"Crystals and Light" includes materials to assemble a simple microscope, polarizing filters, sample crystals, a book of experiments, and a more comprehensive book about crystals.

"Energy from the Sun" contains raw materials and instructions for making your own solar cell, experiments for determining solar-cell characteristics, and details for building a light-powered pendulum, a light-commutated motor, and a radio receiver.

"Speech Synthesis" enables you to assemble a simple battery-powered circuit to artificially produce the vowel sounds. A booklet describes similarities between the circuit and human voice production, and discusses early attempts to create artificial voice machines.

"From Sun to Sound" contains a ready-made solar cell, a booklet, and materials for building a solar-powered radio.

Good Reading

Several good paperbacks in the science Study Series (Anchor Books, Doubleday and Co.) are appropriate for Unit 4, including *The Physics of Television,* by Donald G. Fink and David M. Lutyens; *Waves and Messages* by John R. Pierce; *Quantum Electronics* by John R. Pierce; *Electrons and Waves* by John R. Pierce; *Computers and the Human Mind,* by Donald G. Fink. See also "Telephone Switching," *Scientific American,* July, 1962, and the Project Physics *Reader 4.*

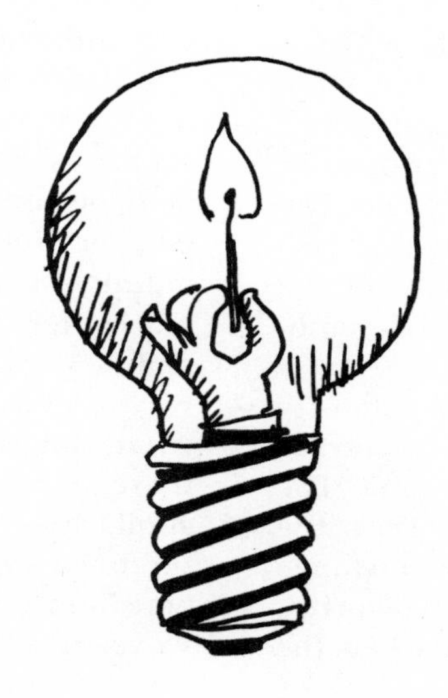

FILM LOOP

FILM LOOP 45: STANDING ELECTROMAGNETIC WAVES

Standing waves are not confined to mechanical waves in strings or in gas. It is only necessary to reflect the wave at the proper distance from a source so that two oppositely moving waves superpose in just the right way. In this film, standing electromagnetic waves are generated by a radio transmitter.

The transmitter produces electromagnetic radiation at a frequency of 435×10^6 cycles/sec. Since all electromagnetic waves travel at the speed of light, the wavelength is $\lambda = c/f = 0.69$ m. The output of the transmitter oscillator (Fig. 16-10) passes through a power-indicating meter, then to an antenna of two rods each $\frac{1}{4}\lambda$ long.

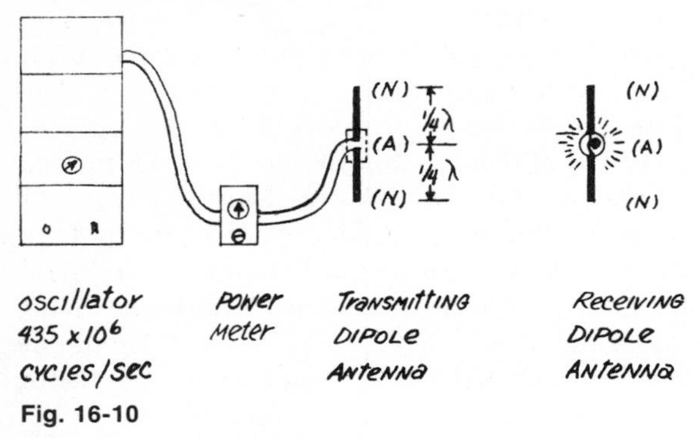

Fig. 16-10

The receiving antenna (Fig. 16-11) is also $\frac{1}{2}\lambda$ long. The receiver is a flashlight bulb connected between two stiff wires each $\frac{1}{4}\lambda$ long. If the electric field of the incoming wave is parallel to the receiving antenna, the force on the electrons in the wire drives them back and forth through the bulb. The brightness of the bulb indicates the intensity of the electromagnetic radiation at the antenna. A rectangular aluminum cavity, open toward the camera, confines the waves to provide sufficient intensity.

Initial scenes show how the intensity depends on the distance of the receiving antenna from the transmitting antenna. The radiated power is about 20 watts. Does the received intensity decrease as the distance increases? The radiation has vertical polarization, so the response falls to zero when the receiving antenna is rotated to the horizontal position.

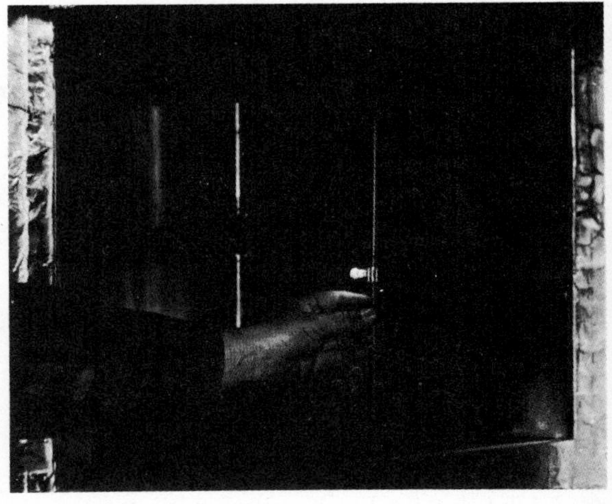

Fig. 16-11

Standing waves are set up when a metal reflector is placed at the right end of the cavity. The reflector can't be placed just anywhere; it must be at a node. The distance from source to reflector must be an integral number of half-wavelengths plus $\frac{1}{4}$ of a wavelength. The cavity length must be "tuned" to the wavelength. Nodes and antinodes are identified by moving a receiving antenna back and forth. Then a row of vertical receiving antennas is placed in the cavity, and the nodes and antinodes are shown by the pattern of brilliance of the lamp bulbs. How many nodes and antinodes can be seen in each trial?

Standing waves of different types can all have the same wavelength. In each case a source is required (tuning fork, loudspeaker, or dipole antenna). A reflector is also necessary (support for string, wooden piston, or sheet aluminum mirror). If the frequencies are 72 vib/sec for the string, 505 vib/sec for the gas, and 435×10^6 vib/sec for the electromagnetic waves and all have the same wavelength, what can you conclude about the speeds of these three kinds of waves? Discuss the similarities and differences between the three cases. What can you say about the "medium" in which the electromagnetic waves travel?

The Project Physics Course

Handbook 5

Models of the Atom

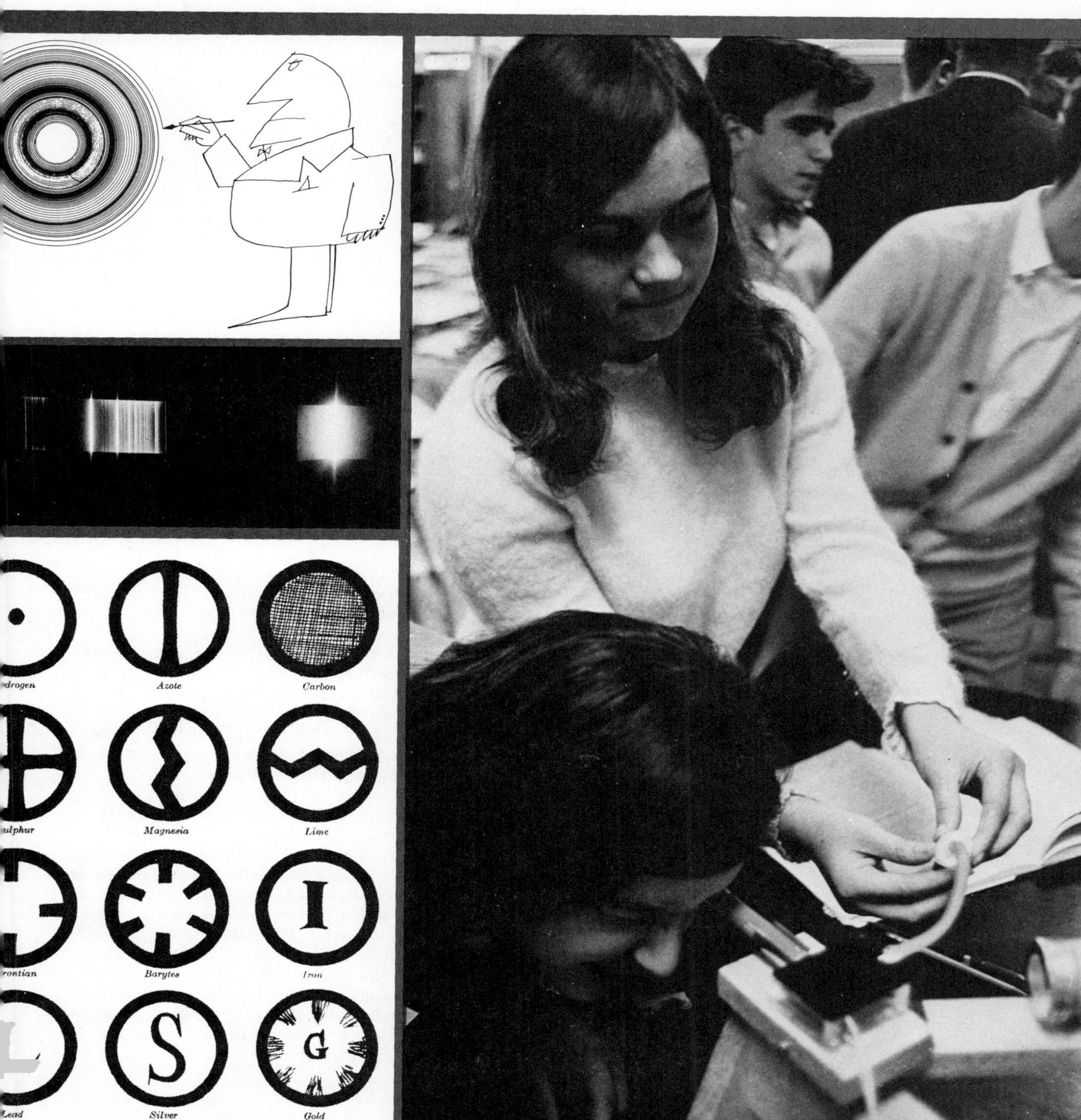

Picture Credits, Unit 5

Cover: Drawing by Saul Steinberg, from *The Sketchbook for 1967*, Hallmark Cards, Inc.

P. 286 These tables appear on pp. 122, 157 and 158 of *Types of Graphic Representation of the Periodic System of Chemical Elements* by Edmund G. Mazurs, published in 1957 by the author. They also appear on p. 8 of *Chemistry* Magazine, July 1966.

P. 292 Courtesy L. J. Lippie, Dow Chemical Company, Midland, Michigan.

P. 305 From the cover of *The Science Teacher*, Vol. 31, No. 8, December 1964, illustration for the article, "Scientists on Stamps; Reflections of Scientists' Public Image," by Victor Showalter, *The Science Teacher*, December 1964, pp. 40–42.

All photographs used with film loops courtesy of National Film Board of Canada.

Photographs of laboratory equipment and of students using laboratory equipment were supplied with the cooperation of the Project Physics staff and Damon Corporation.

Contents HANDBOOK, Unit 5

Chapter 17 The Chemical Basis of Atomic Theory

EXPERIMENT 40 ELECTROLYSIS

Volta and Davy discovered that electric currents created chemical changes never observed before. As you have already learned, these scientists were the first to use electricity to break down apparently stable compounds and to isolate certain chemical elements.

Later Faraday and other experimenters compared the *amount* of electric charge used with the *amount* of chemical products formed in such electrochemical reactions. Their measurements fell into a regular pattern that hinted at some underlying link between electricity and matter.

In this experiment you will use an electric current just as they did to decompose a compound. By comparing the charge used with the mass of one of the products, you can compute the mass and volume of a single atom of the product.

Theory Behind the Experiment

A beaker of copper sulfate ($CuSO_4$) solution in water is supported under one arm of a balance (Fig. 17-1). A negatively charged copper electrode is supported in the solution by the balance arm so that you can measure its mass without removing it from the solution. A second, positively charged copper electrode fits around the inside wall of the beaker. The beaker, its solution and the positive electrode are *not* supported by the balance arm.

If you have studied chemistry, you probably know that in solution the copper sulfate comes apart into separate charged particles, called ions, of copper (Cu^{++}) and sulfate ($SO_4^{=}$), which move about freely in the solution.

When a voltage is applied across the copper electrodes, the electric field causes the $SO_4^{=}$ ions to drift to the positive electrode (or anode) and the Cu^{++} ions to drift to the negative electrode (or cathode). At the cathode the Cu^{++} particles acquire enough negative charge to form neutral copper atoms which deposit on the cathode and add to its weight. The motion of charged particles toward the electrodes is a continuation of the electric current in the wires and the rate of transfer of charge (coulombs per second) is equal to it in magnitude. The electric current is provided by a power supply that converts 100-volt alternating current into low-voltage direct current. The current

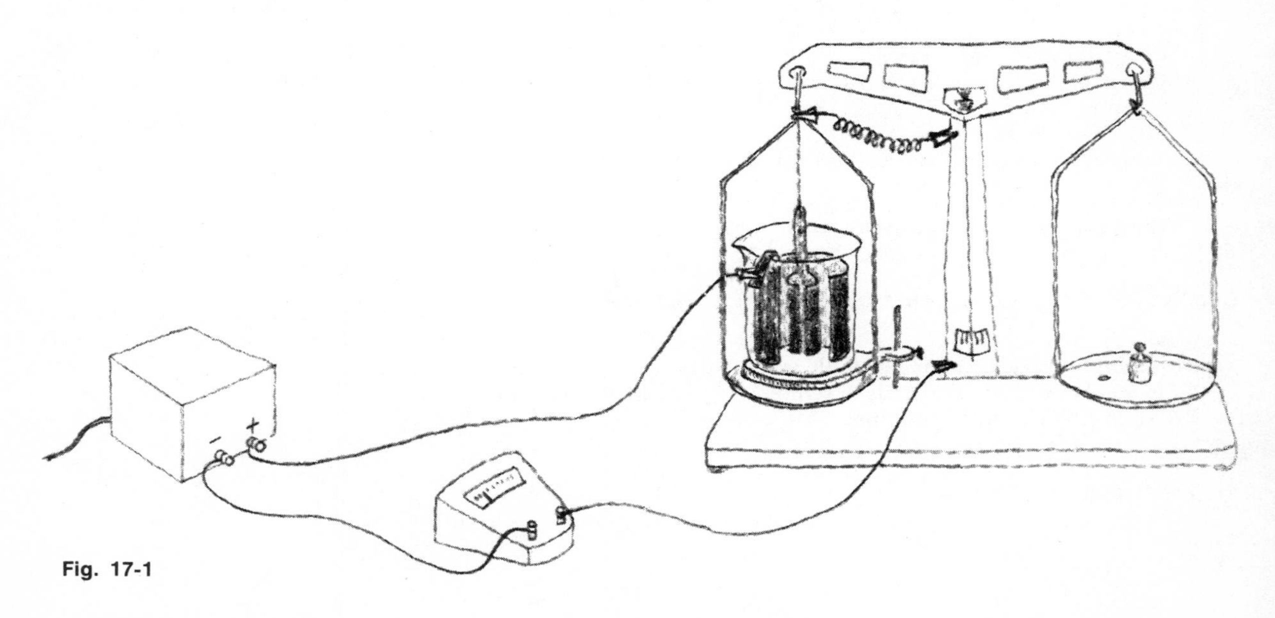

Fig. 17-1

is set by a variable control on the power supply (or by an external rheostat) and measured by an ammeter in series with the electrolytic cell as shown in Fig. 17-1.

With the help of a watch to measure the time the current flows, you can compute the electric charge that passed through the cell. By definition, the current I is the rate of transfer or charge: $I = \Delta Q/\Delta t$. It follows that the charge transferred is the product of the current and the time.

$$\Delta Q = I \times \Delta t$$

$$(\text{coulombs} = \frac{\text{coulombs}}{\text{sec}} \times \text{sec})$$

Since the amount of charge carried by a single electron is known ($q_e = 1.6 \times 10^{-19}$ coulombs), the number of electrons transferred, N_e, is

$$N_e = \frac{\Delta Q}{q_e}$$

If n electrons are needed to neutralize each copper ion, then the number of copper atoms deposited, N_{Cu}, is

$$N_{Cu} = \frac{N_e}{n}$$

If the mass of each copper atom is m_{Cu}, then the total mass of copper deposited, M_{Cu}, is

$$M_{Cu} = N_{Cu} m_{Cu}$$

Thus, if you measure I, Δt and M_{Cu}, and you know q_e and n, you can calculate a value for m_{Cu}, the mass of a single copper atom!

Setup and Procedure

Either an equal-arm or a triple-beam balance can be used for this experiment. First arrange the cell and the balance as shown in Fig. 17-1. The cathode cylinder must be supported far enough above the bottom of the beaker so that the balance arm can move up and down freely when the cell is full of the copper sulfate solution.

Next connect the circuit as illustrated in the figure. Note that the electrical connection from the negative terminal of the power supply to the cathode is made through the balance beam. The knife-edge and its seat *must* be by-passed by a short piece of thin flexible wire, as shown in Fig. 17-1 for equal-arm balances, or in Fig. 17-2 for triple-beam balances. The positive terminal of the power supply is connected directly to the anode in any convenient manner.

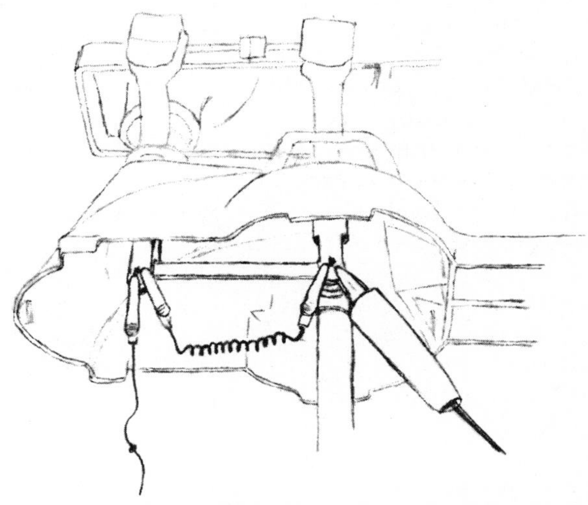

Fig. 17-2 This cutaway view shows how to by-pass the knife-edge of a typical balance. The structure of other balances may differ.

Before any measurements are made, operate the cell long enough (10 or 15 minutes) to form a preliminary deposit on the cathode—unless this has already been done. In any case, run the current long enough to set it at the value recommended by your teacher, probably about 5 amperes.

When all is ready, adjust the balance and record its reading. Pass the current for the length of time recommended by your teacher. Measure and record the current I and the time interval Δt during which the current passes. Check the ammeter occasionally and, if necessary, adjust the control in order to keep the current set at its original value.

At the end of the run, record the new reading of the balance, and find by subtraction the increase in mass of the cathode.

Calculating Mass and Volume of an Atom

Since the cathode is buoyed up by a liquid, the masses you have measured are not the true masses. Because of the buoyant force exerted by the liquid, the mass of the cathode and its increase in mass will both appear to be less than they would be in air. To find the true mass increase, you must divide the observed mass increase by the factor $(1 - D_s/D_c)$, where D_s is the density of the solution and D_c is the density of the copper.

Your teacher will give you the values of these two densities if you cannot find values for them yourself. He will also explain how the correction factor is derived. The important thing for you to understand here is why a correction factor is necessary.

Q1 How much positive or negative charge was transferred to the cathode?

In the solution this positive charge is carried from anode to cathode by doubly charged copper ions, Cu^{++}. At the cathode the copper ions are neutralized by electrons and neutral copper atoms are deposited: $Cu^{++} + 2e^{-}Cu$.

Q2 How many electrons were required to neutralize the total charge transferred? (Each electron carries -1.6×10^{-19} coulombs.)

Q3 How many electrons (single negative charge) were required to neutralize each copper ion?

Q4 How many copper atoms were deposited?

Q5 What is the mass of each copper atom?

Q6 The mass of a penny is about 3 grams. If it were made of copper only, how many atoms would it contain? (In fact modern pennies contain zinc as well as copper.)

Q7 The volume of a penny is about 0.3 cm^3. How much volume does each atom occupy?

ACTIVITIES

DALTON'S PUZZLE

Once Dalton had his theory to work with, the job of figuring out relative atomic masses and empirical formulas boiled down to nothing more than working through a series of puzzles. Here is a very similar kind of puzzle with which you can challenge your classmates.

Choose three sets of objects, each having a different mass. Large ball bearing with masses of about 70, 160, and 200 grams work well. Let the smallest one represent an atom of hydrogen, the middle-sized one an atom of nitrogen, and the large one an atom of oxygen.

From these "atoms" construct various "molecules." For example, NH_3 could be represented by three small objects and one middle-sized one, N_20 by two middle-sized ones and one large, and so forth.

Conceal one molecule of your collection in each one of a series of covered Styrofoam cups (or other light-weight, opaque containers). Mark on each container the symbols (but not the formula!) of the elements contained in the compound. Dalton would have obtained this information by qualitative analysis.

Give the covered cups to other students. Instruct them to measure the "molecular" mass of each compound and to deduce the relative atomic masses and empirical formulas from the set of masses, making Dalton's assumption of simplicity. If the objects you have used for "atoms" are so light that the mass of the styrofoam cups must be taken into account, you can either supply this information as part of the data or leave it as a complication in the problem.

If the assumption of simplicity is relaxed, what other atomic masses and molecular formulas would be consistent with the data?

ELECTROLYSIS OF WATER

The fact that electricity can decompose water was an amazing and exciting discovery, yet the process is one that you can easily demonstrate with materials at your disposal. Fig. 17-3 provides all the necessary information. Set up an electrolysis apparatus and demonstrate the process for your classmates.

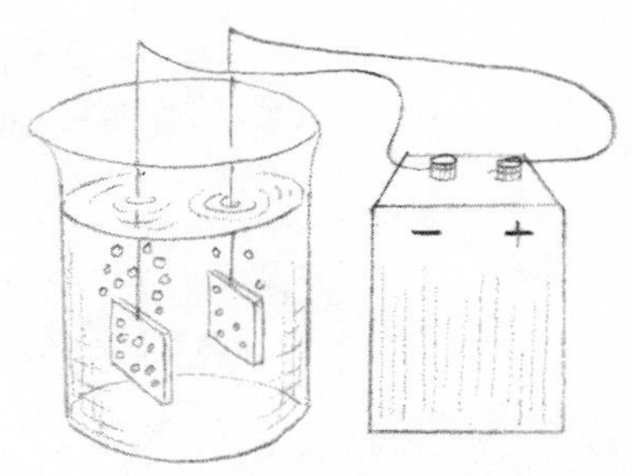

Fig. 17-3

In Fig. 17-3 it looks as if about twice as many bubbles were coming from one electrode as from the other. Which electrode is it? Does this happen in your apparatus? Would you expect it to happen?

How would you collect the two gases that bubble off the electrodes? How could you prove their identity?

If water is really just these two gases "put together" chemically, you should be able to put the gases together again and get back the water with which you started. Using your knowledge of physics, predict what must then happen to all the electrical energy you sent flowing through the water to separate it.

PERIODIC TABLE

You may have seen one or two forms of the periodic table in your classroom, but many others have been devised to emphasize various relationships among the elements. Some, such as the ones shown on the next page, are more visually interesting than others. Check various sources in your library and prepare an exhibit of the various types. An especially good lead is the article, "Ups and Down of the Periodic Table" in *Chemistry,* July 1966, which shows many different forms of the table, including those in Fig. 17-4.

It is also interesting to arrange the elements in order of discovery on a linear time chart. Periods of intense activity caused by breakthroughs in methods of extended work by a certain group of investigators show up in groups of names. A simple way to do this is to use a typewriter, letting each line represent one year (from 1600 on). All the elements then fit on six normal typing pages which can be

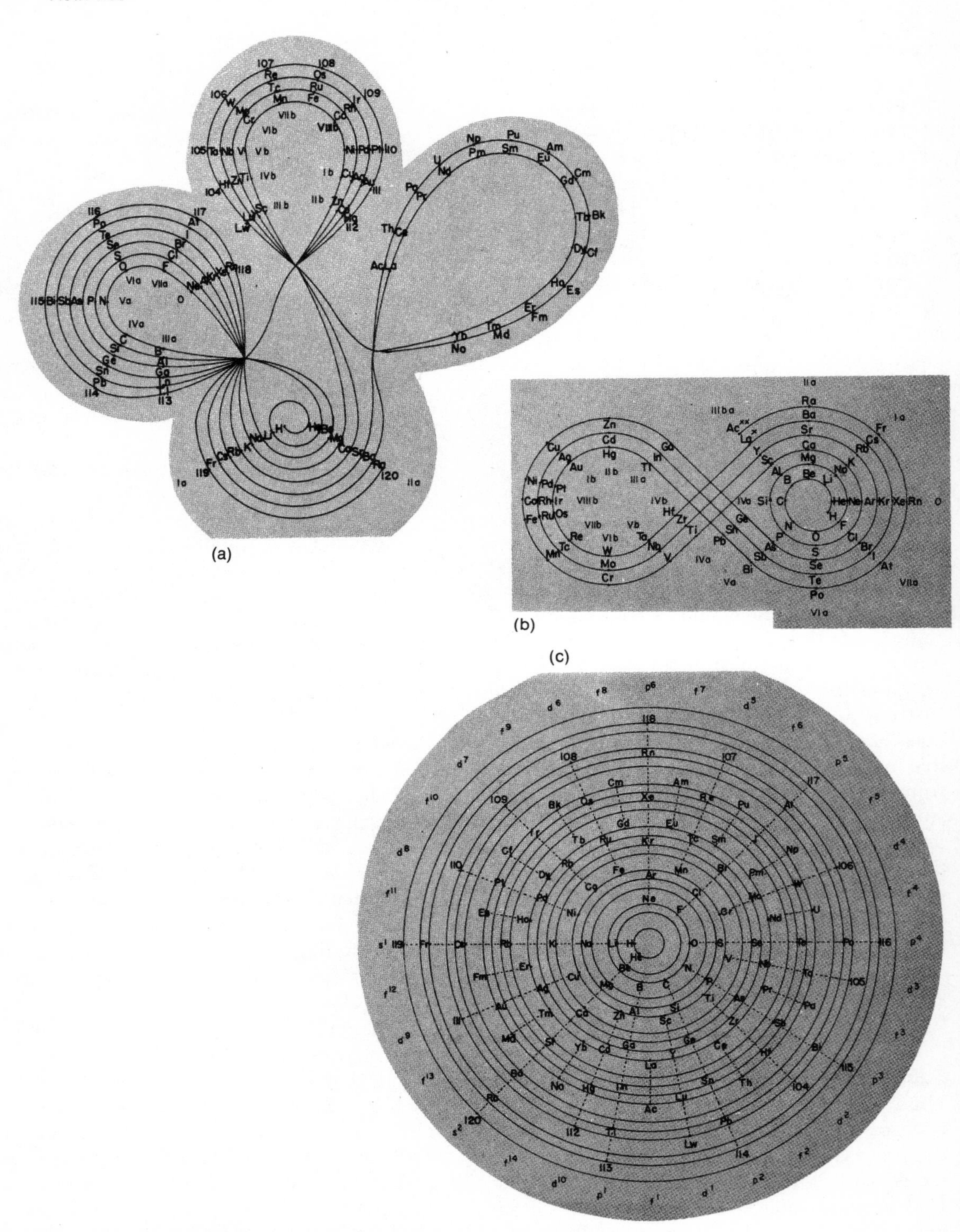

Three two-dimensional spiral forms. (a) Janet, 1928. (b) Kipp, 1942. (c) Sibaiua, 1941.

fastened together for mounting on a wall. A list of discovery dates for all elements appears at the end of Chapter 21 in the Text.

SINGLE-ELECTRODE PLATING

A student asked if copper would plate out from a solution of copper sulfate if only a negative electrode were placed in the solution. It was tried and no copper was observed even when the electrode was connected to the negative terminal of a high voltage source for five minutes. Another student suggested that only a very small (invisible) amount of copper was deposited since copper ions should be attracted to a negative electrode.

A more precise test was devised. A nickel-sulfate solution was made containing several microcuries of radioactive nickel (no radio-copper was available). A single carbon electrode was immersed in the solution, and connected to the negative terminal of the high voltage source again for five minutes. The electrode was removed, dried, and tested with a Geiger counter. The rod was slightly radioactive. A control test was run using identical test conditions, except that *no* electrical connection was made to the electrode. The control showed *more* radioactivity.

Repeat these experiments and see if the effect is true generally. What explanation would you give for these effects? (Adapted from *Ideas for Science Investigations,* N. S.-T. A. 1966).

ACTIVITIES FROM *SCIENTIFIC AMERICAN*

The following articles from the "Amateur Scientist" section of *Scientific American* relate to Unit 5. They range widely in difficulty.

Accelerator, electron, Jan. 1959, p. 138.
Beta ray spectrometer, Sept. 1958, p. 197.
Carbon 14 dating, Feb. 1957, p. 159.
Cloud chamber, diffusion, Sept. 1952, p. 179.
Cloud chamber, plumber's friend, Dec. 1956, p. 169.
Cloud chamber, Wilson, Apr. 1956, p. 156.
Cloud chamber, with magnet, June 1959, p. 173.
Cyclotron, Sept. 1953, p. 154.
Gas discharge tubes, how to make, Feb, 1958, p. 112.
Geiger counter, how to make, May 1960, p. 189.
Isotope experiments, May 1960, p. 189.
Magnetic resonance spectrometer, Apr. 1959, p. 171.
Scintillation counter, Mar. 1953, p. 104.
Spectrograph, astronomical, Sept. 1956, p. 259.
Spectrograph, Bunsen's, June 1955, p. 122.
Spinthariscope, Mar. 1953, p. 104.
Spectroheliograph, how to make, Apr. 1958, p. 126.
Subatomic particle scattering, simulating, Aug. 1965, p. 102.

FILM LOOP

FILM LOOP 46: PRODUCTION OF SODIUM BY ELECTROLYSIS

In 1807, Humphry Davy produced metallic sodium by electrolysis of molten lye—sodium hydroxide.

In the film, sodium hydroxide (NaOH) is placed in an iron crucible and heated until it melts, at a temperature of 318°C. A rectifier connected to a power transformer supplies a steady current through the liquid NaOH through iron rods inserted in the melt. Sodium ions are positive and are therefore attracted to the negative electrode; there they pick up electrons and become metallic sodium, as indicated symbolically in this reaction:

$$Na^{+} + e^{-} = Na.$$

The sodium accumulates in a thin, shiny layer floating on the surface of the molten sodium hydroxide.

Sodium is a dangerous material which combines explosively with water. The experimenter in the film scoops out a little of the metal and places it in water. (Fig. 17-5.) Energy is released rapidly, as you can see from the violence of the reaction. Some of the sodium is vaporized and the hot vapor emits the yellow light characteristic of the spectrum of sodium. The same yellow emission is easily seen if common salt, sodium chloride, or some other sodium compound, is sprinkled into an open flame.

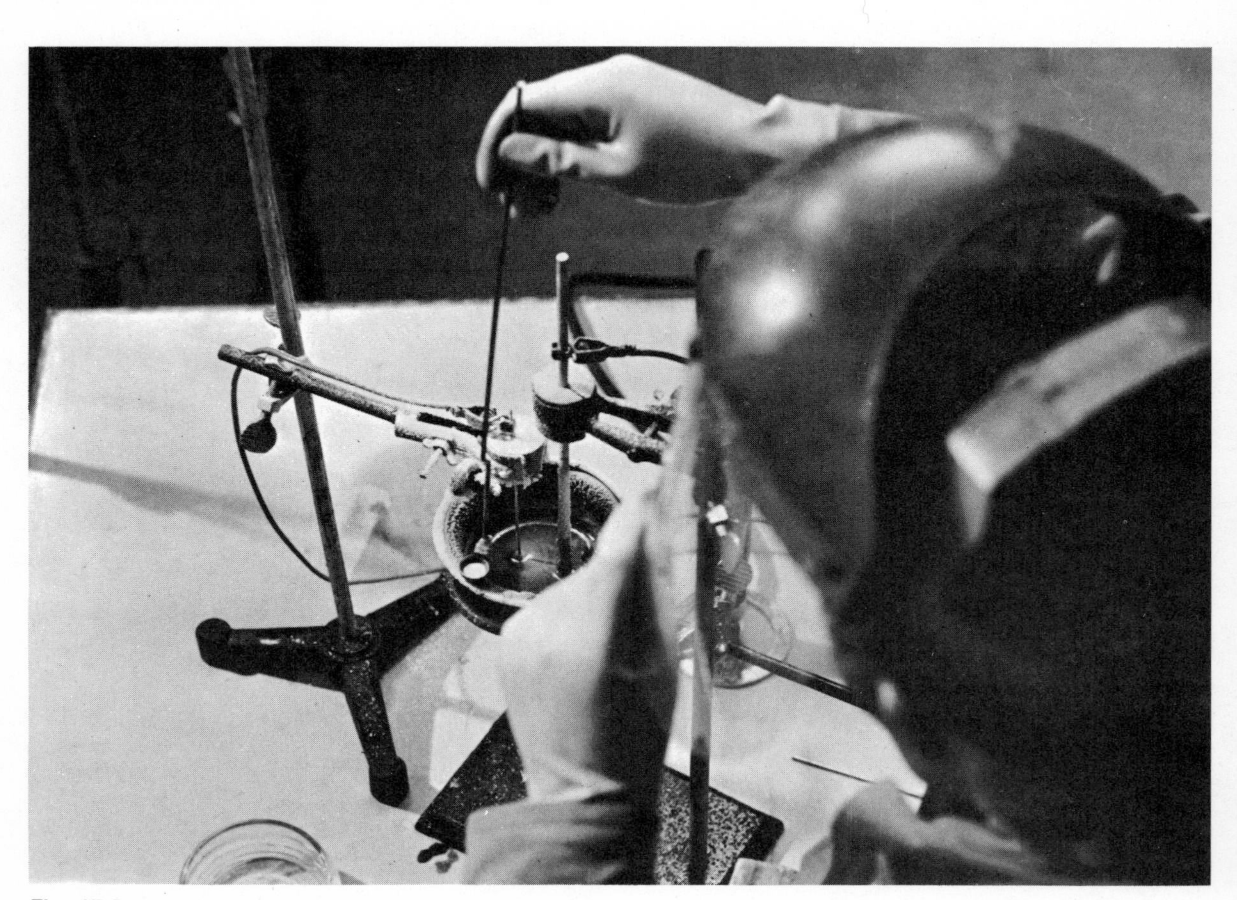

Fig. 17-5

Chapter 18 Electrons and Quanta

EXPERIMENT 41 THE CHARGE-TO-MASS RATIO FOR AN ELECTRON

In this experiment you make measurements on cathode rays. A set of similar experiments by J. J. Thomson convinced physicists that these rays are not waves but streams of identical charged particles, each with the same ratio of charge to mass. If you did experiment 38 in Unit 4, "Electron-Beam Tube," you have already worked with cathode rays and have seen how they can be deflected by electric and magnetic fields.

Thomson's use of this deflection is described on page 36 of the Unit 5 *Text*. Read that section of the text before beginning this experiment.

Theory of the experiment

The basic plan of the experiment is to measure the bending of the electron beam by a known magnetic field. From these measurements and a knowledge of the voltage accelerating the electrons, you can calculate the electron charge-to-mass ratio. The reasoning behind the calculation is illustrated in Fig. 18-1. The algebraic steps are described below.

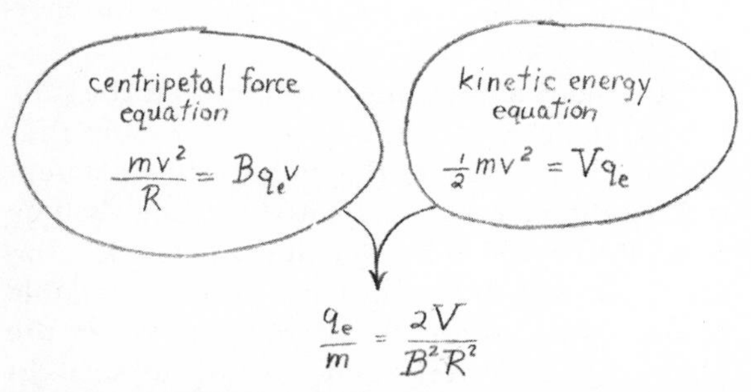

Fig. 18-1 The combination of two relationships, for centripetal and kinetic energy, with algebraic steps that eliminate velocity, v, lead to an equation for the charge-to-mass ratio of an electron.

When the beam of electrons (each of mass m and charge q_e) is bent into a circular arc of radius R by a uniform magnetic field B, the centripetal force mv^2/R on each electron is supplied by the magnetic force Bq_ev. Therefore

$$\frac{mv^2}{R} = Bq_ev,$$

or, rearranging to get v by itself,

$$v = \frac{Bq_er}{m}.$$

The electrons in the beam are accelerated by a voltage V which gives them a kinetic energy

$$\frac{mv^2}{2} = Vq_e.$$

If you replace v in this equation by the expression for v in the preceding equation, you get

$$\frac{m}{2} = \left(\frac{Bq_eR}{m}\right)^2 = Vq_e$$

or, after simplifying,

$$\frac{q_e}{m} = \frac{2V}{B^2R^2}.$$

You can measure with your apparatus all the quantities on the right-hand side of this expression, so you can use it to calculate the charge-to-mass ratio for an electron.

Preparing the apparatus

You will need a tube that gives a beam at least 5 cm long. If you kept the tube you made in Experiment 38, you may be able to use that. If your class didn't have success with this experiment, it may mean that your vacuum pump is not working well enough, in which case you will have to use another method.

In this experiment you need to be able to adjust the strength of the magnetic field until the magnetic force on the charges just balances the force due to the electric field. To enable you to change the magnetic field, you will use a pair of coils instead of permanent magnets. A current in a pair of coils, which are separated by a distance equal to the coil radius, produces a nearly uniform magnetic field in the central region between the coils. You can vary the magnetic field by changing the current in the coils.

Into a cardboard tube about 3″ in diameter and 3″ long cut a slot $1\frac{1}{4}$″ wide. (Fig. 18-2.) Your electron-beam tube should fit into this slot as shown in the photograph of the completed set-up. (Fig. 18-4.) Current in the pair of coils will create a magnetic field at right angles to the axis of the cathode rays.

Now wind the coils, one on each side of the slot, using a single length of insulated copper wire (magnet wire). Wind about 20 turns of wire for each of the two coils, one coil on each side of the slot, leaving 10″ of wire free at both ends of the coil. Don't cut the wire off the reel until you have found how much you will need. Make the coils as neat as you can and keep them close to the slot. Wind both coils in the same sense (for example, make both clockwise).

When you have made your set of coils, you must "calibrate" it; that is, you must find out what magnetic field strength B corresponds to what values of current I in the coils. To do

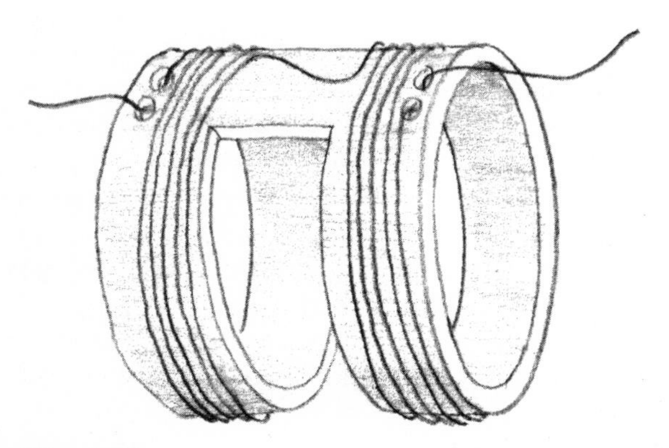

Fig. 18-2

Fig. 18-3

this, you can use the current balance, as you did in Experiment 36. Use the shortest of the balance "loops" so that it will fit inside the coils as shown in Fig. 18-3.

Connect the two leads from your coils to a power supply capable of giving up to 5 amps direct current. There must be a varable control on the power supply (or a rheostat in the circuit) to control the current; and an ammeter to measure it.

Measure the force F for a current I in the loop. To calculate the magnetic field due to the current in the coils, use the relationship $F = BI\ell$ where ℓ is the length of short section of the loop. Do this for several different values of current in the coil and plot a calibration graph of magnetic field B against coil current I.

Set up your electron-beam tube as in Experiment 38. Reread the instructions for operating the tube.

Connect a shorting wire between the pins for the deflecting plates. This will insure that the two plates are at the same electric potential, so the electric field between them will be zero. Pump the tube out and adjust the filament current until you have an easily visible beam. Since there is no field between the plates, the electron beam should go straight up the center of the tube between the two plates. (If it does not, it is probably because the filament and the hole in the anode are not properly aligned.)

Turn down the filament current and switch off the power supply. Now, without releasing

Fig. 18-4 The magnetic field is parallel to the axis of the coils; the electric and magnetic fields are perpendicular to each other and to the electron beam.

the vacuum, mount the coils around the tube as shown in Fig. 18-4.

Connect the coils as before to the power supply. Connect a voltmeter across the power supply terminals that provide the accelerating voltage V.

Your apparatus is now complete.

Performing the experiment

Turn on the beam, and make sure it is travelling in a straight line. The electric field remains off throughout the experiment, and the deflecting plates should still be connected together.

Turn on and slowly increase the current in the coils until the magnetic field is strong enough to deflect the electron beam noticeably.

Record the current I in the coils.

Using the calibration graph, find the magnetic field B.

Record the accelerating voltage V between the filament and the anode plate.

Finally you need to measure R, the radius of the arc into which the beam is bent by the magnetic field. The deflected beam is slightly fan-shaped because some electrons are slowed by collisions with air molecules and are bent into a curve of smaller R. You need to know the largest value of R (the "outside" edge of the curved beam), which is the path of electrons that have made no collisions. You won't be able to measure R directly, but you can find

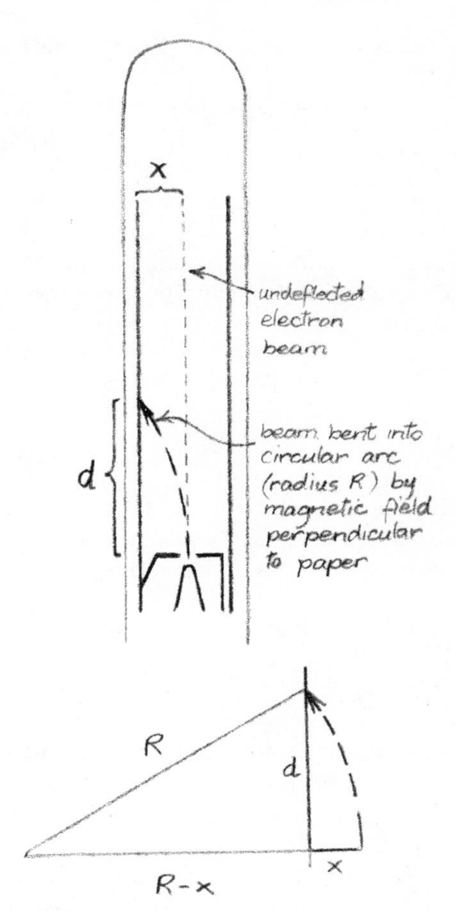

Fig. 18-5

it from measurements that are easy to make. (Fig. 18-5.)

You can measure x and d. It follows from Pythagoras' theorem that $R^2 = d^2 + (R - x)^2$,

so $R = \dfrac{d^2 + x^2}{2x}$.

Q1 What is your calculation of R on the basis of your measurements?

Now that you have values for V, B and R, you can use the formula $q_e/m = 2V/B^2R^2$ to calculate your value for the charge-to-mass ratio for an electron.

Q2 What is your value for q_e/m, the charge-to-mass ratio for an electron?

EXPERIMENT 42 THE MEASUREMENT OF ELEMENTARY CHARGE

In this experiment, you will investigate the charge of the electron, a fundamental physical constant in electricity, electromagnetism, and nuclear physics. This experiment is substantially the same as Millikan's famous oil-drop experiment, described on page 39 of the Unit 5 *Text*. The following instructions assume that you have read that description. Like Millikan, you are going to measure very small electric charges to see if there is a limit to how small an electric charge can be. Try to answer the following three questions before you begin to do the experiment in the lab.

Q1 What is the electric field between two parallel plates separated by a distance d meters, if the potential difference between them is V volts?

Q2 What is the electric force on a particle carrying a charge of q coulombs in an electric field of E volts/meter?

Q3 What is the gravitational force on a particle of mass m in the earth's gravitational field?

Background

Electric charges are measured by measuring the forces they experience and produce. The extremely small charges that you are seeking require that you measure extremely small forces. Objects on which such small forces can have a visible effect must also in turn be very small.

Millikan used the electrically charged droplets produced in a fine spray of oil. The varying size of the droplets complicated his measurements. Fortunately you can now use suitable objects whose sizes are accurately known. You use tiny latex spheres (about 10^{-4} cm diatmeter), which are almost identical in size in any given sample. In fact, these spheres, shown magnified (about 5000 ×) in Fig. 18-6, are used as a convenient way to find the magnifying power of electron microscopes. The spheres can be bought in a water suspension, with their diameter recorded on the bottle. When the suspension is sprayed into the air, the water quickly evaporates and leaves

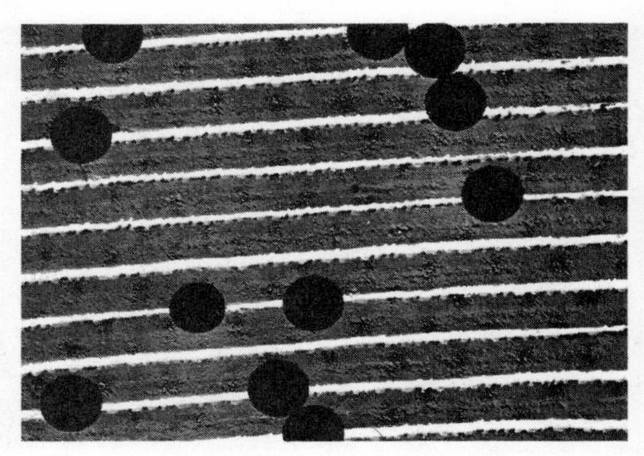

Fig. 18-6 Electron micrograph of latex spheres 1.1×10^{-4}cm, silhouetted against diffracting grating of 28,800 lines/inch. What magnification does this represent?

a cloud of these particles, which have become charged by friction during the spraying. In the space between the plates of the Millikan apparatus they appear through the 50-power microscope as bright points of light against a dark background.

You will find that an electric field between the plates can pull some of the particles upward against the force of gravity, so you will know that they are charged electrically.

In your experiment, you adjust the voltage producing the electric field until a particle hangs motionless. On a balanced particle carrying a charge q, the upward electric force Eq and the downward gravitational force ma_g are equal, so

$$ma_g = Eq.$$

The field $E = V/d$, where V is the voltage between the plates (the voltmeter reading) and d is the separation of the plates. Hence

$$q = \frac{ma_g d}{V}.$$

Notice that $ma_g d$ is a constant for all measurements and need be found only once. Each value of q will be this constant $ma_g d$ times $1/V$ as the equation above shows. That is, the value of q for a particle is proportional to $1/V$: the greater the voltage required to balance the weight of the particle, the smaller the charge of the particle must be.

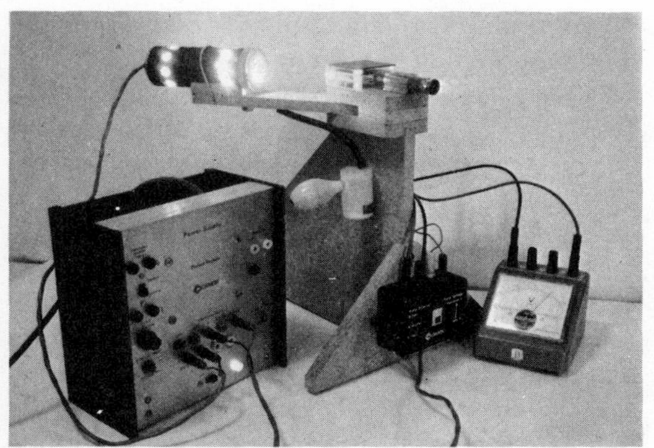

Fig. 18-7 A typical set of apparatus. Details may vary considerably.

Using the apparatus

If the apparatus is not already in operating condition, consult your teacher. Study Figs. 18-7 and 18-8 until you can identify the various parts. Then switch on the light source and look through the microscope. You should see a series of lines in clear focus against a uniform gray background.

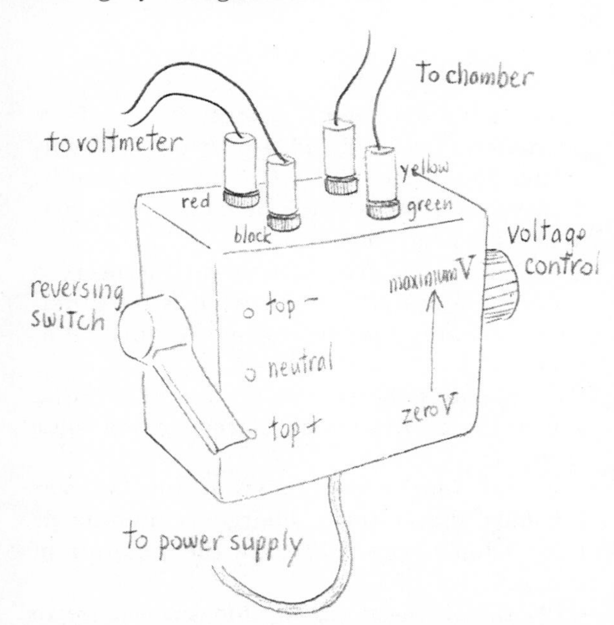

Fig. 18-8 A typical arrangement of connections to the high-voltage reversing switch.

The lens of the light source may fog up as the heat from the lamp drives moisture out of the light-source tube. If this happens, remove the lens and wipe it on a clean tissue. Wait for the tube to warm up thoroughly before replacing the lens.

Squeeze the bottle of latex suspension two or three times until five or ten particles drift into view. You will see them as tiny bright spots of light. You may have to adjust the focus slightly to see a specific particle clearly. Notice how the particle appears to move upward. The view is inverted by the microscope—the particles are actually falling in the earth's gravitational field.

Now switch on the high voltage across the plates by turning the switch up or down. Notice the effect on the particles of varying the electric field by means of the voltage-control knob.

Notice the effect when you reverse the electric field by reversing the switch position. (When the switch is in its mid-position, there is zero field between the plates.)

Q4 Do all the particles move in the same direction when the field is on?

Q5 How do you explain this?

Q6 Some particles move much more rapidly in the field than others. Do the rapidly moving particles have larger or smaller charges than the slowly moving particles?

Sometimes a few particles cling together, making a clump that is easy to see—the clump falls more rapidly than single particles when the electric field is off. Do not try to use these for measuring q.

Try to balance a particle by adjusting the field until the particle hangs motionless. Observe it carefully to make sure it isn't slowly drifting up or down. The smaller the charge, the greater the electric field must be to hold up the particle.

Taking data

It is not worth working at voltages much below 50 volts. Only highly charged particles can be balanced in these small fields, and you are interested in obtaining the smallest charge possible.

Set the potential difference between the plates to about 75 volts. Reverse the field a few times so that the more quickly moving particles (those with greater charge) are swept out of the field of view. Any particles that remain have low charges. If *no* particles remain, squeeze in some more and look again for some with small charge.

When you have isolated one of these particles carrying a low charge, adjust the voltage carefully until the particle hangs motionless. Observe it for some time to make sure that it isn't moving up or down very slowly, and that the adjustment of voltage is as precise as possible. (Because of uneven bombardment by air molecules, there will be some slight, uneven drift of the particles.)

Read the voltmeter. Then estimate the precision of the voltage setting by seeing how little the voltage needs to be changed to cause the particle to start moving just perceptibly. This small change in voltage is the greatest amount by which your *setting* of the balancing voltage can be uncertain.

When you have balanced a particle, make sure that the voltage setting is as precise as you can make it before you go on to another particle. The most useful range to work in is 75–150 volts, but try to find particles that can be brought to rest in the 200–250 volt range too, if the meter can be used in that range. Remember that the higher the balancing field the smaller the charge on the particle.

In this kind of an experiment, it is helpful to have large amounts of data. This usually makes it easier to spot trends and to distinguish main effects from the background scattering of data. Thus you may wish to contribute your findings to a class data pool. Before doing that, however, arrange your values of V in a vertical column of increasing magnitude.

Q7 Do the numbers seem to clump together in groups, or do they spread out more or less evenly from the lowest to the highest values?

Now combine your data with that collected by your classmates. This can conveniently be done by placing your values of V on a class histogram. When the histogram is complete, the results can easily be transferred to a transparent sheet for use on an overhead projector. Alternatively, you may wish to take a Polaroid photograph of the completed histogram for inclusion in your laboratory notebook.

Q8 Does your histogram suggest that all values of q are possible and that electric charge is therefore endlessly divisible, or the converse?

If you would like to make a more complete quantitative analysis of the class results, calculate an average value for each of the highest three or four clumps of V values in the class histogram. Next change those to values of $1/V$ and list them in order. Since q is proportional to 1/V, these values represent the magnitude of the charges on the particles.

To obtain actual values for the charges, the $1/V$'s must be multiplied by $ma_g d$. The separation d of the two plates, typically about 5.0 mm, or 5.0×10^{-3}m, is given in the specification sheets provided by the manufacturer. You should check this.

The mass m of the spheres is worked out from a knowledge of their volume and the densitiy D of the material they are made from.

Mass = volume × density, or

$$m = \tfrac{4}{3} r^3 \times D$$

The sphere diameter (careful: 2) has been previously measured and is given on the supply bottle. The density D is 1077 kg/m³ (found by measuring a large batch of latex before it is made into little spheres).

Q9 What is the spacing between the observed average values of $1/V$ and what is the difference in charge that corresponds to this difference in $1/V$?

Q10 What is the smallest value of $1/V$ that you obtained? What is the corresponding value of q?

Q11 Do your experimental results support the idea that electric charge is quantized? If so, what is *your value* for the quantum of charge?

Q12 If you have already measured q_e/m in Experiment 39, compute the mass of an electron. Even if your value differs the accepted value by a factor of 10, perhaps you will agree that its measurement is a considerable intellectual triumph.

EXPERIMENT 43 THE PHOTOELECTRIC EFFECT

In this experiment you will make observations on the effect of light on a metal surface; then you will compare the appropriateness of the wave model and the particle model of light for explaining what you observe.

Before doing the experiment, read text Sec. 18.4 (Unit 5) on the photoelectric effect.

How the apparatus works

Light that you shine through the window of the phototube falls on a half-cylinder of metal called the emitter. The light drives electrons from the emitter surface.

Along the axis of the emitter (the center of the tube) is a wire called the collector. When the collector is made a few volta positive with respect to the emitter, practically all the emitted electrons are drawn to it, and will return to the emitter through an external wire. Even if the collector is made slightly negative, some electrons will reach it and there will be a measurable current in the external circuit.

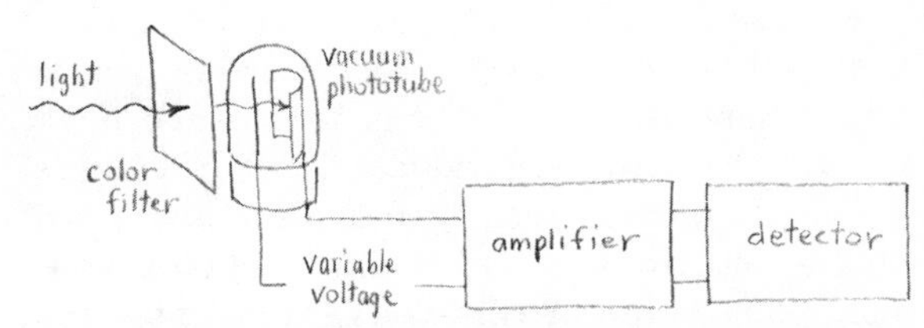

However much the details may differ, any equipment for the photoelectric effect experiment will consist of these basic parts.

The small current can be amplified several thousand times and detected in any of several different ways. One way is to use a small loudspeaker in which the amplified photoelectric current causes an audible hum; another is to use a cathode ray oscilloscope. The following description assumes that the output current is read on a microammeter (Fig. 18-9).

The voltage control knob on the phototube unit allows you to vary the voltage between emitter and collector. In its full counterclockwise position, the voltage is zero. As you turn the knob clockwise the "photocurrent" decreases. You are making the collector more

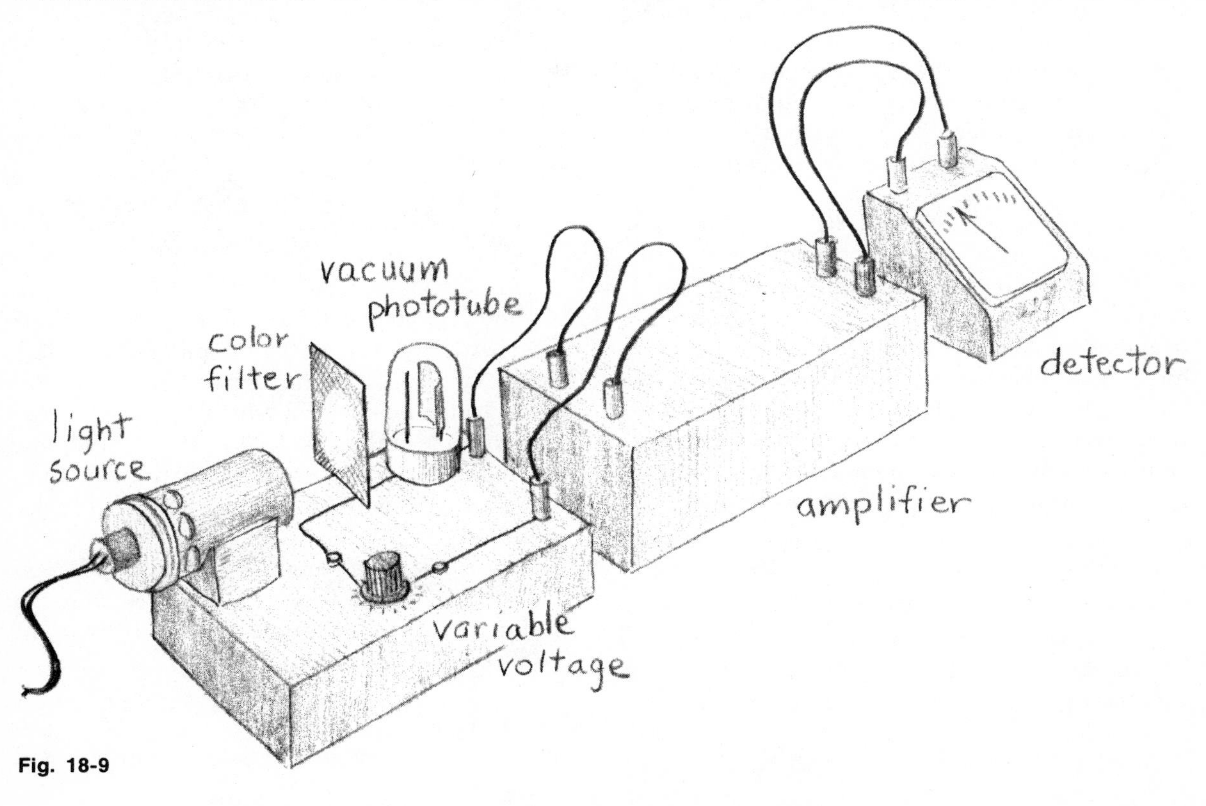

Fig. 18-9

and more negative and fewer and fewer electrons get to it. Finally the photocurrent ceases altogether—all the electrons are turned back before reaching the collector. The voltage between emitter and collector that just stops all the electrons is called the "stopping voltage." The value of this voltage indicates the maximum kinetic energy with which the electrons leave the emitter. To find the value of the stopping voltage precisely you will have to be able to determine precisely when the photocurrent is reduced to zero. Because there is some drift of the amplifier output, the current indicated on the meter will drift around the zero point even when the actual current remains exactly zero. Therefore you will have to adjust the amplifier *offset* occasionally to be sure the zero level is really zero. An alternative is to ignore the precise reading of the current meter and adjust the collector voltage until *turning the light off and on causes no detectable change in the current*. Turn up the negative collector voltage until blocking the light from the tube (with black paper) has no effect on the meter reading—the exact location of the meter pointer isn't important.

The position of the voltage control knob at the current cutoff gives you a rough measure of stopping voltage. To measure it more precisely, connect a voltmeter as shown in Fig. 18-10.

In the experiment you will measure the stopping voltages as light of different frequencies falls on the phototube. Good colored filters will allow light of only a certain range of frequencies to pass through. You can use a hand spectroscope to find the highest frequency line passed by each filter. The filters select frequencies from the mercury spectrum emitted by an intense mercury lamp. Useful frequencies of the mercury spectrum are:

Yellow	5.2×10^{14}/sec
Green	5.5×10^{14}/sec
Blue	6.9×10^{14}/sec
Violet	7.3×10^{14}/sec
(Ultraviolet)	8.2×10^{14}/sec

DOING THE EXPERIMENT

Part I

The first part of the experiment is qualitative. To see if there is *time delay* between light falling on the emitter and the emission of photoelectrons, cover the phototube and then quickly remove the cover. Adjust the light source and filters to give the smallest photocurrent that you can conveniently notice on the meter.

Q1 Can you detect any time delay between the moment that light hits the phototube and the moment that motion of the microammeter pointer (or a hum in the loudspeaker or deflection of the oscilloscope trace) signals the passage of photoelectrons through the phototube?

To see if the *current* in the phototube depends on the intensity of incident light, vary the distance of the light source.

Q2 Does the *number* of photoelectrons emitted from the sensitive surface vary with light intensity—that is, does the output current of the amplifier vary with the intensity of the light?

To find out whether the *kinetic energy* of the photoelectrons depends on the *intensity* of the incident light, measure the stopping voltage with different intensities of light falling on the phototube.

Q3 Does the *kinetic energy* of the photoelectrons depend on *intensity*—that is, does the stopping voltage change?

Finally, determine how the *kinetic energy* of photoelectrons depends on the *frequency* of incident light. You will remember (Text Sec. 18.5) that the maximum kinetic energy of the photoelectrons is $V_{stop}q_e$, where V_{stop} is the stopping voltage and $q_e = 1.60 \times 10^{-19}$ coulombs, the charge on an electron. Measure the stopping voltage with various filters over the window.

Q4 How does the stopping voltage and hence the kinetic energy change as the light is changed from red through blue or ultraviolet (no filters)?

Part II

In the second part of the experiment you will

make more precise measurements of stopping voltage. To do this, adjust the voltage control knob to the cutoff (stopping voltage) position and then measure V with a voltmeter (Fig. 18-10.) Connect the voltmeter only after the cutoff adjustment is made so that the voltmeter leads will not pick up any ac voltage induced from other conducting wires in the room.

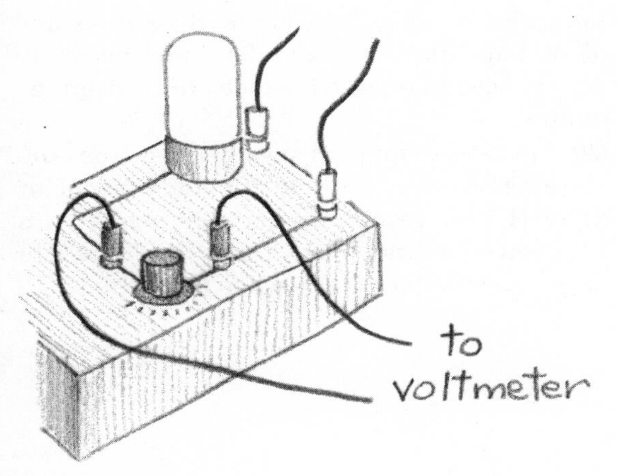

Fig. 18-10

Measure the stopping voltage V_{stop} for three or four different light frequencies, and plot the data on a graph. Along the vertical axis, plot electron energy $V_{stop}q_e$. When the stopping voltage V is in volts, and q_e is in coulombs, Vq_e will be energy, in joules.

Along the horizontal axis plot frequency of light f.

Interpretation of Results

As suggested in the opening paragraph, you can compare the wave model of light and the particle model in this experiment. Consider, then, how these models explain your observations.

Q5 If the light striking your phototube acts as *waves*—

a) Can you explain why the stopping voltage should depend on the *frequency* of light?

b) Would you expect the stopping voltage to depend on the *intensity* of the light? Why?

c) Would you expect a delay between the time that light first strikes the emitter and the emission of photoelectrons? Why?

Q6 If the light is acting as a stream of *particles*, what would be the answer to questions a, b and c above?

If you drew the graph suggested in the Part II of the experiment, you should now be prepared to interpret the graph. It is interesting to recall that Einstein predicted its form in 1905, and by experiments similar to yours, Millikan verified Einstein's prediction in 1916.

Einstein's photoelectric equation (Text Sec. 18.5) describes the energy of the most energetic photoelectrons (the last ones to be stopped as the voltage is increased), as

$$\begin{aligned}\tfrac{1}{2}mv^2_{max} &= V_{stop}q_e \\ &= hf - W.\end{aligned}$$

This equation has the form

$$y = kx - c.$$

In this equation $-c$ is a constant, the value of y at the point where the straight line cuts the vertical axis; and k is another constant, namely the slope of the line. (See Fig. 18-11.) Therefore, the slope of a graph of $V_{stop}q_e$ against f should be h.

Q7 What is the value of the slope of *your* graph? How well does this value compare with

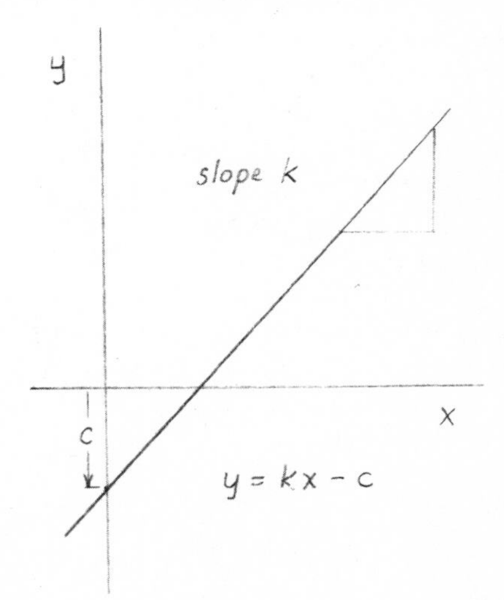

Fig. 18-11

CHAPTER 18

the value of Planck's constant, $h = 6.6 \times 10^{-34}$ joule-sec? (See Fig. 18-12).

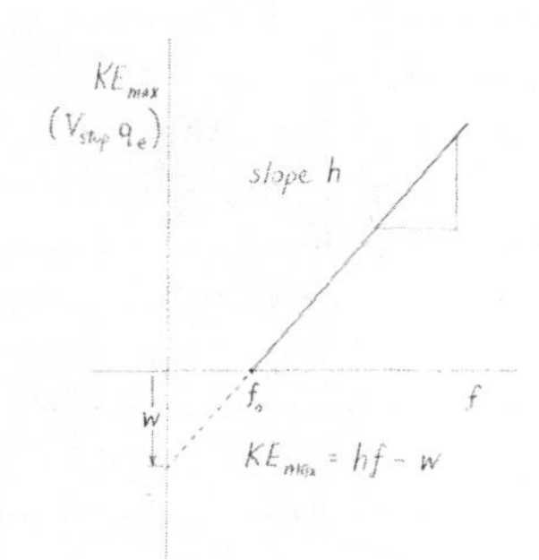

Fig. 18-12

With the equipment you used, the slope is unlikely to agree with the accepted value of h (6.6×10^{-34} joule-sec) more closely than an order of magnitude. Perhaps you can give a few reasons why your agreement cannot be more approximate.

Q8 The lowest frequency at which any electrons are emitted from the cathode surface is called the ***threshold frequency***, f_0. At this frequency $\frac{1}{2}mv_{max} = 0$ and $hf_0 = W$, where W is the "work function." Your experimentally obtained value of W is not likely to be the same as that found for very clean cathode surfaces, more carefully filtered light, etc. The important thing to notice here is that there *is* a value of W, indicating that there is a minimum energy needed to release photoelectrons from the emitter.

Q9 Einstein's equation was derived from the assumption of a particle (photon) model of light. If your results do not fully agree with Einstein's equation, does this mean that your experiment supports the wave theory?

ACTIVITIES

WRITINGS BY OR ABOUT EINSTEIN

In addition to his scientific works, Einstein wrote many perceptive essays on other areas of life which are easy to read, and are still very current. The chapter titles from *Out of My Later Years* (Philosophical Library, N.Y. 1950) indicate the scope of these essays: Convictions and Beliefs; Science; Public Affairs; Science and Life; Personalities; My People. This book includes his writings from 1934 to 1950. *The World As I See It* includes material from 1922 to 1934. *Albert Einstein: Philosopher-Scientist,* Vol. I. (Harper Torchbook, 1959) contains Einstein's autobiographical notes, left-hand pages in German and right hand pages in English, and essays by twelve physicist contemporaries of Einstein about various aspects of his work. See also the three articles, "Einstein," "Outside and Inside the Elevator," and "Einstein and Some Civilized Discontents" in *Reader* 5.

MEASURING q/m FOR THE ELECTRON

With the help of a "tuning eye" tube such as you may have seen in radio sets, you can measure the charge-to-mass ratio of the electron in a way that is very close to J. J. Thomson's original method.

Complete instructions appear in the PSSC *Physics Laboratory Guide,* Second Edition, D. C. Heath Company, Experiment IV-12, "The Mass of the Electron," pp. 79-81.

CATHODE RAYS IN A CROOKES TUBE

A Crookes tube having a metal barrier inside it for demonstrating that cathode rays travel in straight lines may be available in your classroom. In use, the tube is excited by a Tesla coil or induction coil.

Use a Crookes tube to demonstrate to the class the deflection of cathode rays in magnetic fields. To show how a magnet focuses cathode rays, bring one pole of a strong bar magnet toward the shadow of the cross-shaped obstacle near the end of the tube. Watch what happens to the shadow as the magnet gets closer and closer to it. What happens when you switch the poles of the magnet? What do you think would happen if you had a stronger magnet?

Can you demonstrate deflection by an electric field? Try using static charges as in Experiment 34, "Electric Forces I," to create a deflecting field. Then if you have an electrostatic generator, such as a small Van de Graaff or a Wimshurst machine, try deflecting the rays using parallel plates connected to the generator.

X RAYS FROM A CROOKES TUBE

To demonstrate that x rays penetrate materials that stop visible light, place a sheet of 4″ × 5″ 3000-ASA-speed Polaroid Land film, still in its protective paper jacket, in contact with the end of the Crookes' tube. (A film pack cannot be used, but any other photographic film in a light-tight paper envelope could be substituted.) Support the film on books or the table so that it doesn't move during the exposure. Fig. 18-13 was a 1-minute exposure using a hand-held Tesla coil to excite the Crookes tube.

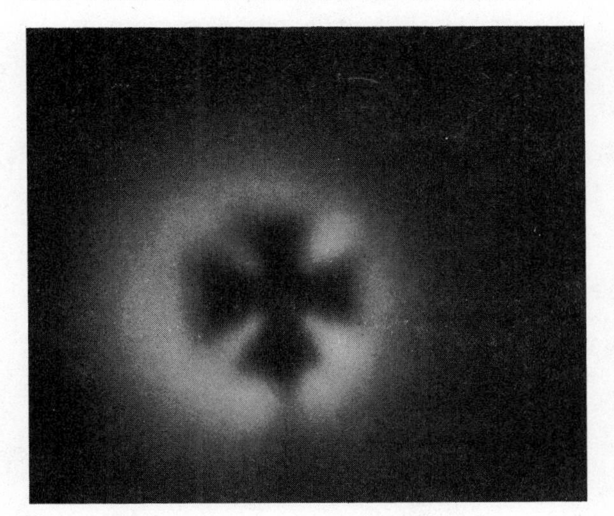

Fig. 18-13

LIGHTING AN ELECTRIC LAMP WITH A MATCH

Here is a trick with which you can challenge your friends. It illustrates one of the many amusing and useful applications of the photoelectric effect in real life. You will need the phototube from Experiment 42, "The Photoelectric Effect," together with the Project Physics Amplifier and Power Supply. You will also need a $1\frac{1}{2}$V dry cell or power supply and a 6V light source such as the one used in the Millikan Apparatus. (If you use this light source, remove the lens and cardboard tube and use only the 6V lamp.) Mount the lamp on the Photoelectric Effect apparatus and connect it to the 0-5V, 5 amps variable output on the power supply. Adjust the output to maximum. Set the *transistor switch input* switch to *switch*.

Connect the Photoelectric Effect apparatus to the Amplifier as shown in Fig. 18-14. Notice that the polarity of the 1.5V cell is reversed and that the output of the Amplifier is connected to the *transistor switch input*.

Advance the gain control of the amplifier to maximum, then adjust the offset control in a positive direction until the filament of the 6V lamp ceases to glow. Ignite a match near the apparatus (the wooden type works the best) and bring it quickly to the window of the phototube while the phosphor of the match is still glowing brightly. The phosphor flare of the match head will be bright enough to cause sufficient photocurrent to operate the transistor switch which turns the bulb on. Once the bulb is lit, it keeps the photocell activated by its own light; you can remove the match and the bulb will stay lit.

When you are demonstrating this effect, tell your audience that the bulb is really a candle and that it shouldn't surprise them that you can light it with a match. And of course one way to put out a candle is to moisten your fingers and pinch out the wick. When your fingers pass between the bulb and the photocell, the bulb turns off, although the filament may glow a little, just as the wick of a freshly snuffed candle does. You can also make a "candle-snuffer" from a little cone of any reasonable opaque material and use this instead of your fingers. Or you can "blow out" the bulb: It will go out obediently if you take care to remove it from in front of the photocell as you blow it out.

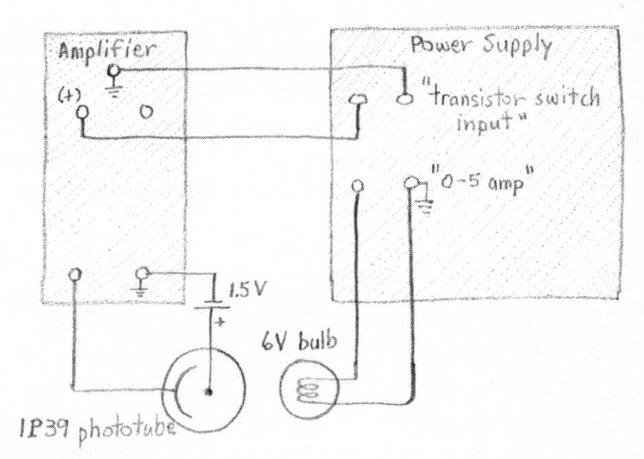

Fig. 18-14

FILM LOOP

FILM LOOP 47 THOMSON MODEL OF THE ATOM

Before the development of the Bohr theory, a popular model for atomic structure was the "raisin pudding" model of J. J. Thomson. According to this model, the atom was supposed to be a uniform sphere of positive charge in which were embedded small negative "corpuscles" (electrons). Under certain conditions the electrons could be detached and observed separately, as in Thomson's historic experiment to measure the charge/mass ratio.

The Thomson model did not satisfactorily explain the stability of the electrons and especially their arrangement in "rings," as suggested by the periodic table of the elements. In 1904 Thomson performed experiments which to him showed the *possibility* of a ring structure within the broad outline of the raisin-pudding model. Thomson also made mathematical calculations of the various arrangements of electrons in his model.

In the Thomson model of the atom, the cloud of positive charge created an electric field directed along radii, strongest at the surface of the sphere of charge and decreasing to zero at the center. You are familiar with a gravitational example of such a field. The earth's downward gravitational field is strongest at the surface and it decreases uniformly toward the center of the earth.

For his model-of-a-model Thomson used still another type of field—a magnetic field caused by a strong electromagnet above a tub of water. Along the water surface the field is "radial," as shown by the pattern of iron filings sprinkled on the glass bottom of the tub. Thomson used vertical magnetized steel needles to represent the electrons; these were stuck through corks and floated on the surface of the water. The needles were oriented with like poles pointing upward; their mutual repulsion tended to cause the magnets to spread apart. The outward repulsion was counteracted by the radial magnetic field directed inward toward the center. When the floating magnets were placed in the tub of water, they came to equilibrium configurations under the combined action of all the forces. Thomson saw in this experiment a partial verification of his calculation of how electrons (raisins) might come to equilibrium in a spherical blob of positive fluid.

In the film the floating magnets are 3.8 cm long, supported by ping pong balls (Fig. 18-15). Equilibrium configurations are shown for various numbers of balls, from 1 to 12. Perhaps you can interpret the patterns in terms of rings, as did Thomson.

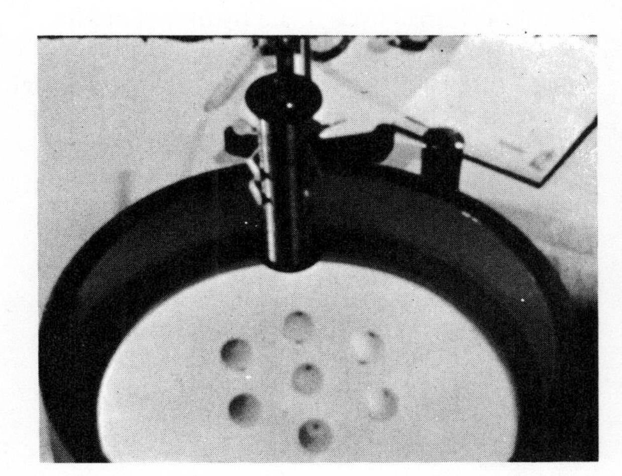

Fig. 18-15

Thomson was unable to make an exact correlation with the facts of chemistry. For example, he knew that the eleventh electron is easily removed (corresponding to sodium, the eleventh atom of the periodic table), yet his floating magnet model failed to show this. Instead, the patterns for 10, 11 and 12 floating magnets are rather similar.

Thomson's work with this apparatus illustrates how physical theories may be tested with the aid of analogies. He was disappointed by the failure of the model to account for the details of atomic structure. A few years later the Rutherford model of a nuclear atom made the Thomson model obsolete, but in its day the Thomson model received some support from experiments such as those shown in the film.

Chapter 19 The Rutherford-Bohr Model of the Atom

EXPERIMENT 44 SPECTROSCOPY

In text Chapter 19 you learn of the immense importance of spectra to our understanding of nature. You are about to observe the spectra of a variety of light sources to see for yourself how spectra differ from each other and to learn how to measure the wavelengths of spectrum lines. In particular, you will measure the wavelengths of the hydrogen spectrum and relate them to the structure of the hydrogen atom.

Before you begin, review carefully Sec. 19.1 of text Chapter 19.

Creating spectra

Materials can be made to give off light (or be "excited") in several different ways: by heating in a flame, by an electric spark between electrodes made of the material, or by an electric current through a gas at low pressure.

The light emitted can be dispersed into a spectrum by either a prism or a diffraction grating.

In this experiment, you will use a diffraction grating to examine light from various sources. A diffraction grating consists of many very fine parallel grooves on a piece of glass or plastic. The grooves can be seen under a 400-power microscope.

In experiment 33 (Young's Experiment) you saw how two narrow slits spread light of different wavelengths through different angles, and you used the double slit to make approximate measurements of the wavelengths of light of different colors. The distance between the two slits was about 0.2 mm. The distance between the lines in a diffraction grating is about 0.002 mm. And a grating may have about 10,000 grooves instead of just two. Because there are more lines and they are closer together, a grating diffracts more light and separates the different wavelengths more than a double-slit, and can be used to make very accurate measurements of wavelength.

Observing spectra

You can observe diffraction when you look at light that is reflected from a phonegraph record. Hold the record so that light from a distant source is almost parallel to the record's surface, as in the sketch below. Like a diffraction grating, the grooved surface disperses light into a spectrum.

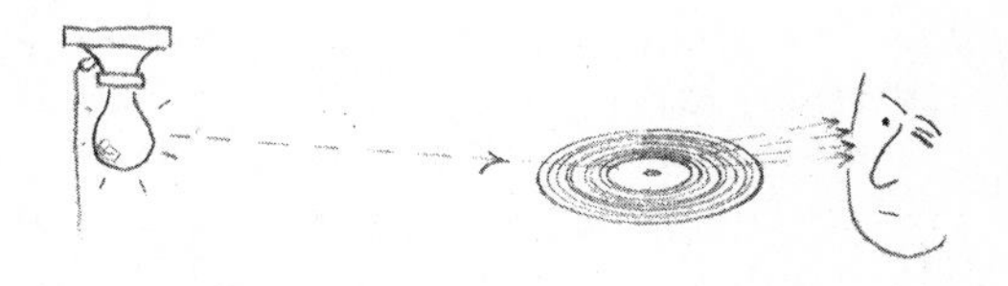

Use a real diffraction grating to see spectra simply by holding the grating close to your eye with the lines of the grating parallel to a distant light source. Better yet, arrange a slit about 25 cm in front of the grating, as shown below, or use a pocket spectroscope.

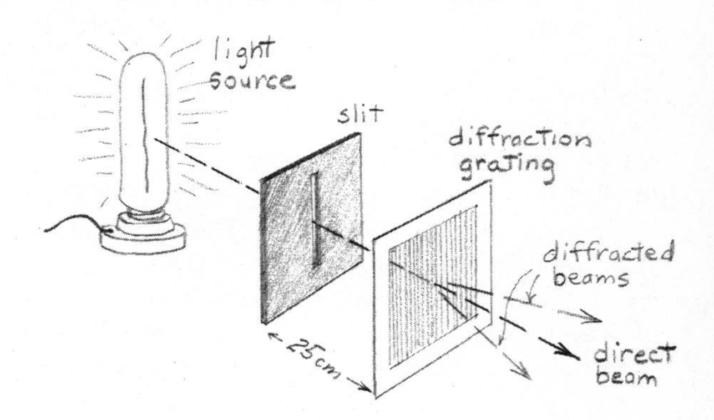

Look through the pocket spectroscope at a fluorescent light, at an ordinary (incandescent) light bulb, at mercury-vapor and sodium-vapor street lamps, at neon signs, at light from the sky (but *don't* look directly at the sun), and at a flame into which various compounds are introduced (such as salts of sodium, potassium, strontium, barium, and calcium).

Q1 Which color does the grating diffract into the widest angle and which into the narrowest? Are the long wavelengths diffracted at a wider angle than the short wavelengths, or vice-versa?

Q2 The spectra discussed in the *Text* are (a) either emission or absorption, and (b) either line or continuous. What different *kinds* of spectra have you observed? Make a table showing the type of spectrums produced by each of the light sources you observed. Do you detect any relationship between the nature of the source and the kind of spectra it produces?

Photographing the spectrum

A photograph of a spectrum has several advantages over visual observation. A photograph reveals a greater range of wavelengths; also it allows greater convenience for your measurement of wavelengths.

When you hold the grating up to your eye, the lens of your eye focuses the diffracted rays to form a series of colored images on the retina. If you put the grating in front of the camera lens (focused on the source), the lens will produce sharp images on the film.

The spectrum of hydrogen is particularly interesting to measure because hydrogen is the simplest atom and its spectrum is fairly easily related to a model of its structure. In this experiment, hydrogen gas in a glass tube is excited by an electric current. The electric discharge separates most of the H_2 molecules into single hydrogen atoms.)

Set up a meter stick just behind the tube (Fig. 19-1). This is a scale against which to observe and measure the position of the spectrum lines. The tube should be placed at about the 70-cm mark since the spectrum viewed through the grating will appear nearly 70 cm long.

From the camera position, look through the grating at the glowing tube to locate the positions of the visible spectral lines against the meter stick. Then, with the grating fastened over the camera lens, set up the camera with its lens in the same position your eye was. The lens should be aimed perpendicularly at the 50 cm mark, and the grating lines must be parallel to the source.

Now take a photograph that shows both the scale on the meter stick and the spectral

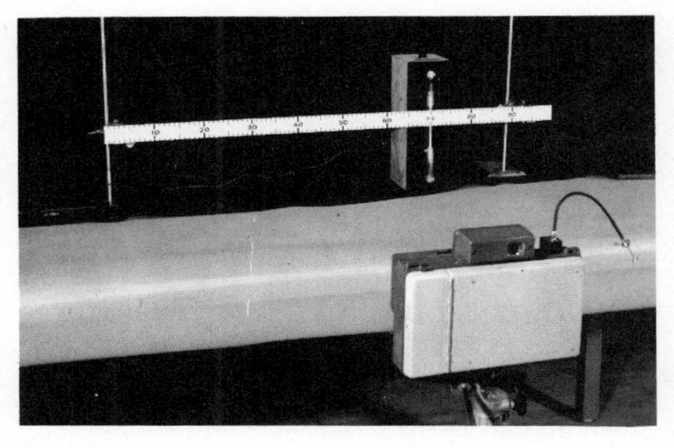

Fig. 19-1

lines. You may be able to take a single exposure for both, or you may have to make a double exposure—first the spectrum, and then, with more light in the room, the scale. It depends on the amount of light in the room. Consult your teacher.

Analyzing the spectrum

Count the number of spectral lines on the photograph, using a magnifier to help pick out the faint ones.

Q3 Are there more lines than you can see when you hold the grating up to your eye? If you do see additional lines, are they located in the visible part of the spectrum (between red and violet) or in the infrared or ultraviolet part?

The angle θ through which light is diffracted by a grating depends on the wavelength λ of the light and the distance d between lines on the grating. The formula is a simple one:

$$\lambda = d \sin \theta.$$

To find θ, you need to find $\tan \theta = x/\ell$ as shown in Fig. 19-2. Here x is the distance of the spectral line along the meter stick from the source, and ℓ is the distance from the source to the grating. Use a magnifier to read x from your photograph. Calculate $\tan \theta$, and then look up the corresponding values of θ and $\sin \theta$ in trigonometric tables.

To find d, remember that the grating space is probably given as lines per inch. You must convert this to the distance between lines in meters. One inch is 2.54×10^{-2} meters, so if there are 13,400 lines per inch, then d is

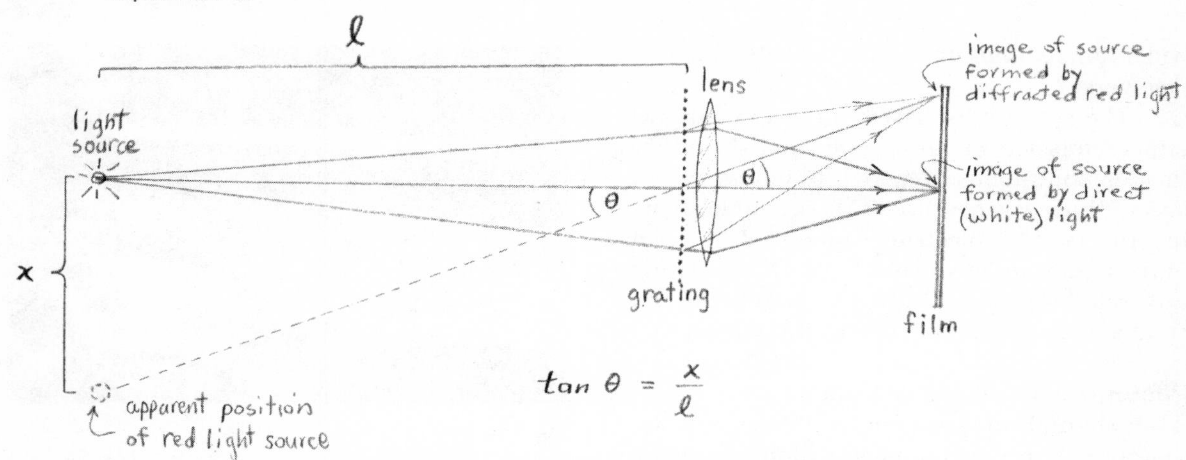

Fig. 19-2 Different images of the source are formed on the film by different colors of diffracted light. The angle of diffraction is equal to the apparent angular displacement angle of the source in the photograph so $\tan\theta = \frac{x}{\ell}$

$(2.54 \times 10^{-2}) / (1.34 \times 10^{4}) = 1.89 \times 10^{-6}$ meters.

Calculate the values of λ for the various spectral lines you have measured.

Q4 How many of these lines are visible to the eye?

Q5 What would you say is the shortest wave length to which your eye is sensitive?

Q6 What is the shortest wavelength that you can measure on the photograph?

Compare your values for the wavelengths with those given in the text, or in a more complete list (for instance, in the *Handbook of Chemistry and Physics*). The differences between your values and the published ones should be less than the experimental uncertainty of your measurement. Are they?

This is not all that you can do with the results of this experiment. You could, for example, work out a value for the Rydberg constant for hydrogen (mentioned in *Text* Sec. 19.2).

More interesting perhaps is to calculate some of the energy levels for the excited hydrogen atom. Using Planck's constant ($h = 6.6 \times 10^{-34}$), the speed of light in vacuum ($c = 3.0 \times 10^{8}$ m/sec), and your measured value of the wavelength λ of the separate lines, you can calculate the energy of photons' various wavelengths, $E / hf = hc/\lambda$, emitted when hydrogen atoms change from one state to another. The energy of the emitted photon is the difference in energy between the initial and final states of the atom.

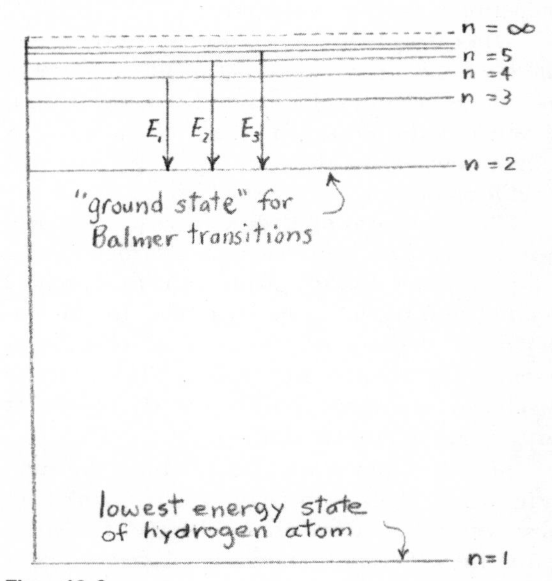

Fig. 19-3

Make the assumption (which is correct) that for all lines of the series you have observed the final energy state is the same. The energies that you have calculated represent the energy of various excited states above this final level.

Draw an energy-level diagram something like the one shown here (Fig. 19-3.). Show on it the energy of the photon emitted in transition from each of the excited states to the final state.

Q7 How much energy does an excited hydrogen atom lose when it emits red light?

ACTIVITIES

SCIENTISTS ON STAMPS

As shown here, scientists are pictured on the stamps of many countries, often being honored by other than their homeland. You may want to visit a stamp shop and assemble a display for your classroom.

See also "Science and the Artist," in the Unit 4 *Handbook*.

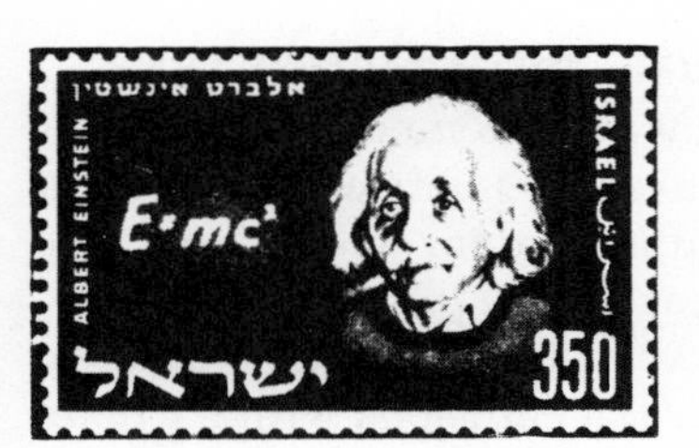

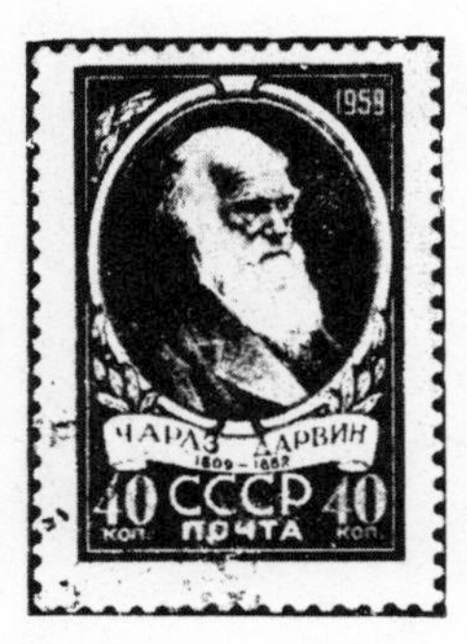

MEASURING IONIZATION, A QUANTUM EFFECT

With an inexpensive thyratron 885 tube, you can demonstrate an effect that is closely related to the famous Franck-Hertz effect.

Theory

According to the Rutherford-Bohr model, an atom can absorb and emit energy only in certain amounts that correspond to permitted "jumps" between states.

If you keep adding energy in larger and larger "packages," you will finally reach an amount large enough to separate an electron entirely from its atom—that is, to ionize the atom. The energy needed to do this is called the *ionization energy*.

Now imagine a beam of electrons being accelerated by an electric field through a region of space filled with argon atoms. This is the situation in a thyratron 884 tube with its grid and anode both connected to a source of variable voltage, as shown schematically in Fig. 19-4).

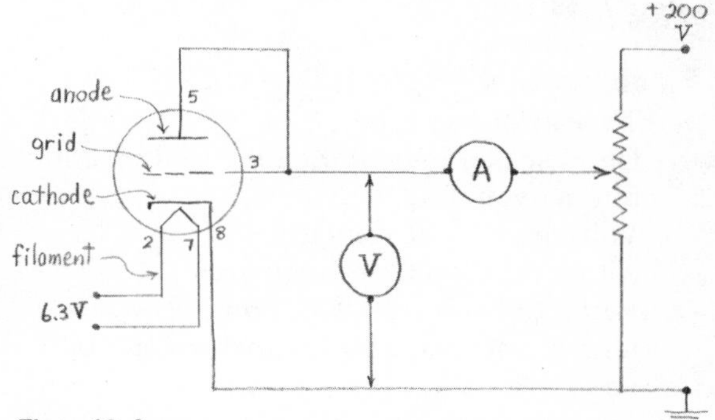

Fig. 19-4

In the form of its kinetic energy each electron in the beam carries energy in a single "package." The electrons in the beam collide with argon atoms. As you increase the accelerating voltage, the electrons eventually become energetic enough to excite the atoms, as in the Franck-Hertz effect. However, your equipment is not sensitive enough to detect the resulting small energy absorptions. So nothing seems to happen. The electron current from cathode to anode appears to increase quite linearly with the voltage, as you would expect—until the electrons get up to the ionization energy of argon. This happens at the *ionization potential* V_i, which is related to the ionization energy E_i and to the charge q_e on the electron as follows:

$$E_i = q_e V_i$$

As soon as electrons begin to ionize argon atoms, the current increases sharply. The argon is now in a different state, called an ionized state, in which it conducts electric current much more easily than before. Because of this sudden decrease in electrical resistance, we may use the thyratron tube as an "electronic switch" in such devices as stroboscopes. (A similar process ionizes the air so that it can conduct lightning.) As argon ions recapture electrons, they emit photons of ultraviolet and of visible violet light. When you see this violet glow, the argon gas is being ionized.

For theoretical purposes, the important point is that ionization takes place in any gas at a particular energy that is characteristic of that gas. This is easily observed evidence of one special case of Bohr's postulated discrete energy states.

CHAPTER 19

Equipment

Thyratron 884 tube
Octal socket to hold the tube (not essential but convenient)
Voltmeter (0-30 volts dc)
Ammeter (0-100 milliamperes)
Potentiometer (10,000 ohm, 2 watts or larger) or variable transformer, 0-120 volts ac
Power supply, capable of delivering 50-60 mA at 200 volts dc
Connect the apparatus as shown schematically in Fig. 19-7.

Procedure

With the potentiometer set for the lowest available anode voltage, turn on the power and wait a few seconds for the filament to heat. Now increase the voltage by small steps. At each new voltage, call out to your partner the voltmeter reading. Pause only long enough to permit your partner to read the ammeter and to note both readings in your data table. Take data as rapidly as accuracy permits: Your potentiometer will heat up quickly, especially at high currents. If it gets too hot to touch, turn the power off and wait for it to cool before beginning again.

Watch for the onset of the violet glow. Note in your data table the voltage at which you first observe the glow, and then note what happens to the glow at higher voltages.

Plot current versus voltage, and mark the point on your graph where the glow first appeared. From your graph, determine the first ionization potential of argon. Compare your experimental value with published values, such as the one in the *Handbook of Chemistry and Physics*.

What is the energy an electron must have in order to ionize an argon atom?

MODELING ATOMS WITH MAGNETS

Here is one easy way to demonstrate some of the important differences between the Thomson "raisin pudding" atom model and the Rutherford nuclear model.

To show how alpha aprticles would be expected to behave in collisions with a Thomson atom, represent the spread-out "pudding" of positive charge by a roughly circular arrangement of small disc magnets, spaced four or five inches apart, under the center of a smooth tray, as shown in Fig. 19-5. Use tape

Fig. 19-5 The arrangement of the magnets for a "Thomson atom".

or putty to fasten the magnets to the underside of the tray. Put the large magnet (representing the alpha particle) down on top of the tray in such a way that the large magnet is repelled by the small magnets and sprinkle onto the tray enough tiny plastic beads to make the large magnet slide freely. Now push the "alpha particle" from the edge of the tray toward the "atom." As long as the "alpha particle" has enough momentum to reach the other side, its deflection by the small magnets under the tray will be quite small—never more than a few degrees.

For the Rutherford model, on the other hand, gather all the small magnets into a vertical stack under the center of the tray, as shown in Fig. 19-6. Turn the stack so that it

Fig. 19-6 The arrangement of the magnets for a "Rutherford atom."

repels "alpha particles" as before. This "nucleus of positive charge" now has a much greater effect on the path of the "alpha particle."

Have a partner tape an unknown array of magnets to the bottom of the tray—can you determine what it is like just by scattering the large magnet?

With this magnet analogue you can do some quantitative work with the scattering relationships that Rutherford investigated. (See text Sec. 19.3 and Film Loop 48, "Rutherford Scattering" at the end of this *Handbook* chapter.) Try again with different sizes of magnets. Devise a launcher so that you can control the velocity of your projectile magnets and the distance of closest approach.

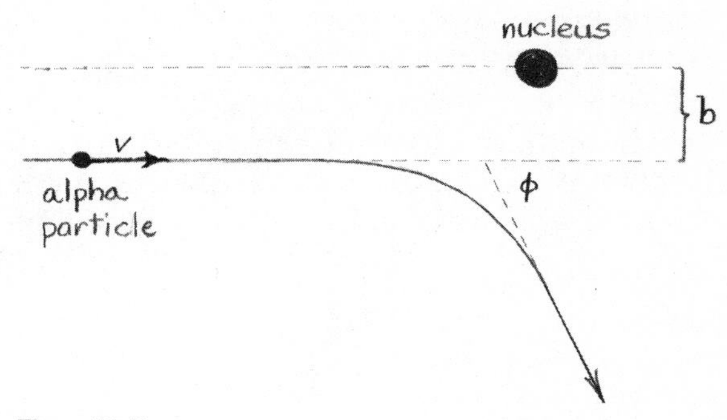

Fig. 19-7

1) Keep the initial projectile velocity v constant and vary the distance b (see Fig. 19-7); then plot the scattering angle ϕ versus b.
2) Hold b constant and carry the speed of the projectile, then plot ϕ versus v.
3) Try scattering hard, nonmagnetized discs off each other. Plot ϕ versus b and ϕ versus v as before. Contrast the two kinds of scattering-angle distributions.

"BLACK BOX" ATOMS

Place two or three different objects, such as a battery, a small block of wood, a bar magnet, or a ball bearing, in a small box. Seal the box, and have one of your fellow students try to tell you as much about the contents as possible, without opening the box. For example, sizes might be determined by tilting the box, relative masses by balancing the box on a support, or whether or not the contents are magnetic by checking with a compass.

The object of all this is to get a feeling for what you can or cannot infer about the structure of an atom purely on the basis of secondary evidence. It may help you to write a report on your investigation in the form you may have used for writing a proof in plane geometry, with the property of the box in one column and your reason for asserting that the property is present in the other column. The analogy can be made even better if you are exceptionally brave: Don't let the guesser open the box, ever, to find out what is really inside.

ANOTHER SIMULATION OF THE RUTHERFORD ATOM

A hard rubber "potential-energy hill" is available from Stark Electronics Instruments, Ltd., Box 670, Ajax, Ontario, Canada. When you roll steel balls onto this hill, they are deflected in somewhat the same way as alpha particles are deflected away from a nucleus. The potential-energy hill is very good for quantitative work such as that suggested for the magnet analogue in the activity "Modeling atoms with magnets."

FILM LOOPS

FILM LOOP 48: RUTHERFORD SCATTERING

This film simulates the scattering of alpha particles by a heavy nucleus, such as gold, as in Ernest Rutherford's famous experiment. The film was made with a digital computer.

The computer program was a slight modification of that used in film loops 13 and 14, on program orbits, concerned with planetary orbits. The only difference is that the operator selected an inverse-square law of *repulsion* instead of a law of attraction such as that of gravity. The results of the computer calculation were displayed on a cathode-ray tube and then photographed. Points are shown at equal time intervals. Verify the law of areas for the motion of the alpha particles by projecting the film for measurements. Why would you expect equal areas to be swept out in equal times?

All the scattering particles shown are near a nucleus. If the image from your projector is 1 foot high, the nearest adjacent nucleus would be about 500 feet above the nucleus shown. Any alpha particles moving through this large area between nuclei would show no appreciable deflection.

We use the computer and a mathematical model to tell us what the result will be if we shoot particles at a nucleus. The computer does not "know" about Rutherford scattering. What it does is determined by a program placed in the computer's memory, written in this particular instance in a language called Fortran. The programmer has used Newton's laws of motion and has assumed an inverse-square repulsive force. It would be easy to change the program to test another force law, for example $F = K/r^3$. The scattering would be computed and displayed; the angle of deflection for the same distance of closest approach would be different than for inverse-square force.

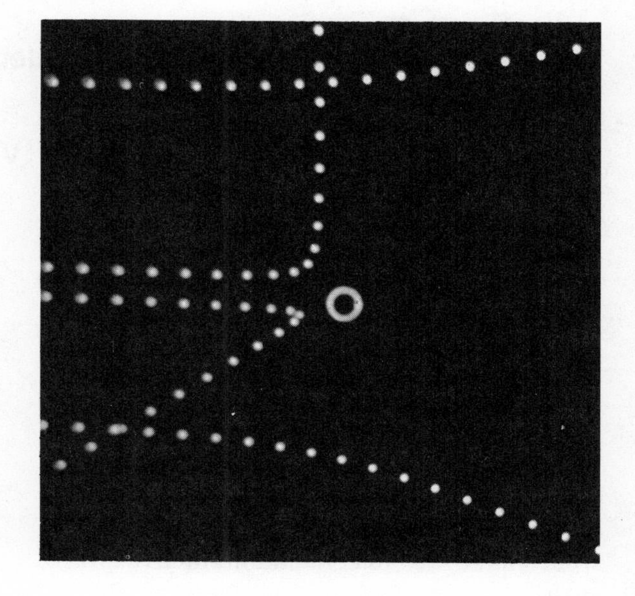

Working backward from the observed scattering data, Rutherford deduced that the inverse-square Coulomb force law is correct for all motions taking place at distances greater than about 10^{-14}m from the scattering center, but he found deviations from Coulomb's law for closer distances. This suggested a new type of force, called nuclear force. Rutherford's scattering experiment showed the size of the nucleus (supposedly the same as the range of the nuclear forces) to be about 10^{-14}m, which is about 1/10,000 the distance between the nuclei in solid bodies.

Chapter 20 Some Ideas from Modern Physical Theories

ACTIVITIES

STANDING WAVES ON A BAND-SAW BLADE

Standing waves on a ring can be shown by shaking a band-saw blade with your hand. Wrap tape around the blade for about six inches to protect your hand. Then gently shake the blade up and down until you have a feeling for the lowest vibration rate that produces reinforcement of the vibration. Then double the rate of shaking, and continue to increase the rate of shaking, watching for standing waves. You should be able to maintain five or six nodes.

TURNTABLE OSCILLATOR PATTERNS RESEMBLING DE BROGLIE WAVES

If you set up two turntable oscillators and a Variac as shown in Fig. 20-1, you can draw pictures resembling de Broglie waves, like those shown in Chapter 20 of your text.

Place a paper disc on the turntable. Set both turntables at their lowest speeds. Before starting to draw, check the back-and-forth motion of the second turntable to be sure the pen stays on the paper. Turn both turntables on and use the Variac as a precise speed control on the second turntable. Your goal is to get the pen to follow exactly the same path each time the paper disc goes around. Try higher frequencies of back-and-forth motion to get more wavelengths around the circle.

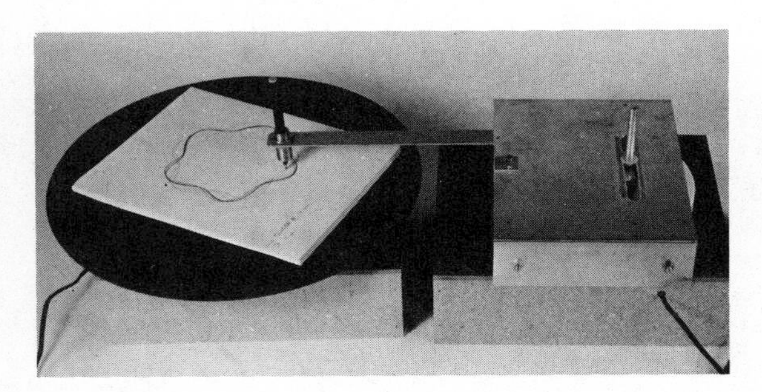

Fig. 20-1

For each stationary pattern that you get, check whether the back-and-forth frequency is an integral multiple of the circular frequency.

STANDING WAVES IN A WIRE RING

With the apparatus described below, you can set up circular waves that somewhat resemble the de Broglie wave models of certain electron orbits. You will need a strong magnet, a fairly stiff wire loop, a low-frequency oscillator, and a power supply with a transistor chopping switch.

The output current of the oscillator is much too small to interact with the magnetic field enough to set up visible standing waves in the wire ring. However, the oscillator current can operate the transistor switch to control ("chop") a much larger current from the power supply (see Fig. 20-2).

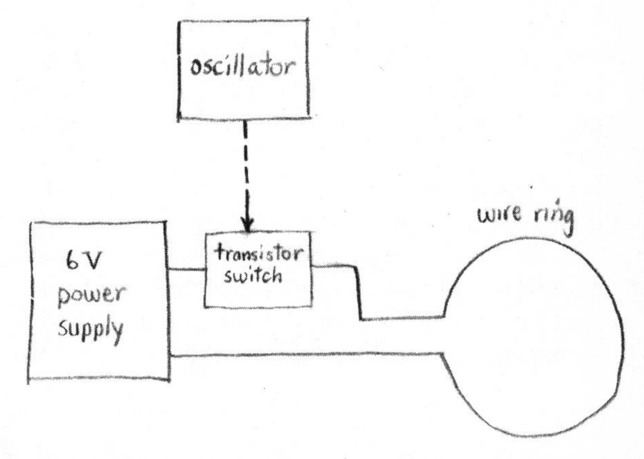

Fig. 20-2 The signal from the oscillator controls the transistor switch, causing it to turn the current from the power supply on and off. The "chopped" current in the wire ring interacts with the magnetic field to produce a pulsating force on the wire.

The wire ring must be of non-magnetic metal. Insulated copper magnet wire works well: Twist the ends together and support the ring at the twisted portion by means of a binding post, Fahnestock clip, thumbtack, or ring-stand clamp. Remove a little insulation from each end for electrical connections.

A ring 4 to 6 inches in diameter made of 22-guage enameled copper wire has its lowest rate of vibration at about 20 cycles/sec. Stiffer wire or a smaller ring will have higher characteristic vibrations that are more difficult to see.

Position the ring as shown, with a section of the wire passing between the poles of the magnet. When the pulsed current passes through the ring, the current interacts with the magnetic field, producing alternating forces which cause the wire to vibrate. In Fig. 20-2, the magnetic field is vertical, and the vibrations are in the plane of the ring. You can turn the magnet so that the vibrations are perpendicular to the ring.

Because the ring is clamped at one point, it can support standing waves that have any integral number of half wavelengths. In this respect they are different from waves on a *free* wire ring, which are restricted to integral numbers of *whole* wavelengths. Such waves are more appropriate for comparison to an atom.

When you are looking for a certain mode of vibration, position the magnet between expected nodes (at antinodes). The first "characteristic, or state" "mode of vibration," that the ring can support in its plane is the first harmonic, having two nodes: the one at the point of support and the other opposite it. In the second mode, three nodes are spaced evenly around the loop, and the best position for the magnet is directly opposite the support, as shown in Fig. 20-3.

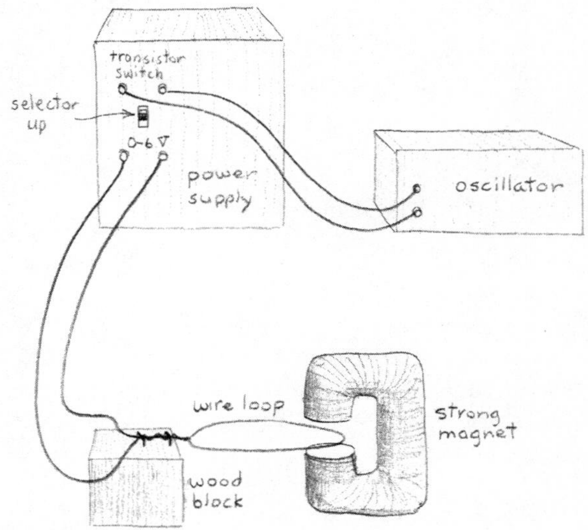

Fig. 20-3

You can demonstrate the various modes of vibration to the class by setting up the magnet, ring, and support on the platform of an overhead projector. Be careful not to break the glass with the magnet, especially if the frame of the projector happens to be made of a magnetic material.

The Project Physics *Film Loop* "Vibrations of a Wire," also shows this.

The Project Physics Course

Handbook 6

The Nucleus

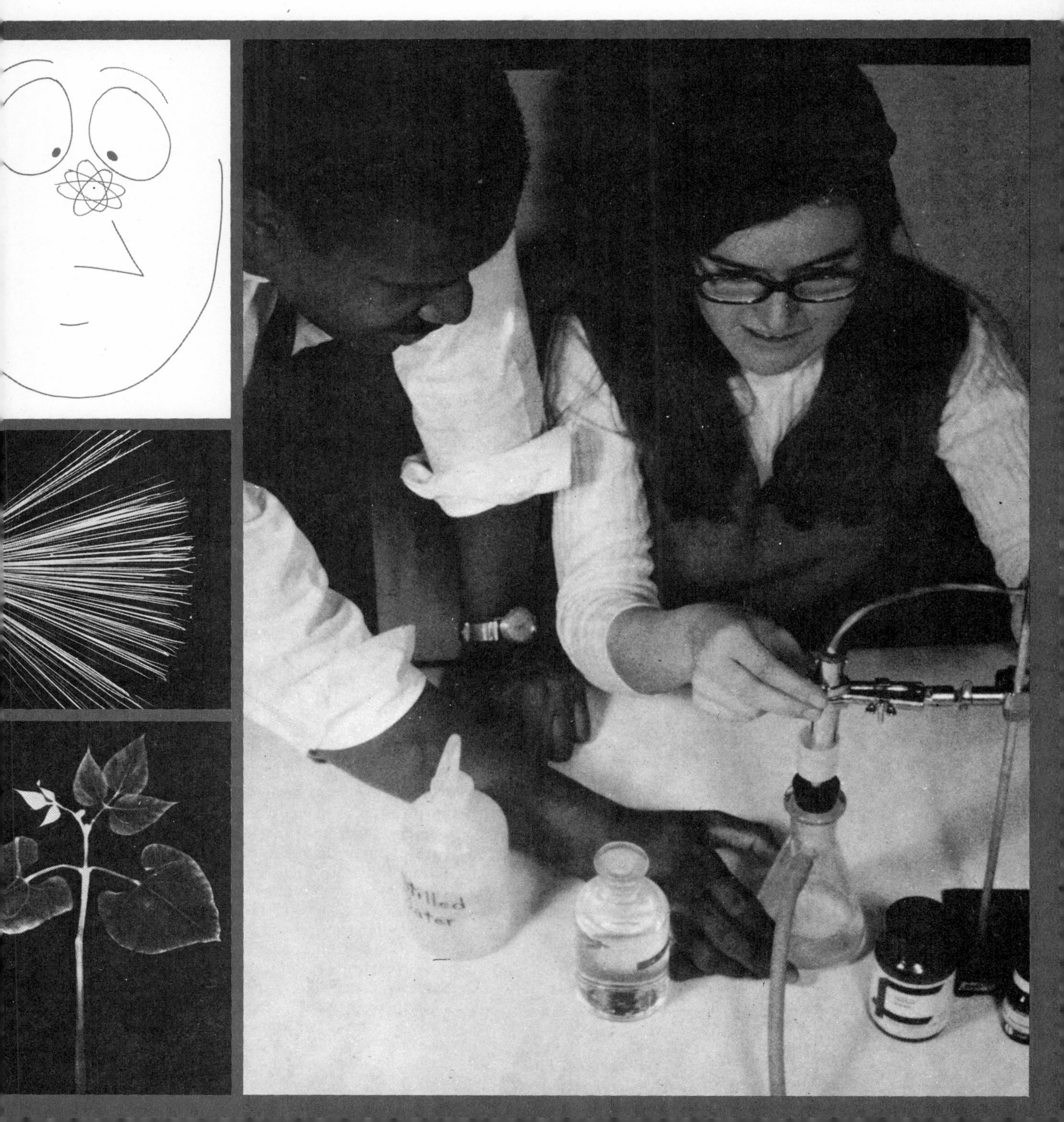

Picture Credits, Unit 6

Cover: (cartoon) Andrew Ahlgren; (alpha particle tracks in cloud chamber) Professor J. K. Bøggild, Niels Bohr Institute, Copenhagen; (autoradiograph of leaves) from *Photographs of Physical Phenomena*, Kodanska, Tokyo, 1968.

P. 333 Atomic Energy Commission photograph.

P. 336 Alan Dunn, *New Yorker* Magazine 1965.

All photographs used with film loops courtesy of National Film Board of Canada

Photographs of laboratory equipment and of students using laboratory equipment were supplied with the cooperation of the Project Physics staff and Damon Corporation.

Contents HANDBOOK, Unit 6

Chapter 21 Radioactivity

EXPERIMENT 45 RANDOM EVENTS

In Unit 6, after having explored the random behavior of gas molecules in Unit 3, you are learning that some atomic and nuclear events occur in a random manner. The purpose of this experiment is to give you some firsthand experience with random events.

What is a random event?

Dice are useful for studying random behavior. You cannot predict with certainty how many spots will show on a single throw. But you are about to discover that you *can* make useful predictions about a large number of throws. If the behavior of the dice is truly random, you can use probability theory to make predictions. When, for example, you shake a box of 100 dice, you can predict with some confidence how many will fall with one spot up, how many with two spots up, and so on. Probability theory has many applications. For example, it is used in the study of automobile traffic flow, the interpretation of faint radar echoes from the planets, the prediction of birth, death, and accident rates, and the study of the breakup of nuclei. An interesting discussion of the rules and uses of probability theory is found in George Gamow's article, "The Law of Disorder," in *Reader 3*.

The theory of probability provides ways to determine whether a set of events are random. An important characteristic of all truly *random* events is that each event is independent of the others. For example, if you throw a legitimate die four times in a row and find that a single spot turns up each time, your chance of observing a single spot on the fifth throw is no greater or smaller than it was on the first throw.

If events are to be independent, the circumstances under which the observations are made must never favor one outcome over another. This condition is met in each of the following three parts of this experiment. You are expected to do only one of these parts, (a), (b), or (c). The section "Recording your data" that follows the three descriptions applies to all parts of the experiment. Read this section in preparing to do any part of the experiment.

(a) Twenty-sided dice

A tray containing 120 dice is used for this experiment. Each die has 20 identical faces (the name for a solid with this shape is *icosahedron*). One of the 20 faces on each die should be marked; if it is not, mark one face on each die with a felt-tip pen.

Q1 What is the probability that the marked face will appear at the top for any one throw of one die? To put it another way, *on the average* how many marked faces would you expect to see face up if you roll all 120 dice?

Now try it, and see how well your prediction holds. Record as many trials as you can in the time available, shaking the dice, pouring them out onto the floor or a large tabletop, and counting the number of marked faces showing face up. (See Fig. 21-1.)

Fig. 21-1 Icosahedral dice in use.

The counting will go faster if the floor area or tabletop is divided into three or four sections, with a different person counting each section and another person recording the total count. Work rapidly, taking turns with others in your group if you get tired, so that you can count at least 100 trials.

(b) Diffusion cloud chamber

A cloud chamber is a device that makes visible the trail left by the particles emitted by radioactive atoms. One version is a transparent box filled with supercooled alcohol vapor. When an α particle passes through, it leaves a trail of ionized air molecules. The alcohol molecules are attracted to these ions and they condense into tiny droplets which mark the trail.

Your purpose in this experiment is not to learn about the operation of the chamber, but simply to study the randomness with which the α particles are emitted. A barrier with a narrow opening is placed in the chamber near a radioactive source that emits α particles. Count the number of tracks you observe coming through the opening in a convenient time interval, such as 10 seconds. Continue counting for as many intervals as you can during the class period.

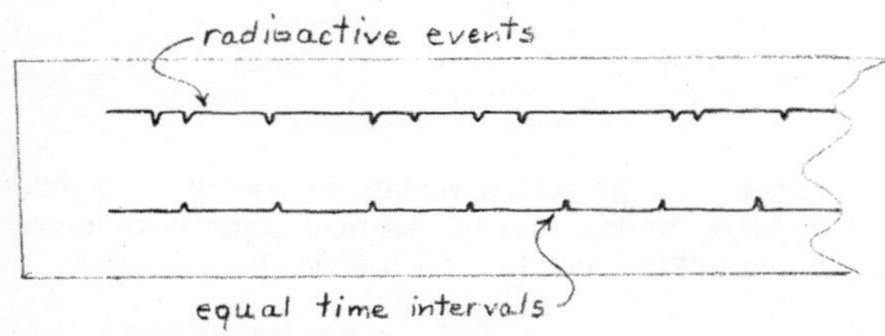

A convenient method of counting events in successive time intervals is to mark them in one slot of the "drag-strip" recorder, while marking seconds (or ten second intervals) in the other slot.

(c) Geiger counter

A Geiger counter is another device that detects the passage of invisible particles. A potential difference of several hundred volts is maintained between the two electrodes of the Geiger tube. When a β particle or a γ ray ionizes the gas in the tube, a short pulse of electricity passes through it. The pulse may be heard as an audible click in an earphone, seen as a "blip" on an oscilloscope screen, or read as a change in a number on an electronic scaling device. When a radioactive source is brought near the tube, the pulse rate goes up rapidly. But even without the source, an occasional pulse still occurs. These pulses are called "background" and are caused by cosmic radiation and by a slight amount of radioactivity always present in objects around the tube.

Use the Geiger counter to determine the rate of background radiation, counting over and over again the number of pulses in a convenient time interval, such as 10 seconds.

Recording your data

Whichever of the three experiments you do, prepare your data record in the following way:

Down the left-hand edge of your paper write a column of numbers from 0 to the highest number you ever expect to observe in one count. For example, if your Geiger counts seem to range from 3 to 20 counts in each time interval, record numbers from 0 to 20 or 25.

Number of events observed in one time interval (n)	(frequency) (f)	Total number of events observed ($n \times f$)
0	I	0
1		0
2	I	2
3	𝍸 𝍸	30
4	𝍸 𝍸 III	52
5	𝍸 𝍸 II	65
6	𝍸 𝍸 𝍸 𝍸 I	126
7	𝍸 𝍸 𝍸 I	112
8	𝍸 𝍸 II	96
9	IIII	36
10	IIII	40
11	IIII	44
12	I	12
13	I	13
		623

Fig. 21-2 A typical data page.

To record your data, put a tally mark opposite each number in the column for each time this number occurred. Continue making tally marks for as many trial observations as you can make during the time you have. When you are through, add another column in which you multiply each number in the first column by the number of tallies opposite it. Whichever experiment you did, your data sheet will look something like the sample in Fig. 21-2. The third column shows that a total of 623 marked faces (or pulses or tracks) were observed in the 100 trials. The *average* is 623 divided by 100, or about 6. You can see that most of the counts cluster around the mean.

This arrangement of data is called a *distribution table*. The distribution shown was obtained by shaking the tray of 20-sided dice 100 times. Its shape is also typical of Geiger-counter and cloud-chamber results.

A graph of random data

The pattern of your results is easier to visualize if you display your data in the form of a bar graph, or *histogram*, as in Fig. 21-3.

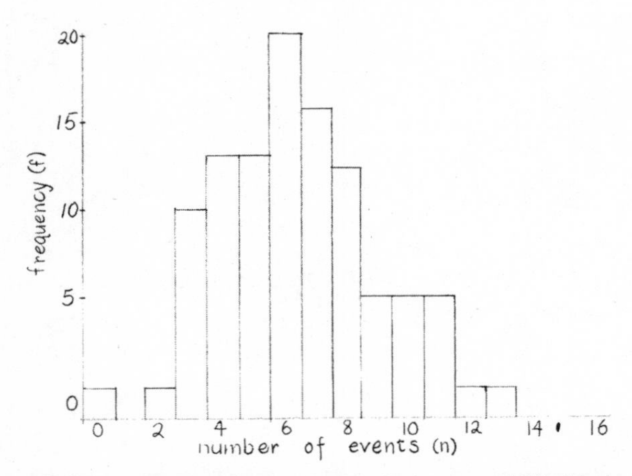

Fig. 21-3 The results obtained when a tray of 20-sided dice (one side marked) were shaken 100 times.

If you were to shake the dice another set of 100 times, your distribution would not be exactly the same as the first one. However, if sets of 100 trials were repeated several times, the *combined* results would begin to form a smoother histogram. Fig. 21-4 shows the kind of result you could expect if you did 1,000 trials.

Compare this with the results for only ten trials shown in Fig. 21-5. As the number of trials increases, the distribution generally becomes smoother and more like the distribution in Fig. 21-4.

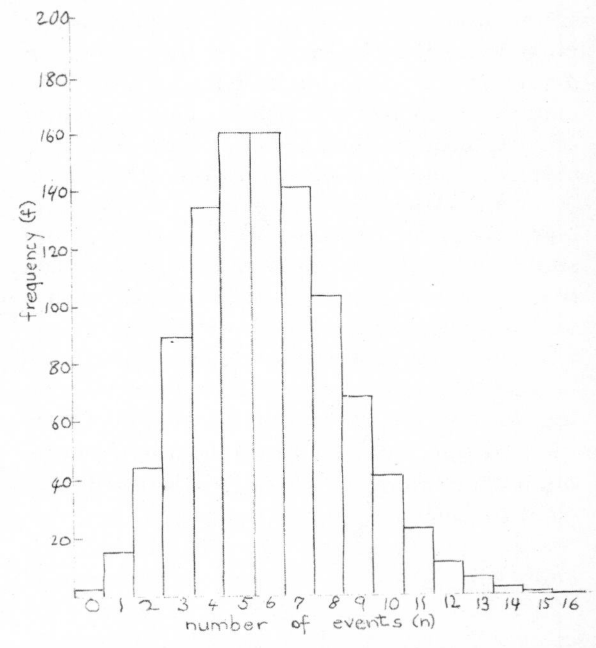

Fig. 21-4 The predicted results of shaking the dice 1000 times. Notice that the vertical scale is different from that in Fig. 21-3. Do you see why?

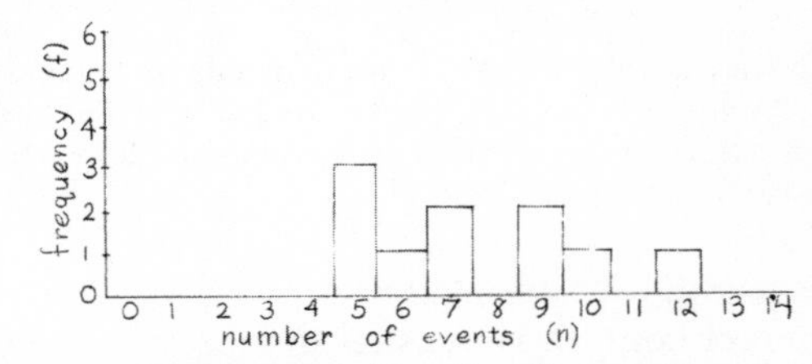

Fig. 21-5 Results of shaking the dice ten times.

Predicting random events

How can data like these be used to make predictions?

On the basis of Fig. 21-4, the best prediction of the number of marked faces turning up would be 5 or 6 out of 120 rolls. Apparently the chance of a die having its marked face up is about 1 in 20—that is, the probability is $\frac{1}{20}$.

But not all trials had 5 or 6 marked faces showing. In addition to the average of a distribution, you also need to know something about how the data spread out around the average. Examine the histogram and answer the following questions:

Q2 How many of the trials in Fig. 21-4 had from 5 to 7 counts?

Q3 What fraction is this of the total number of observations?

Q4 How far, going equally to the left and right of the average, must you go to include half of all the observations? to include two-thirds?

For a theoretical distribution like this (which your own results will closely approximate as you increase the number of trials), it turns out that there is a simple rule for expressing the spread: If the average count is A, then $\frac{2}{3}$ of the counts will be between $A - \sqrt{A}$ and $A + \sqrt{A}$. Putting it another way, about $\frac{2}{3}$ of the values will be in the range of $A \pm \sqrt{A}$.

Another example may help make this clear. For example, suppose you have been counting cloud-chamber tracks and find that the average of a large number of one-minute counts is 100 tracks. Since the square root of 100 is 10, you would find that about two-thirds of your counts would lie between 90 and 110.

Check this prediction in Fig. 21-4. The average is 6. The square root of 6 is about 2.4. The points along the base of the histogram corresponding to 6 ± 2.4 are between 3.6 and 8.4. (Of course, it doesn't really make sense to talk about a fraction of a marked side. One would need to round off to the nearest whole numbers, 4 and 8.) Therefore the chances are about two out of three that the number of marked sides showing after any shake of the tray will be in the range 4 to 8 out of 120.

Q5 How many of the trials did give results in the range 4 to 8? What fraction is this of the total number of trials?

Q6 Whether you rolled dice, counted tracks, or used the Geiger counter, inspect your results to see if $\frac{2}{3}$ of your counts do lie in the range $A \pm \sqrt{A}$.

If you counted for only a *single* one-minute trial, the chances are about two out of three that your single count C will be in the range $A \pm \sqrt{A}$, where A is the true average count (which you would find over many trials). This implies that you can predict the true average value fairly well even if you have made only a single one-minute count. The chances are about two out of three that the single count C will be within $\sqrt{A}$ of the true average A. If we assume C is a fairly good estimate of A, we can use $\sqrt{C}$ as an estimate of $\sqrt{A}$ and conclude that the chances are two out of three that *the value obtained for C is within $\pm\sqrt{C}$ of the true average.*

You can decrease the uncertainty in predicting a true average like this by counting for a longer period. Suppose you continued the count for ten minutes. If you counted 1,000 tracks the expected "two-thirds range" would be about $1000 \pm \sqrt{1000}$ or 1000 ± 32. The result is 1000 ± 32 counts in *ten* minutes, which gives an average of 100 ± 3.2 counts per minute. If you counted for still longer, say 100 minutes, the range would be $10{,}000 \pm \sqrt{10{,}000}$ or $10{,}000 \pm 100$ counts in *100* minutes. Your estimate of the average count rate would be 100 ± 1 counts per minute. The table below lists these sample results.

Notice that although the range of uncertainty in the *total* count increases as the count goes up, it becomes a smaller *fraction* of the total count. Therefore, the uncertainty in the *average* count rate (number of counts per minute) decreases.

SAMPLE RESULTS AND ESTIMATED "TWO-THIRDS RANGES"

TIME min	TOTAL COUNT	EXPECTED UNCERTAINTY	AVERAGE COUNT per min	EXPECTED UNCERTAINTY per min
1	100	±10	100	±10
10	1000	±32	100	±3.2
100	10000	±100	100	±1.0

(The percent uncertainty can be expressed as $\frac{\sqrt{C}}{C}$, which is equal to $\frac{1}{\sqrt{C}}$. In this expression, you can see clearly that the percent uncertainty goes down as C increases.)

You can see from these examples that the higher the total count (the longer you count or the more dice-rolling trials you do) the more precisely you can estimate the true average. This becomes important in the measurement of the activity of radioactive samples and many other kinds of random events. To get a precise measure of the activity (the average count rate), you must work with large numbers of counts.

Q7 If you have time, take more data to increase the precision of your estimate of the mean.

Q8 If you count 10 cosmic ray tracks in a cloud chamber during one minute, for how long would you expect to have to go on counting to get an estimate of the average with a "two-thirds range" that is only 1% of the average value.

This technique of counting over a longer period to get better estimates is fine as long as the true count rate remains constant. But it doesn't always remain constant. If you were measuring the half-life of a short-lived radioactive isotope, the activity rate would change appreciably during a ten-minute period. In such a case, the way to increase precision is still to increase the number of observations—by having a larger sample of material or putting the Geiger tube closer to it—so that you can record a large number of counts during a short time.

Q9 In a small town it is impossible to predict whether there will be a fire next week. But in a large metropolitan area, firemen know with remarkable accuracy how many fires there will be. How is this possible? What assumption must the firemen make?

EXPERIMENT 46
RANGE OF α AND β PARTICLES

An important property of particles from radioactive sources is their ability to penetrate solid matter. In this experiment you will determine the distances α and β particles can travel in various materials.

α particles are most easily studied in a cloud chamber, a transparent box containing super-cooled alcohol vapor. Since the α particles are relatively massive and have a double positive charge, they leave a thick trail of ionized air molecules behind them as they move along. The ions then serve as centers about which alcohol condenses to form tracks of visible droplets.

β particles also ionize air molecules as they move. But because of their smaller mass and smaller charge, they form relatively few ions, which are farther apart than those formed by α's. As a result, the trail of droplets in the chamber is much harder to see.

A Geiger counter, on the other hand detects β particles better than α particles. This is because α particles, in forming a heavy trail, lose all their energy long before they get through even the thin window of an ordinary Geiger tube. β particles encounter the atoms in the tube window also, but they give up relatively less energy so that their chances of getting through the wall are fairly good.

For these reasons you count α particles using a cloud chamber and β particles with a Geiger counter.

Observing α particles

Mark off a distance scale on the bottom of the cloud chamber so that you will be able to estimate, at least to the nearest $\frac{1}{2}$ cm, the lengths of the tracks formed (Fig. 21-6). Insert a source of α radiation and a barrier (as in the preceding experiment on random events) with a small slot opening at such a height that the tracks form a fairly narrow beam moving parallel to the bottom of the chamber. Put the cloud chamber into operation according to the instructions supplied with it.

Practice watching the tracks until you can report the length of any of the tracks you see.

Fig. 21-6

When you are ready to take data, count and record the number of α's that come through the opening in the barrier in one minute. Measure the opening and calculate its area. Measure and record the distance from the source to the barrier.

Actually you have probably not seen all the particles coming through the opening, since the sensitive region in which tracks are visible is rather shallow and close to the chamber floor. You will probably miss the α's above this layer.

The range and energy of α particles

The maximum range of radioactive particles as they travel through an absorbing material depends on several factors, including the density and the atomic number of the absorber. The graph (Fig. 21-7) summarizes the results

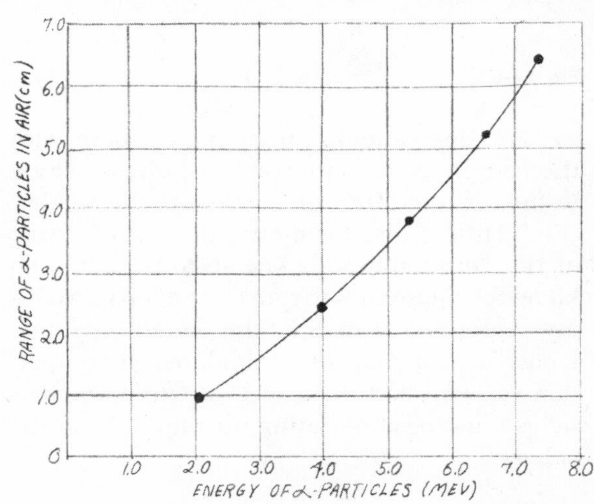

Fig. 21-7 Range of α-particles in air as a function of their energy.

of many measurements of the range of α particles traveling through air. The range-energy curve for particles in air saturated with alcohol vapor, as the air is in your chamber, does not differ significantly from the curve shown. You are therefore justified in using Fig. 21-7 to get a fair estimate of the kinetic energy of the α particles you observed.

Q1 Was there a wide variation in α-particle energies, or did most of the particles appear to have about the same energy? What was the energy of the α particle that caused the longest track you observed?

Now calculate the rate at which energy is being carried away from the radioactive source. Assume that the source is a point. From the number of α particles per minute passing through an opening of known area at a known distance from the source, estimate the number of α particles per minute leaving the source in *all* directions.

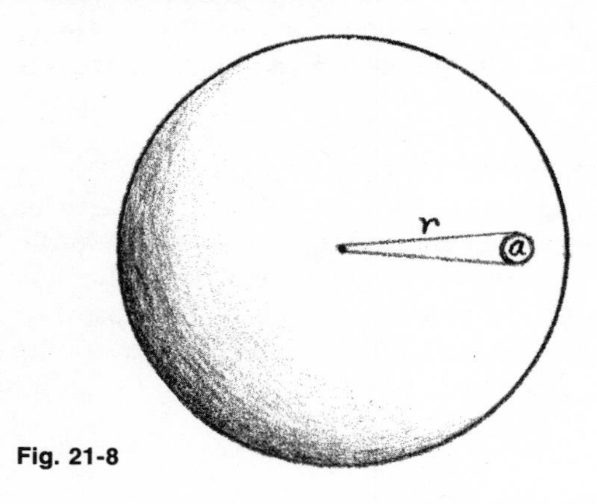

Fig. 21-8

For this estimate, imagine a sphere with the source at its center and a radius r equal to the distance from the source to the barrier. (Fig. 21-8.) From geometry, the surface area of the entire sphere is known to be $4\pi r^2$. You know the approximate rate c at which particles are emerging through the small opening, whose area a you have calculated. By proportion you can find the rate C at which the particles must be penetrating the total area of the sphere:

$$\frac{C}{c} = \frac{4\pi r^2}{a}$$

(The α-particle source is not a point, but probably part of a cylinder. This discrepancy, combined with a failure to count those particles that pass above the active layer, will introduce an error of as much as a factor of 10.)

The total number of particles leaving the source per minute, multiplied by the average energy of the particles, is the total energy lost per minute.

To answer the following questions, use the relationships

$$1 \text{ MeV} = 1.60 \times 10^{-13} \text{ joules}$$

$$1 \text{ calorie} = 4.18 \text{ joules}$$

Q2 How many joules of energy are leaving the source per minute?

Q3 How many calories per minute does this equal?

Q4 If the source were placed in one gram of water in a perfectly insulated container, how long would it take to heat the water from 0° C to 100° C?

Q5 How many joules per *second* are leaving the source? What is the power output in watts?

Observing β particles

After removing all radioactive sources from near the Geiger tube, count the number of pulses caused by background radiation in several minutes. Calculate the average background radiation in counts per minute. Then place a source of β radiation near the Geiger tube, and determine the new count rate. (Make sure that the source and Geiger tube are not moved during the rest of the experiment.) Since you are concerned only with the particles from the source, subtract the average background count rate.

Next, place a piece of absorbing material (such as a sheet of cardboard or thin sheet metal) between the source and the tube, and count again. Place a second, equally thick sheet of the same material in front of the first, and count. Keep adding absorbers and recording counts until the count rate has dropped nearly to the level of background radiation.

Plot a graph on which the horizontal scale is the total thickness (number of) absorbers and the vertical scale is the number of β's getting through the absorber per minute.

In addition to plotting single points, show the uncertainty in your estimate of the count rate for each point plotted. You know that because of the random nature of radioactivity, the count rate actually fluctuates around some average value. You do not know what *true* average value is; it would ideally take an infinite number of one-minute counts to determine the "true" average. But you know that the distribution of a great number of one-minute counts will have the property that two-thirds of them will differ from the average by less than the square root of the average. (See Experiment 45.)

For example, suppose you have observed 100 counts in one given minute. The chances are two out of three that, if you counted for a very long time, the mean count rate would be between 90 and 110 counts (between $100 - \sqrt{100}$ and $100 + \sqrt{100}$ counts). For this reason you would mark a vertical line on your graph extending from 90 counts up to 110. In this way you avoid the pitfall of making a single measurement and assuming you know the "correct" value. (For an example of this kind of graph see notes for Film Loop 9 in Unit 1 *Handbook*.)

If other kinds of absorbing material are available, repeat the experiment with the same source and another set of absorbers. For sources that emit very low-energy β rays, it may be necessary to use very thin materials, such as paper or household aluminum foil.

Range and absorption of β particles

Examine your graph of the absorption of particles.

Q6 Is it a straight line?

Q7 What would the graph look like if (as is the case for α particles) all β particles from the source were able to penetrate the same thickness of a given absorber material before giving up all their energy?

Q8 If you were able to use different absorbing materials, how did the absorption curves compare?

Q9 What might you conclude about the kinetic energies of β particles?

EXPERIMENT 47 HALF-LIFE—I

The more people there are in the world, the more people die each day. The less water there is in a tank, the more slowly water leaks out of a hole in the bottom.

In this experiment, you will observe three other examples of quantities that change at a rate that depends on the total amount of the quantity present. The objective is to find a common principle of change. Your conclusions will apply to many familiar growth and decay processes in nature.

If you experimented earlier with rolling dice and with radioactive decay (Experiment 45), you were studying random events you could observe one at a time. You found that the fluctuations in such small numbers of random events were relatively large. But this time you will deal with a large number of events, and you will find that the outcome of your experiments is therefore more precisely predictable.

Part A. Twenty-sided Dice

Mark any two sides of each 20-sided die with a (washable) marking pen. The chances will therefore be one in ten that a marked surface will be face up on any one die when you shake and roll the dice. When you have rolled the 120 dice, *remove* all the dice that have a marked surface face up. Record the number of dice you removed (or line them up in a column). With the remaining dice, continue this process of shaking, rolling, and removing the marked dice at least twenty times. Record the number you remove each time (or line them up in a series of columns).

Plot a graph in which each roll is represented by one unit on the horizontal axis, and the number of dice *removed* after each roll is plotted on the vertical axis. (If you have lined up columns of removed dice, you already have a graph.)

Plot a second graph with the same horizontal scale, but with the vertical scale representing the number of dice *remaining* in the tray after each roll.

You may find that the numbers you have recorded are too erratic to produce smooth curves. Modify the procedure as follows: Roll the dice and count the dice with marked surfaces face up. Record this number but do not remove the dice. Shake and count again. Do this five times. Now find the mean of the five numbers, and remove that number of dice. The effect will be the same as if you had actually started with 120×5 or 600 dice. Continue this procedure as before, and you will find that it is easier to draw smooth curves which pass very nearly through all the points on your two graphs.

Q1 How do the shapes of the two curves compare?

Q2 What is the ratio of the number of dice removed after each shake to the number of dice shaken in the tray?

Q3 How many shakes were required to reduce the number of dice in the tray from 120 to 60? from 60 to 30? from 100 to 50?

Part B. Electric Circuit

A capacitor is a device that stores electric charge. It consists of two conducting surfaces placed very close together, but separated by a thin sheet of insulating material. When the two surfaces are connected to a battery, negative charge is removed from one plate and added to the other so that a potential difference is established between the two surfaces. (See Sec. 14.6 of Unit 4 *Text*.) If the conductors are disconnected from the battery and connected together through a resistor, the charge will begin to flow back from one side to the other. The charge will continue to flow as long as there is a potential difference between the sides of the capacitor. As you learned in Unit 4, the rate of flow of charge (the current) through a conducting path depends both on the resistance of the path and the potential difference across it.

Fig. 21-9 An analogy: The rate of flow of water depends upon the difference in height of the water in the two tanks and upon the resistance the pipe offers to the flow of water.

To picture this situation, think of two partly filled tanks of water connected by a pipe running from the bottom of one tank to the bottom of the other (Fig. 21-9). When water is transferred from one tank to the other, the additional potential energy of the water is given by the difference in height, just as the potential difference between the sides of a charged capacitor is proportional to the potential energy stored in the capacitor. Water flows through the pipe at the bottom until the water levels are the same in the two tanks. Similarly, charge flows through the conducting path connecting the sides of the capacitor until there is no potential difference between the two plates.

Connect the circuit as in Fig. 21-10, close the switch, and record the reading on the voltmeter. Now open the switch and take a series of voltmeter readings at regular intervals. Plot a graph, using time intervals for the horizontal axis and voltmeter readings for the vertical.

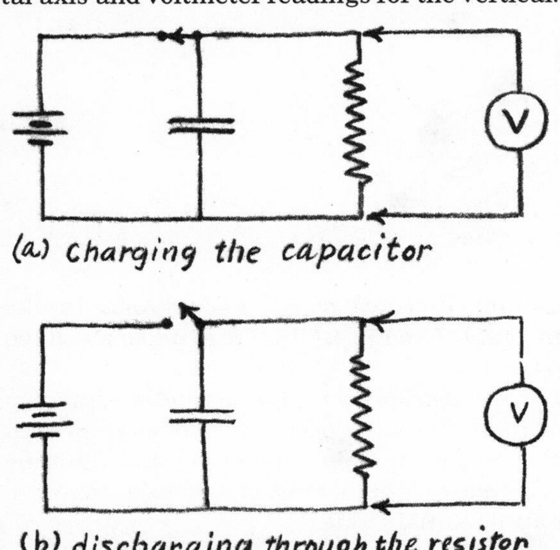

Fig. 21-10

Q4 How long does it take for the voltage to drop to half its initial value? from one-half to one-fourth? from one-third to one-sixth?

Repeat the experiment with a different resistor in the circuit. Find the time required for the voltage to drop to half its initial value. Do this for several resistors.

Q5 How does the time required for the voltage to drop to half its initial value change as the resistance in the circuit is changed?

Part C. Short-lived Radioisotope

Whenever you measure the radioactivity of a sample with a Geiger counter, you must first determine the level of background radiation. With no radioactive material near the Geiger tube, take a count for several minutes and calculate the average number of counts per minute caused by background radiation. This number must be subtracted from any count rates you observe with a sample near the tube, to obtain what is called the *net count rate* of the sample.

The measurement of background rate can be carried on by one member of your group while another prepares the sample according to the directions given below. Use this measurement of background rate to become familiar with the operation of the counting equipment. You will have to work quite quickly when you begin counting radiation from the sample itself.

First, a sample of a short-lived radioisotope must be isolated from its radioactive parent material and prepared for the measurement of its radioactivity.

Although the amount of radioactive material in this experiment is too small to be considered at all dangerous (unless you drink large quantities of it), it is a very good idea to practice caution in dealing with the material. Respect for radioactivity is an important attitude in our increasingly complicated world.

The basic plan is to (1) prepare a solution which contains several radioactive substances, (2) add a chemical that absorbs only one of the radioisotopes, (3) wash most of the solution away leaving the absorbing chemical on a piece of filter paper, (4) mount the filter paper close to the end of the Geiger counter.

(1) Prepare a funnel-filter assembly by placing a small filter paper in the funnel and wetting it with water.

Pour 12 cc of thorium nitrate solution into one graduated cylinder, and 15 cc of dilute nitric acid into another cylinder.

(2) Take these materials to the filter flask

which has been set up in your laboratory. Your teacher will connect your funnel to the filter flask and pour in a quantity of ammonium phosphomolybdate precipitate, $(NH_4)_3PMo_{12}O_{40}$. The phosphomolybdate precipitate adsorbs the radioisotope radioactive elements present in the thorium nitrate solution.

(3) Wash the precipitate by sprinkling several cc of distilled water over it, and then *slowly* pour the thorium nitrate solution onto the precipitate (Fig. 21-11). Distribute the solution over the whole surface of the precipitate. Wash the precipitate again with 15 cc of dilute nitric acid and wait a few moments while the pump attached to the filter flask dries the sample. By the time the sample is dry, the nitric acid should have carried all the thorium nitrate solution through the filter. Left behind on the phosphomolybdate precipitate should be the short-lived daughter product whose radioactivity you wish to measure.

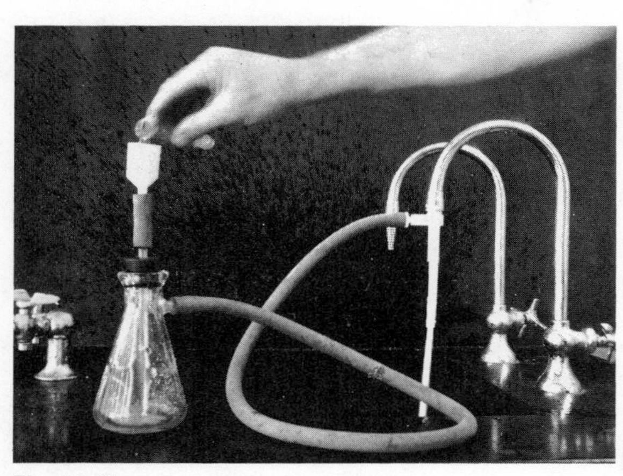

Fig. 21-11

(4) As soon as the sample is dry, remove the upper part of the funnel from the filter flask and take it to the Geiger counter. Make sure that the Geiger tube is protected with a layer of thin plastic food wrapping. Then lower it into the funnel carefully until the end of the tube almost touches the precipitate (Fig. 21-12).

You will probably find it convenient to count for one period of 30 seconds in each minute. This will give you 30 seconds to record the count, reset the counter, and so on, before be-

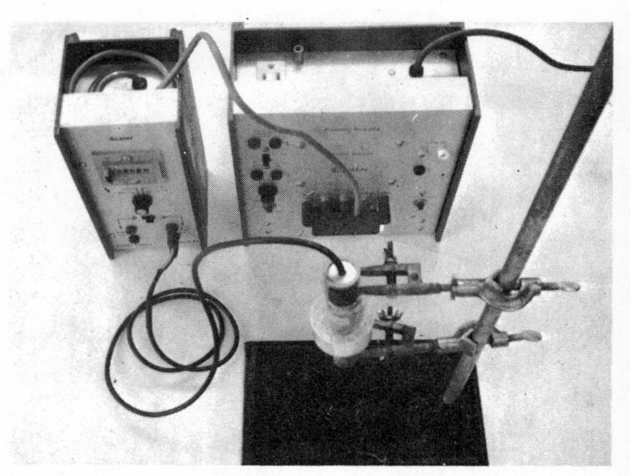

Fig. 21-12

Background = 12 counts per minute
= 6 counts per ½ minute

time (mins)	count	net count rate (counts per ½ min)
0 - ½	803	797
1 - 1½	627	621
2 - 2½	.	.
3 - 3½	.	.
4 - 4½	.	.

Fig. 21-13

ginning the next count. Record your results in a table like Fig. 21-13. Try to make about ten trials.

Plot a graph of *net* count rate as a function of time. Draw the best curve you can through all the points. From the curve, find the time required for the net count rate to decrease to half its initial value.

Q6 How long does it take for the net count rate to decrease from one-half to one-fourth its initial value? one-third to one-sixth? one-fourth to one-eighth?

Q7 The half-life of a radioisotope is one of the important characteristics which helps to identify it. Using the *Handbook of Chemistry and Physics,* or another reference source, identify which of the decay products of thorium is present in your sample.

Q8 Can you tell from the curve you drew whether your sample contains only one radio-isotope or a mixture of isotopes?

Discussion

It should be clear from your graphs and those of your classmates that the three kinds of quantities you observed all have a common property: It takes the same time (or number of rolls of the dice) to reduce the quantity to half its initial value as it does to reduce from a half to a fourth, from a third to a sixth, from a fourth to an eighth, etc. This quantity is the *half-life*.

In the experiments on the "decay" of twenty-sided dice with two marked faces, you knew beforehand that the "decay rate" was one-tenth. That is, over a large number of throws an average of one-tenth of the dice would be removed for each shake of the tray.

The relationship between the half-life of a process and the decay constant λ is discussed on the gray page *Mathematics of Decay* in Chapter 21 of the *Text*. There you learned that for a large number of truly random events, the half life $T_{\frac{1}{2}}$ is related to the decay constant λ by the equation:

$$T_{\frac{1}{2}} = \frac{0.693}{\lambda}$$

Q9 From the known decay constant of the dice, calculate the half-life of the dice and compare it with the experimental value found by you or your classmates.

Q10 If you measured the half-life for capacitor discharge or for radioactive decay, calculate the decay constant for that process.

EXPERIMENT 48 HALF LIFE—II

Look at the thorium decay series in the table below. One of the members of the series, radon 220, is a gas. In a sealed bottle containing thorium or one of its salts, some radon gas always gathers in the air space above the thorium. Radon 220 has a very short half-life (51.5 sec). The subsequent members of the series (polonium 214, lead 210, etc.) are solids. Therefore, as the radon 220 decays, it forms a solid deposit of radioactive material in the bottle. In this experiment you will measure the half-life of this radioactive deposit.

Although the amount of radioactive material in this experiment is too small to be considered at all dangerous (unless you drank large quantities of it), it is a very good idea to practice caution in dealing with the material.

THE THORIUM DECAY SERIES

NAME	SYMBOL	MODE OF DECAY	HALF-LIFE
Thorium 232	${}_{90}Th^{232}$	α	1.39×10^{10} yrs.
Radium 228	${}_{88}Ra^{228}$	β	6.7 years
Actinium 228	${}_{89}Ac^{228}$	β	6.13 hours
Thorium 228	${}_{90}Th^{228}$	α	1.91 years
Radium 224	${}_{88}Ra^{224}$	α	3.64 days
Radon 220	${}_{86}Rn^{220}$	α	51.5 sec
Polonium 216	${}_{84}Po^{216}$	α	0.16 sec
Lead 212	${}_{82}Pb^{212}$	β	10.6 hours
Bismuth 212	${}_{83}Bi^{212}$	α or β*	60.5 min
Polonium 212	${}_{84}Po^{212}$	α	3.0×10^{-7} sec
Thallium 208	${}_{81}Tl^{208}$	β	3.10 min
Lead 208	${}_{82}Pb^{208}$	Stable	3.10 min

*Bismuth 212 can decay in two ways: 34 per cent decays by α emission to thallium 208; 66 per cent decays by β emission to polonium 212. Both thallium 208 and polonium 212 decay to lead 208.

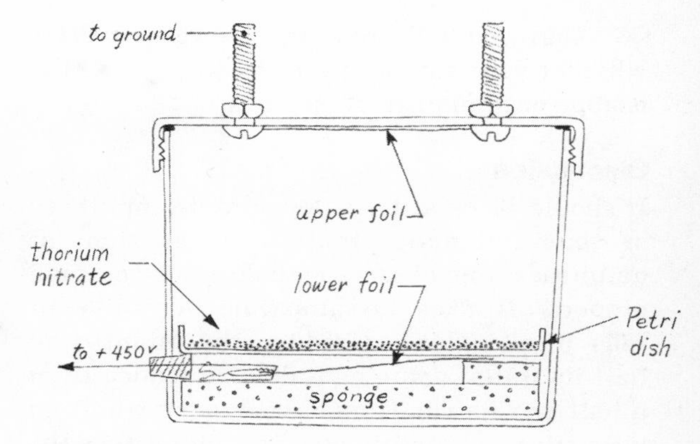

Fig. 21-14

Respect for radioactivity is an important attitude in our increasingly complicated world.

The setup is illustrated in Fig. 21-14. The thorium nitrate is spread on the bottom of a sealed container. (The air inside should be kept damp by moistening the sponge with water.) Radon gas escapes into the air of the container, and some of its decay products are deposited on the upper foil.

When radon disintegrates in the nuclear reaction

$$ {}_{86}Rn^{220} \rightarrow {}_{84}Po^{216} + {}_{2}He^{4} $$

the polonium atoms formed are ionized, apparently because they recoil fast enough to lose an electron by inelastic collision with air molecules.

Because the atoms of the first daughter element of radon are ionized (positively charged), you can increase the amount of deposit collected on the upper foil by charging it negatively to several hundred volts. Although the electric field helps, it is not essential; you will get some deposit on the upper foil even if you don't set up an electric field in the container.

After two days, so much deposit has accumulated that it is decaying nearly as rapidly as the constant rate at which it is being formed.

Therefore, to collect a sample of maximum activity, your apparatus should stand for about two days.

Before beginning to count the activity of the sample, you should take a count of the background rate. Do this far away from the vessel containing the thorium. Remove the cover, place your Geiger counter about one mm above the foil, and begin to count. Make sure, by adjusting the distance between the sample and the window of the Geiger tube, that the initial count rate is high—several hundred per minute. Fix both the counter and the foil in position so that the distance will not change. To get fairly high precision, take a count over a period of at least ten minutes (see Experiment 45). Because the deposit decays rather slowly, you can afford to wait several hours between counts, but you will need to continue taking counts for several days. Make sure that the distance between the sample and the Geiger tube stays constant.

Record the net count rate and its uncertainty (the "two-thirds" range discussed in Experiment 45). Plot the net count rate against time.

Remember that the deposit contains several radioactive isotopes and each is decaying. The net count rate that you measure is the sum of the contributions of all the active isotopes. The situation is not as simple as it was in Experiment 46, in which the single radioactive isotope decayed into a *stable* isotope.

Q1 Does your graph show a constant half-life or a changing half-life?

Look again at the thorium series and in particular at the half-lives of the decay products of radon. Try to interpret your observations of the variation of count rate with time.

Q2 Which isotope is present in the greatest amount in your sample? Can you explain why this is so? Make a sketch (like the one on page 22 of Unit 6 *Text*) to show approximately how the relative amounts of the different isotopes in your sample vary with time.

Ignore the isotopes with half-lives of less than one minute.

You can use your measurement of count rate and half-life to get an estimate of the amount of deposit on the foil. The activity, $\frac{\Delta N}{\Delta t}$, depends on the number of atoms present, N:

$$\frac{\Delta N}{\Delta t} = \lambda N.$$

The decay constant λ is related to the half-life $T_{\frac{1}{2}}$ by

$$\lambda = \frac{0.693}{T_{\frac{1}{2}}}$$

Use your values of counting rate and half-life to estimate N, the number of atoms present in the deposit. What mass does this represent? (1 amu = 1.7×10^{-27} kg.) The smallest amount of material that can be detected with a chemical balance is of the order of 10^{-6} gram.

Discussion

It is not too difficult to calculate the speed and hence the kinetic energy of the polonium atom. In the disintegration

$$_{86}Rn^{220} \rightarrow \ _{84}Po^{216} + {}_2He^4$$

the α particle is emitted with kinetic energy 6.8 MeV. Combining this with the value of its mass, you can calculate v^2 and, therefore, v. What is the momentum of the α particle? Momentum is of course conserved in the disintegration. So what is the momentum of the polonium atom? What is its speed? What is its kinetic energy?

The ionization energy—the energy required to remove an outer electron from the atom—is typically a few electron volts. How does your value for the polonium atom's kinetic energy compare with the ionization energy? Does it seem likely that most of the recoiling polonium atoms would ionize?

EXPERIMENT 49
RADIOACTIVE TRACERS

In this group of experiments, you have the opportunity to invent your procedures yourself and to draw your own conclusions. Most of the experiments will take more than one class period and will require careful planning in advance. You will find below a list of books and magazine articles that can help you.

A Caution

All these experiments take cooperation from the biology or the chemistry department, and require that safety precautions be observed very carefully so that neither you nor other students will be exposed to radiation.

For example, handle radioisotopes as you would a strong acid; if possible, wear disposable plastic gloves, and work with all containers in a tray lined with paper to soak up any spills. Never draw radioactive liquids into a pipette by mouth as you might do with other chemical solutions; use a mechanical pipette or a rubber bulb. Your teacher will discuss other safety precautions with you before you begin.

None of these activities is suggested just for the sake of doing tricks with isotopes. You should have a question clearly in mind before you start, and should plan carefully so that you can complete your experiment in the time you have available.

Tagged Atoms

Radioactive isotopes have been called tagged atoms because even when they are mixed with stable atoms of the same element, they can still be detected. To see how tagged atoms are used, consider the following example.

A green plant absorbs carbon dioxide (CO_2) from the air and by a series of complex chemical reactions builds the carbon dioxide (and water) into the material of which the plant is made. Suppose you tried to follow the steps in the series of reactions. You can separate each compound from the mixture by using ordinary chemical methods. But how can you trace out the chemical steps by which each compound is transformed into the next when they are all jumbled together in the same place? Tagged atoms can help you.

Put the growing green plant in an atmosphere containing normal carbon dioxide, to which has been added a tiny quantity of CO_2 molecules which contain the radioactive isotop carbon 14 in place of normal carbon 12. Less than a minute later the radioactivity can be detected within some, but not all, of the molecules of complex sugars and amino acids being synthesized in the leaves. As time goes on, the radioactive carbon enters step by step into each of the carbon compounds in the leaves.

With a Geiger counter, in effect, one can watch each compound in turn to detect the moment when radioactive molecules begin to be added to it. In this way, the mixture of compounds in a plant can be arranged in their order of formation, which is obviously a useful clue to chemists studying the reactions. Photosynthesis, long a mystery, has been studied in detail in this way.

Radioactive isotopes used in this manner are called *tracers*. The quantity of tracer material needed to do an experiment is astonishingly small. For example, compare the amount of carbon that can be detected by an analytical balance with the amount needed to do a tracer experiment. Your Geiger counter may, typically, need 100 net counts per minute to distinguish the signal from background radiation. If only 1% of the particles emitted by the sample are detected, then in the smallest detectable sample, 10,000 or 10^4 atoms are decaying each minute. This is the number of atoms that decay each minute in a sample of only 4×10^{-4} micrograms of carbon 14. Under ideal conditions, a chemical balance might detect one microgram.

Thus, in this particular case, measurement by radioactivity is over ten thousand times more sensitive than the balance.

In addition, tracers give you the ability to find the precise location of a tagged substance *inside* an undisturbed plant or animal. Radiation from thin sections of a sample placed on photographic film produces a visible spot. (Fig. 21-14.) This method can be made so precise that scientists can tell not only which *cells* of an organism have taken in the tracer, but also which *parts* of the cell (nucleus, mitochondria, etc.).

Choice of Isotope

The choice of which radioactive isotope to use in an experiment depends on many factors, only a few of which are suggested here.

Carbon 14, for example, has several properties that make it a useful tracer. Carbon compounds are a major constituent of all living organisms. It is usually impossible to follow the fate of any one carbon compound that you inject into an organism, since the added molecules and their products are immediately lost in the sea of identical molecules of which the organism is made. Carbon 14 atoms, however, can be used to tag the carbon compounds, which can then be followed step by step through complex chains of chemical processes in plants and animals. On the other hand, the carbon 14 atom emits only β particles of rather low energy. This low energy makes it impractical to use carbon 14 inside a large liquid or solid sample since all the emitted particles would be stopped inside the sample.

The half-life of carbon 14 is about 6000 years, which means that the activity of a sample will remain practically constant for the duration of an experiment. But sometimes the experimenter prefers to use a short-lived isotop so that it will rapidly drop to negligibly low activity in the sample—or on the laboratory table if it gets spilled.

Some isotopes have chemical properties that make them especially useful for a specific kind of experiment. Phosphorus 32 (half-life 14.3 days) is especially good for studying the growth of plants, because phosphorous is used by the plant in many steps of the growth process. Practically all the iodine in the human body is used for just one specific process—the manufacture of a hormone in the thyroid gland which regulates metabolic rate. Radioactive iodine 131 (half-life 8.1 days) has been immensely useful as a tracer in unravelling the steps in that complex process.

The amount of tracer to be used is determined by its activity, by how much it will be diluted during the experiment, and by how much radiation can be safely allowed in the laboratory. Since even very small amounts of radiation are potentially harmful to people, safety precautions and regulations must be carefully followed. The Atomic Energy Commission has established licensing procedures and regulations governing the use of radioisotopes. As a student you are permitted to use only limited quantities of certain isotopes under carefully controlled conditions. However, the variety of experiments you can do is still so great that these regulations need not discourage you from using radioactive isotopes as tracers.

One unit used to measure radioactivity of a source is called the *curie*. When 3.7×10^{10} atoms within a source disintegrate or decay in one second, its activity is said to be one curie (c). (This number was chosen because it was the approximate average activity of 1 gram of pure radium 226.) A more practical unit for tracer experiments is the microcurie (μc) which is 3.7×10^4 disintegrations per second or 2.2×10^6 per minute. The quantity of radioisotope that students may safely use in experiments, without special license, varies from 0.1 μc to 50 μc depending on the type and energy of radiation.

Notice that even when you are restricted to 0.1 μc for your experiments, you may still expect 3700 disintegrations per second, which would cause 37 counts a second in a Geiger counter that recorded only 1% of them.

Q1 What would be the "$\frac{2}{3}$ range" in the activity (disintegrations per minute) of a 1 μc source?

Q2 What would be the "$\frac{2}{3}$ range" in counts per minute for such a source measured with a Geiger counter that detects only 1% of the disintegrations?

Q3 Why does a Geiger tube detect such a small percentage of the β particles that leave the sample? (Review that part of Experiment 46 on the range of β particles.)

Part A. Autoradiography

One rather simple experiment you can almost certainly do is to re-enact Becquerel's original discovery of radioactivity. Place a radioactive object—lump of uranium ore, luminous watch dial with the glass removed, etc.—on a Polaroid film packet or on a sheet of x-ray film in a light-tight envelope. A strong source of radiation will produce a visible image on the film within an hour, even through the paper wrapping. If the source is not so strong, leave it in place overnight. To get a very sharp picture, you must use unwrapped film in a completely dark room and expose it with the radioactive source pressed firmly against the film.

(Most Polaroid film can be developed by placing the packet on a flat surface and passing a metal or hard-rubber roller firmly over the pod of chemicals and across the film. Other kinds of film are processed in a darkroom according to the directions on the developer package.)

This photographic process has grown into an important experimental technique called *autoradiography*. The materials needed are relatively inexpensive and easy to use, and there are many interesting applications of the method. For example, you can grow plants in soil treated with phosphorus 32, or in water to which some phosphorus 32 has been added, and make an autoradiograph of the roots, stem, and leaves (Fig. 21-15). Or each day take a leaf from a fast-growing young plant and show how the phosphorous moves from the roots to the growing tips of the leaves. Many other simple autoradiograph experiments are described in the source material listed at the end of this experiment.

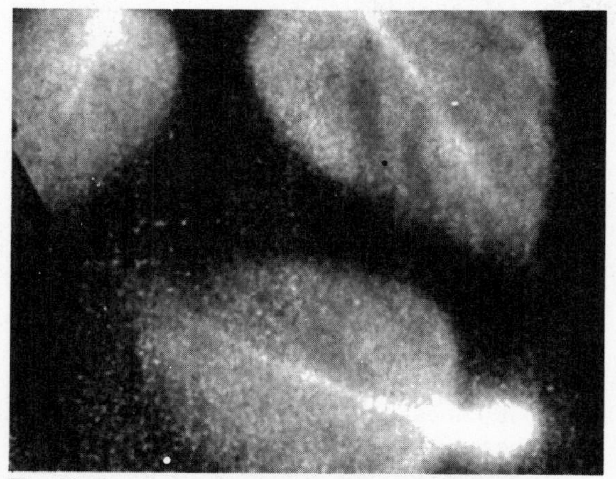

Fig. 21-15 Autoradiograph made by a high school student to show uptake of phosphorus-32 in coleus leaves.

Part B. Chemical Reactions and Separations

Tracers are used as sensitive indicators in chemical reactions. You may want to try a tracer experiment using iodine 131 to study the reaction between lead acetate and potassium iodide solutions. Does the radioactivity remain in the solute or is it carried down with the precipitate? How complete is the reaction?

When you do experiments like this one with liquids containing β sources, transfer them carefully (with a special mechanical pipette or a disposable plastic syringe) to a small, disposable container called a planchet, and evaporate them so that you count the dry sample. This is important when you are using β sources since otherwise much of the radiation would be absorbed in the liquid before it reached the Geiger tube.

You may want to try more elaborate experiments involving the movement of tracers through chemical or biological systems. Students have grown plants under bell-jars in an atmosphere containing radioactive carbon dioxide, fed radioactive phosphorus to earthworms and goldfish, and studied the metabolism of rats with iodine 131.

Some Useful Articles

"Laboratory Experiments with Radioisotopes for High School Demonstrations," edited by S. Schenberg; U.S. Atomic Energy Commission, 1958. Order from Superintendent of Documents, Government Printing Office, Washington, D.C. 20402 for thirty-five cents.

"Radioactive Isotopes: A Science Assembly Lecture." Illustrated. Reprints of this article available from *School Science and Mathematics*, P.O. Box 246, Bloomington, Indiana 47401 for twenty-five cents.

"Radioisotope Experiments for the Chemistry Curriculum" (student manual 17311) prepared by U.S. Atomic Energy Commission.

Order from Office of Technical Services, Washington, D.C. 20545, for two dollars. (A companion teacher's guide is also available at one dollar from the same source.)

American Biology Teacher, August 1965, Volume 27, No. 6. This special issue of the magazine is devoted to the use of radioisotopes and contains several articles of use in the present exercise on tracers. Order single copies from Mr. Jerry Lightner, P.O. Box 2113, Great Falls, Montana 59401 for seventy-five cents.

Scientific American, May 1960. The Amateur Scientist section (by C. L. Stong), page 189, is devoted to a discussion of "how the amateur scientist can perform experiments that call for the use of radioactive isotopes." Copies of the magazine are available in many libraries or can be obtained from Scientific American, 415 Madison Ave., New York, New York 10017. (Reprints of this article are not available).

Scientific American, March 1953. The Amateur Scientist section is on "scintillation counters and a home-made spinthariscope for viewing scintillations."

"Low Level Radioisotope Techniques," John H. Woodburn, *The Science Teacher* magazine, November 1960. Order from The Science Teacher, 1201 16th Street, N.W., Washington, D.C. 20036. Single copies are one dollar.

Safe disposal of radioactive wastes with long half-lives is becoming a significant problem. Here steel cases containing dangerously large amounts of radioactive wastes from nuclear reactors are being buried.

ACTIVITIES

MAGNETIC DEFLECTION OF β RAYS

Clamp a radioactive β source securely a distance of about a foot from a Geiger tube. Place a sheet of lead at least 1 mm thick between source and counter to reduce the count to background level. Hold one end (pole) of a strong magnet above or to the side of the sheet, and change its position until the count rate increases appreciably. By what path do the β rays reach the counter? Try keeping the magnet in the same position but reversing the two poles; does the radiation still reach the counter? Determine the polarity of the magnet by using a compass needle. If β rays are particles, what is the sign of their charge? (See Experiment 37 for hints.)

MEASURING THE ENERGY OF β RADIATION

With a device called a β-ray spectrometer, you can sort out the β particles emitted by a radioactive source according to their energy just as a grating or prism spectroscope spreads out the colors of the visible spectrum. You can make a simple β-ray spectrometer with two disk magnets and a packet of 4″ × 5″ Polaroid film. With it you can make a fairly good estimate of the average energy of the β particles emitted from various sources by observing how much they are deflected by a magnetic field of known intensity.

Mount two disk magnets as shown in the Activity, "Measuring Magnetic Field Intensity," in the Unit 4 *Handbook,* Chapter 14. Be sure the faces of the magnets are parallel and opposite poles are facing each other.

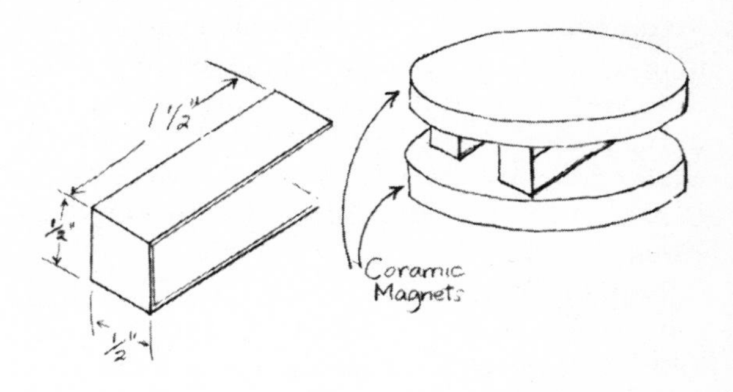

Bend a piece of sheet metal into a curve so that it will hold a Polaroid film packet snugly around the magnets. Place a β source behind a barrier made of thin sheet lead with two narrow slits that will allow a beam of β particles to enter the magnetic field as shown in Fig. 21-16. Expose the film to the β radiation for two days. Then carefully remove the magnets without changing the relative positions of the film and β source. Expose the film for two more days. The long exposure is necessary because the collimated beam contains only a small fraction of the β given off by the source, and because Polaroid film is not very sensitive to β radiation. (You can shorten the exposure time to a few hours if you use x-ray film.)

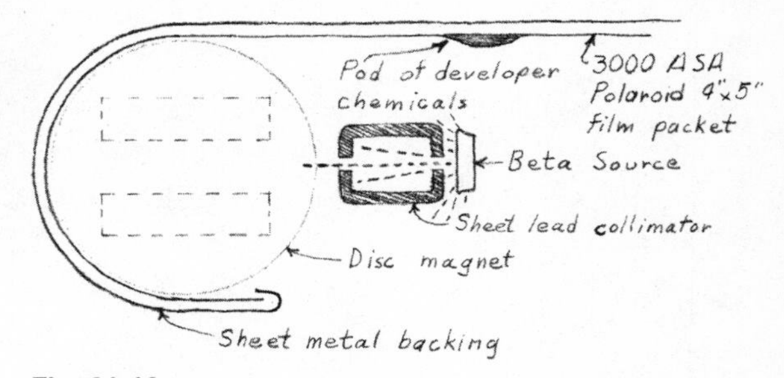

Fig. 21-16

When developed, your film will have two blurred spots on it; the distance between their centers will be the arc length a in Fig. 21-17.

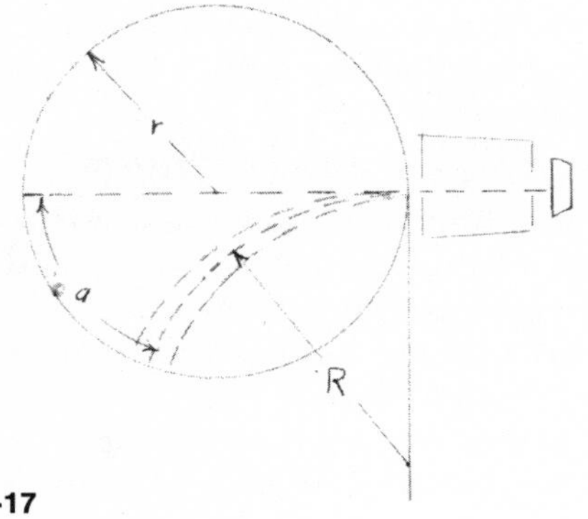

Fig. 21-17

An interesting mathematical problem is to find a relationship between the angle of deflection, as indicated by *a,* and the average energy of the particles. It turns out that you can calculate the *momentum* of the particle fairly easily. Unfortunately, since the β particles from radioactive sources are traveling at nearly the speed of light, the simple relationships between momentum, velocity, and kinetic energy (which you learned about in Unit 3) cannot be used. Instead, you would need to use equations derived from the special theory of relativity which, although not at all mysterious, are a little beyond the scope of this course. (The necessary relations are developed in the supplemental unit, "Elementary Particles.") A graph (Fig. 21-19) that gives the values of kinetic energy for various values of momentum is provided.

First, you need an expression which will relate the deflection to the momentum of the particle. The relationship between the force on a charged particle in a magnetic field and the radius of the circular path is derived in Sec. 18.2 of Unit 5 *Text.* Setting the magnetic force equal to the centripetal force gives

$$Bqv = \frac{mv^2}{R}$$

which simplifies to

$$mv = BqR$$

If you know the magnetic field intensity B (measured with the current balance as described in the Unit 4 *Handbook*), and the charge on the electron, and can find R, you can compute the momentum. A little geometry will enable you to calculate R from a, the arc length, and r, the radius of the magnets. A detailed solution will not be given here, but a hint is shown in Fig. 21-18.

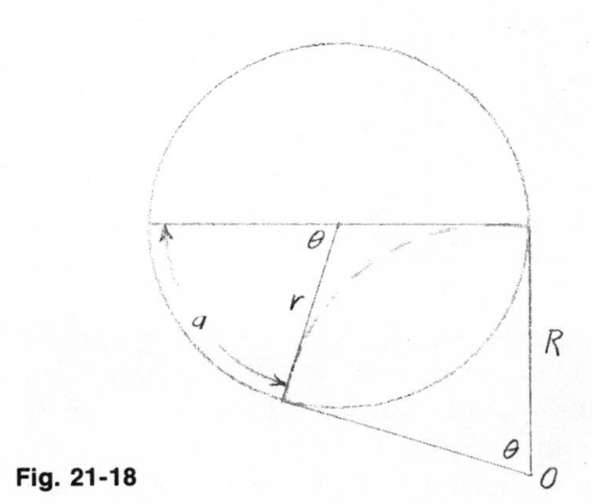

Fig. 21-18

The angle θ is equal to $\frac{a}{2\pi r} \times 360°$, and you should be able to prove that if tangents are drawn from the center of curvature 0 to the points where the particles enter and leave the field, the angle between the tangents at 0 is also θ. With this as a start, see if you can calculate R.

The relationship between momentum and kinetic energy for objects traveling at nearly the speed of light

$$E = \sqrt{p^2c^2 + m_o{}^2c^4}$$

is discussed in most college physics texts. The graph in Fig. 21-19 was plotted using data calculated from this relationship.

From the graph, find the average kinetic energy of the β particles whose momentum you have measured. Compare this with values given in the *Handbook of Chemistry and Physics,* or another reference book, for the particles emitted by the source you used.

You will probably find a value listed which is two to three times higher than the value you found. The value in the reference book is

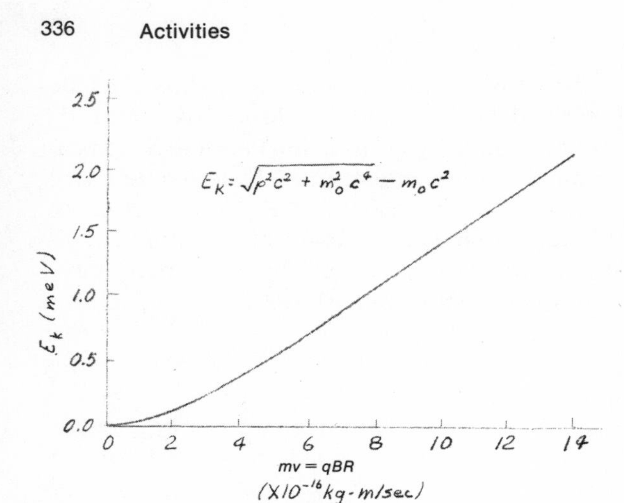

Fig. 21-19 Kinetic energy versus momentum for electrons ($m_oc^2 = 0.511$ meV).

the *maximum* energy that any one β particle from the source can have, whereas the value you found was the *average* of all the β's reaching the film. This discrepancy between the maximum energy (which all the β's should theoretically have) and the average energy puzzled physicists for a long time. The explanation, suggested by Enrico Fermi in the mid-1930's, led to the discovery of a strange new particle called the neutrino which you will want to find out about.

A SWEET DEMONSTRATION

In Experiment 46, "Half-Life I," it is difficult to show that the number of dice "decaying" is directly proportional to the initial number of dice, because statistical fluctuations are fairly large with only 120 dice. An inexpensive way to show that ΔN is directly proportional to N is to use at least 400 sugar cubes (there are 198 in the commonly available 1 pound packages). Mark one face with edible food coloring. Then shake them and record how many decayed as described in Experiment 46.

IONIZATION BY RADIOACTIVITY

Place a different radioactive sample inside each of several identical electroscopes. Charge the electroscopes negatively (as by rubbing a hard, rubber comb on wool and touching the comb to the electroscope knob). Compare the times taken for the electroscopes to completely lose their charges, and interpret your observations.

Place no sample in one electroscope so that you can check how fast it discharges without a sample present. What causes this type of discharge?

EXPONENTIAL DECAY IN CONCENTRATION

Stir 10 drops of food coloring into 1000 cc of water. Pour off 100 cc into a beaker. Add 100 cc of water, stir up the mixture, and pour off a second 100-cc sample. Keep repeating until you have collected 10 to 15 samples.

Questions:

The original concentration was 10 drops/1000 cc or 1 drop/100 cc. What is the concentration after one removal and the addition of pure water (one dilution cycle)? What is the concentration after two cycles? after three cycles? and after n cycles? [Answer: $(0.9)^n$ drops/100 cc.]

What is the number of cycles required to reduce the concentration to approximately $\frac{1}{2}$ of its original concentration?

How many times would you have to repeat the process to get rid of the dye completely?

Chapter 23 The Nucleus

ACTIVITY

NEUTRON DETECTION PROBLEM ANALOGUE (CHADWICK'S PROBLEM)

It is impossible to determine both the mass and the velocity of a neutron from measurements of the mass and the final velocity of a target particle which the neutron has hit. To help you understand this, try the following:

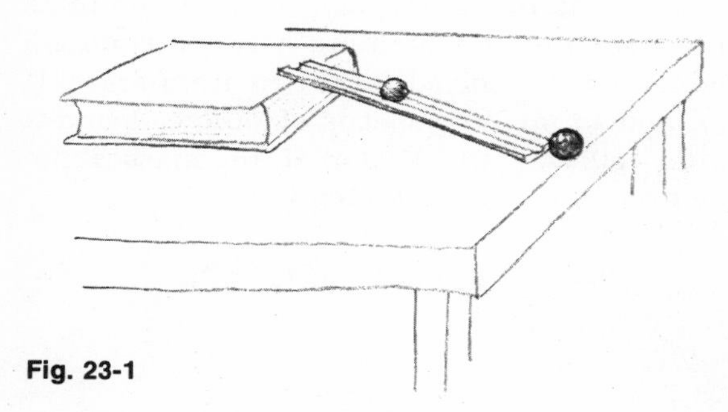

Fig. 23-1

Set up an inclined groove on a table as shown in Fig. 23-1. Let a small ball bearing roll part way down the groove, hitting the larger target ball and knocking it off the table. Note the point where the target ball strikes the floor. Now use another smaller ball as the projectile. Can you adjust the point of release until the target ball strikes the *same spot* on the floor as it did when you used the large projectile? If so, then two different combinations of mass and velocity for the projectile cause the same velocity of the target ball. Are there more combinations of mass and velocity of the target ball. Are there more combinations of mass and velocity of the "neutron" that will give the same result?

Now repeat the experiment, but this time have the same projectile collide in turn with two different target balls of different masses, and measure the velocities of the targets.

Use these velocity values to calculate the mass of the incoming neutron. (Hint: Refer to Sec. 23.4, *Text*. You need only the ratio of the final velocities achieved by the different targets; therefore, you can use the ratio of the two distances measured along the floor from directly below the edge of the table, since they are directly proportional to the velocities.) See also *Film Loop 49*.

"Incredible as it may seem to those of us who live in the world of anti-matter, a mirror image exists—the reverse of ourselves—which we can only call the world of matter."

FILM LOOP 49: COLLISIONS WITH AN OBJECT OF UNKNOWN MASS

In 1932, Chadwick discovered the neutron by analyzing collision experiments. This film allows a measurement similar to Chadwick's, using the laws of motion to deduce the mass of an unknown object. The film uses balls rather than elementary particles and nuclei, but the analysis, based on conservation laws, is remarkably similar.

The first scene shows collisions of a small ball with stationary target balls, one of similar mass and one a larger ball. The incoming ball always has the same velocity, as you can see.

The slow-motion scenes allow you to measure the velocity acquired by the targets. The problem is to find the mass and velocity of the incoming ball without measuring it directly. The masses of the targets are $M_1 = 352$ grams, $M_2 = 4260$ grams.

Chadwick used hydrogen and nitrogen nuclei as targets and measured their recoil velocities. The target balls in the film do not have the same mass ratio, but the idea is the same.

The analysis is shown in detail on a grey page in Chapter 23 of the *Text*. For each of the two collisions, equations can be written expressing conservation of energy and conservation of momentum. These four equations contain three quantities which Chadwick could not measure, the initial neutron velocity and the two final neutron velocities. Some algebraic manipulation allows us to eliminate these quantities, obtaining a single equation which can be solved for the neutron mass. If v_1' and v_2' are the speeds of targets 1 and 2 after collision, and M_1 and M_2 the masses, the neutron mass m can be found from

$$m(v_1' - v_2') = M_2v_2' - M_1v_1'$$

or

$$m = \frac{M_2v_2' - M_1v_1'}{v_1' - v_2'}$$

Make measurements only on the targets, as the incoming ball (representing the neutron) is supposed to be unobservable both before and after the collisions. Measure v_1' and v_2' in any convenient unit, such as divisions per second. (Why is the choice of units not important here?) Calculate the mass m of the invisible, unknown particle. In what ways might your result differ from Chadwick's?

Chapter 24 Nuclear Energy; Nuclear Forces

ACTIVITIES

TWO MODELS OF A CHAIN REACTION

Mousetraps

Carefully put six or more set mousetraps in a large cardboard box. Place two small corks on each trap in such a position that they will be thrown about violently when the trap is sprung. Place a sheet of clear plastic over the top. Then drop one cork in through the corner before you slide the cover completely on. Can you imagine the situation with trillions of tiny mousetraps and corks in a much smaller space?

Questions: What in the nucleus is represented by the potential energy of the mousetrap spring? What do the corks represent? Does the model have a critical size? How might you control the reaction? Describe the effect of the box cover.

Match Heads

Break off the heads of a dozen wooden matches about $\frac{1}{8}$ inch below the match head. Arrange the match heads as shown in the drawing. Place wads of wet paper at certain points. Light a match and place it at point A.

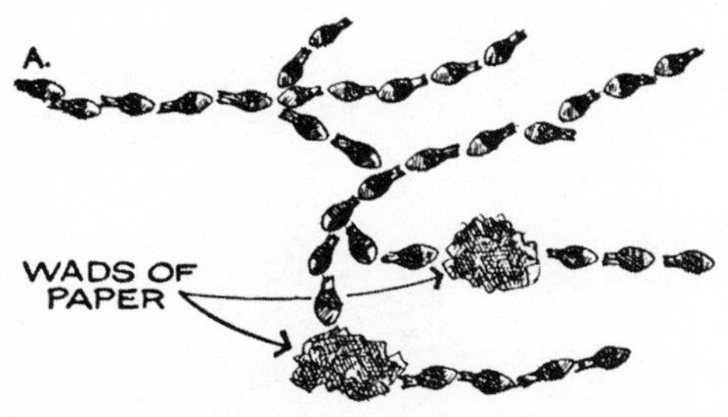

Observe what happens to the right and left sides of the arrangement. What component of a nuclear reactor is represented by the wet paper? How could you modify this model to demonstrate the function of a moderator?

Comment on how good an analogue this is of a nuclear chain reaction. (Adapted from *A Physics Lab of Your Own,* Steven L. Mark, Houghton Mifflin Co., Boston, 1964.)

MORE INFORMATION ON NUCLEAR FISSION AND FUSION

The U.S. Atomic Energy Commission has issued the following booklets on the practical applications of nuclear fission and fusion:

"Nuclear Reactors"
"Power Reactors in Small Packages"
"Nuclear Power and Merchant Shipping"
"Atomic Fuel"
"Direct Conversion of Energy"
"Power from Radioisotopes"
"Atomic Power Safety"
"Controlled Nuclear Fusion"

All are available free by writing USAEC, P.O. Box 62, Oak Ridge, Tenn. 37831.

PEACEFUL USES OF RADIOACTIVITY

Some of the uses of radioactive isotopes in medicine or in biology can be studied with the help of simple available equipment. See Experiment 48, "Radioactive Tracers," in Chapter 21 of this *Handbook*.

A few USAEC booklets that may provide useful information are:

"Food Preservation by Irradiation"
"Whole Body Counters"
"Fallout from Nuclear Tests"
"Neutron Activation Analysis"
"Plowshare"
"Atoms, Nature and Man"
"Radioisotopes in Industry"
"Nuclear Energy for Desalting"
"Nondestructive Testing"

For experiments see: "Laboratory Experiments with Radioisotopes," U.S. Government Printing Office, Washington, D.C. twenty-five cents.

For necessary safety precautions to be taken in working with radioactive materials, see "Radiation Protection in Educational Institutions," NCRP Publication, P.O. Box 4867, Washington, D.C. 20008. Seventy-five cents.

Additional Books and Articles

On the following pages are three separate bibliographies:

Science and Literature
Collateral Reading for Physics Courses
Technology, Literature and Art since World War II

Skim through them to see the variety of kinds of books and articles that have been written on these topics. If you find an item that looks particularly interesting, see if you can find it in the library—or if you can get the library to order it.

Resource Letter SL-1 on Science and Literature *

MARJORIE NICOLSON
Institute for Advanced Study, Princeton, New Jersey

I. INTRODUCTION

AS a field for scholarly research, the interrelation of science and literature is comparatively recent, extending back little more than thirty-five years. A pioneer study was **The Background of the Battle of the Books,** RICHARD F. JONES (Washington University, St. Louis, 1920). Until the appearance of this work, SWIFT's **Battle of the Books** had been interpreted as reflecting only a literary quarrel, an English chapter in the French **Querelle des anciens et des modernes.** Jones showed that a most important background for the mock-epic lay in growing antagonism between the humanists and the propounders of the New Science in England, who had become self-conscious of themselves as a party after the establishment of the Royal Society in 1662. The humanists were becoming the Ancients, the scientists the Moderns. This was followed by Jones' more comprehensive work, **Ancients and Moderns: A Study of the Rise of the Scientific Movement in Seventeenth Century England** (Washington University Studies, St. Louis, 1936; revised ed. with index, 1961).

In the meantime, Jones also published a series of papers in which he suggested the New Science as one of the main forces lying behind the remarkable change in prose style that occurred in England during the seventeenth century, a change from the earlier sonorous, rhetorical, elaborate styles (familiar in Sir Thomas Browne and Milton) to a plainer, more straightforward, conversational style, familiar in Dryden, Addison and Steele, and others. In the first book of **The Advancement of Learning** (1605) FRANCIS BACON had warned against the prevailing literary emphasis on "manner" rather than "matter," particularly in what we would call "scientific writing." Descartes' emphasis on clarity and mathematical plainness merged with Bacon's influence in the program of the Royal Society which, shortly after its inception, declared that its members would "reject all amplifications, digressions and swellings of style," and "exacted from all their members . . . positive expressions, clear senses, bringing all things as near the Mathematical plainness as they can." (THOMAS SPRAT, **The History of the Royal Society of London,** 1667, Part II, Sec. xx.) More than one well-known man of letters was refused membership in the Society because of his over-elaborate style.

Early students in the field of science and literature concerned themselves chiefly with the literature of the seventeenth and early eighteenth centuries, when the New Philosophy, as it was called, both startled and enthralled laymen. The Copernican hypothesis had had little effect upon literary imagination. Galileo's **Sidereus Nuncius** (1610) first brought science "home to men's business and bosoms." My own articles on **"The Telescope and Imagination," "The New Astronomy and English Imagination,"** and **"Milton and the Telescope,"** published in 1935 [collected in *Science and Imagination* (Cornell University Press, Ithaca, New York, 1956)] only scratched the surface. They were followed shortly by two more extensive and important books: **John Donne and the New Philosophy,** C. M. COFFIN (Columbia University Press, New York, 1937) and **Astronomical Thought in Renaissance England,** F. R. JOHSON (Johns Hopkins University, Baltimore, 1937).

During the last twenty years, the field of science and literature has become increasingly popular among scholars, teachers, and students. Since 1939 a group has met anually under the title, "MLA General Topics VII: Literature and Science" at the meetings of the Modern Language Association of America. Papers are read and discussed. An annual bibliography is com-

*Prepared at the request of the Committee on Resource Letters of the American Association of Physics Teachers, supported by a grant from the National Science Foundation and published in the *American Journal of Physics,* Vol. 33, No. 3, March 1965, pp. 175-183.

piled for members. In recent years this has appeared in Symposium (Department of Modern Foreign Languages and Literatures, Syracuse University). The field has broadened to science and the humanities, and includes important treatments of the effect of science on music, the plastic arts, and the social sciences.

II. SCOPE AND LIMITATIONS OF THIS LETTER

I have limited myself to works written in English about English or American literature, making no attempt to include foreign literatures or critics, which would require another list. Except in Sec. X, Contemporary, I have referred chiefly to books, introducing articles only when they seemed to have particular interest for students of physics. I have not attempted, nor would it have been possible, to emphasize physics alone. In the earlier periods, it would be impossible to isolate any one science with the exception of astronomy, so far as lay imagination was concerned. Astronomy (still astrology) was the science most familiar to laymen, emphasized in the schools, because, until well after the invention of the mariner's compass, knowledge of the heavens was essential for travellers by sea or land, and because astrology was basic to all medical treatment. Chaucer, for example, undoubtedly knew more about descriptive astronomy than do most modern liberal arts students, unless they have taken courses in the field. Chaucer wrote a treatise on the astrolabe. It is interesting to note that one of his major works, **Troilus and Criseyde,** has been dated in part on the basis of Chaucer's reference to a conjunction of Jupiter, Saturn, and the crescent moon in the sign Cancer, a conjunction that had not occurred for six hundred years before May 1385, when Chaucer must have been engaged on the poem. [R. K. Root and H. N. Russell, **"A Planetary Date for Chaucer's Troilus,"** Publications of the Modern Language Association **39**, 48–63 (1924)].

After the publication of the **Sidereus Nuncius** (1610), the chief popular interest was, for some years, in astronomy. The telescope stimulated imagination, as did the later microscope. At the end of the seventeenth century, laymen became greatly interested in the findings of microbiology. Physics, like other sciences, remained part of natural history or natural philosophy well down into the eighteenth century. Indeed, the word "science" itself was comparatively rare until almost the nineteenth century. It should be said, too, that although men of letters read some scientific works at first hand (e.g., **Sidereus Nuncius** and Newton's **Opticks**), much of their science was gleaned from popular encyclopedias which afforded comprehensive surveys of natural history as a whole.

During the nineteenth century, physics was of little interest to men of letters, in comparison with the conflicting emotions aroused by geology and the Darwinian theory. Evolution was debated *pro* and *con*, and ideas of development and progress greatly affected literary as well as other history. At the turn of our century, physics may be said to have returned as a literary force with Henry Adams, who said that his "historical neck was broken by the inruption of forces totally new," when he saw the great dynamo at the Paris Exposition in 1900. In **The Rule of Phase Applied to History** (1909), he wrote: "The future of thought, and therefore of History, lies in the hands of the physicists. A new generation must be brought up to think by new methods, and if our historical department in the Universities cannot enter the next phase, the physical department will have to assume the task alone." He felt a basic conflict between the supposed "lessons" of physics and those of biological evolution. In 1910 he addressed **A Letter to American Teachers of History** (J. S. Furst, Baltimore, 1910), in which he emphasized the conflict. Assuming the general validity of the second law of thermodynamics, he claimed, teachers of history faced an insoluble problem if they continued to postulate progressive evolution toward a state of perfection. Both these papers may be found in **The Degeneration of the Democratic Dogma** (The Macmillan Company, New York, 1920), pp. 137–311. During and after the World War, pessimism implicit in Adams' works, particularly **The Education of Henry Adams** (Houghton Mifflin Company, Boston and New York, 1918; privately printed, 1907) merged in the popular mind with such interpretations of history as that of Oswald Spengler, **The Decline of the West** (published in German in 1918, in English in 1926–28).

During recent years, the influence of Einstein, as interpreted by Eddington, Jeans, and Whitehead, has been pervasive in literature, although there is no one definitive treatment of the subject. The most obvious effect of science upon contemporary literature is found in the widespread interest of novelists, poets, dramatists and critics in psychology, chiefly psychoanalysis, as developed by Freud, Jung, Adler. For an indication of the merging of these two forces in contemporary literature see the description below (#72) of the thesis of LAWRENCE DURRELL, **A Key to Modern British Poetry.**

I suggest what seems to me one important aspect of this new field, so far as college teaching is concerned. From 1936–1941 I offered at Smith College a course called "Science and Imagination," which was elected almost as largely by science majors as by students in the humanities. The term-papers written by the young scientists were as original and ingenious as any I received, and their teachers told me that the students returned to the observatory or the laboratories with new interest, particularly in astronomy, biology, and physics. During my early years at Columbia University, I offered the course on the graduate level. Many students found new and interesting approaches for their Master's essays, and several were able to publish parts of their essays in learned or popular journals. I know from colleagues in various fields of science that their students have profited by such study in reverse. In this connection, I call particular attention to the last item listed in this letter, #96, **"Physics and Culture,"** by GERALD HOLTON. The author offers a most interesting discussion of one kind of joint-area course, "a connective approach to the teaching of physics." On p. 9 he gives titles of some of the papers written by students in such a course. In a period when all our disciplines are criticized for over-specialization, the introduction of such area-study into college courses seems one way in which we can help break down barriers which seem to have been increasing between our fields.

III. BASIC HISTORICAL-PHILOSOPHICAL REFERENCES

I preface the readings in science and literature by listing six volumes which, while dealing with the subject only by indirection, are of importance in indicating how our ancestors thought and how they looked at Nature in various periods.

*1. **The Making of the Modern Mind.** J. H. RANDALL. (Houghton Mifflin Company, Boston, 1926; revised ed., 1940.) One of the earliest books to attempt such a survey, this has been widely used as secondary reading for students in English, philosophy, and the social sciences. While inevitably superseded in some respects, it remains a valuable and very readable introduction to modern thought, by a philosopher who has contributed important articles to the history of science.

2. **The Idea of Progress.** J. B. BURY. (The Macmillan Company, New York, 1932; paperback, Dover Publishers, New York.) Although the idea of progress has been pushed much farther back in human history since this appeared, and other materials have been added by scholars, this still remains a standard work in showing why the idea of progress was so long delayed, how it developed, and the extent to which it dominated thinking for almost three hundred years.

3. **The Idea of Nature.** R. G. COLLINGWOOD. (Oxford University Press, 1945, 1960; paperback, Galaxy Books.) A brief and illuminating account of attitudes toward Nature from the Greeks to the twentieth century. The old idea of Nature as animate gave way in the seventeenth century to the theory of Nature as mechanism, which was replaced in the nineteenth century by the concept of development.

4. **The Great Chain of Being: A Study in the History of an Idea.** A. O. LOVEJOY. (Harvard University Press, Cambridge, Massachusetts, 1936; paperback, Harper Torchbooks.) A monumental study of one of the longest-lived of all ideas: graduation, the chain, ladder, or scale of Nature, which, originating with Plato, continued as one of the most basic ideas in human thought through the eighteenth century.

5. **Science and the Modern World.** A. N. WHITEHEAD. (The Macmillan Company, New York, 1925, eleventh printing, 1962; paperback, Mentor Books.) A work of great importance for students in many fields of thought. Chapter 3, **"A Century of Genius,"** is an admirable background for the literature of the seventeenth century.

6. **Metaphysical Foundations of Modern Physical Science: A Historical and Critical Essay.** E. A. BURTT. (Harcourt, Brace and Company, New York, 1925; paperback, Anchor Books.) An important study of the philosophies from which Newton drew and which have been deduced from the **Principia.**

IV. GENERAL WORKS ON SCIENCE AND LITERATURE

(dealing with more than one century)

***7. Science and English Poetry: An Historical Sketch, 1590-1950.** DOUGLAS BUSH. (Oxford University Press, New York, 1950.) These Patten Lectures at the University of Indiana afford the best possible introduction to the subject, indicating attitudes of poets toward science from the Elizabethan period to our own.

***8. Literature and Science.** B. IFOR EVANS. (George Allen and Unwin, London, 1954.) Covers much the same ground as Bush, from the point of view of a thesis: "My aim is to explain the position of the artist, and more particularly the writer, in our modern scientific society."

9. Scientific Thought in Poetry. R. B. CRUM. (Columbia University Press, New York, 1931.) One of the earliest works in the field.

10. Science and the Creative Spirit. Edited by HARCOURT BROWN. (University of Toronto Press, Toronto, Ontario, 1958.) "This volume is the fruit of five years of meetings of the Committee on the Humanistic Aspects of Science, held under the auspices of the American Council of Learned Societies." Contains essays by Harcourt Brown, Karl Deutsch, F. E. L. Priestley, David Hawkins.

***11. The Common Sense of Science.** JACOB BRONOWSKI. (Harvard University Press, Cambridge, Massachusetts, 1953; paperback, Vintage Books.) A very interesting discussion of scientific attitudes in the seventeenth, eighteenth, and nineteenth centuries as well as in the present.

12. Pilgrims through Space and Time: Trends and Patterns in Scientific and Utopian Fiction. J. O. BAILEY. (Argus Books, Inc., New York, 1947.) Scientific fiction from the seventeenth to the twentieth century. Appendix lists many scientific romances of various periods.

13. Mountain Gloom and Mountain Glory: The Development of the Aesthetics of the Infinite. M. H. NICOLSON. (Cornell University Press, Ithaca, New York, 1959; paperback, Norton Library, 1963.) The transfer of a new sense of the vast to great terrestrial objects, particularly mountains. Profound changes caused by astronomy and geology brought about the modern glorification of grand scenery.

14. The Orphic Voice: Poetry and Natural History. ELIZABETH SEWELL. (Yale University Press, New Haven, Connecticut, 1960.) A distinguished British critic and novelist discusses "the biological function of poetry in the natural history of mankind as symbolized by the myth of Orpheus in the works of major British writers from Bacon and Shakespeare, to Erasmus Darwin and Goethe, to Wordsworth and Rilke."

***15. Literature and Science: An Anthology from English and American Literature, 1600-1900.** Edited by GRANT McCOLLEY. (Packard and Company, Chicago, 1940.) Originally designed as a textbook for students in an Institute of Technology, this anthology includes passages from 53 authors. About one-half are from works of science, the others from poets and prose-writers commenting on science. Although the book has not been reprinted (the editor is deceased), it is frequently available in college libraries.

***16. A Book of Science Verse: The Poetic Relations of Science and Technology.** Edited by E. EASTWOOD. (MacMillan Company Ltd., London, 1961.) A pleasant anthology, containing poems or parts of poems from Lucretius to the present, all related directly to either science or technology.

***17. Watchers of the Skies** and **The Torchbearers.** ALFRED NOYES. (Sneed, London, 1937.) In these long poems, a modern poet retells, often dramatically, important moments in the history of science.

V. THE RENAISSANCE AND SEVENTEENTH CENTURY

***18. The Seventeenth Century Background: Studies in the Thought of the Age in Relation to Poetry and Religion.** BASIL WILLEY. (Chatto Windus, London, 1934, 1953; paperback, Anchor Books.) Professor Willey, of Cambridge University, orients the literature of the period against the philosophical and scientific background.

19. Science and Religion in Elizabethan England. P. H. KOCHER. (Huntington Library, San Marino, California, 1953.) Science, religion, and literature are inextricably interwoven in this particular period, so that the effect of science upon religion is important for understanding the literature.

20. The Scientific Renaissance, 1450-1630. MARIE BOAS. (Harper and Row, New York, 1962.) A comprehensive survey, valuable for the literary or other student, who wishes to understand the background of the New Philosophy of the seventeenth century.

21. The Platonic Renaissance in England. ERNST CASSIRER, translated by J. P. Pettegrove. (University of Texas Press, Austin, 1953.) An indispensable background for the study of Renaissance literature, philosophy or science, by one of the most important and original philosophers of modern times. By no means "easy reading" but immensely rewarding.

***22. The Elizabethan World Picture.** E. M. W. TILLYARD. (Chatto Windus, London, 1943.) An excellent, brief, readable account of concepts and attitudes or Renaissance ancestors took for granted (analogy, correspondence, microcosm–macrocosm, etc.) before the age of mechanism.

***23. The Star-Crossed Renaissance.** D. C. ALLEN. (Duke University Press, Durham, North Carolina, 1941.) The quarrel about astrology and its influence in England.

23. A Handbook of Renaissance Meteorology, with Particular Reference to Elizabethan and Jacobean

Literature. S. K. HENINGER. (Duke University Press, Durham, North Carolina, 1960.) The term "meteorology" was used in a broad sense to cover all atmospheric phenomena. The author discusses the general background of the science, and the meteorological imagery of Spenser, Marlowe, Jonson, Chapman, Donne, Shakespeare.

***24. Francis Bacon, Philosopher of Industrial Science.** BENJAMIN FARRINGTON. (Henry Schuman, New York, 1949; paperback, Collier Books, 1961.) **Francis Bacon: The First Statesman of Science.** J. C. CROWTHER. (Cresset Press, London, 1960.) **Francis Bacon: His Career and Thought.** F. H. ANDERSON. (University Publishers, New York, 1962.) **Francis Bacon and the Modern Dilemma.** L. C. EISELEY. (University of Nebraska Press, Lincoln, 1962.)

25. The Psychiatry of Robert Burton. BERGEN EVANS, in consultation with GEORGE J. MOHR, M.D. (Columbia University Press, New York, 1944.) The Anatomist of Melancholy recognized, accepted, and used many of the basic pressuppositions we associate with modern psychoanalysis.

VI. THE NEW PHILOSOPHY

***26. From the Closed World to the Infinite Universe.** ALEXANDRE KOYRÉ. (Johns Hopkins Press, Baltimore, 1952.) An outstanding, comprehensive treatment of the change from the old to the new cosmos by a distinguished historian of science, mathematics, and philosophy, whose **Études galiléennes** (Hermann et Cie., Paris, 1939) is a definitive study of Galileo's place in thought.

27. Galileo as a Critic of the Arts. ERWIN PANOFSKY. (Martinus Nijhoff, The Hague, 1954.) Galileo, son of a distinguished musician, grew up in an environment humanistic and artistic rather than scientific. This (illustrated) volume not only discusses Galileo's attitude toward painting and sculpture but offers a remarkably interesting contrast between Galileo and Kepler, particularly in their attitudes toward planetary circularity. A little classic in the study of the interrelationship of science and the arts.

***28. "Science and Literature."** DOUGLAS BUSH. In *Seventeenth Century Science and the Arts*, edited by H. H. RHYS. (Princeton University Press, Princeton, 1961.) Also contains essays on **"Seventeenth Century Science and the Arts,"** STEPHEN TOULMIN; **"Science and Visual Art,"** J. S. ACKERMAN; **"Scientific Empiricism in Musical Thought,"** C. V. PALISCA.

***29. The Breaking of the Circle: Studies in the Effect of the New Science on Seventeenth Century Poetry.** M. H. NICOLSON. (Northwestern University Press, Evanston, Illinois, 1950; revised ed. Columbia University Press, 1960; paperback, Columbia, 1962.) The breakdown of older conceptions of Nature and the rise of the mechanistic philosophy as reflected in the poetry of the period.

***30. Voyages to the Moon.** M. H. NICOLSON. (The Macmillan Company, New York, 1948; Macmillan paperback, 1960.) Galileo's observations on the topography of the moon stimulated important experimentation in aerostatics, and also led many authors to write imaginary moon-voyages, fanciful, satiric, realistic.

31. "Milton and the Telescope," In *Science and Imagination*, M. H. NICOLSON. (Cornell University Press, Ithaca, New York, 1956), pp. 80–109. The effect of telescopic astronomy on Milton's imagination in **Paradise Lost. Milton and Science.** KESTER SVENDSEN. (Harvard University Press, Cambridge, Massachusetts, 1956.) A thorough study of Milton's knowledge of an interest in various aspects of science.

32. Science and Imagination in Sir Thomas Browne. E. S. MERTON. (King's Crown Press, New York, 1949.) Discusses Browne's knowledge of physical science but feels that his chief interest was in the biological sciences.

VII. THE RESTORATION AND EIGHTEENTH CENTURY

***33. The Eighteenth Century Background.** BASIL WILLEY. (Chatto Windus, London, 1940.) Emphasizes attitudes toward "Nature" as the important clue to the thought of the century.

***34. The Heavenly City of the Eighteenth Century Philosophers.** C. L. BECKER. (Yale University Press, New Haven, Connecticut, 1932, eleventh printing, 1957; Yale Paperback, 1959.) This small volume of four admirably written lectures analyzes the attitudes of the eighteenth century Philosophers and *philosophes* in such a way as to indicate that, in spite of their revolt from revealed religion, they were really closer to the universe of Dante and Aquinas than to that of Einstein.

35. The Philosophy of the Enlightenment. ERNST CASSIRER. (Princeton University Press, Princeton 1951; paperback, Beacon, 1955.) Important for this period as #21 for the Renaissance.

***36. "The Microscope and English Imagination."** In *Science and Imagination*. M. H. NICOLSON. (Cornell University Press, Ithaca, New York, 1956), pp. 155–234. The great popular interest aroused by the microscope, which became a popular plaything for ladies as well as gentlemen. Serious and satiric attitudes toward the small, complementing the interest in the vast aroused by the telescope.

***37. Newton Demands the Muse: Newton's Opticks and the Eighteenth Century Poets.** M. H. NICOLSON. (Princeton University Press, Princeton, 1946; reprinted, Archon Books, Hamden, Connecticut, 1963.) The **Opticks** was widely read by laymen to whom it was more readily comprehensible than the **Principia,** since it was written in English and dealt with light and color, always the stuff of poetry. The popular reception of the work, attitudes toward Newton, changes in poetic treatment of light and color, the aesthetics and metaphysics read into the **Opticks.**

*38. **Scientists and Amateurs. A History of the Royal Society.** DOROTHY STIMSON. (Henry Schuman, New York, 1948.) Emphasizes the difference between the original and the modern Royal Society. Many members were amateurs, gentlemen, asistocrats. divines, authors. As Sprat suggests, one group wished to make the Society into a British Academy. An interesting and amusing way of charting changing attitudes toward the Royal Society is to read by index (entries under "Gresham College" as well as "Royal Society") **The Diary of Samual Pepys,** which almost parallels the first decade of the Society. Pepys began with laughter. After his election, he attended meetings and reported experiments. A climax of his career was his installation as President of the Royal Society.

39. **Swift's Satire on Learning in A Tale of a Tub.** M. K. STARKMAN. (Princeton University Press, Princeton, 1950.) A thorough study of Swift's satiric attitudes toward various kinds of learning, including the New Science.

*40. **"The Scientific Background of Swift's Voyage to Laputa."** N. M. MOHLER AND M. H. NICOLSON. In *Science and Imagination.* M. H. Nicolson (Cornell University Press, Ithaca, New York, 1956), pp. 110–154. The third voyage of **Gulliver's Travels** is largely satire on science ranging from mathematics to astronomy, and particularly to experimentation in the Royal Society. The absurd experiments in the Grand Academy are satires on actual experiments performed by members of the Royal Society.

41. **"Swift's Flying Island in the Voyage to Laputa."** N. M. MOHLER and M. H. NICOLSON. Ann. Sci. 2, 405–430 (1937). The scientific background of the structure of the "Flying Island" and its principle of flight by terrestrial magnetism.

42. **Edward Tyson, M.D., F.R.S., and the Rise of Comparative Anatomy in England.** M. F. ASHLEY MONTAGU. (American Philosophical Society, Philadelphia, 1943.) In 1698 Tyson dissected a chimpanzee and discovered the close parallels between its structure and organs and those of man. His work attracted wide attention among writers as well as scientists.

43. **"Description and Science."** In *The Background of Thomson's Seasons.* A. D. MCKILLOP. (University of Minnesota Press, Minneapolis, 1942), Chap. 2. James Thomson's interest in science and his revision of *The Seasons* on the basis of his reading in physics, meteorology, astronomy.

44. **"Scientific Verse."** In *English Literature in the Early Eighteenth Century.* BONAMY DOBRÉE. (Clarendon Press, Oxford, 1959.) Discussion of a group of writers who prided themselves upon being "scientific poets."

45. **Doctor Johnson on Ballooning and Flight.** J. E. HODGSON. (Oxford University Press, London, 1925.) Originally published in London Mercury **10,** 63 ff., (1924). Samuel Johnson pessimistically discussed the future of flight in **Rasselas.** Like his contemporaries, however, he was greatly interested in the experiments with balloons in his time.

46. **Doctor Darwin.** HESKETH PEARSON. (Walker and Company, New York, 1963.) **Erasmus Darwin.** DESMOND KING-HELE. (Charles Scribner's Sons, New York, 1963). The grandfather of Charles Darwin was widely known for his long poem, **The Botanic Garden,** based upon Linnean botany, in which he developed a form of evolutionism later expounded by Lamarck.

47. **William Blake: A Man Without a Mask.** JACOB BRONOWSKI. (Secker-Warburg, London, 1948.) **Tracks in the Snow: Studies in English Science and Art.** RUTHVEN TODD. (Charles Scribner's Sons, New York, 1947.) Essays on William Blake, as artist rather than as poet.

48. **The New England Mind.** PERRY MILLER. (The Macmillan Company, New York, 1 vol. 1939; Beacon, Boston, 2 vols. 1961; paperback, Beacon.) Recognized as the standard work upon the intellectual background of the colonists. Various sections on attitudes toward science in New England during the seventeenth and eighteenth centuries.

49. **The Pursuit of Science in Revolutionary America, 1735-1789.** (University of North Carolina Press, Chapel Hill, North Carolina, 1956.) Shows significant connections between the Colonies and Europe, particularly Britain. **Franklin and Newton: An Inquiry into Speculative Newtonian Experimental Science and Franklin's Work in Electricity.** I. B. COHEN. (American Philosophical Society, Philadelphia, 1956.)

VIII. THE NINETEENTH CENTURY

*50. **The Concept of Nature in Nineteenth Century English Poetry.** J. W. BEACH. (The Macmillan Company, New York, 1936.) A discussion of changing concepts of Nature as basic to the poetry of the period.

*51. **Ideas and Beliefs of the Victorians.** Edited by HARMAN GRISEWOOD. (Sylvan Press, London, 1950.) An interesting series of British Broadcasting Lectures by distinguished scientists, clergymen, writers, including **"Unbelief in Science,"** by J. BRONOWSKI, 164–169, in a group of lectures on **"Man and Nature,"** 164–243.

52. **The Metaphysical Society.** A. W. BROWN. (Columbia University Press, New York, 1947.) During the period 1869–1881, meetings of this society were attended by nearly all important leaders of thought in theology, philosophy, science, and literature.

*53. **Science and Literary Criticism.** HERBERT DINGLE. (Thomas Nelson and Sons, New York, 1949.) Essays on various nineteenth century writers. Dingle has also edited **A Century of Science, 1851-1951.** (Hutchinson, London, 1951.)

54. **The Road to Xanadu.** J. L. LOWES. (Houghton Mifflin Company, Boston, 1927, rev. ed., 1930; Vintage paperback, 1959.) While not primarily about

science, this extraordinary study of Coleridge's imagination makes use of every possible kind of background. Coleridge's scientific reading is frequently mentioned.

55. **A Newton among Poets: Shelley's Use of Science in Prometheus Unbound.** CARL GRABO. (University of North Carolina Press, Chapel Hill, North Carolina, 1930.) Whitehead had said in **Science and the Modern World** that if Shelley had not been determined to be a poet he might have been a great chemist. In this study, **Prometheus Unbound** is discussed as "a drama of the chemical elements." Grabo followed this with **The Magic Plant: The Growth of Shelley's Thought.** (University of North Carolina Press, Chapel Hill, North Carolina, 1936.)

56. **Keats as Doctor and Patient.** SIR WILLIAM HALE-WHITE. (Oxford University Press, London, 1938.) **A Doctor's Life of John Keats.** W. A. WELLS. (Vantage Press, New York, 1959.)

57. **Walt Whitman: Poet of Science.** JOSEPH BEAVER. (King's Crown Press, New York, 1951.) A rather slight treatment of Whitman's interest in science.

58. **Emerson's Angle of Vision: Man and Nature in American Experience.** SHERMAN PAUL. (Harvard University Press, Cambridge, Massachusetts, 1952.) Not primarily about science, though science in part determines the "angle of vision."

59. **Matthew Arnold the Ethnologist.** F. E. FAVERTY. (Northwestern University Press, Evanston, Illinois, 1951.) A good study of the effect of anthropology upon a leading man of letters.

DARWIN AND EVOLUTION

*60. **"A Liberal Education and Where to Find It."** T. H. HUXLEY. This famous essay, delivered at the South London Working Man's College on 4 January 1868, published in **Science and Education,** provoked much discussion. It may be found in various modern editions of Huxley's works.

*61. **"Literature and Science."** MATTHEW ARNOLD. Arnold's rejoinder to Huxley was originally given as the Rede Lecture at Cambridge, and later recast for presentation in the United States. It was published among **Discourses in America** (MacMillan Company Ltd., London, 1885, 1889) and is reprinted in modern editions of Arnold's prose works.

62. **Darwin Among the Poets.** LIONEL STEVENSON. (University of Chicago Press, Chicago, 1932; reprinted, Russell and Russell, New York, 1963.)

63. **Darwinism in the English Novel: The Impact of Evolution on Victorian Fiction.** J. L. HENKIN. (Corporate Press, New York, 1940; reprinted, Russell and Russell, New York, 1963.)

*64. **Apes, Angels and Victorians: Darwin, Huxley and Evolution.** WILLIAM IRVINE. (McGraw-Hill Book Company, New York, 1955; paperback, Meridian Books, 1962.) Chief emphasis on Huxley; some discussion of Tennyson, Arnold, Carlyle and other writers.

65. **Cosmic Evolution: A Study of the Interpretation of Evolution by American Poets from Emerson to Robinson.** F. W. CONNER. (University of Florida Press, Gainesville, Florida, 1949.) Pre-Darwinian evolution in Emerson, Whitman, Poe; Darwinian evolution shown in "rejection, discomfort, indifference" on the part of Bryant, Longfellow, Lowell.

66. **Evolution and Poetic Belief: A Study in Some Victorian and Modern Writers.** GEORG ROPPEN. (Oslo University Press, Oslo, 1956; Basil Blackwell, Oxford, 1956.) An extensive study, dealing with Tennyson, Browning, Swinburne, Meredith, Hardy, Butler Wells, Shaw.

67. **Darwin and Butler: Two Versions of Evolution.** BASIL WILLEY. (Harcourt, Brace & World, Inc., New York, 1960.) Attitudes of Darwin as scientist and Samuel Butler as novelist and satirist.

*68. **Darwin and His Critics.** B. R. KOGAN. (Wadsworth Publishing Company, Belmont, California, 1960); **Darwin's Century: Evolution and the Men who Discovered It.** L. C. EISELEY. (Doubleday & Company, Inc., New York, 1958.)

69. **Principles of Geology. Being an Attempt to Explain the Former Changes of the Earth's Surface, by Reference to Causes Now In Operation.** SIR CHARLES LYELL. (J. Murray, London, 1830–1833.) Widely read by laymen (a twelfth edition appeared in 1875), this volume put an end to older catastrophic theories, establishing ideas of development. Its influence merged with that of Darwin's **Origin of Species.** Lyell later published **The Geological Evidence of the Antiquity of Man. With Remarks on Theories of the Origin of Species by Variation.** (J. Murray, London, 1863.)

70. **Genesis and Geology: A Study in the Relation of Scientific Thought, Natural Theology, and Social Opinion in Great Britain, 1790-1854.** C. C. GILLESPIE. (Harvard University Press, Cambridge, Massachusetts, 1951.) The long title clearly indicates the scope of the work.

IX. THE TWENTIETH CENTURY

*71. **Forces in Modern British Literature.** W. Y. TINDALL. (Alfred A. Knopf, Inc., New York, 1947.) Among the forces discussed are science, with relation particularly to naturalism, and the impact of Freud and other psychologists.

72. **A Key to Modern British Poetry.** LAWRENCE DURRELL. (University of Oklahoma Press, Norman, Oklahoma, 1952.) The British novelist develops the thesis that the impact of relativity and psychology has changed the world view of all important writers. Discusses Eliot, Joyce, V. Woolf, Proust, and others, arguing that the principle of indeterminancy throws the structure of the external world in doubt, and the discoveries of psychologists throw the internal world in doubt, resulting in drastic changes in character-

ization, plot, metaphor, even sentence structure. Durrell's four recent novels, which constitute the **Alexandria Quartet,** involve various aspects of his thesis. See **"Durrell and Relativity."** A. M. BORK. The Centennial Review **7,** 191–203 (Spring, 1963) for a very interesting critique of Durrell's treatment of relativity in this book and in the **Alexandria Quartet.**

73. **Science and Poetry.** I. A. RICHARDS. (W. W. Norton and Company, Inc., New York, 1926); **Principles of Literary Criticism** (Harcourt, Brace and Company, Inc., fifth ed., 1934); paperback, Harvest Book); **Practical Criticism: A Study of Literary Judgment** (Harvard University Press, Cambridge, Massachusetts, 1950). Three works by one of the most influential modern critics, dealing with behaviorist psychology in relation to literature.

74. **Archetypal Patterns in Poetry.** MAUD BODKIN. (H. Milford, London and Oxford, 1934; Vintage paperback.) The psychology of Jung applied to poetry.

75. **Philosophy in a New Key: A Study in the Symbolism of Reason, Rite, and Art.** SUSANNE LANGER. (Harvard University Press, Cambridge, Massachusetts, 1957.) In part, the approach is influenced by Whitehead's **Symbolism**; more directly by Ernst Cassirer.

76. **The Wound and the Bow.** EDMUND WILSON. (Houghton Mifflin Company, Boston, 1941; Oxford University Press, New York, 1947.) An influential critic uses the classical story of Philoctetes to discuss the idea of the wounded artist healing society.

77. **Hamlet and Orestes.** ERNEST JONES. (Victor Gollancz, London, 1949.) Dr. Jones, student and editor of Freud, applies his method to **Hamlet.** Other examples of the Freudian approach to literature may be found in **The Life and Works of Sigmund Freud,** edited and abridged by LIONEL TRILLING and STEVEN MARCUS. (Hogarth Press, London, 1961; Anchor paperback.)

78. **Freudianism and the Literary Mind.** FREDERICK HOFFMAN. (Louisiana State University Press, Baton Rouge, Louisiana, 1945.)

79. **Psychoanalysis and American Literary Criticism.** LOUIS FRAIBERG. (Wayne State University Press, Detroit, 1959.)

80. **The Tangled Bank: Darwin, Marx, Freud as Imaginative Writers.** S. E. HYMAN. (Atheneum, New York, 1962.)

81. **The Heel of Elohim: Science and Values in Modern American Poetry.** (University of Oklahoma Press, Norman, Oklahoma, 1950.) Discussion of Robinson, Frost, Eliot, Jeffers, MacLeish, Hart Crane.

82. **Space, Time and Architecture.** S. GIEDION. (Harvard University Press, Cambridge, Massachusetts, 1938; 4 ed. enl. 1962). The thesis is that the development of modern physics and of modern art proceed from unrest with older theories of space and time.

83. **Time in Literature.** HANS MEYERHOFF. (University of California Press, Berkeley, 1955; paperback, 1960.) A comparison between the scientific concept of time in physics and the literary treatment of time in Proust, Scott Fitzgerald, Thomas Wolfe, and others.

X. CONTEMPORARY: CONTROVERSY AND SYNTHESIS

During the last few years, controversialists have become particularly vocal in defense of or attack on C. P. Snow's Rede lecture of 1959. (#84). I select some titles—#85 through #87—from the quarrel, followed by other papers on the general theme of science and the humanities, some by scientists, some by humanists.

84. **The Two Cultures: A Second Look.** C. P. SNOW. (Cambridge University Press, Cambridge, England, second ed., 1964; first ed., 1959.) The theme of humanism versus science is that of a majority of Snow's novels. In this lecture, he discussed what he considers the growing division of intellectuals in our time into two groups which no longer communicate.

85. **"Look in Thy Glass: Science Looks at Itself."** PAUL WEISS. The Graduate Journal **5,** 43–59 (Spring, 1962). In part, this is based on **"The Message of Science,"** delivered at the Brussels World's Fair by Dr. Weiss, as representative of the United States, which preceded Snow's lecture. Dr. Weiss concludes here with a discussion of Snow's position.

86. **Two Cultures? The Significance of C. P. Snow. With a new preface for the American reader.** F. R. LEAVIS. **And an essay on Sir Charles Snow's Rede Lecture by Michael Yudkin.** (Pantheon Books, New York, 1963.) The most vehement of all replies to Snow, this provoked the four items under #87.

87. **"Science, Literature and Culture: A Commentary on the Snow-Leavis Controversy."** LIONEL TRILLING. Commentary **33,** 461–477 (June 1962); **"The Snow-Leavis Rumpus,"** S. G. PUTT. Antioch Review **23,** 299–312 (1963); **C. P. Snow: A Spectrum.** STANLEY WEINTRAUB. (Charles Scribner's Sons, New York, 1963); **Literature and Science.** ALDOUS HUXLEY. (Harper & Row, New York, 1963.)

88. **The History of Science and the New Humanism.** GEORGE SARTON. (Indiana University Press, Bloomington, Indiana, 1963.) A new and revised form of the Colver Lectures at Brown University, 1930, by a distinguished historian of science.

***89.** **Science and the Modern Mind. A Symposium.** Edited by GERALD HOLTON. (Beacon Press, Boston, 1958.) (Originally published in Daedalus **87** (Winter, 1958) as a result of a Conference of the American Academy of Arts and Sciences.) "The nine contributors in this volume present a spectrum of points of view on the influence which science has had upon our image of the universe within and around us." In the first three essays, Henry Guerlac, Harcourt Brown, and Giorgio de Santillana look back to consider how the rise of science affected succeeding culture. Philipp Frank, Robert Oppenheimer, and Jerome S. Bruner discuss the effect of key concepts of science upon contemporary attitudes. The last three, P. W. Bridgman, Charles Morris, and Howard Mumford Jones, look to the future.

*90. **Science and Human Values.** JACOB BRONOWSKI. (Julian Messner Inc., New York, 1956.) Three illuminating essays on **"The Creative Mind," "The Habit of Truth," "The Sense of Human Dignity,"** by a distinguished scientist. See, also, a very interesting essay by the same author, **"The Educated Man in 1984,"** Science 710 ff. (1956).

*91. **Science and the Humanities.** MOODY PRIOR. (Northwestern University Press, Evanston, Illinois, 1962.) Five judicious and thoughtful essays by a humanist, which had their origin in an Aspen Conference between scientists and humanists.

92. **"Closing the Gap Between the Scientists and the Others."** MARGARET MEAD. Daedalus **88**, 139–146 (Winter 1959). A well-known anthropologist cautions that, with the acceleration in specialized areas of learning, we must find new educational and communication devices which "will protect society from the schismatic effects of too great a separation of thought patterns, language and interest between the specialized practitioners of a scientific or humane discipline and those who are laymen."

93. **The Sciences and the Arts: A New Alliance.** H. G. CASSIDY. (Harper & Row, New York, 1962.) **"The Muse and the Axiom."** H. G. CASSIDY. Am. Sci. **51**, 315–326 (1963). The latter begins with Henry Adams as "an epitome of the nonscientist faced with science he does not understand."

*94. **Science: The Glorious Entertainment.** JACQUES BARZUN. (Harper & Row, New York, 1964.) "It is Barzun's thesis that science, whose true glories he eloquently celebrates, has imposed itself as a single, imperialistic mode of though upon all experience—partly by right of conquest, but chiefly through our own inertia."

95. **Science and Literature: Toward a Synthesis.** J. J. CADDEN and P. R. BROSTOWIN. (D. C. Heath, Boston, 1964; a paperback original.)

96. **"Physics and Culture."** GERALD HOLTON. Bull. Inst. Phys. & Phys. Soc., 321–329 (December 1963). Reprinted in **Why Teach Physics?** S. C. BROWN and N. CLARKE. (M.I.T. Press, Cambridge, Massachusetts, 1964.) The opening address to the Second International Conference on Physics Education held in Rio de Janeiro, July, 1963. Gives a rationale for building science courses that introduce explicit connections to humanistic and other fields of study.

Resource Letter ColR-1 on Collateral Reading for Physics Courses *

ALFRED M. BORK
Department of Physics, Reed College, Portland, Oregon
AND
ARNOLD B. ARONS
Department of Physics, Amherst College, Amherst, Massachusetts

I. INTRODUCTION

THIS resource letter lists materials outside of physics suitable for use with physics classes as reading assignments, perhaps combined with class discussion. Most are for beginning courses, both for science majors and for nonscience majors.

One could easily construct a list 10 times as long; hence a large element of personal predilection is involved, and a reading list compiled by others might well include very different items. We should like to present the references listed here as suggestions and not as prescriptions. Each listing opens the door to a variety of other items that an individual teacher may find more to his own liking. Many of the given references have been used by us, as well as by many other teachers, in classroom situations.

There is some overlap between this resource letter and a few earlier ones which also consider similar material, particularly those dealing with the history of science and the relation of science to other aspects of our culture. We list other useful resource letters below.

Because almost all the selections are intended for beginning courses, the notation of levels E, I, and A, and the use of the asterisk, followed in other resource letters, are not employed.

Another arbitrary boundary condition on the present list is that it does not concern itself with original scientific papers. Many such sources are referred to in other resource letters or anthologies mentioned. The present list is restricted to articles which are not the raw material of science itself. In most cases specific articles rather than full books are listed, since we wish to confine ourselves to relatively short selections that lend themselves to assignment as collateral reading.

*Prepared at the request of the Committee on Resource Letters of the American Association of Physics Teachers, supported by a grant from the National Science Foundation and published in the *American Journal of Physics*, Vol. 35, No. 2, February 1967, pp. 71-78.

II. RELEVANT RESOURCE LETTERS

1. **"Philosophical Foundations of Classical Mechanics" (PhM-1).** MARY HESSE. Am. J. Phys. **32**, 905 (1964).
2. **"Science and Literature" (SL-1).** MARJORIE NICOLSON. Am. J. Phys. **33**, 175 (1965).
3. **"Evolution of Energy Concepts from Galileo to Helmholtz" (EEC-1).** T. M. BROWN. Am. J. Phys. **33**, 759 (1965).

A resource letter on History of Physics is currently being prepared by H. Woolf of The Johns Hopkins University, and, when published, it will also form a useful item in this sequence.

III. COLLECTIONS OF COLLATERAL READING AND BIBLIOGRAPHIES

4. **Science and Ideas.** ARNOLD B. ARONS AND A. M. BORK. (Prentice-Hall, Inc., Englewood Cliffs, N. J., 1964.) Readings in the history, philosophy, and sociology of science: excerpts from such authors as Butterfield, Gillispie, Feynman, Bridgman, Einstein, Jammer, Conant, Merton, Frank, and others.
5. **A Stress Analysis of a Strapless Evening Gown.** ROBERT A. BAKER. (Prentice-Hall, Inc., Englewood Cliffs, N. J., 1963). Occasionally a point can best be made by the use of humor. This volume gathers many well-known satires on science.
6. **The Sociology of Science.** B. BARBER AND W. HIRSCH, EDS. (The Free Press of Glencoe, New York, 1962.) Articles by Parsons, Merton, Kubie, Kuhn, Price, Barber, Nagel, and others.
7. **Science and Language.** ALFRED M. BORK. (D. C. Heath Co., Boston, Mass., 1966.) This collection, not restricted to physics, is meant for freshman English classes; however, some of the articles should be useful with physics classes.
8. **Science and Literature—a Reader.** JOHN J. CADDEN AND PATRICK R. BROSTOWIN. (D. C. Heath Co., Boston, Mass., 1964.) The beginning contains critical essays about the relations of science to literature; the last part has excerpts from literature influenced by science. There is a brief bibliography.
9. **The Scientific Endeavor.** CENTENNIAL CELEBRATION OF THE NATIONAL ACADEMY OF SCIENCES. (The

Rockefeller Institute, New York, 1963.) The last section, "The Scientific Endeavor," is particularly relevant. E. P. Wigner's article on "Symmetry and Conservation Laws" is for beginning classes.

10. **Science and Society.** THOMAS D. CLARESON. (Harper and Brothers, New York, 1961.) Intended for freshman composition classes.
11. **The Validation of Scientific Theories.** PHILIPP FRANK, ED. (Collier Books, New York, 1961), Paperback A5101. Articles by Frank, Margenau, Lindsay, Bridgman, Skinner, Reshevsky, Guerlac, Koyré, Boring, and others.
12. **The Voices of Time: A Cooperative Survey of Man's Views of Time as Expressed by the Sciences and by the Humanities.** J. J. FRASER, ED. (George Braziller, New York, 1966.)
13. **Scientists as Writers.** J. HARRISON. (MIT Press, Cambridge, Mass., 1965.) Includes a useful bibliography.
14. **Studies in Explanation: A Collection of 27 Explanations from Physics, Biology, Psychology, Sociology, and History.** RUSSELL KAHL, ED. (Prentice–Hall, Inc., Englewood Cliffs, N. J., 1963).
15. **The New Scientist: Essays on the Methods and Values of Modern Science.** P. C. OBLER AND H. A. ESTRIN, EDS. (Doubleday–Anchor, Garden City, New York, 1962), Paperback A319.
16. **The New Treasury of Science.** HARLOW SHAPLEY, SAMUEL RAPPORT, AND HELEN WRIGHT, EDS. (Harper and Row, New York, 1965.) Contains selections different from those in their earlier book. The first two sections concern science in general and physics in particular.
17. **Readings in the Physical Sciences.** HARLOW SHAPLEY, SAMUEL RAPPORT, AND HELEN WRIGHT. (Appleton–Century–Crofts, New York, 1948.) Particularly relevant sections are Part I, "Science and the Scientific Method," and Part V, "Physics."
18. **Science and Society.** A. VAROULIS AND A. W. COLVER. (Holden-Day Inc., San Francisco, Calif., 1966.) Essays useful for physical science classes.
19. **Exploring the Universe.** LOUISE B. YOUNG, ED. (McGraw–Hill Book Co., Inc., New York, 1963.)
20. **The Mystery of Matter.** LOUISE B. YOUNG, ED. (McGraw–Hill Book Co., Inc., New York, 1964.) Both this and the above are published by the American Foundation for Continuing Education. Some material is suitable for beginning classes in physics; the bibliographies are useful.
21. **"Books on Science for the Nonscience Student."** COMPILED BY KENNETH FORD. In *The Proceedings of the Boulder Conference on Physics for the Nonscience Major.* M. Correll, Ed. (Commission on College Physics, Ann Arbor, Mich., 1965), p. 237.

IV. THE TWO-CULTURES CONTROVERSY

22. **"Literature and Science."** MATTHEW ARNOLD. In *Discourses in America* (Lecture 2) (1883–1884). This article, stimulated by a comment of T. H. Huxley concerning the lack of teaching of science, is an early discussion of the two cultures, the humanistic and the scientific.
23. **"Liberal Education in a Scientific Age."** BENTLEY GLASS. Long version in *Science and Liberal Education.* (Louisiana State University Press, Louisiana, n.d.), pp. 54–86; short version in Bull. Atomic Sci. **14**, 346–353 (1958). An eminent biologist argues the importance of science in the modern curriculum, showing how it can be infused into almost every area of the curriculum, suggesting that science should be taught in the humanities.
24. **"Split Personality in the Universities."** E. ASHBY. In *Technology and the Academics.* (The MacMillan Co., Ltd., London, 1959). Ashby sees technology as the bridge between the sciences and the humanities.
25. **The Two Cultures and the Scientific Revolution.** C. P. SNOW. (Cambridge University Press, Cambridge, England, 1959). The most famous of all the articles concerning the two cultures. It contains three lectures; perhaps the first is the most useful. A later edition, published in 1963, contains a reply of Snow to his critics, an interesting restatement and expansion of his original position.
26. **"Modern Science and the Intellectual Tradition."** GERALD HOLTON. Science **131**, 1187–1193 (1960). Also in *Science and Ideas* [4] and *The New Scientist* [15] and in several other collections. This is perhaps the most careful study arguing that there *is* a problem involving the interaction of scientists and humanists. The focus is on intellectual issues rather than the sociological ones which tend to cloud Snow's analysis. The middle section discussing false views of science in our society is particularly interesting.
27. **"Faintly Macabre's Story."** NORTON JUSTER. In *The Phantom Tollbooth.* (Epstein and Carroll, New York, 1961), pp. 71–77. Although this allegorical description of two cities, one where numbers are considered the most important and the other where letters are most important, was done without knowledge of Snow, it is concerned with a similar problem. Particularly useful because of its allegorical humor. Short excerpt in *Science and Ideas* [4].
28. **Two Cultures? The Significance of C. P. Snow.** F. R. LEAVIS. (Chatto and Windus, London, 1962.) A violent attack on Snow both as a writer and as an individual, and on the thesis of the two cultures. This talk was not intended for publication, but it was published because of the notoriety it received.
29. **"A Comment on the Leavis–Snow Controversy."** LIONEL TRILLING. In Commentary (June 1962). Also in L. TRILLING, *Beyond Culture* (Viking Press, New York, 1965). A more moderate rebuttal of the Snow position.
30. **"The Battle of the Books."** MARJORIE HOPE NICOLSON. Brown University Papers, No. XLI, address before the graduate convocation, Brown University, 11 July 1964. Although this article starts with the interactions of science and literature, its thrust at

the end is toward the two-cultures problem. "We are not antagonists. We supplement and complement one another." And again, "Perhaps what we really need to do is to laugh together."

31. **Science and the Shabby Curate of Poetry.** Martin Green. (W. W. Norton, New York, 1965.) A defense of Snow, particularly against the attacks of Leavis and Trilling, by a former student of Leavis. The writer felt it necessary to go back and study more science. There are interesting comments concerning popularization.

V. SCIENCE AND ART

32. **"The Relation to Literature."** H. J. Muller. In *Science and Criticism.* (Yale University Press, New Haven, Conn., 1934), pp. 256–268. This passage, in a book with other interesting material for beginning science classes, concerns the interactions of science and the arts. Although examples are from literature, the discussion has wide general applicability.
33. **"The Representation of Motion."** Gyorgy Kepes. In *Language of Vision.* (Paul Theobalz, Chicago, Ill., 1947), pp. 170–185. The theme is one that Kepes was to expand in the Vision+Values Series, the artist's reaction to the problem of motion.
34. **"Modern Art and the Humanities."** G. H. Forsyth. In *Man and Learning in Modern Society,* papers and addresses delivered at the inaguration of Charles E. Odegaard as President of the University of Washington, 6 and 7 November 1958 (University of Washington Press, Seattle, Wash., 1959), pp. 141–155. The author's purpose is to stress "the importance of mutual respect in close understanding between the modern humanists and the modern scientist and to suggest that the visual arts provide a most effective channel between the two." His main concern is with the concept of space in contemporary art and science.
35. **"Initial Manifesto of Futurism, 1909."** J. C. Taylor. In *Futurism.* (Museum of Modern Art, New York, 1961), pp. 124–125. This might be interesting to read with physics classes to show the excitement that the prosaic topic of motion can generate in contemporary artists. It is a wild document.
36. **"The Cubist Perspective–The New World of Relationships: Camera and Cinema."** Wylie Sypher. In *Rococo to Cubism in Art and Literature.* (Random House, Vintage Books, New York, 1963), Chap. 1, Pt. 4. Although primarily concerned with cubism, this article refers to related developments in literature, philosophy, and science, particularly relativity. One persistent interest is the change in the concept of reality and another is the manipulation of time in both cinema and cubist views of perspective.
37. **"The Vision of Our Age."** J. Bronowski. In *Insight—Ideas of Modern Science.* (Harper and Row, New York, 1964), Chap. 15. This is an attempt to place science in contemporary culture. Bronowski discussed the situation with four individuals: an artist, Eduardo Paoluzzi; an architect, Eero Saarinen; a physicist, Abdus Salam; and a novelist, Lawrence Durrell.
38. **The Nature and Art of Motion.** Gyorgy Kepes, Ed. (George Braziller, New York, 1965.) This is in the Vision+Values Series. Like all of Kepes's works, it is beautifully done and illustrated with material from both the arts and the sciences. It is listed, as a whole, because of the material from the artists—particularly Stanley William Hayter, George Rickey, and Katherine Kuh—showing how motion can be exciting to the modern artist.
39. **The Feynman Lectures on Physics.** R. P. Feynman, R. B. Leighton, and M. Sands. (Addison–Wesley Publ. Co., Reading, Mass., 1965), Vol. 2, pp. 20–9 to 20-11. Feynman asks, in the middle of the second year of his course, "How do I imagine the electric and magnetic field?" This leads to a discussion of scientific imagination and to comments on beauty in scientific thought and experience.
40. **"Art in Science."** Albany Institute of History and Art, September 1965. The artistic value of scientific photographs has been noted by contemporary artists and contemporary scientists. Several exhibitions, with different organization, have been based on this theme, usually stressing the similarity between scientific photographs and contemporary abstract art. This is a catalog of one such recent exhibition.
41. **"Physical Science and the Temper of the Age."** Erwin Schrödinger. In *Science and the Human Temperament.* Translated by J. Murphy and W. H. Johnston. (W. W. Norton, New York, n.d.), Chap. 5, pp. 106–132. Schrödinger begins, "I shall discuss the question of how far the picture of the physical universe as presented to us by modern science has been outlined under the influence of certain contemporary trends which are not peculiar to science at all."
42. **"The Esthetic Experience of the Machine."** Lewis Mumford. In *Technics and Civilization.* (George Routledge and Sons, London, 1947), pp. 335–344. Mumford is concerned with the relation between modern art and modern science. He ends with an interesting discussion on the film, viewing it as a Twentieth Century art form with close affinity to modern physics.
43. **"The Identity of Methods."** S. Giedion. In *Space, Time and Architecture.* (Harvard University Press, Cambridge, Mass., 1963), 4th ed., pp. 11–17. Giedion argues that science and art are essentially interrelated.
44. **"The New Space-Conception: Space–Time."** S. Giedion. In *Space, Time and Architecture.* (Harvard University Press, Cambridge, Mass., 1963), 4th ed., pp. 426–446. This could be read together with the previously mentioned selection of this book; it concerns the possible interaction between relativity and modern art.
45. **"The Impact of Recent Scientific Trends on Art."** Armand Siegel. In *Boston Studies in the Philosophy of Science.* M. W. Wartofsky, Ed. (R. Reidel Publ. Co., Dordrecht, The Netherlands, 1963), pp. 168–173. A physicist is unhappy with the effect of science on art.

VI. PHILOSOPHY OF SCIENCE

46. **"System-Connectedness, Completeness, and Logical Order."** M. R. COHEN. In *Reason and Nature—An Essay on the Meaning of the Scientific Method.* (Harcourt, Brace and Co., New York, 1931), pp. 106–114. A lucid account of the importance of a logical system in theory, discussing the nature of such a system and why it is useful for scientific purposes.

47. **"Newtonian Physics and Aviation Cadets."** ANATOLE RAPOPORT. In *Language, Meaning and Maturity.* S. I. Hayakawa, Ed. (Harper and Brothers, New York, n.d.). The main point is that the language often used in teaching classical mechanics can cause difficulties for the student because of its anthropomorphic overtones. An easy article on the problems associated with language.

48. **"Merits of the Quantitative Method."** RUDOLF CARNAP. In *Philosophical Foundations of Physics.* (Basic Books, Inc., New York, 1966), Chap. II, pp. 105–114. A leading philosopher of science illustrates, using historical examples such as Goethe's work on colors, the usefulness of quantitative methods over qualitative procedures.

49. **"The Unreasonable Effectiveness of Mathematics in the Natural Sciences."** E. P. WIGNER. Commun. Pure Appl. Math. **13,** 1–14 (1960). This is a personal discussion characterized by the title. While Wigner's views are not representative, they stem from a carefully thought-out position. Wigner calls the accuracy of mathematical laws in physics the "empirical law of epistemology." Toward the end of the article, he discusses the uniqueness of physical theory and whether independently developed physical theories are necessarily consistent with each other.

50. **"The Experimental Method."** RUDOLF CARNAP. In *Philosophical Foundations of Physics.* M. Gardner, Ed. (Basic Books, Inc., New York, 1966), Chap. 4, pp. 40–47. This short and lucidly written chapter explains the difference between experimentation and observation and tries to show why experimentation has been such a powerful tool.

51. **"Suggestions from Physics."** P. W. BRIDGMAN. In *The Intelligent Individual and Society.* (Macmillan Co., New York, 1938), pp. 10–47. A chapter on operationalism from one of Bridgman's lesser known books. Also reprinted in *Science and Ideas* [4].

52. **"Geometry: An Example of a Science."** PHILIPP FRANK. In *Philosophy of Science.* (Prentice–Hall, Inc., Englewood Cliffs, N. J., 1957), pp. 48–89. A comprehensive introduction to positivistic philosophy of science, based on the example of geometry. Also reprinted in *Science and Ideas* [4].

53. **"Physics and Reality."** ALBERT EINSTEIN. In *Out of My Later Years.* (Philosophical Library, New York, 1950), pp. 59–65. A beautiful, very brief discussion of the nature of scientific knowledge.

54. **"Geometry and Experience."** ALBERT EINSTEIN. In *Ideas and Opinions.* (Crown Publishers, Inc., New York, 1954.) Considerations of reality and uncertainty in the experimental sciences, with geometry as an example: "···as far as the propositions of mathematics refer to reality, they are not certain; and as far as they are certain, they do not refer to reality."

55. **"The Value of Science."** R. P. FEYNMAN. In *Frontiers in Science—A Survey.* E. Hutchins, Ed. (Basic Books, Inc., New York, 1958), pp. 260–267. This short article develops the theme that science is an open thing and that the admission that "we do not know" is a vital concomitant.

56. **"Malicious Philosophies of Science."** ERNEST NAGEL. In *Sovereign Reason and Other Studies in the Philosophy of Science.* (Free Press of Glencoe, New York, 1954), pp. 17–35. Also, in Barber and Hirsch [6]. Considers some common criticisms of science and attempts to show that they are without basis.

57. **"The Positivistic and the Metaphysical Conception of Physics."** PHILIPP FRANK. In *Between Physics and Philosophy.* (Harvard University Press, Cambridge, Mass., 1941), Chap. 5, pp. 127–138. A statement and defense of a moderate positivistic position, contrasted with a more metaphysical view held by Planck.

58. **"Philosophical Misinterpretations of the Quantum Theory."** PHILIPP FRANK. In *Between Physics and Philosophy.* (Harvard University Press, Cambridge, Mass., 1941), Chap. 7, pp. 151–171. This classic article warns against rushing to draw philosophical conclusions from a scientific theory. Frank shows lucidly how dangerous and foolish such an effort can be; quantum mechanics has been particularly fruitful for generating associated nonsense.

59. **"The Meaning of Reduction in the Natural Sciences."** ERNEST NAGEL. In *Science and Civilization.* Robert C. Stauffer, Ed. (University of Wisconsin Press, Madison, Wisc., 1949), pp. 99–135. Discusses the meanings attached to the idea that one scientific theory has been "reduced" to another. Most of the examples are from physics.

60. **"The Model in Physics."** H. J. GROENWALD. In *The Concept and the Role of the Model in Mathematics and Natural and Social Sciences—Proceedings of the Colloquium sponsored by the Division of Philosophy of Sciences of the Internation Union of History and Philosophy of Sciences organized at Utrecht, January, 1960.* Hans Freudenthal Ed. (D. Reidel Publishing Co., Dordrecht, The Netherlands, 1961), pp. 98–103. This is a brief and readable account by a contemporary physicist of the ways the concept of the model is used in physics. If nothing else, it indicates that there is no standard use of the term in physics.

61. **Quest, the Evolution of A Scientist.** LEOPOLD INFELD. (Doubleday Doran and Co., Inc., New York, 1941.) The material toward the end of Infeld's book is an extremely personal statement about the nature and struggles of science.

62. **"The Reasonableness of Science."** W. H. DAVIS. Sci.

Monthly **15**, 193–214 (1922). Reprinted in abridged form in *Readings in the Physical Sciences* [17]. This easy-to-read article starts with a pleasant fable to illustrate the interaction between observation, invention, and deduction in the development of scientific theory.

63. "Conditions of Scientific Discovery." I. B. COHEN. In *Science, Servant of Man.* (Little, Brown and Co., Boston, Mass., 1948), pp. 16–35. Cohen studies two examples of scientific developments: the discovery of penicillin and the invention of the battery.

VII. SOCIOLOGY OF SCIENCE

64. "Resistance by Scientists to Scientific Discovery." BERNARD BARBER. Science **134**, 596–602 (1961). Also in Barber and Hirsch [6]. A compendium of examples of resistance to scientific discovery and an attempt to sort out factors leading to this resistance.

65. "The College Student's Image of the Scientist." D. C. BEARDSLEY AND DONALD C. O'DOWD. In Barber and Hirsch [6]. A continuation of the Mead and Metraux study [69] to the college level, with similar results.

66. "Reflections on Horror Movies." ROBERT BRUSTEIN. Partisan Rev. **25**, 288–296 (1958). Supplements and reinforces the studies of high school and college students' attitudes. The unconscious attitudes revealed in horror films are analyzed, particularly the association of theoretical science with evil.

67. "Structural Change: Functional Differentiation." WARREN O. HAGSTROM. In *The Scientific Community.* (Basic Books, Inc., New York, 1965), Chap. 5, pp. 244–253. In spite of sociological terminology such as that in the title, this is an interesting brief resume of the roles of the theorist and the experimentalist in physics. The emphasis is on the relative status value. This topic is often mentioned, but there seem to be few reading assignments usable with beginning classes.

68. "The Image of the Scientist in Science Fiction: A Content Analysis." WALTER HIRSCH. Am. J. Sociol. **63**, 506–512 (1958). Also in Barber and Hirsch [6]. In this too-brief paper, science fiction becomes a source of sociological information on attitudes about the scientist. Hirsch divides the period from 1926 to 1950 into six subperiods and arbitrarily selects 50 works from each; the main emphasis is on the trends during successive periods.

69. "Image of the Scientist Among High School Students." MARGARET MEAD AND RHODA METRAUX. Science, 384–390 (1957). Also in Barber and Hirsch [6]. A statistical study of the views that high-school students have about scientists. Every teacher of science should be aware of the discouraging conclusions of the study.

70. "Priorities in Scientific Discovery: A Chapter in the Sociology of Science." ROBERT K. MERTON. Am. Sociol. Rev. **22**, 635–659 (1957). Also in Barber and Hirsch [6]. One of a series of continuing studies by Merton concerning this problem. He argues that multiple discoveries are more common than we have previously suspected and may be the rule in scientific work. In this paper he is interested in the violent quarrels over multiple discoveries, using the institutional norms of science as a basis.

71. "Science in the Social Order." ROBERT K. MERTON. In *Social Theory and Social Structure.* R. K. Merton Ed. (Free Press of Glencoe, New York, 1962), rev. ed., Chap. 15, pp. 537–561. Also in Barber and Hirsch [6]. Merton starts with science in Nazi Germany and then notes other ways in which science interacts with society. The internal goals of science create problems for it in the general society. It might be interesting to combine this article in a reading assignment with Gerald Holton's "Modern Science and the Intellectual Tradition [26]."

72. "Singletons and Multiples in Scientific Discovery: A Chapter in the Sociology of Science." ROBERT K. MERTON. Proc. Am. Phil. Soc. **105**, 470–486 (1961). The first section concerns Francis Bacon's views; it can be omitted. Merton's theme is that multiple discoveries are more common than might be expected in science and are often hidden.

73. "Water Witching as Magical Divination." E. Z. VOGT AND R. HYMAN. In *Water Witching U. S. A.* (University of Chicago Press, Chicago, Ill., 1959.) The authors point out that a nonscientific activity survives in contemporary society and study the sociological implications.

74. There is a selected bibliography on sociology of science in Barber and Hirsch [6], covering the period from 1952 to 1961. There is an earlier bibliography [in *Science and the Social Order.* BERNARD BARBER. (Free Press, New York, 1952. Also Collier Books, Paperback B382X, New York, 1962)] for the work up to that time. A third bibliography is in *Sociology of Science: A Trend Report and Bibliography of Current Sociology.* BERNARD BARBER. (UNESCO, New York, 1956).

VIII. HISTORY OF SCIENCE

75. "Natural Science in the Fourteenth Century." E. J. DIJKSTERHUIS. In *The Mechanization of the World Picture.* (Oxford University Press, Oxford, England, 1961), pp. 164–219. This is a long chapter in an important book. It might be advisable to use only a part of it. The conventional view that nothing happened in science before Copernicus is far from correct.

76. "Medieval Mechanics in Retrospect." M. CLAGETT. In *The Science of Mechanics in the Middle Ages.* (University of Wisconsin Press, Madison, Wisc., 1959), pp. 673–682. A brief summary of an important but neglected area of the history of physics, written by a master.

77. "The Conservatism of Copernicus." H. BUTTERFIELD. *The Origins of Modern Science.* (The Macmillan Co., New York, 1957. Also Collier Books, Paperback

AS259V). Copernicus is closer to Greek astronomy than he is often pictured; he uses most of Ptolemy's mathematical devices.

78. **"Johannes Kepler's Universe: Its Physics and Metaphysics."** G. HOLTON. Am. J. Phys. **24**, 340–351 (1956). Kepler's astronomy and the problem of reality.

79. **"The Galilean Revolution in Physics."** A. RUPERT HALL. In *From Galileo to Newton.* (Harper and Row, New York, 1963), pp. 36–77. An account of Galileo's contributions by a well-known historian of science.

80. **"Newton with His Prism and Silent Face."** C. C. GILLISPIE. In *The Edge of Objectivity.* (Princeton University Press, Princeton, N. J., 1960.) Of all the brief biographies, this is one of the most readable and one of the better for tracing the development of Newton's ideas. The author perhaps overemphasizes the difficulty of reading the *Principia.*

81. **"The Newtonian World Machine."** J. H. RANDALL, JR. In *Making of the Modern Mind.* (Houghton Mifflin Co., Boston, Mass., 1940), pp. 253–279. Randall surveys the overwhelming influence of Newtonian physics in many nonscientific areas. Bibliography. Also in *Science and Ideas* [**4**].

82. **"The Significance of the Newtonian Synthesis."** A. KOYRÉ. In *Newtonian Studies.* (Harvard University Press, Cambridge, Mass., 1965), pp. 3–24. An important statement by a leading historian of ideas. It is not easy to read.

83. **"How the Scientific Revolution of the Seventeenth Century Affected Other Branches of Thought."** BASIL WILLEY. In *A Short History of Science. Origins and Results of the Scientific Revolution.* (Doubleday–Anchor Books, New York, 1951), pp. 61–68. Main emphasis is on literature, because of Willey's background. Quite easy to read.

84. **"The Development of Modern Science."** ERNEST NAGEL. In *Chapters in Western Civilization.* Selected and edited by the Contemporary Civilization Staff of Columbia College. (Columbia University Press, New York, 1954), 2nd ed., Vol. 1, Chap. 8, pp. 282–324. This is perhaps one of the most readable accounts of the rise of modern science, considering the interaction of science with other phases of our society and emphasizing the history of ideas.

85. **The Discovery of Neptune.** M. GROSSER. (Harvard University Press, Cambridge, Mass., 1962.) This entire book can be covered in a series of reading assignments. It gives a very interesting view of scientific development. The ending will surprise many scientists. Useful when coupled with *The Black Cloud* [**103**].

86. **"James Clerk Maxwell."** J. R. NEWMAN. In *Science and Sensibility.* (Simon and Schuster, New York, 1961), Vol. 1, pp. 139–193. This biography of Maxwell, like all the available ones, is based on the Campbell–Garnett *Life* (a new biography is currently under preparation). It also tries to review Maxwell's developing thought.

87. **"Physics Just before Einstein."** A. M. BORK. Science **152**, 597–603 (1966). Primarily concerned with the history of electromagnetic theory after Maxwell.

88. **"The Solvay Meetings and the Development of Quantum Physics."** N. BOHR. In *Atomic Physics and Human Knowledge.* (John Wiley & Sons, Inc., New York, 1963), pp. 79–100. The Solvay Meetings were very important in the evolution of quantum physics; Bohr's account is useful as a historical introduction.

89. **"The Fundamental Idea of Wave Mechanics."** E. SCHRÖDINGER. *Nobel Lectures—Physics 1922–1941.* (Elsevier Publishing Co., Amsterdam, 1965), pp. **304–316.** The historical development of wave mechanics, a history which is often mistold.

90. **"The Function of Dogma in Scientific Research."** T. S. KUHN. In *Scientific Change.* A. C. Crombie, Ed. (William Heinemann, London, 1963). Kuhn argues that scientific development has two very different stages, normal science and revolutions. This thesis is developed in more detail in *The Structure of Scientific Revolutions* (University of Chicago Press, Chicago, Ill., 1962).

IX. SCIENCE AND GOVERNMENT

91. The most useful reference on the detailed interaction between science and government in the United States is the "News and Comments" section of Science. Several reporters follow carefully the hearings of Congressional committees and the work of executive branches of the government, giving a day-by-day account of this interaction. This is a must for anyone interested in the area.

92. **"Behind the Decision to Use the Atomic Bomb: Chicago, 1944–45."** ALICE KIMBALL SMITH. Bull. Atomic Sci. **14**, 280–312 (1958). A long and detailed account of how one group of atomic scientists, those at the metallurgical laboratory at the University of Chicago, reacted as time for using the atomic bomb appeared to be coming closer and closer. It is done with considerable care.

93. **"Responsibilities of Scientists in the Atomic Age."** EUGENE RABINOWITZ. Bull. Atomic Sci. **15**, 2–7 (1959.) Starts with the conflicting loyalties of the scientist to his nation and to society. It discusses the attitudes that various scientists have had toward the question of arms development and ends with the Pugwash program.

94. **"The Scientific Establishment."** DON K. PRICE. Proc. Am. Phil. Soc. **106**, 235–245 (1962). Argues that we have hardly come to grips with the fact that science "has become the major Establishment in the American political system" (see item **101**).

95. **"Federal Expenditures and the Quality of Education."** HAROLD ORLANS. Science **142**, 1625–1629 (1963). The effect of federal support on education.

96. **"Science Goes to Washington."** MEG GREENFIELD. The Reporter **29**, 20–26 (1963). Scientists work in many different areas of the federal government. Often

a scientist will have several different attachments, making it unclear when he is acting as a scientist and when he is not. Some revealing examples are mentioned.

97. **"Central Scientific Organization in the United States Government."** A. HUNTER DUPREE. Minerva **1**, 453–469 (Summer 1963). A historical account of the interactions between the American government and scientific organizations, going back to the earliest period in American history.

98. **"Technology and Society."** JEROME B. WEISNER. In *Science as a Cultural Force.* Introduction by Harry Woolf, Ed. (The Johns Hopkins Press, Baltimore, Md., 1964), Chap. 3, pp. 35–53. Weisner's theme is the extreme importance of technology in contemporary society and the implications this has for the government.

99. **"Scientists and Politics: The Rise of an Apolitical Elite."** R. C. WOOD. In *Scientists and National Policy-Making.* R. Gilpin and C. Wright, Eds. (Columbia University Press, New York, 1964), pp. 41–72. Wood reviews and comments on the attitudes commonly expressed on the interaction of scientists and politics. Many of the other articles in the volume could also be used in readings on this topic.

100. **"The Nationalization of U. S. Science."** S. KLAW. Fortune **70**, 158 (1964). Primarily concerned with the financial support of science and technology by the national government.

101. **"The Established Dissenters."** DON K. PRICE. In *The Scientific Estate.* (Harvard University Press, Cambridge, Mass., 1965.)

102. **"Where the Brains Are."** R. E. LAPP. Fortune **73**, 154 (1966). Scientists as a financial asset for their state; the geographical distribution of scientists and the distribution of federal funds.

X. MISCELLANEOUS

103. **The Black Cloud.** FRED HOYLE. (Harper and Bros., New York, 1957.) The portrait of scientific work in this science fiction book is very good, particularly in the early material. The entire book makes an exciting reading assignment.

104. **"Style in Science."** JOHN RADER PLATT. In *The Excitement of Science.* (Houghton-Mifflin Co., Boston, Mass., 1962.) Originally in Harper's Magazine (October 1956). Points out that the personality of the individual scientist is of importance in determining what kind of science he produces, emphasizing the human elements in scientific developments.

105. **"Passion and Controversy in Science."** MICHAEL POLANYI. Bull. Atomic Sci. (April 1957). Polanyi's purpose is to counteract the idea that scientists are not passionately involved in what they do. He gives historical examples of emotional involvement.

106. **"The Importance of Science."** LEWIS MUMFORD. In *Technics and Civilization.* (George Routledge and Sons, London, 1947), pp. 215–221. The dependence of modern engineering on the results of science.

107. **"Scientific Concepts and Cultural Change."** HARVEY BROOKS. Daedalus, pp. 66–83 (Winter, 1965.) Also in *Science and Culture.* G. Holton, Ed. (Houghton Mifflin Co., Boston, Mass., 1965), pp. 70–87. Suggests how "a number of important themes from the physical and biological sciences have found their way into our general culture, or have the potential for doing so." The themes are relativity, indeterminancy, feedback, and noise. Near the beginning there is some discussion of the ways science and culture interact.

108. **"Society and the Intelligent Physicist."** P. W. BRIDGMAN. Am. Phys. Teacher **7**, 109 (1939). Talking just before the beginning of World War II, Bridgman asks how the contemporary intelligent physicist should behave in the light of national and international social problems.

109. **"The Real Responsibilities of the Scientist."** J. BRONOWSKI. Bull. Atomic Sci. **12**, 10–20 (1956). Discusses the responsibility of the scientist to the society in which he lives. The principal conclusion is, "The sense of intellectual heresy is the life-blood of our civilization."

110. **"Visionaries and the Era of Fulfillment."** RENÉ DUBOS. In *The Dreams of Reason—Science and Utopias.* (Columbia University Press, New York, 1961), Chap. 3, pp. 40–62. Science increasingly has had enormous practical value and consequence in everyday life, making many of the dreams of past utopias practical realities; but this creates new social responsibilities for the scientist.

111. **"Science and the Deallegorization of Motion."** GERALD HOLTON. In *The Nature and Art of Motion.* Gyorgy Kepes, Ed. (George Braziller, New York, 1965). Also in Scientia **57**, 1–10 (1963). Motion is a rich concept, with ramifications far beyond that in the sciences themselves. This beautiful paper is concerned with the "process of separation from the generalized meaning of motion during the rise of modern scientific conceptions···." This might be useful in connection with an elementary treatment of mechanics, since it stresses what is usually left out in the latter.

112. **"The Ethical Basis of Science."** BENTLEY GLASS. *Science and Ethical Values.* (University of North Carolina Press, Chapel Hill, N. C., 1965), pp. 69–101. Glass holds that science and ethics are intimately interwoven. This long article gives a point of view too often missing from an elementary physics course.

113. **"Quo Vadis."** P. W. BRIDGMAN. In *Science and the Modern Mind.* G. Holton, Ed. (Beacon Press, Boston, Mass., 1958), pp. 83–91. Bridgman asks what consequences one would expect if one used the method of intellect more in the future society than we have already done. Also in *Science and Ideas* [4].

ACKNOWLEDGMENTS

We are indebted to Daniel S. Greenberg and Gerald Holton for various helpful suggestions in connection with this Resource Letter.

Resource Letter TLA-1 on Technology, Literature, and Art * since World War II

William H. Davenport
Department of Humanities, Harvey Mudd College, Claremont, California 91711
(Received 10 December 1969)

I. INTRODUCTION

This resource letter lists materials for collateral reading in classes in physics and other sciences as well as in new cross-discipline courses; it also offers professors and students alike an opportunity to see how modern science and technology appear to artists and writers—in other words, to see themselves as others see them. A sampling of books and articles in the increasingly publicized area of cross relationships between technology and society in general would require a separate resource letter, which may one day materialize. The reader will note that the proper distinctions between science and technology are often blurred, as indeed they are daily by the public. The basic hope in this letter is to promote the mutual communication and understanding between disciplines so necessary for personal growth and so vital in such areas as top-level decision making.

As an earlier letter has said, the following listings are suggestions, not prescriptions. They are samplings guided by personal taste and experience, offered with the notion of tempting readers to go farther and deeper on their own.

By agreement with the editors, the listings contain a minimum of overlap with earlier letters, two of which are cited below; indeed, they are intended to pick up where the former left off. Anyone wishing bibliography on the "Two Cultures" argument, missing the presence of earlier "must" items by Mumford, Giedion, Holton, Bronowski, Barzun *et al.*, or seeking historical material can probably find what he wants in the earlier letters. This bibliography is arranged alphabetically by author or editor, except where no such identification exists; in the latter instances, titles are listed in the appropriate alphabetical sequence.

*Prepared at the request of the Committee on Resource Letters of the American Association of Physics Teachers, supported by a grant from the National Science Foundation and published in the *American Journal of Physics,* Vol. 38, No. 4, April 1970, pp. 407-414.

II. RELEVANT RESOURCE LETTERS

1. **"Science and Literature" (SL-1).** Marjorie Nicolson. Amer. J. Phys. **33,** 175 (1965).
2. **"Resource Letter ColR-1 on Collateral Reading for Physics Courses,"** Alfred M. Bork and Arnold B. Arons. Amer. J. Phys. **35,** 71 (1967).

III. TECHNOLOGY, LITERATURE, AND ART SINCE WORLD WAR II: INTERPLAY AND CROSS RELATIONS

1. **"Two Cultures in Engineering Design,"** Anonymous. Engineering **197,** 373 (13 Mar. 1964). A model essay demonstrating that dams and highway-lighting standards can be and should be both useful and beautiful.
2. **Poetics of Space.** Gaston Bachelard. (Orion Press, New York, 1964). A physicist–philosopher justifies poetry as an answer to technology and formulas. In a provocative discussion of the "spaceness" of cellars, attics, and closets and of their relative effects on us, of which we are generally unaware, the author makes us see the familiar in a new light. He offers stimulating contrasts with common notions of space in physics and in the public mind, as influenced by Apollo missions.
3. **"Science: Tool of Culture,"** Cyril Bibby. Saturday Rev. **48,** No. 23, 51 (6 June 1964). Pits the scientists and creative artists against the purely verbal scholars, asks for more science (better taught) in schools, and appeals to administrators to change their methods of training teachers, so that science will appear not as an ogre but as a fairy godmother.
4. **Voices from the Crowd (Against the H-Bomb).** David Boulton, Ed. (Peter Owen, London, 1964.) This anthology of poetry and prose, stemming from the Campaign for Nuclear Disarmament, is a good example of direct testimony of the effect of the Bomb on thinking and writing. Among the literary people included: Priestley, Comfort, Russell, Read, Osborne, Braine (*Room at the Top*).
5. **Poetry and Politics: 1900–1960.** C. M. Bowra. (University Press, Cambridge, England, 1966.) Discusses in part the effect of Hiroshima on poets, notably Edith Stilwell, whose form and vision were radically affected by the event, and the Russian, Andrei Voznesensky, whose poem on the death of Marilyn Monroe foresaw a universal disaster.

6. **"Science as a Humanistic Discipline,"** J. Bronowski. Bull. Atomic Scientists **24**, No. 8, 33 (1968). The author of *Science and Human Values* here covers the history of humanism, values, choice, and man as a unique creature. It is the duty of science to transmit this sense of uniqueness, to teach the world that man is guided by self-created values and thus comfort it for loss of absolute purpose.
7. **"Artist in a World of Science,"** Pearl Buck. Saturday Rev. **41**, No. 38, 15–16, 42–44 (1958). Asks for artists to be strong, challenges writers to use the findings of science and illuminate them so that "human beings will no longer be afraid."
8. **The Novel Now.** Anthony Burgess. (W. W. Norton and Co., New York, 1967.) Prominent British novelist discusses the aftermath of nuclear war as a gloomy aspect of fictional future time and advances the thesis that comparatively few good novels came out of the war that ended with Hiroshima, although a good deal of ordinary fiction has the shadow of the Bomb in it.
9. **Beyond Modern Sculpture: Effects of Science & Technology on the Sculpture of this Century.** Jack Burnham. (George Braziller Inc., New York, 1969.) "Today's sculpture is preparing man for his replacement by information-processing energy" Burnham sees an argument for a mechanistic teleological interpretation of life in which culture, including art, becomes a vehicle for qualitative changes in man's biological status. [See review by Charlotte Willard in Saturday Rev. **52**, No. 2, 19 (1969).]
10. **Cultures in Conflict.** David K. Cornelius and Edwin St. Vincent, Eds. (Scott, Foresman and Company, Glenview, Ill., 1964). A useful anthology of primary and secondary materials on the continuing C. P. Snow debates.
11. **"The Computer and the Poet,"** Norman Cousins. Saturday Rev. **49**, No. 30, 42 (23 July 1966). Suggests editorially (and movingly) that poets and programmers should get together to "see a larger panorama of possibilities than technology alone may inspire" and warns against the "tendency to mistake data for wisdom."
12. **Engineers and Ivory Towers.** Hardy Cross. Robert C. Goodpasture, Ed. (McGraw–Hill Book Co., New York, 1952). A sort of common-sense bible covering the education of an engineer, the full life, and concepts of technological art.
13. **Engineering: Its Role and Function in Human Society.** William H. Davenport and Daniel Rosenthal, Eds. (Pergamon Press, Inc., New York, 1967). An anthology with four sections on the viewpoint of the humanist, the attitudes of the engineer, man and machine, and technology and the future. Many of the writers in this bibliography are represented in an effort to present historical and contemporary perspectives on technology and society.
14. **"Art and Technology—The New Combine,"** Douglas M. Davis. Art in Amer. **56,** 28 (Jan.–Feb. 1968). Notes a new enthusiasm among many modern artists because of the forms, effects, and materials made possible by the new technology. Envisions full partnership between artist and machine in the creative process.
15. **So Human an Animal.** René Dubos. (Charles Scribner's & Sons, New York, 1968). Dubos, a prominent microbiologist, won a Pulitzer Prize for this work, and it deserves wide reading. Motivated by humanistic impulses, writing now like a philosopher and again like a poet, he discusses man's threatened dehumanization under technological advance. Man can adjust, Dubos says—at a price But first he must understand himself as a creature of heredity and environment and then learn the science of life, not merely science.
16. **The Theatre of the Absurd.** Martin Esslin. (Anchor Books–Doubleday and Co., Inc., Garden City, N. J., 1961). The drama director for the British Broadcasting Company explains the work of Beckett, Ionesco, Albee, and others as a reaction to loss of values, reason, and control in an age of totalitarianism and of that technological development, the Bomb.
17. **Engineering and the Liberal Arts.** Samuel C. Florman. (McGraw–Hill Book Co., New York, 1968). The subtitle tells the story: *A Technologist's Guide to History, Literature, Philosophy, Art, and Music.* Explores the relationships between technology and the liberal arts—historical, aesthetic, functional. Useful reading lists are included.
18. **The Creative Process.** Brewster Ghiselin, Ed. (University of California Press, Berkeley, 1952; Mentor Books, The New American Library, Inc., New York, paperback, 1961). Mathematicians, musicians, painters, and poets, in a symposium on the personal experience of creativity. Of use to those interested in the interplay between science and art.
19. **Postwar British Fiction: New Accents and Attitudes.** James Gindin. (University of California Press, Berkeley, 1963). Traces the comic or existentialist view of the world in recent British novels as resulting in part from the threat of the hydrogen bomb.
20. **The Poet and the Machine.** Paul Ginestier. Martin B. Friedman, Transl. (University of North Carolina Press, Chapel Hill, 1961; College and University Press, New Haven, Conn., paperback, 1964). Considers through analysis of generous examples from modern and contemporary poetry the effect of the machine on subject matter, form, and attitude. An original approach to the value, meaning, and influence, as the author puts it, of the poetry of our technology-oriented era.
21. **"Nihilism in Contemporary Literature,"** Charles I. Glicksberg. Nineteenth Century **144,** 214 (Oct., 1948). An example of the extreme view that man is lost in a whirlpool of electronic energy, that cosmic doubts, aloneness, and fear of cataclysmic doom have led to a prevailing mood of nihilism in writing.
22. **"Impact of Technological Change on the Humanities,"**

MAXWELL H. GOLDBERG. Educational Record **146,** No. 4, 388–399 (1965). It is up to the humanities to soften the impact of advancing technology upon the pressured individual. One thing they can do is to help us pass the almost unlimited leisure time prophesied for the near future under automation and, thus, to avoid Shaw's definition of hell.

23. **"A Poet's Investigation of Science,"** ROBERT GRAVES. Saturday Rev. **46,** 82 (7 Dec. 1963). The dean of English poets in a lecture at the Massachusetts Institute of Technology takes technologists to task good-humoredly but with a sting, too. He is concerned about the upset of nature's balance, the weakening of man's powers through labor-saving devices, synthetic foods, artificial urban life, dulling of imagination by commercialized art, loss of privacy—all products or results of technology. Graves finds no secret mystique among advanced technologists, only a sense of fate that makes them go on, limited to objective views and factual accuracy, forgetting the life of emotions and becoming diminished people.

24. **Social History of Art.** ARNOLD HAUSER. (Vintage Books, Random House, Inc., 4 vols., New York, 1951). In the fourth volume of this paperback edition of a standard work, considerable space is given to the cultural problem of technics and the subject of film and technics.

25. **"Automation and Imagination,"** JACQUETTA HAWKES. Harper's **231,** 92 (Oct. 1965). Prominent archaeologist fears loss of man's imaginative roots under years of technical training. While the technological revolution sweeps on toward a total efficiency of means, she says, we must control the ends and not forget the significance of the individual.

26. **The Future as Nightmare.** MARK R. HILLEGAS. (Oxford University Press, New York, 1967). A study that begins with Wells and ends with recent science fiction by Ray Bradbury, Kurt Vonnegut, and Walter Miller, Jr. The latter three are worried about the mindless life of modern man with his radio, TV, and high-speed travel; the need to learn nothing more than how to press buttons; the machine's robbing man of the pleasure of working with his hands, leaving him nothing useful to do, and lately making decisions for him; and, of course, the coming nuclear holocaust.

27. **Science and Culture.** GERALD HOLTON, Ed. (Beacon Press, Boston, 1967). Almost all of the 15 essays in this outstanding collection appeared, several in different form, in the Winter 1965 issue of Daedalus. Of particular relevance to the area of this bibliography are Herbert Marcuse's view of science as ultimately just technology; Gyorgy Kepes' criticism of modern artists for missing vital connections with technological reality; René Dubos' contention that technological applications are becoming increasingly alienated from human needs; and Oscar Handlin's documentation of the ambivalent attitude of modern society toward technology.

28. **"The Fiction of Anti-Utopia,"** IRVING HOWE. New Republic **146,** 13 (23 Apr. 1962). An analysis of the effect on modern fiction of the splitting apart of technique and values and the appearance of technical means to alter human nature, both events leading to the American dream's becoming a nightmare.

29. **The Idea of the Modern.** IRVING HOWE, Ed. (Horizon Press, New York, 1967). A perspective on post-Hiroshima literature and its relation to technology calls for a frame of reference on modernism in art and literature in general. A useful set of ideas is contained in this volume, the summing-up of which is that "nihilism lies at the center of all that we mean by modernist literature."

30. **The Machine.** K. G. P. HULTÉN, Ed. (Museum of Modern Art, New York, 1968). A metal-covered book of pictorial reproductions with introduction and running text, actually an exhibition catalogue, offering clear visual evidence of the interplay of modern art and modern technology in forms and materials.

31. **Literature and Science.** ALDOUS HUXLEY. (Harper & Row, Publishers, New York, 1963). A literary and highly literate attempt to show bridges between the two cultures. Technological know-how tempered by human understanding and respect for nature will dominate the scene for some time to come, but only if men of letters and men of science advance together.

32. **The Inland Island.** JOSEPHINE JOHNSON. (Simon & Schuster, Inc., New York, 1969). One way to avoid the evils of a technological society is to spend a year on an abandoned farm, study the good and the cruel aspects of nature, and write a series of sketches about the experience. Escapist, perhaps, but food for thought.

33. **The Sciences and the Humanities.** W. T. JONES. (University of California Press, Berkeley, 1965). A professor of philosophy discusses conflict and reconciliation between the two cultures, largely in terms of the nature of reality and the need to understand each other's language.

34. **"The Literary Mind,"** ALFRED KAZIN. Nation **201,** 203 (20 Sept. 1965). Advances the thesis that it is more than fear of the Bomb that produces absurdist and existentialist writing, it is dissatisfaction that comes from easy self-gratifications: "Art has become too easy."

35. **"Imagination and the Age,"** ALFRED KAZIN. Reporter **34,** No. 9, 32 (5 May 1966). Analyzes the crisis mentality behind modern fiction, the guilt feelings going back to Auschwitz and Hiroshima. Salvation from the materialism of modern living lies in language and in art.

36. **New Landscape in Science and Art.** GYORGY KEPES. (Paul Theobald, Chicago, 1967). Like the earlier *Vision in Motion* by L. Moholy-Nagy (Paul Theobald, Chicago, 1947), this work will make the reader see more, better, and differently. Essays and comments by Gabo, Giedion, Gropius, Rossi, Wiener, and others

plus lavish illustration assist Kepes, author of the influential *Language of Vision* and head of the program on advanced visual design at the Massachusetts Institute of Technology, to discuss morphology in art and science, form in engineering, esthetic motivation in science—in short, to demonstrate that science and its applications belong to the humanities, that, in Frank Lloyd Wright's words, "we must look to the artist brain . . . to grasp the significance to society of this thing we call the machine."

37. **"If You Don't Mind My Saying So . . .,"** JOSEPH WOOD KRUTCH. Amer. Scholar **37,** 572 (Autumn, 1968). Expresses fear over extending of experimentation with ecology and belief that salvation does not lie with manipulation and conditioning but may come from philosophy and art.
38. **The Scientist vs the Humanist.** GEORGE LEVINE AND OWEN THOMAS, Eds. (W. W. Norton, New York, 1963). Among the most relevant items are I. I. Rabi's "Scientist and Humanist"; Oppenheimer's "The Tree of Knowledge"; Howard Mumford Jones's "The Humanities and the Common Reader" (which treats technological jargon); and P. W. Bridgman's "Quo Vadis."
39. **Death in Life: Survivors of Hiroshima.** ROBERT J. LIFTON. (Random House, Inc., New York, 1967). Chapter 10, "Creative Response: A-Bomb Literature," offers samples of diaries, memoirs, and poems by survivors, running the gamut from protest to reconstruction. See, also, Lifton's "On Death and Death Symbolism: The Hiroshima Disaster" in Amer. Scholar 257 (Spring, 1965).
40. **"The Poet and the Press,"** ARCHIBALD MACLEISH. Atlantic **203,** No. 3, 40 (March, 1959). Discusses the "divorce between knowing and feeling" about Hiroshima as part of a social crisis involving the decay of the life of the imagination and the loss of individual freedom. The danger here is growing acquiescence to a managed order and satisfaction with the car, TV, and the material products of our era.
41. **"The Great American Frustration,"** ARCHIBALD MACLEISH. Saturday Rev. **51,** No. 28, 13 (13 July 1968). Prior to Hiroshima, it seemed that technology would serve human needs; after the event, it appeared that technology is bound to do what it can do. We are no longer men, but consumers filled with frustrations that produce the satirical novels of the period. We must try to recover the management of technology and once more produce truly educated men.
42. **"The New Poetry,"** FRANK MACSHANE. Amer. Scholar **37,** 642 (Autumn, 1968). Frequently, the modern poet writes of confrontation of man and machine. He is both attracted and repelled by technological change, which both benefits and blights.
43. **The Machine in the Garden: Technology and the Pastoral Ideal in America.** LEO MARX. (Oxford University Press, New York, 1964; Galaxy, Oxford Univ. Press, New York, paperback, 1967). One of the three most significant contemporary works on the interplay of literature and technology [along with Sussman (69) and Sypher (72)], this study concentrates on 19th-century American authors and their ambivalent reactions to the sudden appearance of the machine on the landscape. Whitman, Emerson, Thoreau, Hawthorne, Melville, and others reveal, under Marx's scrutiny, the meaning inherent in productivity and power. Whitman assimilated the machine, Emerson welcomed it but disliked ugly mills, Thoreau respected tools but hated the noise and smoke, Hawthorne and Melville noted man's growing alienation with the green fields gone, Henry Adams set the theme for the "ancient war between the kingdom of love and the kingdom of power...waged endlessly in American writing ever since." The domination of the machine has divested of meaning the older notions of beauty and order, says Marx, leaving the American hero dead, alienated, or no hero at all. Aptly used quotations, chronological order, and clarity of perspective and statement (with which all may not agree) make this a "must" for basic reading in this special category. Furthermore, there are links to Frost, Hemingway, Faulkner, and other modern writers.
44. **Technology and Culture in Perspective.** ILENE MONTANA, Ed. (The Church Society for College Work, Cambridge, 1967). Includes "Technology and Democracy" by Harvey Cox, "The Spiritual Meaning of Technology and Culture," by Walter Ong, and "The Artist's Response to the Scientific World," by Gyorgy Kepes.
45. **"Science, Art and Technology,"** CHARLES MORRIS. Kenyon Rev. **1,** No. 4, 409 (Autumn, 1939). A tight study of three forms of discourse—scientific, aesthetic, and technological and a plea that the respective users acquire vision enough to see that each complements the other and needs the other's support.
46. **"Scientist and Man of Letters,"** HERBERT J. MULLER. Yale Rev. **31,** No. 2, 279 (Dec., 1941). Similarities and differences again. Science has had some bad effects on literature, should be a co-worker; literature can give science perspective on its social function.
47. **The Myth of the Machine.** LEWIS MUMFORD. (Harcourt, Brace & World, Inc., New York, 1967). Important historical study of human cultural development, that shows a major shift of emphasis from human being to machine, questions our commitment to technical progress, and warns against the downplaying of literature and fine arts so vital to complete life experience. See also his earlier *Art and Technics* (Columbia University Press, New York, 1952).
48. **"Utopia, the City, and the Machine,"** LEWIS MUMFORD. Daedalus **94,** No. 2, of the Proceedings of the American Academy of Arts and Sciences, 271 (Spring, 1965). The machine has become a god beyond challenge. The only group to understand the dehumanizing effects and eventual price of technology are the ***avant-garde*** artists, who have resorted to caricature.
49. **"Utopias for Reformers,"** FRANCOIS BLOCH-LAINE.

Daedalus **94,** No. 2, 419 (1965). Discusses the aim of two utopias—technological and democratic—to increase man's fulfillment by different approaches, which must be combined.

50. **"Utopia and the Good Life,"** GEORGE KATEB, Daedalus **94,** No. 2, 454 (1965). Describes the thrust of technology in freeing men from routine drudgery and setting up leisure and abundance, with attendant problems, however.
51. **Utopia and Utopian Thought.** FRANK MANUEL, Ed. (Houghton Mifflin Co., Boston, 1966). A gathering of the preceding three, and other materials in substantially the same form.
52. **Aesthetics and Technology in Building.** PIER LUIGI NERVI. (Harvard University Press, Cambridge, Mass., 1965). "Nervi's thesis is that good architecture is a synthesis of technology and art," according to an expert review by Carl W. Condit, in Technol. and Culture **7,** No. 3, 432 (Summer, 1966), which we also recommend.
53. **Liberal Learning for the Engineer.** STERLING P. OLMSTED. (Amer. Soc. Eng. Educ., Washington, D. C., 1968). The most recent and comprehensive report on the state of liberal studies in the engineering and technical colleges and institutes of the U. S. Theory, specific recommendations, bibliography.
54. **Road to Wigan Pier.** GEORGE ORWELL. (Berkley Publishing Corporation, New York, 1967). A paperback reissue of the 1937 work by the author of *1984*. Contains a 20-page digression, outspoken and controversial, on the evils of the machine, which has made a fully human life impossible, led to decay of taste, and acquired the status of a god beyond criticism.
55. **"Art and Technology: 'Cybernetic Serendipity',"** S. K. OVERBECK. The Alicia Patterson Fund, 535 Fifth Ave., New York, N. Y., 10017, SKO-1 (10 June 1968). The first of a dozen illustrated newsletter-articles by Overbeck that are published by the Fund. The series (space limitations forbid separate listings) describes various foreign exhibitions of computer music, electronic sculpture, and sound and light, which, in turn, recall "Nine Evenings: Theater and Engineering" staged in New York, fall 1966, by Experiments in Art and Technology, with outside help, "to familiarize the artist with the realities of technology while indulging the technician's penchant to transcend the mere potentialities of his discipline."
56. **"Myths, Emotions, and the Great Audience,"** JAMES PARSONS. Poetry **77,** 89 (Nov., 1950). Poetry is important to man's survival because it is a myth maker at a time when "it is the developing rationale of assembly-line production that all society be hitched to the machine."
57. **"Public and Private Problems in Modern Drama,"** RONALD PEACOCK. Tulane Drama Rev. **3,** No. 3, 58 (March, 1959). The dehumanizing effects of technocratic society as seen in modern plays going back as far as Georg Kaiser's *Gas* (1918).
58. **"The American Poet in Relation to Science,"** NORMAN HOLMES PEARSON. Amer. Quart. **1,** No. 2, 116 (Summer, 1949). Science and technology have done a service to poets by forcing them into new modes of expression; however, the poet remains the strongest force in the preservation of the freedom of the individual.
59. **Science, Faith and Society.** MICHAEL POLANYI. (University of Chicago Press, Chicago, 1964). Originally published by Oxford University Press, London, in 1946, this work appears in a new format with a new introduction by the author, which fits the present theme, inasmuch as it considers the idea that all great discoveries are beautiful and that scientific discovery is like the creative act in the fine arts.
60. **Avant-Garde: The Experimental Theater in France.** LEONARD C. PRONKO. (University of California Press, Berkeley, 1966). A keen analysis of the work of Beckett, Ionesco, Genêt, and others, which no longer reflects a rational world but the irrational world of the atom bomb.
61. **"Scientist and Humanist: Can the Minds Meet?"** I. I. RABI. Atlantic **197,** 64 (Jan., 1956). Discusses modern antiintellectualism and the urge to keep up with the Russians in technology. Calls for wisdom, which is unobtainable as long as sciences and humanities remain separate disciplines.
62. **"Integral Science and Atomized Art,"** EUGENE RABINOWITCH. Bull. Atomic Scientists **15,** No. 2, 65 (February, 1959). Through its own form and expression, art could help man find the harmony now threatened by the forces of atomism and fear of nuclear catastrophe.
63. **"Art and Life,"** SIR HERBERT READ. Saturday Evening Post **232,** 34 (26 Sept. 1959). Modern violence and restlessness stem in great part from a neurosis in men who have stopped making things by hand. Production, not grace or beauty, is the guiding force of technological civilization. Recommends the activity of art to release creative, rather than destructive, forces.
64. **The New Poets: American and British Poetry Since World War II.** M. L. ROSENTHAL. (Oxford University Press, New York, 1967). Detects a dominant concern among contemporary poets with violence and war and links it to a general alienation of sensibility, due in great part to the fact that human values are being displaced by technology.
65. **"The Vocation of the Poet in the Modern World,"** DELMORE SCHWARTZ. Poetry **78,** 223 (July, 1951). The vocation of the poet today is to maintain faith in and love of poetry, until he is destroyed as a human being by the doom of a civilization from which he has become alienated.
66. **"Science and Literature,"** ELIZABETH SEWELL. Commonweal **73,** No. 2, 218 (13 May 1966). Myth and the simple affirmation of the human mind and body are the only two forms of imagination capable of facing modern enormities. The two terminal points of our technological age were Auschwitz and

Hiroshima; in literature about them, we may yet see "the affirmation of simple humanity."

67. **"Is Technology Taking Over?"** Charles E. Silberman. Fortune **73,** No. 2, 112 (Feb., 1966). A brisk discussion of familiar topics: art as defense; technology as an end; dehumanization and destruction; mass idleness; meaninglessness. Technology may not determine our destiny, but it surely affects it and, in enlarging choice, creates new dangers. As the author points out, however, borrowing from Whitehead, the great ages have been the dangerous and disturbed ones.

68. **"One Way to Spell Man,"** Wallace Stegner. Saturday Rev. **41,** No. 21, 8; 43 (24 May 1958). Finds a real quarrel between the arts and technology but not between the arts and science, the latter two being open to exploitation by the technology of mass production. Reminds us that nonscientific experience is valid, and nonverifiable truth important.

69. **Victorians and the Machine: The Literary Response to Technology.** Herbert L. Sussman. (Harvard University Press, Cambridge, Mass., 1968). Does for English writers of the 19th century what Leo Marx [43] did for the Americans, with substantially similar conclusions. Writers stressed are Carlyle, Butler, Dickens, Wells, Ruskin, Kipling, and Morris, whose thought and art centered on the effects of mechanization on the intellectual and aesthetic life of their day. A major study of the machine as image, symbol, servant, and god—something feared and respected, ugly and beautiful, functional and destructive—as seen by the significant Victorian literary figures, this work also helps explain the thrust of much contemporary writing.

70. **"The Poet as Anti-Specialist,"** May Swenson. Saturday Rev. **48,** No. 5, 16 (30 Jan. 1965). A poet tells how her art can show man how to stay human in a technologized age, compares and contrasts the languages of science and poetry, wonders about the denerving and desensualizing of astronauts "trained to become a piece of equipment."

71. **"The Poem as Defense,"** Wylie Sypher. Amer. Scholar **37,** 85 (Winter, 1967). The author is not worried about opposition between science and art, but about opposition between both of them together and technology. Technique can even absorb criticism of itself. Technological mentality kills the magic of surprise, grace, and chance. If Pop art, computer poetry, and obscene novels are insolent, society is even more so in trying to engineer people.

72. **Literature and Technology.** Wylie Sypher. (Random House, Inc., New York, 1968). The best, almost the only, general study of its kind, to be required reading along with Leo Marx [43] and Herbert Sussman [69]. Develops the thesis that technology dreads waste and, being concerned with economy and precaution, lives by an ethic of thrift. The humanities, including art, exist on the notion that every full life includes waste—of virtue, intention, thinking, and work. The thesis is illustrated by examples from literature and art. Although, historically, technology minimizes individual participation and resultant pleasure, Sypher concedes that lately "technology has been touched by the joy of finding in its solutions the play of intellect that satisfies man's need to invent."

73. **Dialogue on Technology.** Robert Theobald, Ed. (The Bobbs–Merrill Co., Inc., Indianapolis, 1967). Contains essays on the admiration of technique, human imagination in the space age, educational technology and value systems, technology and theology, technology and art.

74. **Science, Man and Morals.** W. H. Thorpe. (Cornell University Press, Ithaca, N. Y., 1965). Brings out interplay among science, religion, and art, accenting an over-all tendency toward wholeness and unity. Traces modern plight in some degree to the Bomb.

75. **"Modern Literature and Science,"** I. Traschen. College English **25,** 248 (Jan., 1964). Explores the common interests of scientist and poet in their search for truth as well as their differences, which produce alienation and literary reaction.

76. **"The New English Realism,"** Ossia Trilling. Tulane Drama Rev. **7,** No. 2, 184 (Winter, 1962). Ever since Osborne's "Look Back in Anger," the modern British theatre has shown a realism based on revolt against class structure and the dilemma of threatening nuclear destruction, although scarcely touching on the new technology itself.

77. **"The Poet in the Machine Age,"** Peter Viereck. J. History Ideas **10,** No. 1, 88 (Jan., 1949). A classification of antimachine poets, who for esthetic, pious, instinctual, or timid reasons have backed away, and promachine poets, who, as materialists, cultists, or adapters, have used the new gadgets to advantage. We must try to unite the world of machinery and the world of the spirit, or "our road to hell will be paved with good inventions."

78. **The Industrial Muse.** With introduction by Jeremy Warburg, Ed. (Oxford University Press, New York, 1958). An amusing and informative anthology of verse from 1754 to the 1950's dealing in all moods with engines, factories, steamboats, railways, machines, and airplanes.

79. **"Poetry and Industrialism,"** Jeremy Warburg. Modern Language Rev. **53,** No. 2, 163 (1958). Treats the problem of imaginative comprehension as the modern poet strives to assimilate the new technology, make statements, and find terms for a new form of expression.

80. **Reflections on Big Science.** Alvin Weinberg. (The MIT Press, Cambridge, Mass., 1967). The director of Oak Ridge National Laboratory devotes his first chapter, "The Promise of Scientific Technology; The New Revolutions," to nuclear energy, cheap electricity, technology of information, the Bomb, and dealing with nuclear garbage. He calls upon the humanists to restore meaning and purpose to our lives.

81. **The Theater of Protest and Paradox.** George Wellwarth. (New York University Press, 1964). A discussion of contemporary playwrights, e.g., Ionesco, who finds a machine-made preplanned city essentially drab; and Dürrenmatt, whose "The Physicists" teaches the lesson that mankind can be saved only through suppression of technical knowledge.
82. **Flesh of Steel: Literature and the Machine in American Culture.** Thomas Reed West. (Vanderbilt University Press, Nashville, Tenn., 1967). A consideration of the writings of Sherwood Anderson, Dos Passos, Sandburg, Sinclair Lewis, Mumford, and Veblen which, while conceding that most of them are antimachine most of the time, preaches the positive virtues of the Machine: law, order, energy, discipline, which, at a price, produce a city like New York, where artists and writers may live and work on their own terms who could not exist if the machine stopped.
83. **"The Discipline of the History of Technology,"** Lynn White, Jr. Eng. Educ. **54**, No. 10, 349 (June, 1964). Technologists have begun to see that they have an intellectual need for the knowledge of the tradition of what they are doing. Engineers, too, must meet the mark of a profession, namely, the knowledge of its history. Even the humanists are realizing what this explosive new discipline can contribute to their personal awareness.
84. **Drama in a World of Science.** Glynne Wickham. (University of Toronto Press, Toronto, Canada, 1962). Treats the renascence of English theatre in the 50's, the Bomb as topic, the individual confused by technology and its tyranny as protagonist, and mass conformity, violence, or apathy as themes.
85. **"The Scientist and Society."** J. Tuzo Wilson. Imperial Oil Rev., 20–22 (Dec., 1963). The humanist who pretends to have no interest in science and the technocrat who relies completely on science are equally deluded. Calls for tolerance and understanding among all intellectual disciplines. The scientist must reconsider his position *vis-à-vis* the humanities and the arts.
86. **"Science is Everybody's Business,"** J. Tuzo Wilson. Amer. Scientist **52**, 266A (1964). Includes new directions in technology.
87. **"On the History of Science,"** J. Tuzo Wilson, Saturday Rev. **47**, No. 18, 50 (2 May 1964). Suggests new university departments to train scientifically literate humanists.
88. **"The Long Battle between Art and the Machine."** Edgar Wind. Harper's **228**, 65 (Feb., 1964). Contemplation of fake-modern buildings, dehumanized music, and mass-produced furniture raises once more the old question of whether the artist uses the machine or becomes its slave.

Postscript to Sec. III

Since most of the foregoing material is critical or expository, except for quoted illustration, readers may wish to make a start with firsthand creative literary pieces. Here are some suggestions (unless otherwise indicated, items are available in various paperback editions; see the current issue of *Paperbound Books in Print*, R. R. Bowker Co., New York).

Plays

On the theme of machine replacing man, there are two early modern classics for background:

89. **R.U.R.** Karel Capek.
90. **The Adding Machine.** Elmer Rice.

Three British plays deal directly with the Bomb, and the fourth, the only one available in paper, alludes to it:

91. **The Tiger and the Horse.** Robert Bolt. In *Three Plays* (Mercury Books, London, 1963).
92. **The Offshore Island.** Marghanita Laski. (Cresset Press, London, 1959).
93. **Each His Own Wilderness.** Doris Lessing. In *New English Dramatists*, E. Martin Browne, Ed. (Penguin Plays, London).
94. **Look Back in Anger.** John Osborne.

Two recent plays dealing with physicists:

95. **The Physicists.** Friedrich Durrenmatt.
96. **In the Matter of J. Robert Oppenheimer.** Heinar Kipphardt.

Fiction

A quartet of Utopian or anti-Utopian novels:

97. **Brave New World.** Aldous Huxley.
98. **Nineteen Eighty-Four.** George Orwell.
99. **Walden II.** B. F. Skinner.
100. **We.** E. Zamiatan.

A quartet of science fiction:

101. **Fahrenheit *451*.** Ray Bradbury.
102. **Canticle for Leibowitz.** Walter Miller, Jr.
103. **Player Piano.** Kurt Vonnegut, Jr.
104. **Cat's Cradle.** Kurt Vonnegut, Jr.

A trio of short stories:

105. **"By the Waters of Babylon,"** Stephen V. Benet.
106. **"The Portable Phonograph,"** Walter Van Tilburg Clark. In *The Art of Modern Fiction*, R. West and R. Stallman, Eds., alternate ed. (Holt, Rinehart, & Winston, Inc., New York, 1949).
107. **"The Machine Stops,"** E. M. Forster. In *Modern Short Stories*, L. Brown, Ed. (Harcourt, Brace & World, Inc., New York, 1937).

Poetry

See Ginestier [20], Warburg [78], and Boulton [4] above. Also:

108. The Modern Poets. John M. Brinnin and Bill Read, Eds. (McGraw-Hill Book Co., New York, 1963). Contains poems by Hoffman, Lowell, Moss, and Nemerov pertaining to the Bomb.

109. Weep Before God. John Wain. (The Macmillan Company, London, 1961). Sections VI-VII consider the Machine.

110. Wildtrack. John Wain. (The Macmillan Company, London, 1965). Pages 10-12 satirize Henry Ford and the assembly line.

111. Today's Poets. Chad Walsh, Ed. (Charles Scribner's Sons, New York, 1964). The Introduction mentions the Bomb, and a poem by Gil Orlovitz spoofs the computer.

ACKNOWLEDGMENTS

I wish to thank Professor Gerald Holton of Harvard for suggesting that I prepare this letter and for helping it on its way.

For help in research or bibliography, I owe much to Dr. Emmanuel Mesthene (Director), Charles Hampden-Turner, and Tom Parmenter of the Harvard Program on Technology and Society, where I spent the sabbatical year 1968-69; to Professor Leo Marx of Amherst College; and to Professor Wylie Sypher of Simmons College.

In the later stages, I received helpful advice from Professor Arnold Arons of the University of Washington and from Professor Joel Gordon of Amherst.

None of the above is responsible for errors, omissions, or final choices.

Table of Some "Elementary" Particles

Family name	Particle name	Symbol	Rest mass*	Electric charge	Antiparticle	Average lifetime (seconds)
Photon	photon	γ (gamma ray)	0	neutral	same particle	infinite
Leptons	neutrino	ν, ν_μ	0	neutral	ν, ν_μ	infinite
	electron	e^-	1	negative	e^+ (positron)	infinite
	μ-meson (muon)	μ^-	207	negative	μ^+	10^{-6}
Mesons	π-mesons (pions)	π^+	273	positive	π^- same as	10^{-8}
		π^-	273	negative	π^+ the	10^{-8}
		π°	264	neutral	π° particles	10^{-16}
	K-mesons (Kaons)	K^+	966	positive	K^- (negative)	
		K°	974	neutral	K°	10^{-10} and 10^{-7}
	η-meson (eta)	η°	1073	neutral	η°	10^{-18}
Baryons	proton	p	1836	positive	p (antiproton)	infinite
	neutron	n	1839	neutral	n (antineutron)	10^3
	lambda	Λ°	2182	neutral	Λ°	10^{-10}
	sigma	Σ^+	2328	positive	Σ^- (negative)	10^{-10}
		Σ^-	2341	negative	Σ^+ (positive)	10^{-10}
		Σ°	2332	neutral	Σ°	10^{-20}
	xi	Ξ^-	2580	negative	Ξ^+ (positive)	10^{-10}
		Ξ^+	2570	neutral	Ξ°	10^{-10}
	omega	Ω^-	3290	negative	Ω^+	10^{-10}

* Mass of electron is 1 unit on this scale

INDEX/HANDBOOK